50 Years Lost in Medical Advance

50 Years Lost in Medical Advance

The discovery of Hans Selye's stress mechanism

By Lewis S. Coleman, MD

The American Institute of Stress Press

First Edition

For more information, contact:
Lewis S. Coleman
PO Box 3263
Visalia, CA 93278-3263
coleman006@gmail.com
www.stressmechanism.com

ISBN: 978-0-578-82260-0

Dedication

Johannes Rhodin, MD, PhD 1922-2004
(Photo courtesy Dr. Anders Rhodin)

This book is dedicated to the memory of Johannes “John” Arne Gosta Rhodin, MD., Ph.D., (1922-2004) who was an internationally recognized research scientist and pioneer in electron microscopy. During his career, Dr. Rhodin delivered sophisticated stress theory lectures to more than 5,000 medical students in hopes of promoting the discovery of a testable *stress mechanism* that would confirm Hans Selye's revolutionary Unified Theory of Medicine. He was recruited by New York Medical College to upgrade its core curriculum, and by happy coincidence his sojourn at the school coincided with my basic sciences medical school training from 1970-1972. I thus enjoyed the benefit of his revised curriculum and his compelling lectures on stress theory. Years later, these lectures enabled my discovery of the long-sought stress mechanism after clarification of the chimeric character of coagulation Factor VIII inspired me to embark on what became an extensive review of research literature, which is the subject of this book. Were it not for Dr. Rhodin's selfless efforts, Selye's mechanism would surely have remained undiscovered for decades to come.

Dr. Rhodin was born in Lund, Sweden, and graduated from college in Gothenburg. He received his MD degree from the Karolinska Institute in Stockholm, Sweden, and practiced as a general surgeon for several years before returning to Karolinska to obtain a Ph.D. in anatomy. He was invited to New York University, where he became an associate professor. He was the recipient of numerous honors and awards, and he authored three renowned medical texts, "An Atlas of Ultrastructure" in 1963, "Histology: A Text and Atlas" in 1974 and "An Atlas of Histology" in 1975. His books were translated into several languages and continue to be used in medical schools around the world. In addition, he published hundreds of scientific research articles, dealing most notably with electron microscopy of microcirculation. In 1970 he received the Eugene M. Landis research award, the highest national award for excellence in microvascular research. That same year he became Chairman of the Department of Anatomy at New York Medical College. He left NYMC in 1974 to become Chairman of the Anatomy Department at the University of Michigan medical school. He returned to Sweden as Chairman and Professor of Anatomy at the Karolinska Institute in 1977. From 1977-1979 he served on the Selection Committee for the Nobel Prize in Physiology and Medicine. He became the Chairman of the Department of Anatomy at the University of South Florida in 1977. He retired from USF after 25 years of service on June 1, 2004.

Regrettably, Dr. Rhodin succumbed to cancer-related illness in 2004 at the age of 82, even as his dream was being fulfilled. Had he lived but a few years longer, he could have known that his vision was attained. May his memory endure.

Dr. William Collins (1934-2017)
Photo courtesy of his daughter, Cheri Schultz

Dr. William Collins (1934-2017) was my undergraduate mentor at the Ohio State University. His training paved the path to the discovery of Selye's mechanism. He served as my faculty advisor and guided my undergraduate research project. He introduced me to the classical scientific method, which shuns statistics and tests hypotheses under controlled conditions. Today statistics sometimes identify fruitful relationships,[1] but they are often manipulated for unscientific purposes. He emphasized that amateurs working outside the field can legitimately propose fresh scientific theories founded on the published research of others, regardless of their personal lack of research experience. He warned me that medical journals are notoriously reluctant to publish controversial discoveries, as exemplified by the Ohio State researchers Bruce Glick and Timothy Chang who were obliged to publish their discovery of the Bursa of Fabricius, a key element of the immune mechanism, in a poultry journal after it was repeatedly rejected by medical journals. It was discovered years later, whereupon it revolutionized immunology.[2] He prophetically warned me that his best friend in high school had idealistically sought a career in medicine, only to be surrounded by *circumferuntur venales* obsessives bereft of interest in science or progress.

Dr. Collins was the sole toxicologist at Ohio State. He gave his life to science, for as this was written he died of progressive brain damage that was most simply explained by progressive, irreversible carbamate insecticide exposure, though this remains unconfirmed. His fate accentuated his warning of the danger of toxic environmental substances, including pharmaceuticals.[3] I honor his memory.

Dr. Theodore Stanley (1940-2017)

Dr. Ted Stanley provided invaluable support during my review of research literature that led to the discovery of Selye's mechanism. His annual anesthesia conference introduced me to the chimeric nature of factor VIII, which inspired that adventure. He encouraged my otherwise solitary exploration of published literature. He coached and critiqued my fledging publication efforts. He arranged for two researchers at the University of Utah to review and critique my earliest description of Selye's mechanism. He invited me to present those early findings before the anesthesia research committee at the University of Utah. He helped me to refine my opioid-based anesthesia technique. He cautioned that I would find difficulty returning to the academic environment, but nevertheless supported my (unsuccessful) efforts to find a venue for clinical research. He helped me to organize and present a "point-counterpoint" session at the American Society of Anesthesiology that discussed CO_2 pathophysiology. He introduced me to prominent professors. He was especially intrigued by the extended biological implications of stress theory and its relevance to anesthesia, analgesia and allostasis. In my last conversation with him, shortly before his death, he warned that purposely drawing attention to prevailing carbon dioxide mythology would create politically potent enemies. He was an asset to the anesthesiology profession, and a warm, wonderful friend. Many will share his loss.

Last, but not least, I'd like to acknowledge Dr. C. Richard Chapman, a professor at the University of Utah, who reviewed and critiqued my earliest description of the stress mechanism, offered invaluable advice, and contributed the forward; my friend Jeff Walden, who served as my editor; Dr. Carl Cummings, my roommate who enabled me to survive the gauntlet of medical school; and my Mother, to whom I owe everything. All of them were indispensable.

EPIGRAPH

Hitch your wagon to a star; keep your seat, and there you are!
— Ralph Waldo Emerson

Here's to my friends, I wish them well;
and all my enemies can go to Hell.

NOTES

1 S. Jamal, S. Goyal, A. Shanker, A. Grover, Machine learning and molecular dynamics based insights into mode of actions of insulin degrading enzyme modulators. *Comb Chem High Throughput Screen*, (2017).

2 D. Ribatti, E. Crivellato, A. Vacca, The contribution of Bruce Glick to the definition of the role played by the bursa of Fabricius in the development of the B cell lineage. *Clinical and experimental immunology* 145, 1-4 (2006).

3 E. G. Vallianatos, M. Jenkins, *Poison spring : the secret history of pollution and the EPA*. (Bloomsbury Press, New York, ed. First U.S. edition., 2014), pp. xviii, 284 pages.

TABLE OF CONTENTS

Table of Figures

Foreword

C. Richard Chapman, PhD
Professor Emeritus
Department of Anesthesiology
University of Utah
Salt Lake City, Utah.

More than eight decades have passed since the young Hans Selye published his 1936 paper on the General Adaptation Syndrome. It preceded his adoption of the term "stress" but nonetheless laid the groundwork for the field of stress research. The General Adaptation Syndrome, as he conceived it, was "A syndrome produced by diverse nocuous signals" manifested as autonomic and hormonal responses. The concept came from clinical observations of patients who responded to challenging events in three sequential stages: alarm, resistance, and exhaustion. Selye proposed that privations, losses, traumatic events, and other demanding life experiences can evoke resource-depleting physiological coping responses that ultimately lead to disability, disease, or death. Some years would pass before he would define the burden of challenging events and conditions as stress, the term that now defines the field. Selye pursued and refined his topic relentlessly, ultimately publishing 32 books and 1500 journal articles. He sought to identify the physiological mechanisms of stress until the end of his career.

Selye was an endocrinologist, and so he conceptualized the stress response as fundamentally neuroendocrine. The mediators of the stress response were primarily autonomic changes, glucocorticoids, and catecholamines, and this still characterizes most contemporary thinking in the stress field. Although his original ideas are now obsolete, recent years have witnessed a renewed interest in the fundamental premises of Selye's work and extensions of his ideas. *Allostasis* refers to an individual's capacity to adapt to a stressor, including the ability to turn on or off emergency responses. *Allostatic load* refers to the burden of cumulative negative effects, including organ and tissue-specific damage. Of course, progress in the fields of endocrinology, immunology, and neurology has opened many potential insights into the physiological mechanisms of stress that were not in Selye's reach in the 20th century. A broader understanding of stress and its impact on health can now emerge.

In this book Lewis Coleman revisits Selye's vision and introduces an unprecedented extension of stress theory. The fundamental building blocks of Coleman's reasoning are not the standard neuroendocrine processes, although they remain intrinsic components, but rather regulation of hemodynamic physiology and tissue repair. He introduces the Mammalian Stress Mechanism, a systemic mechanism that continuously regulates hemodynamic physiology and maintains structure in mammals. Coleman's focus on hemodynamic physiology represents a radical departure from contemporary theories of stress and adaptation. He proposes a new understanding of the physiological mechanisms of stress that gives the vascular endothelium, rather than the hypothalamo-pituitary-adrenocortical axis, center stage.

The first chapter of the book provides historical background on stress theory and a critique of classical thinking in the field. Importantly, it provides the groundwork necessary for understanding the central role of hemodynamics. This chapter provides a basic overview of the Mammalian Stress Mechanism and makes the argument that this line of thinking goes well beyond accounting for the stress response. It explains most manifestations of physiology and, Coleman contends, it provides a Unified Theory of Biology.

The second chapter comprises a detailed description of the Mammalian Stress Mechanism. The vascular endothelium plays a central role, acting through the capillary gate component and tissue repair component. Its three key products are thrombin, soluble fibrin, and insoluble fibrin. The chapter covers the various agents of the mechanism in detail. It further explains Mammalian Stress Mechanism activation through three synergistic pathways: Tissue disruption, nociception, and cognitive pathways. These pathways are essentially categories of stressors in classical stress language. Coleman provides examples of each of these. Hyperactivity in the Mammalian Stress Mechanism can lead to disease. This is essentially a greatly revised view of Selye's General Adaptation Syndrome.

The third chapter addresses Selye's goal of defining a "Unified Theory of Medicine." Here Coleman explains the normal beneficial processes of the Mammalian Stress Mechanism and then addresses the pathological conditions that can arise from prolonged hyperactivity in this mechanism. These include accelerated capillary senescence, which can explain myriad diseases and syndromes. Coleman accounts for the generation of numerous seemingly unrelated pathologies. For example, Disseminated Intravascular Coagulation occurs when tissue factor abnormally enters flowing blood. Cancer results from an abnormal state of self-sustaining tissue repair activity induced by excessive and prolonged thrombin elevations. This perspective suggests new interventions based on the Mammalian Stress Mechanism for the prevention and treatment of various pathological conditions, and Coleman discusses some of them at the end of the chapter.

In fourth and final chapter, Coleman again pursues one of Selye's goals and offers the Mammalian Stress Mechanism as the basis for a Unified Theory of Biology. Evolution has produced speciation, and all living multicellular organisms face the challenges of adaptation and survival. Every class of complex multicellular animals and plants possesses a unique extracellular stress mechanism. Coleman discusses the role of stress mechanisms in evolution and the evolution of such mechanisms themselves. The distinction between mammals and reptiles receives substantial treatment. Of course, any discussion of evolution involves speculation, and this chapter is no exception. With that in mind, the material in this chapter is insightful and engaging, and the author succeeds in making a case for the importance of stress mechanisms in evolution.

In summary, this book represents a bold advancement of Selye's stress theory. Coleman departs from the assumption of a fundamental neuroendocrine mechanism and proposes instead an alternative based on tissue repair and regulation of hemodynamic physiology. The Mammalian Stress Mechanism, Coleman contends, has much greater explanatory power than Selye's original neuroendocrine concept and subsequent extensions of that concept. It provides the basis for improved diagnosis of stress related disease and suggests ways to prevent such disease when stressors are unavoidable. To validate the explanatory power of his theory, Coleman extends it to offer a Unified Theory of Medicine and a Unified Theory of Biology.

What impact will the Mammalian Stress Mechanism have? Science itself undergoes evolution, and only time will reveal the viability of these ideas. In the meanwhile, this book provides an unprecedented, provocative perspective on stress theory and many valuable, thought provoking ideas.

Preface

"It must be considered that there is nothing more difficult to carry out nor more doubtful of success nor more dangerous to handle than to initiate a new order of things; for the reformer has enemies in all those who profit by the old order, and only lukewarm defenders in all those who would profit by the new order; this lukewarmness arising partly from the incredulity of mankind who does not truly believe in anything new until they actually have experience of it.
— Niccolo Machiavelli

Fortuitous serendipity paved my path to the discovery of Selye's mechanism. My undergraduate advisor, Dr. William Collins, understood the danger of excessive specialization, and guided me to courses in cell biology, ethology, taxonomy, and botany. He was the only toxicologist at the university, and his impromptu lectures conferred an awareness of pharmaceutical as well as insecticide toxicity. I'll never forget his presentation of carbamate pesticides, which can destroy the brain. They undoubtedly caused his death years later. He stressed the importance of eliminating variables and developing testable hypotheses, and he explained that amateurs could legitimately propose fresh scientific theories founded on published research. He supervised my undergraduate honors research, which involved anesthetizing cockroaches with carbon dioxide. This inspired my enduring interest in carbon dioxide pathophysiology, because I suspected that CO_2 suffocated the cockroaches rather than anesthetizing them.

Miraculously, my basic medical sciences education at New York Medical College coincided with the sojourn of Dr. Johannes Rhodin, an internationally renowned researcher, who reformed the curriculum. His curriculum included lectures that confirmed my suspicions that carbon dioxide is an asphyxiant gas that mimics anesthesia and toxicity but is essential to life. In addition, Dr. Rhodin's lectures summarized the shortcomings of conventional medical theory, and left me convinced that stress theory must be correct. However, Selye's theory could not be confirmed without a testable mechanism, and soon thereafter, as Dr. Rhodin had feared, Selye's theory was largely abandoned and forgotten for failure to find such a mechanism.[1, 2, 3] Years later, I learned that Dr. Rhodin delivered these lectures to some 5,500 medical students during his career. His efforts paved the path to the discovery of Selye's mechanism, as he had hoped. Sadly, he died in 2004, even as I was discovering the long-sought stress mechanism.

Fortunately, I pursued a surgical internship followed by anesthesia residency. These placed me on the path to discover Selye's stress mechanism, because anesthesia *is* surgical stress control. During my residency I found myself surrounded by professors and residents who assumed that carbon dioxide has both toxic and narcotic properties. This conflicted with my medical school training and heightened my skeptical view of medical theory and practice.

For several years in private practice I tried unsuccessfully to alter conventional anesthetic technique to improve surgical outcome. Then, around 1990, capnography became available and critical care researchers simultaneously discovered *permissive hypercarbia.* I soon discovered that I could utilize capnography and permissive hypercarbia to enhance opioid dosage and optimize surgical outcome. I believed that I had discovered an important advance, but I had merely re-discovered the forgotten principles of Crile, Henderson, and Lundy.

Around 2001, in hopes of performing clinical research to confirm the safety and utility of my anesthetic technique, I sought the wisdom of Dr. Ted Stanley, the ranking anesthesiology expert in opioid use. Thus I attended his annual anesthesia conference in Salt Lake City. Coincidentally, this conference presented fresh information indicating that Factor VIII is a chimera, which is rare in biology. I immediately suspected that Factor VIII functions as an important physiological signaling mechanism. My intrigue evolved into an intense six-year computerized review of published literature, and I temporarily forgot clinical research. The unique characteristics of Factor VIII served as a "Rosetta Stone" that led first to fresh explanations of coagulation and atherosclerosis,[4, 5, 6] followed by a description of the capillary gate mechanism,[7, 8] and then the tissue repair mechanism. Ultimately I realized that these seemingly disparate mechanisms were elements of Selye's long forgotten stress mechanism.[9, 10] Next I realized that the mechanism conferred a fresh theory of anesthesia, analgesia, and allostasis that corroborated my clinical experience.[11] Still later I realized that Selye's mechanism exceeds the bounds of medicine and confers a Unified Theory of Biology that explains embryology, evolution, anatomy, ethology, intelligence, taxonomy, and other aspects of botany and zoology. Having comprehended the significance of my discovery, I resumed my efforts to perform clinical research to test Selye's mechanism, but I confronted obstacles at every turn.

As long anticipated, Selye's theory is prepared to complement DNA, extend its meaning, and provide a fresh paradigm of medical and biological research. It portends a new era of health, longevity and prosperity, free of the eternal scourge of disease and premature death. Beyond that, it may yet reveal the very secret of life, with implications that presently reside in the realm of science fiction. First, however, the mechanism must be acknowledged, tested, and confirmed. Sadly, in an era when commercial considerations obstruct scientific advance, that will literally require an Act of Congress.

Lewis S. Coleman, MD, FACA

NOTES

1 S. Fink, in *New York Academy of Sciences Magazine*. (New York Academy of Sciences, 2 E 63 Street, New York, N.Y. 10021, 2005), chap. 18, pp. 2.

2 R. M. Yehuda, B, Behavioral Stress Response: Protective and Damaging Effects. (*New York Academy of Science*, ed. Annals, 2005), pp. 331.

3 B. S. McEwen, E. Stellar, Stress and the individual. Mechanisms leading to disease. Arch Intern Med 153, 2093-2101 (1993).

4 L. S. Coleman, An Improved Explanation of Atherosclerosis. *PLoS Medicine* 2(4):e98, (2005).

5 L. S. Coleman, Atherosclerosis may be caused by inadequate levels of turbulence and mixing. *World J Surg* 30, 638-639 (2006).

6 L. S. Coleman, Insoluble fibrin may reduce turbulence and bind blood components into clots. *Med Hypotheses* 65, 820-821 (2005).

7 L. S. Coleman, A capillary hemostasis mechanism regulated by sympathetic tone and activity via factor VIII or von Willebrand's factor may function as a "capillary gate" and may explain angiodysplasia, angioneurotic edema, and variations in systemic vascular resistance. *Med Hypotheses*, (2005).

8 L. S. Coleman, A capillary hemostasis mechanism regulated by sympathetic tone and activity via factor VIII or von Willebrand's factor may function as a "capillary gate" and may explain angiodysplasia, angioneurotic edema, and variations in systemic vascular resistance. *Med Hypotheses* 66, 773-775 (2006).

9 L. S. Coleman, A Stress Repair Mechanism that Maintains Vertebrate Structure during Stress. Cardiovasc Hematol Disord Drug Targets, (2010).

10 L. S. Coleman, in *Hypotheses in Clinical Medicine*, e. a. Shoja MM, Ed. (Nova Biomedical, New York, NY, 2012), chap. 29.

11 L. S. Coleman, 30 Years Lost in Anesthesia Theory. Cardiovasc Hematol *Agents Med Chem* 10, 31-49 (2012).

Chapter 1
Introduction

"It is a popular delusion that the scientific enquirer is under an obligation not to go beyond generalization of observed facts... but anyone who is practically acquainted with scientific work is aware that those who refuse to go beyond the facts rarely get as far."
— Thomas Henry Huxley

Medicine forever suffers for lack of a theory of disease that enables physicians to treat the *cause* of disease rather than its *symptoms*. The purpose of this book is to introduce such a theory. It proposes a radical hypothesis that is at odds with orthodox medical beliefs, but is nevertheless predicated on published, peer-reviewed medical research and is compatible with all aspects of known science. It promises to revolutionize medicine, and introduce a new era of health, longevity, prosperity, and freedom from the eternal curse of disease and premature death.

The term *stress* is confusing because it is used in various unrelated contexts. For example, engineers regard stress as a force that compresses, twists, or otherwise deforms a structure or material. In medicine and biology stress refers to a physical, chemical, or emotional factor that causes injury or disease. Worse yet, the term is sometimes used to describe the effect of a physical, chemical, or emotional factor. In this book, the term *stress* will mean "any environmental factor that increases stress mechanism activity" and *disease* will mean "harmful stress mechanism hyperactivity that consumes body reserves and damages organs and tissues." The term disease will *not* apply to inherited genetic defects, chromosomal damage, or dietary deficiencies. This is elaborated in Chapter 16: The Unified Theory of Medicine, on page 175.

The tale of money wags the dog of medicine. Primitive technology and the death penalty for dissection hindered medical advance until the printing press provided paper money that fueled fighting, improved technology, and promoted medical advance (see Appendix A: Medical History, on page 395). This led physicians to abandon the traditional *Vis medicatrix naturae* (the body heals itself) concept in favor of the modern view that doctors can diagnose diseases and cure them with specialized treatments. Nevertheless, medicine remains an art based on experiment rather than a true science for lack of a theory that explains disease.

During the previous century of worldwide warfare and medical advance, a prominent researcher named Hans Selye postulated a *Unified Theory of Medicine* that attributes disease to environmental stresses (trauma, surgery, excessive radiation, fear, anxiety, toxic substances, and so forth) that induce hyperactivity in a *stress mechanism* that repairs tissues and regulates physiology.[1, 2] Selye's theory arguably represents the zenith of medical advance because it offers the simplest and therefore most likely explanation of embryology, physiology (body function), pathology (disease) and stress (factors that alter body function).[3] His theory, if verified, would reinstate the traditional medical viewpoint that the body heals itself if causative stresses are relieved and homeostasis (normal stress mechanism activity) is restored. Treatments that restore normal stress mechanism activity would theoretically revolutionize health care. However, Selye's theory remained useless for lack of a testable stress mechanism that explains how it works.

The discovery of DNA focused attention on Selye's ideas. Researchers suspected that his hypothetical stress mechanism works closely with DNA to convert the genetic blueprint into embryological development, and then remains active to repair tissues and regulate physiology, while DNA resumes quiescence once embryological development is complete. Some believed that the discovery of the stress mechanism would explain the very secret of life.

Selye's theory became the prevailing paradigm of medical research and inspired an intense international search for the stress mechanism that lasted 30 years. In an effort to facilitate their search, some stress researchers embraced *capillary gate theory*, which postulated the presence of a submicroscopic molecular mechanism that regulates blood flow in organs and tissues.[4] Others embraced *tissue repair theory*, which postulated the presence of a tissue repair mechanism that orchestrates tissue repair, which is known to proceed in an orderly fashion. However, the two concepts seemed incompatible, and nobody believed that a single stress mechanism could simultaneously explain both.

The era of stress research yielded numerous advances. It discovered many hormones, the immune mechanism, and coagulation enzymes, but it failed to find any testable mechanism that could explain any aspect of Selye's ideas, whereupon stress research was gradually abandoned and is now mostly forgotten.[5, 6] Since then, theoretical advance has foundered for lack of an effective research paradigm (see Appendix F: The Structure of Scientific Revolutions, on page 437, and Appendix G: The Central Dogma of Molecular Biology, on page 441), and there is growing frustration with the lack of theoretical progress.[7] Meanwhile, fresh evidence relevant to stress theory accumulated from unrelated research, and this finally revealed the elusive stress mechanism.[8, 9] Stress theory is thus poised to resume its role as the prevailing paradigm of medical research, renew theoretical advance, and revolutionize medicine.[10, 11]

NOTES

1 H. Selye, Sketch for a unified theory of medicine. *Int Rec Med Gen Pract Clin* 167, 181-203 (1954).

2 H. Selye, A syndrome produced by diverse nocuous agents. 1936. *The Journal of neuropsychiatry and clinical neurosciences* 10, 230-231 (1998).

3 https://en.wikipedia.org/wiki/Occam's_razor

4 Tannenberg, Uber die Kapillartatigkeit. *Verh dtsch. path.* Ges., 20:374, (1925).

5 R. M. Yehuda, B, *Behavioral Stress Response: Protective and Damaging Effects.* (New York Academy of Science, ed. Annals, 2005), pp. 331.

6 S. Fink, in *New York Academy of Sciences Magazine*. (New York Academy of Sciences, 2 E 63 Street, New York, N.Y. 10021, 2005), chap. 18, pp. 2.

7 J. Pontin, in *Technology Review,* J. Pontin, Ed. (MIT, MIT, 2010), vol. 114, chap. 8, pp. 1.

8 L. S. Coleman, A Stress Repair Mechanism that Maintains Vertebrate Structure during Stress. *Cardiovasc Hematol Disord Drug Targets*, (2010).

9 L. S. Coleman, in *Hypotheses in Clinical Medicine,* e. a. Shoja MM, Ed. (Nova Biomedical, New York, NY, 2012), chap. 29.

10 L. S. Coleman, A Stress Repair Mechanism that Maintains Vertebrate Structure during Stress. *Cardiovasc Hematol Disord Drug Targets*, (2010).

11 L. S. Coleman, in *Hypotheses in Clinical Medicine,* e. a. Shoja MM, Ed. (Nova Biomedical, New York, NY, 2012), chap. 29.

Chapter 2

Shortcomings in Present Medical and Biological Theory

"Beware of false knowledge; it is more dangerous than ignorance."
— George Bernard Shaw

An effective theory of medicine must explain the nature of physiology, pathology (disease), stress, and their relationships. Ideally, it should be compatible with a theory of biology that explains embryology, evolution, emotion, ethology, intelligence, anatomy, taxonomy, and other aspects of life.[1] This is easier said than done, for life is so overwhelmingly complex, confusing, and counterintuitive that it remains in the realm of religion.

Paleontology crudely tracks evolution and provides mute testimony that extinct species vastly outnumber those still living, but reveals few clues about behavior, intelligence, anatomy, physiology, and environmental circumstances that might explain their origin and extinction. Taxonomy categorizes animals and plants into kingdom, phylum, class, order, family, genus, species, and even sub-species based on observable anatomy and behavior. Mammals evolved directly from reptiles, and both are composed of similar cells, but they are nevertheless distinctly different. Reptiles thrive in warm climates, are covered with scales, and have minimal food requirements, limited exercise tolerance, inferior intelligence, and poor cold tolerance. Mammals suffer from exaggerated food requirements and poor heat tolerance, but they enjoy superior intelligence and exercise capacity, are covered with fur, and thrive in colder climates. The temperature-sensitive metabolism of eukaryotic cells provides only a partial explanation, because the exercise capacity of reptiles remains inferior to that of mammals regardless of temperature.[2]

Cellular mitosis, metabolism, differentiation, and apoptosis proceed at higher levels during embryological development than at any other time of life, but how this is orchestrated remains a mystery. The specialized cells of complex multicellular creatures retain the ability to re-organize themselves to restore organ tissues when they are separated from one another, and they retain the ability to de-differentiate and then re-differentiate into totally different specialized cell types.[3] The nature of these feats remains unknown. The DNA mechanism explains how genetic information is retained, replicated, and conveyed to subsequent generations, but it fails to explain how the chromosomal DNA of eukaryotic cells communicates outside the nucleus, let alone outside the cell, to enable the formation of complex multicellular structures and specialized organs and tissues during embryological development.

Darwin's presently prevailing explanation of evolution is plagued by shortcomings. It cannot explain the Cambrian Explosion where multiple complex animal forms suddenly appeared at the bottom of the oceans. It cannot explain Lamarck's observation that life grows more complex with time. Darwin insisted that evolution proceeds slowly via small gradual changes, so his theory cannot explain the phenomenon of *saltation*, wherein drastically different offspring suddenly and inexplicably appear in both animals and plants. In Darwin's time, the term *mutation* referred to such sudden changes, but since the discovery of DNA the term has come to be defined as "a sudden departure from the parent type in one or more heritable characteristics, caused by a change in a gene or a chromosome." This definition is problematic, because sudden, drastic environmental change precipitates both extinction and the sudden appearance of new species. Such environmentally induced change cannot be explained by DNA mutation, because Max Delbruck and Salvador Luria, the mentors of James Watson, famously proved that mutations occur at random and are not affected by environmental factors.[4] Lastly, Darwin's theory fails to explain the origin of life. Hypotheses of life's origin have thus far failed to gain universal acceptance, partly because human egocentricity insists that life originated on the earth's surface.[5]

Disease adds confusion by disturbing homeostatic mechanisms, but despite decades of intense medical research, modern medical science cannot explain the relationships between and among seemingly unrelated diseases, immune activity, inflammation, tissue repair, hemostasis, hemodynamic physiology, and cellular hormones such as chemokines, cytokines, and prostaglandins. The innate ability of the body to mend itself without assistance remains so baffling to any discussion. Medicine remains riddled with *syndromes* (groups of associated symptoms) that remain unexplained, such as SIRS, MERS, COVID, ARDS, MOFS, eclampsia, and rheumatoid disease.

Cannon's lingering legacy (Walter B. Cannon, 1871-1945, see Appendix A: Medical History, on page 395 for a quick summary of his contributions) likens hemodynamic physiology to the static hydraulic brake system of a car, where force applied to the brake pedal transmits instantaneously and equally throughout all inner surfaces of the hydraulic braking system. This implies that the blood pressure force should appear simultaneously with the onset of cardiac systole (contraction of the heart), and remain equal throughout the arterial tree until the cardiac muscle relaxes with the onset of diastole (relaxation of the heart). Instead, blood pressure appears during diastole, after the heart muscle is at rest, in the form of a burst of turbulence that speeds throughout the arterial tree, momentarily halting blood flow as it travels, and generating turbulent lateral forces that press against the inner walls of arteries as the pulse waves propagate towards the periphery of the arterial tree. These lateral forces manifest as the palpable pulse and enable blood pressure measurement, but they are perturbed by temperature, arterial length, arterial diameter, arterial curvature, arterial bifurcations, arterial strictures, wave reflections, heart rate, blood viscosity, and small variations in cardiac contractility. Their intensity increases with decreasing arterial diameter, and varies from location to location, so that blood pressure measurements obtained at different locations are invariably different, and "accurate" blood pressure is a misnomer. Invasive monitoring is no more accurate, for it must be calibrated to cuff pressure. Meanwhile, cardiac contractility varies little from beat to beat, and flow resistance (viscosity, aka *systemic vascular resistance*) is the primary determinant of cardiac output, cardiac efficiency, tissue perfusion, and tissue oxygenation. Counter intuitively, blood pressure and atherosclerosis resistance increase in concert as arterial diameter declines toward the periphery. Efforts to derive cardiac output from conventional blood pressure and pulse wave measurements have proved futile.

The terms *vasoconstriction*, *vasodilation*, and *vasoactive* reflect Cannon's crude concept that opposing muscular forces of cardiac contractility and vascular contraction regulate cardiac output and tissue perfusion in accord with autonomic balance. This implies that hemodynamic physiology is inherently inefficient, because both muscular contraction and muscular relaxation are energy expensive events. Cannon accordingly regarded the heart as the "Charles Atlas" of the mammalian body that possesses enormous reserves of power and force that enable cardiac output to increase by a factor of ten during athletic endeavor. The available evidence, however, indicates that animal physiology is efficient, and that the mammalian heart is a "90-pound weakling" that consumes less energy than a 40-watt light bulb and generates the same force with each heartbeat, even during intense exercise. Cannon's concept implies that elephants should possess gigantic hearts that generate high blood pressure necessary to propel blood to distant parts of their gigantic bodies, but mammalian hearts consistently weigh only 1% of total body weight and are so weak that they cannot pump blood more than a few inches above their height.[6] Mammalian brains are accordingly located only inches above the heart to facilitate perfusion. The giraffe is an obvious exception, but it requires a disproportionately gigantic heart that compromises other organs.

Were blood pressure to be the driving force of blood flow, then hypertension should be accompanied by increased cardiac output and enhanced tissue perfusion; instead, hypertension is a prelude to congestive heart failure and is characterized by low cardiac output, poor tissue perfusion, impaired exercise capacity, and limited longevity. Cannon's concept implies that exercise conditioning should exaggerate the size and weight of the heart in order to generate higher blood pressure to sustain exercise requirements. Instead, exercise conditioning induces negligible increases in heart size and weight, and trained athletes exhibit capillary proliferation that reduces flow resistance and lowers blood pressure to abnormal levels at rest, maintains normal blood pressure during intense exercise, and enhances longevity.

The blood pressure generated by cardiac bypass pumps is often so low that it cannot be detected by conventional blood pressure cuff technique, and yet it sustains tissue perfusion and oxygenation. Left-ventricular assist devices (LVAD) produce neither palpable pulse nor detectable blood pressure, and yet they not only sustain life but also enable moderate exercise. By conventional standards, these patients should be dead. These observations indicate that conventional instruments do not measure the driving force of blood flow, and blood pressure measurement does not reflect the driving force of tissue perfusion.

Muscular vasoconstriction is energy extravagant and soon followed by muscular exhaustion and obligatory vasodilation, so arteriolar vasoconstriction cannot explain essential hypertension. Instead, essential hypertension is characterized by accelerated microvascular deterioration that undermines tissue perfusion, exaggerates flow resistance, and increases cardiac work. The increased cardiac work induces myocyte hypertrophy that is soon followed by myocyte apoptosis and pathological collagen deposition that progressively exaggerates cardiac wall thickness and undermines cardiac function until heart failure ensues.[7] In contrast, exercise conditioning proliferates capillaries in affected muscles. The resulting reduction in flow resistance causes hypotension in resting athletes, and it enables 10-fold increases in cardiac output during intense exercise with negligible increases in cardiac size, muscle mass, or blood pressure. Such observations indicate that the heart functions on the basis of extreme efficiency rather than power and force.

The shortcomings of blood pressure were recognized from the beginning, but it stubbornly persists as a confusing and counterproductive mainstay of medical monitoring and treatment. Researchers and clinicians intuitively equate hypertension with effective cardiac reserve, and regard hypotension as a harbinger of death and disaster. In reality, tachycardia and hypertension invariably reflect cardiac distress, reduced cardiac output, and tissue ischemia, while bradycardia (slower heartbeat) and hypotension characterize enhanced cardiac efficiency and tissue perfusion, as in normal sleep and resting athletes. Vasodilator drugs such as nitroprusside lower blood pressure even as they enhance cardiac output and tissue perfusion, and alleviate congestive heart failure.[8] Vasopressor drugs such as Ca+ and epinephrine exaggerate blood pressure even as they undermine cardiac output and tissue perfusion.[9] Streptokinase anticoagulation has no effect on muscular contractility, but it restores perfusion to infarcted tissues even as it counter-intuitively reduces blood pressure.[10, 11, 12]

Familiar *Newtonian fluids* such as water, oil, steam, and atmospheric gases exhibit exponential *increases* in flow resistance (viscosity) when they are accelerated in pipes, but *non-Newtonian* blood exhibits exponential *decreases* in flow resistance when it is accelerated in arteries. Medical literature routinely ignores the significance of this observation.

Medical textbooks routinely assume that the sole purpose of red cells is to transport oxygen to peripheral tissues, even though hemoglobin and its congeners circulate freely in the blood of invertebrates, and hemoglobin encapsulation does not enhance oxygen transport and delivery. Hemoglobin encapsulation distinguishes vertebrates from invertebrates, and distinct red cell morphology characterizes each class of vertebrates, but its significance remains unknown. Mammalian red cell mass is far greater than necessary for effective oxygen transport; approximately one-quarter of all cells in the adult human body are blood cells, and stressful nervous activity exaggerates red cell production.[13] Most vertebrates have large, nucleated red cells that retain the Krebs cycle (and thus produce energy), but mammals are characterized by small, enucleated, biconcave red cells that lack the Krebs cycle, have abbreviated life spans, and are so closely similar as to be nearly indistinguishable among species.

Camels, llamas, and alpacas are presently classed as mammalian, but their red cells are oval and retain a nucleus. Compared to other mammals they suffer notoriously inferior exercise capacity. A satisfactory explanation is presently lacking.

Atherosclerosis defies explanation. Down's Syndrome, anemia, exercise, coagulopathies, anticoagulants, and moderate red wine consumption all inhibit atherosclerosis. Atherosclerosis decreases — and blood pressure increases — with decreasing arterial diameter. Advancing age, increased blood viscosity, emotional stress, sedentary life style, hyperthyroidism, hypothyroidism, hypertension, polycythemia, congestive heart failure, obesity, and chronic diseases all accelerate atherosclerosis. Earlier researchers attributed atherosclerosis to inadequate shear stress,[14] but this fails to explain why atherosclerosis decreases with decreasing arterial diameter. Presently prevailing theories that attribute atherosclerosis to lipoproteins or primary lesions of the vascular endothelium are similarly lacking.[15]

The nature of coagulation remains unclear. Capillary hemostasis occurs within minutes even though capillaries cannot contract. Virchow's Triad identifies hypercoagulability of blood, stasis of blood flow, and endothelial injury as essential elements of coagulation but does not explain how these factors enable clot formation or dissolution. The triad is named after the eminent German researcher Rudolf Virchow (1821-1902), but Virchow never suggested the idea of a triad.[16] Instead the Triad represents a consensus that was achieved decades after Virchow's death. Modern research has identified the *coagulation cascade*, which describes the sequence of blood enzyme interactions that enable clot formation. Insoluble fibrin, the end product of the cascade, is essential for both clot formation and capillary hemostasis, but it has no enzymatic properties that might explain the nature of hemostasis. There is considerable evidence that coagulation cascade enzymes influence tissue repair, but the relationship remains unclear.

The nature of tissue repair remains unknown. Tissue injury initiates an orderly sequence of coagulation, inflammation, chemotaxis (movement towards a chemical stimulus), immune activity, cell proliferation, and cell differentiation, but no mechanism has been identified that governs the tissue repair process and explains its nature.

Conventional medical theory cannot explain the nature of disease and disease relationships. Most diseases produce strikingly similar symptoms and manifestations, such as fever, rashes, exudates, bowel disturbance, dementia, tachycardia, hypertension, malaise, fatigue, edema, catabolism (destructive metabolism), and inflammation. Seemingly unrelated diseases have unexplained relationships. Hypertension is typically accompanied by diabetes, and vice-versa. Obesity exaggerates the risk of hypertension, diabetes, and cancer. Cancer increases the risk of infarction. Chronic infections and rheumatoid diseases increase the risk of cancer, diabetes, and hypertension.

Exaggerated immune activity occurs in association with surgery, trauma, tissue repair, rheumatoid disease, and malignancy. Rheumatoid disease is often categorized as an *autoimmune disease* simply because immune activity can be detected, but there is no evidence that immune activity attacks body tissues. There is no convincing evidence that immune activity either causes or prevents cancer.

Anesthesia suffers for lack of an effective theory that enables the alteration of anesthetic technique to optimize surgical outcome.[17] The nature of anesthesia, analgesia (pain relief), allostasis, nociception (the perception of pain), pain, fear, and their relationships remains unclear.

Animal studies suggest that stress is a normal and even necessary part of life.[18] Animals exhibit physiological perturbations in response to numerous forms of environmental stress, including injury, disease, starvation, cold, heat, predation, parasitism, and confrontations over mates and food. They maintain and repair their structural integrity and sustain their internal milieu despite such stresses. The nature of these phenomena remains unclear.

NOTES

1 S. C. Campbell-Staton *et al.*, Winter storms drive rapid phenotypic, regulatory, and genomic shifts in the green anole lizard. *Science* 357, 495-498 (2017).

2 A. F. Bennett, Gleeson, Todd T., Gorman, George C., Anaerobic metabolism in a lizard (Anolis bonairensis) under natural conditions. *Physiol. Zool.* 54, 237-241 (1981).

3 R. O. Becker, G. Selden, *The body electric : electromagnetism and the foundation of life*. (Quill, New York, ed. 1st Quill, 1985), pp. 364 p.

4 S. E. Luria, M. Delbruck, Mutations of Bacteria from Virus Sensitivity to Virus Resistance. *Genetics* 28, 491-511 (1943).

5 S. L. Miller, A production of amino acids under possible primitive earth conditions. *Science* 117, 528-529 (1953).

6 J. Prothero, Heart weight as a function of body weight in mammals. *Growth* 43, 139-150 (1979).

7 V. P. van Empel *et al.*, Myocyte apoptosis in heart failure. *Cardiovasc Res* 67, 21-29 (2005).

8 G. Tatu-Chitoiu *et al.*, Streptokinase-induced hypotension has no detrimental effect on patients with thrombolytic treatment for acute myocardial infarction. A substudy of the Romanian Study for Accelerated Streptokinase in Acute Myocardial Infarction (ASK-ROMANIA). *Rom J Intern Med* 42, 557-573 (2004).

9 L. S. Coleman, A Stress Repair Mechanism that Maintains Vertebrate Structure during Stress. *Cardiovasc Hematol Disord Drug Targets*, (2010).

10 G. Tatu-Chitoiu *et al.*, Streptokinase-induced hypotension has no detrimental effect on patients with thrombolytic treatment for acute myocardial infarction. A substudy of the Romanian Study for Accelerated Streptokinase in Acute Myocardial Infarction (ASK-ROMANIA). *Rom J Intern Med* 42, 557-573 (2004).

11 A. S. Lew, P. Laramee, B. Cercek, P. K. Shah, W. Ganz, The hypotensive effect of intravenous streptokinase in patients with acute myocardial infarction. *Circulation* 72, 1321-1326 (1985).

12 H. D. Neuhof H, Glaser E, Wolf H, Lasch H.G, Streptokinase Therapy in Patients with Acute Myocardial Infarction. *Intensive Care Medicine* Volume 1, 27-30 (1975).

13 M. A. Al-Drees *et al.*, Making Blood: The Haematopoietic Niche throughout Ontogeny. *Stem Cells Int* 2015, 571893 (2015).

14 M. Texon, The Hemodynamic Concept of Atherosclerosis. *Bull N Y Acad Med* 36, 263-274 (1960).

15 P. Libby, Atherosclerosis: the new view. *Sci Am* 286, 46-55 (2002).

16 C. N. Bagot, R. Arya, Virchow and his triad: a question of attribution. *British journal of haematology* 143, 180-190 (2008).

17 L. S. Coleman, 30 Years Lost in Anesthesia Theory. *Cardiovasc Hematol Agents Med Chem* 10, 31-49 (2012).

18 S. M. Korte, J. M. Koolhaas, J. C. Wingfield, B. S. McEwen, The Darwinian concept of stress: benefits of allostasis and costs of allostatic load and the trade-offs in health and disease. *Neurosci Biobehav Rev* 29, 3-38 (2005).

Chapter 3

Prelude to Stress Theory

"The difference between stupidity and genius is that genius has its limits."
— Albert Einstein

At the turn of the 20th century, the impending apocalypse of WWI accelerated medical research in both Europe and North America, and introduced an era of wartime medical advance that produced germ theory, cell theory, penicillin, polio vaccine, improved anesthetic techniques, and other advances. By the end of World War II, it appeared that continuous medical progress would eradicate disease altogether. The story of DNA exemplifies the idealism that characterized postwar medical research during that promising era.

Oswald Avery (1877-1955) originally planned to become a minister, but he concluded that he could better serve mankind by devoting his life to medical research. He began his research career investigating the disastrous 1918 influenza epidemic that killed more people than WWI, and then performed years of meticulous microbial experiments at Rockefeller University which proved that DNA is the medium of genetic storage and transmission. His research set the stage for the discovery of the DNA molecular structure.

Like Avery, Rosalind Franklin (1920-1958) was an idealist who devoted her life to medical research, even though she was heir to a banking fortune. She developed unique expertise at taking X-ray diffraction photographs and measurements of the DNA molecule that enabled Watson and Crick to decipher its structure. Like Avery, she never married, and she died tragically of ovarian cancer at the age of 37.

James Watson (1928–) and Francis Crick (1916-2004) became friends when they were assigned adjacent desks at Cambridge University. Watson was a protégé of Salvador Luria, who with Max Delbruck had famously proved that genetic mutations occur at random and are not affected by environmental factors.[1] Crick was a mathematics genius who had helped to develop magnetic mines that devastated German shipping during WWII.[2] Their research experience was nil, but they understood the significance of DNA, and they combined their efforts to describe its structure. Franklin's X-ray photographs of the DNA molecule enabled Crick to calculate the reverse angles of the atoms of the DNA molecule, and his calculations enabled Watson to construct a model of the DNA molecule that explains how genetic information is retained and reproduced. They published their discovery in

1953 and were awarded the Nobel Prize in 1962.[3] Both Avery and Franklin lived to witness the triumph, but died too soon to share the prize.

Meanwhile, Hans Selye, a talented endocrinology researcher, had serendipitously discovered that a wide variety of stresses produce a striking triad of adrenal enlargement, thymus shrinkage, and stomach ulceration in rats. He summarized his observations in a classic paper entitled "A syndrome produced by diverse nocuous agents" in 1936.[4] His evolving ideas about the role of stress attracted research funds, and in 1954, immediately after the DNA discovery, Selye published a paper entitled "Sketch for a unified theory of medicine."[5] The DNA discovery focused intense interest on Selye's ideas, because the DNA mechanism by itself cannot explain how chromosomal DNA exerts its influence outside the eukaryotic cell. Many therefore believed that Selye's hypothetical mechanism would most simply explain embryological development, tissue repair, and hemodynamic physiology, and that the discovery of such a mechanism would potentially revolutionize medicine and biology. These exciting ideas inspired an intense international search for the stress mechanism that lasted for 30 years but ended in frustration and failure to find any clue of any testable mechanism that could confirm Selye's idea.

NOTES

1 S. E. Luria, M. Delbruck, Mutations of Bacteria from Virus Sensitivity to Virus Resistance. *Genetics* 28, 491-511 (1943).

2 F. Crick, *What mad pursuit : a personal view of scientific discovery.* Alfred P Sloan Foundation series (Basic Books, New York, 1988), pp. xiii, 182 p., 188 p. of plates.

3 J. D. Watson, F. H. Crick, The structure of DNA. *Cold Spring Harbor symposia on quantitative biology* 18, 123-131 (1953).

4 H. Selye, A syndrome produced by diverse nocuous agents. 1936. *The Journal of neuropsychiatry and clinical neurosciences* 10, 230-231 (1998).

5 H. Selye, Sketch for a unified theory of medicine. *Int Rec Med Gen Pract Clin* 167, 181-203 (1954).

Chapter 4

Hans Selye and the Birth of Stress Theory

"One curious result of this inertia, which deserves to rank among the fundamental 'laws' of nature, is that when a discovery has finally won tardy recognition it is usually found to have been anticipated, often with cogent reasons and in great detail."
— Ferdinand Canning Scott Schiller

Success comes after a lifetime of dedicated toil soon followed by death, but science is seldom so kind. Powerful scientific theories typically require fresh evidence, and visionary theorists seldom outlast their critics. Such was the fate of Hans Selye, the Founder of stress theory, which represents the best hope for an effective theory of medicine.

Selye was born in Vienna, Austria during the declining days of the Hapsburg Empire. His father was Hungarian, of Magyar descent, and his mother was Austrian, of German descent. He grew up in Komarom, a small Hungarian town located on the Danube midway between Vienna and Budapest. The Magyars were the last Caucasian tribe to fight their way into Europe. Their ferocity is legendary, and their cohesive culture persists to the present.[1] Historically, they were close allies of the Khazars, whose powerful empire embraced Judaism. Numerous Khazars took refuge with the Magyars after their empire collapsed.[2]

As a child Selye was exposed to intense Magyar nationalism, and his autobiography describes how he was despised as an Austrian in Hungary and as a Hungarian in Austria. As a schoolboy in the aftermath of WWI he was arrested and put on trial in a Czechoslovakian court for wearing a Hungarian costume in public. He escaped serious punishment, but one can appreciate his terror, because many revolutionaries, even schoolchildren, were summarily shot for bearing firearms. The intensity of these political passions may be judged in the context of the 1956 Hungarian Revolution, where the civilian population, armed with little more than "Molotov cocktails," spontaneously rebelled against Russian oppression. Russian tanks and artillery quelled the rebellion. By this time Selye was living in Canada, but a stray bullet killed his mother during the fighting.

Selye's family enjoyed modest wealth, for he descended from physicians on his father's side, and from bankers and lawyers on his mother's side. His mother maintained servants even during WWI while his father served as a field surgeon on the Russian front for five years. Governesses, private schools, and personal tutors assisted his childhood education, and he attended a Benedictine monastery. He later envied "self-made men" who succeeded without such preparation. Though he described himself as a mediocre student, he had a photographic memory and could readily recall the details of the numerous journals he consumed voraciously in various languages.

Selye's father returned from the war suffering jaundice and emaciation. Soon thereafter the family lost its fortune due to the collapse of the Empire, leaving Selye wary of the vicissitudes of wealth. Komarom was shelled and invaded by competing communist and Czechoslovakian forces, and Selye became a Czechoslovakian without leaving home. Fortunately, his parents established a profitable surgical clinic that enabled him to enter the German Medical School in Prague at the age of seventeen.

Like that of all modern medical students, Selye's medical education emphasized accurate diagnosis to enable specific treatments. This approach arose in the 19th century, after bacteriological research indicated that individual pathogens cause specific diseases, and founded germ theory. Nutrition and endocrinology studies, cell theory, vaccine development, and antibiotics soon followed. These advances conferred the conviction that individual pathogens cause distinct diseases with characteristic symptoms that necessitate individualized treatments. Selye must have been exceptionally perceptive, for he immediately questioned this conventional reasoning. His autobiography describes the seminal epiphany that was to dominate his life:

"As far as I can remember, I first thought of what I later called 'biologic stress' when I was a second-year student at the University of Prague. During the first two years of our curriculum, we were never confronted with a patient; we followed exclusively pre-clinical courses in anatomy, histology and biochemistry. It was only at the end of the second year that the great moment arrived which is awaited with so much anticipation by all students of medicine, that of being faced for the first time with a patient. I will always remember the first lesson of our eminent internist, Professor von Jaksch, one of the most famous hematologists of that period. The subject of his introductory lecture was diagnosis. He asked his assistants to bring him five patients, especially chosen from different services of the University Hospital, who suffered from completely unrelated maladies. In every case he could arrive at the correct diagnosis by merely asking questions and looking for specific signs and symptoms of disease. I was deeply impressed by the logic of his interrogation and the precision of his observations, based on many years of experience. Almost all the questions he asked concerned diagnostic indexes, the very existence of which was unknown to me. For example, I remember that he looked for the reaction of a patient's pupils to light and for a series of other reflexes. In the case of a child sent over from the pediatrics clinic with a high fever, he examined the inner surface of the cheek, on which he discovered little white spots, and immediately he exclaimed 'Measles!' He then explained to us that these little spots were first described by professor Koplik and are known as 'Koplik spots.' They are neither painful nor likely to turn into anything dangerous, but they are quite characteristic of measles. In a similar manner he managed to identify the cause of the disease in every case. I must admit that he had a little luck, because this is not always possible by mere inspection and questioning; yet it was an awe-inspiring spectacle to see that medicine had progressed so far that it was now possible to for an experienced physician to make a correct diagnosis in so many different cases, even without the use of any complicated instrumentation or chemical examination. However, as soon as I recovered from the first emotional impact of my boundless admiration, I began to wonder why the professor did not say a word about all those signs and symptoms of disease which were perfectly obvious even to me, without any previous knowledge of practical medicine, which even the patients must have recognized, for they were what induced them to seek medical advice. All five patients, whatever their disease (one suffered from cancer of the stomach, another from tuberculosis, yet another from intense burns), had something in common: *they all looked and felt sick.* This may seem ridiculously childish, and self-evident, but it was because I wondered about the obvious that the concept of 'stress' was born in my mind."[3]

Selye immediately appreciated the implications of his epiphany:

"I wondered why the already well-known methods of exact scientific investigation, of looking for quantitatively measurable biochemical, microscopic or functional changes, could not be employed to clarify the mechanism of 'the syndrome of just being sick.' Once this mystery was solved we might even succeed in suppressing the stereotyped manifestations of illness in all patients, whatever the specific underlying cause of their particular malady. We might prevent the loss of weight, the feeling of weakness and other less uniformly displayed but still quite common disease manifestations, such as generalized aches and pains, coated tongue, intestinal disturbances, loss of appetite, fever, mental confusion, enlarged liver and spleen, inflamed tonsils, skin rash, inflammation, weight loss, and loss of energy. After all, we already knew that antipyretics could restore body temperature to normal in different kinds of fever, and we had found that analgesics would diminish pain due to diverse causes. It seemed plausible, therefore, that treatment could be directed against the nonspecific manifestations, rather than against particular disease producers."

Selye's professors and fellow students had little patience for his unorthodox insight. Their hostility caused him to abandon his ideas for ten years while he pursued his career. He completed medical school, obtained a PhD in organic chemistry, and won a prestigious Rockefeller scholarship to study at Johns Hopkins University in 1931. He was repelled by the egalitarian atmosphere at Johns Hopkins, and transferred his Fellowship to McGill University in Montreal, where he became a colleague of James Collip, a famous researcher who had helped to discover insulin. He became a member of the faculty and a Canadian citizen.

Soon after Selye began his research career, he believed that he had discovered a new female sex hormone, but he could find no evidence of estrogen effects when he injected his crude extracts into experimental animals. Instead, the animals became very sick, and some of them died. Autopsies revealed three striking observations: marked enlargement of the adrenal glands, shrinkage of the thymus, and ulcerations of the stomach. Reasoning that formaldehyde contamination might have caused these effects, he injected pure formaldehyde into several animals, which produced identical results. These observations re-awakened his medical school ideas and he wondered if other circumstances could elicit the same reaction, whereupon he exposed rats to toxic chemicals, scorching heat, freezing cold, intense light, deafening noise, strenuous exercise, and psychological frustration. All these produced the same pathology in adrenals, thymus, and stomach. In 1936 he reported these observations in his now-famous letter to Nature entitled "A Syndrome Produced by Diverse Nocuous Agents."[4] He came to believe that he had discovered a universal reaction to stress that could explain the nature of disease and disease relationships.

Research funding by profit-oriented corporations was disparaged in Selye's time, but his promising ideas enabled him to obtain generous funding from non-profit private foundations. In 1945, he moved to his own Institute of Experimental Medicine and Surgery at the University of Montreal. By 1952 Selye had hypothesized that disease represents a "fight to maintain the homeostatic balance of our tissues, despite damage."[5, 6]

James Watson and Francis Crick discovered DNA in 1953, and in 1954 Selye published a paper entitled "Sketch for a unified theory of medicine."[7] The unprecedented discovery of DNA caused intense excitement in the world of medical research. Researchers soon identified the DNA/RNA/transcription/translation mechanism that explains how prokaryotic cells manufacture proteins and enzymes, and identified the amino acids that compose them. However, all DNA research had utilized prokaryotic (bacterial) cells that are vastly simpler than the eukaryotic cells that exclusively compose structurally complex multicellular animals and plants. In bacterial cells, DNA exists as simple strands that float free in the cytoplasm. In contrast, the DNA of animal and plant cells exists in the form of chromosomes that are isolated from the cytoplasm by a thick nuclear membrane. Scientists quickly realized that the DNA mechanism by itself cannot explain embryological development, let alone physiology, pathology and stress. Many therefore reasoned that Selye's putative extracellular stress mechanism works closely with chromosomal DNA to convert genetic information into embryological development, whereupon the DNA mechanism resumes quiescence, while the stress mechanism remains active for the duration of life to regulate tissue repair and hemodynamic physiology. This represented the simplest available explanation, and the discovery of the stress mechanism promised to enable Selye's idea of a revolutionary unified theory of medicine. Some believed that the stress mechanism would reveal the very secret of life. These promising prospects inspired an intense international search for the stress mechanism that became the prevailing paradigm of medical research for more than 30 years, and Selye became the most prominent medical researcher in the world.

Selye struggled to derive terms that described his concept, develop a hypothesis that explained it, and devise techniques to test it. He altered his ideas to accommodate accumulating evidence from stress research. He selected the term *stress*, which he defined as "the nonspecific response of the body to any demand" but this failed to distinguish between cause and effect, and he was obliged to add the term *stressor* to describe the source of *stress*. He proceeded to subject animals to prolonged exposure to noxious agents, whereupon he observed a predictable sequence of pathological manifestations. He subdivided these into a *Stage of Alarm* that was followed by a *Stage of Resistance* in which he presumed the body's defense mechanisms to be maximized, and a *Stage of Exhaustion* that culminated with death. Selye viewed these stages as elements of a *General Adaptation Syndrome* (GAS). Detailed autopsies of these animals revealed gross and microscopic changes identical to those found in human patients suffering from arthritis, diabetes, kidney disease, hypertension, heart disease, and gastrointestinal ulcers. Selye called these *Diseases of Adaptation*. He found that he could produce any and all of these pathological manifestations by sensitizing the animals with dietary or hormonal perturbations and then subjecting them to various forms of stress.

Selye embraced Cannon's concepts of homeostasis, simpatico-adrenal system effects, and *fight or flight*. In his time it was understood that the body releases *stress hormones* in response to stressful stimuli, so he assumed that these hormones were the key to understanding the nature of stress.[8] He assumed that the mammalian body possesses homeostatic mechanisms that produce hormones that "defend" the body against stressful effects. He demonstrated that the adrenal gland produces two types of hormones that he called *syntoxic* (anti-inflammatory glucocorticoids) and *catatoxic* (pro-inflammatory mineralcorticoids). He envisioned inflammation as a "basically useful response that attempts to seal off or quarantine an intruder by a wall of inflammatory tissue, which separates the directly affected area from the surrounding healthy tissue; in this way, it prevents a disease producer from multiplying and spreading throughout the blood." These ideas seem misguided in retrospect, for they provided the conceptual foundation for treating inflammation with steroids, a practice that remains widespread despite little evidence of lasting benefit and considerable evidence of harm. His ideas inspired others to clarify the existence and function of HPA hormones, as they are understood today. Selye also introduced the classical terms *eustress* and *distress*. Eustress is beneficial effects produced by stress, such as enhanced athletic performance induced by exercise training. Distress is the opposite of eustress. It is destructive effects such as social withdrawal, disease, hypertension, inflammation, organ damage, and premature death that are provoked by excessive, unremitting, harmful stress.

Selye introduced the *Local Adaptation Syndrome* (LAS) to account for localized effects that did not involve systemic manifestations, and he developed standardized experimental techniques to assess stressful effects. For example, he discovered that injecting air under the skin caused the formation of a *granuloma pouch* that could be used to assess chemical toxicity, and he devised a quick and simple means to remove a rat pituitary gland without performing a craniotomy.

Meanwhile, stress researchers pursued two seemingly incompatible ideas that might explain the nature of the Selye's hypothetical stress mechanism. One group investigated *capillary gate theory* which postulated the presence of a sub-microscopic molecular mechanism that regulated capillary flow resistance. The other investigated *tissue repair theory* which postulated a physiological mechanism that regulated the tissue repair process, which was known to occur in a series of predictable stages.

Fueled by generous postwar government research subsidies, stress research produced a flood of reports. In an era when computer technology remained primitive and the Internet was unknown, Selye was obliged to invent a new library system to keep track of the fresh information. He ultimately published a summary of stress research in book form.[9]

Numerous important advances were achieved during the era of stress research. The coagulation cascade was described, the immune system was discovered, HPA hormones were clarified, and apoptosis was recognized. However, the search for the stress mechanism was a frustrating failure. No clue could be found of any mechanism that might explain any aspect of Selye's theory. Selye summarized the results of this massive effort as follows:

1. There is not merely an acute emergency response of the body but an entire General Adaptation Syndrome consisting of the alarm reaction followed by the stage of resistance and eventually by the stage of exhaustion.
2. The pituitary and adrenal cortex play a decisive role in this response.
3. The reaction is totally nonspecific and merely represents the body's reply to the common aspects of any kind of demand.
4. The most diverse diseases can result from stress, depending on the simultaneous influence of different conditioning agents.

As Selye ruefully observed, "All this does not help us as yet to define biologic stress or disease in precise terms; still it does leave us with a vague feeling that various types of treatment and many, if not all, diseases have certain things in common, have certain nonspecific, stereotyped features. Could all this vagueness be translated somehow into measurable terms?"

NOTES

1 W. Durant, A. Durant, Rouben Mamoulian Collection (Library of Congress), *The story of civilization*. (Simon and Schuster, New York,, 1935).

2 A. Koestler, *The thirteenth tribe : the Khazar Empire and its heritage*. Picador (Pan Books, London, 1977), pp. 224 p.

3 H. Selye, *The stress of my life : a scientist's memoirs*. (McClelland and Stewart, Toronto, 1977), pp. 272 p.

4 H. Selye, A syndrome produced by diverse nocuous agents. 1936. *The Journal of neuropsychiatry and clinical neurosciences* 10, 230-231 (1998).

5 H. Selye, *Annual Report on Stress*. (Acta, 1951), vol. Volume 1 of The Physiology and Pathology of Exposure to Stress: A Treatise Based on the Concepts of the General-adaptation-syndrome and the Diseases of Adaptation.

6 H. Selye, The general adaptation syndrome as a basis for a unified theory of medicine. *Oral Surg Oral Med Oral Pathol* 5, 408-413 (1952).

7 H. Selye, Sketch for a unified theory of medicine. *Int Rec Med Gen Pract Clin* 167, 181-203 (1954).

8 https://www.ncbi.nlm.nih.gov/pmc/articles/PMC1447286/#r4

9 H. Selye, *Hormones and resistance*. (Springer-Verlag, Berlin, New York,, 1971).

Chapter 5

Abandonment of Stress Theory

"All truth passes through three stages. First, it is ridiculed, second it is violently opposed, and third, it is accepted as self-evident."
— Arthur Schopenhauer, Philosopher, 1788–1860

Like Jules Verne predicting a flight to the moon before rockets were invented, Selye was ahead of his time. After 30 years of fruitless frustration, governments withdrew support for stress research. Scientists reasoned that some clue of Selye's mechanism would have been discovered by the massive search effort if the mechanism actually existed, and besides, no single mechanism could possibly explain the bewildering profusion of stress manifestations.[1, 2] Selye lived to see his ideas abandoned. Though some of his students received the Nobel Prize, he was denied that honor, and he died a forgotten and embittered man.

As Selye and his ideas fell from favor, the surviving stress researchers supplanted his classical terminology with the new terms *allostasis* and *allostatic load*. Allostasis refers to physiological adaptations that presumably protect organisms from harmful environmental stressors and restore homeostasis.[3] Such unidentified mechanisms putatively involve cortical function that pre-emptively mobilizes autonomic activity, releases HPA (hypothalamic-pituitary-adrenal) axis hormones, and activates cardiovascular, metabolic, and immune effects to maintain homeostasis in the presence of stress.[4, 5] Allostatic Load is "wear and tear" on the body that results from ongoing adaptive efforts that supposedly maintain homeostasis in response to unremitting environmental stress.[6, 7, 8]

None of these crude concepts explains the nature of stress, tissue repair, disease, or physiology. Stress reliably elevates HPA hormones, but these subside to normal levels within hours, while stress effects persist for months. For lack of a better explanation, many have attributed stress and disease to "auto-immune" activity, as if the body attacks itself, but there is no evidence that supports this notion. The nature of stress thus remains a mystery. Meanwhile, patient research toil has produced a growing body of evidence that *is* pertinent to stress theory. For example, coagulation and tissue repair are intimately related, and von Willebrand Factor and

Factor VIII activity are more closely associated with stress effects than are the short-lived "classical" HPA stress hormones.[9, 10, 11]

As stress theory was abandoned, it was replaced by the idea that DNA transcribes its information to RNA, which in turn translates the information into proteins that account for all manifestations of physiology, disease, and stress.

Meanwhile, there is overwhelming evidence that environmental stresses cause cancer. Since the discovery of DNA, cancer has been attributed to environmentally induced mutations, but this is at odds with the famous findings of Salvador Luria and Max Delbruck, who proved that DNA mutations occur at random, and are independent of environmental stress.[12] Ironically, Salvador Luria was the mentor of James Watson, but this nuance is ignored.

STRESS RESEARCH IN PERSPECTIVE

In retrospect, Selye's ideas represent a resurrection of *Vis medicatrix naturae*, the venerable Greek viewpoint that combinations of stresses cause disease, and that the body possesses the means to heal itself if the stresses are relieved.

Selye may have been his own worst enemy. First, he promoted *in vivo* animal studies as essential for medical research and disparaged the protests of animal rights activists. Even in Claude Bernard's time there was vigorous antivivisectionist opposition to animal research. From a humane viewpoint, the era of stress research resembles a gruesome nightmare of calloused cruelty. Thousands of animals were wantonly burned, frozen, electrocuted, immobilized, poisoned, mutilated, suffocated, terrified, tortured, dispatched, and discarded in the effort to identify Selye's mechanism. To appreciate the calloused view of these postwar researchers, it is helpful to recall that soldiers and civilians were wantonly subjected to ghastly experiments during the recent era of warfare, often without their knowledge. Such suffering provoked an ever-increasing animal rights backlash that besmirched the reputation of science and eventually inspired legal restrictions that frustrate research.

Second, Selye's ideas focused on endocrinology and organic chemistry, and he lacked expertise in evolution, biochemistry, taxonomy, and histology (the study of tissues). He assumed that HPA hormones were the focus of the mammalian stress response, and ignored the inability of known hormones to explain stress manifestations as well as alternative ideas such as capillary gate theory, tissue repair theory, and the role of the vascular endothelium.

Third, Selye, like Bichet (see Appendix A: Medical History, on page 395), preferred direct visual inspection. He did not trust microscopes, biochemical tests, laboratory values, and results that he was unable to confirm with the naked eye.

Fourth, Selye disparaged the possibility that emotional stress, as demonstrated by Mason and Cannon, could be a significant factor in stress and disease.[13] This seems inconsistent, because his books often emphasized the need to avoid emotional stress for the sake of good health.

Fifth, his teleological notion that the stress mechanism defends the body against stress, as opposed to mechanistically reacting to stress, may have confused and distracted stress research.

Last, but not least, Selye became enamored with money and fame, and prospered from his books and lectures. He accepted funds from the notorious Tobacco Institute. In 1979, during the declining days of stress research, he organized the World Congress on Stress in Monte Carlo at the behest of unscrupulous individuals who promised him stock and cash worth thousands of dollars in return for endorsing their questionable treatments. Serious scientists frown on such endeavors, and several shunned Selye's invitations.[14]

CONCLUSION

"An important scientific innovation rarely makes its way by gradually winning over and converting its opponents: it rarely happens that Saul becomes Paul. What does happen is that its opponents gradually die out and that the growing generation is familiarized with the idea from the beginning."
— Max Planck

Powerful new theories typically appear at least 30 years ahead of their time, and must await supporting evidence before they are accepted. Selye's theory that a single mechanism explains physiology, pathology, and stress was never disproved, and, it remains the most promising prospect to provide an effective theory of medicine.

Despite nearly 50 years of intense research since stress theory was abandoned, the DNA paradigm has failed to clarify biology or produce effective medical treatments. The Genome Project exemplifies this failure. It promised to revolutionize medicine and cost billions, but failed to produce promised results, and efforts to rationalize how DNA activity might explain physiology, pathology, and stress have become increasingly strained.[15] Thus DNA theory appears to have exhausted its utility, and scientific history suggests that a new paradigm must soon supplant or supplement the exhausted DNA paradigm and reinvigorate research.[16]

Meanwhile, patient research toil has accumulated fresh evidence from unrelated research that is relevant to stress theory. Based on this fresh evidence, this book proposes a hitherto unrecognized mammalian stress mechanism that meets and exceeds the predictions and expectations of the original stress theorists.

NOTES

1 R. M. Yehuda, B, *Behavioral Stress Response: Protective and Damaging Effects.* (New York Academy of Science, ed. Annals, 2005), pp. 331.

2 S. Fink, in *New York Academy of Sciences Magazine.* (New York Academy of Sciences, 2 E 63 Street, New York, N.Y. 10021, 2005), chap. 18, pp. 2.

3 B. S. McEwen, E. Stellar, Stress and the individual. Mechanisms leading to disease. *Arch Intern Med* 153, 2093-2101 (1993).

4 S. M. Korte, J. M. Koolhaas, J. C. Wingfield, B. S. McEwen, The Darwinian concept of stress: benefits of allostasis and costs of allostatic load and the trade-offs in health and disease. *Neurosci Biobehav Rev* 29, 3-38 (2005).

5 B. L. Ganzel, P. A. Morris, E. Wethington, Allostasis and the human brain: Integrating models of stress from the social and life sciences. *Psychol Rev* 117, 134-174 (2010).

6 R. P. Juster, B. S. McEwen, S. J. Lupien, Allostatic load biomarkers of chronic stress and impact on health and cognition. *Neurosci Biobehav Rev*, (2009).

7 R. von Kanel *et al.*, Altered blood coagulation in patients with posttraumatic stress disorder. *Psychosom Med* 68, 598-604 (2006).

8 R. von Kanel, J. E. Dimsdale, T. L. Patterson, I. Grant, Acute procoagulant stress response as a dynamic measure of allostatic load in Alzheimer caregivers. *Ann Behav Med* 26, 42-48 (2003).

9 K. Kario, B. S. McEwen, T. G. Pickering, Disasters and the heart: a review of the effects of earthquake-induced stress on cardiovascular disease. *Hypertens Res* 26, 355-367 (2003).

10 D. B. Rubin *et al.*, Elevated von Willebrand factor antigen is an early plasma predictor of acute lung injury in nonpulmonary sepsis syndrome. *The Journal of clinical investigation* 86, 474-480 (1990).

11 P. M. Mannucci, Z. M. Ruggeri, G. Gagnatelli, Nervous regulation of factor-VIII levels in man. *British journal of haematology* 20, 195-207 (1971).

12 S. E. Luria, M. Delbruck, Mutations of Bacteria from Virus Sensitivity to Virus Resistance. *Genetics* 28, 491-511 (1943).

13 https://en.wikipedia.org/wiki/John_Wayne_Mason

14 P. J. Rosch, Reminiscences of Hans Selye, and the birth of "stress." *International journal of emergency mental health* 1, 59-66 (1999).

15 J. Pontin, in *Technology Review,* J. Pontin, Ed. (MIT, MIT, 2010), vol. 114, chap. 8, pp. 1.

16 T. S. Kuhn, I. Hacking, *The structure of scientific revolutions.* (The University of Chicago Press, Chicago ; London, ed. Fourth edition., 2012), pp. xlvi, 217 p.

Chapter 6

A "Helicopter View" of the Stress Mechanism

"The improver of natural knowledge absolutely refuses to acknowledge authority as such. Every great advance in natural knowledge has involved the absolute rejection of authority."
— Thomas Henry Huxley

The old stress researchers would have been gratified to know that the stress mechanism satisfies their insightful predictions and expectations. It incorporates both their capillary gate theory and their tissue repair theory. It appears early during embryological development and converts the genetic blueprint into complex multicellular structures. It then remains active to maintain the *internal milieu* that sustains cell survival regulates blood flow and organ function, and repairs tissues. It consists of a capillary gate mechanism and a tissue repair mechanism that share the same enzymes and substrates, but operate independently, so that the activity of each exaggerates that of the other. In addition, its operation entails ancillary elements that appear to be unrelated but are essential to its function, such as the immune mechanism and the mechanism of oxygen transport and delivery. Its operation is accordingly complex, counterintuitive, confusing, and daunting to describe. This abbreviated introduction will hopefully facilitate the reader's ability to understand the more complex and nuanced presentation that follows.

WHAT IS "STRESS"?

Definitions of stress-related phenomena remain imprecise for lack of an effective theoretical understanding of the nature of stress. For example, the International Association for the Study of Pain defines *pain* as, "an unpleasant sensory or emotional experience associated with actual or potential tissue damage." It defines *analgesia* as, "absence of pain in response to stimulation which would normally be painful," and it defines *nociception* as, "the neural process of encoding noxious stimuli."[1] The American Institute of Stress states, "Stress is difficult for scientists to define because it is a subjective sensation associated with varied symptoms that differ for each of us."[2] The mammalian stress mechanism (MSM) clarifies and simplifies the understanding of physiology, pathology and stress, and enables improved definitions as follows:

Stress is *any force or stimulus that increases MSM activity via tissue disruption and nervous activity*, including sepsis, radiation, trauma, surgery, toxic chemicals, nociception, sight, sound, smell, taste, pain, fear, and exercise training.

Disease is *stress-induced stress mechanism hyperactivity* that produces pathological excesses of thrombin, soluble fibrin, and insoluble fibrin and depletes MSM substrates such as fibrinogen, ATP, and blood enzymes. Such MSM hyperactivity explains catabolism, malaise, edema, fever, inflammation, allodynia, exudates, pus, scabs, abscesses, tachycardia, hypertension, amyloidosis, atherosclerosis, accelerated capillary senescence, infarction, cancer, hypercoagulability, hyperviscosity, hemostasis failure, organ dysfunction, and death.

WHAT IS THE STRESS MECHANISM?

The stress mechanism consists of three elements that are ubiquitous throughout the body:

1. The entire *nervous system*.
2. *Blood enzyme factors VII, VIII, IX,* and *X* that are continuously released from the liver into blood.
3. The *vascular endothelium*, which is a delicate layer of highly specialized cells, one cell thick, that lines the interior surface of all blood vessels and is the sole element of capillaries, which are the smallest blood vessels in the body.

The vascular endothelium is the focus of stress mechanism activity. It isolates flowing blood from extravascular tissues, and thereby insulates blood enzymes from a glycoprotein called *tissue factor* that is produced by extravascular cells. Tissue damage disrupts the vascular endothelium, exposes tissue factor to blood enzymes, and activates a complex enzymatic interaction of factors VII, VIII, IX, and X that enables tissue repair. I call this the *tissue repair mechanism*.

The nervous system innervates the vascular endothelium directly and indirectly, causing it to release hormones that govern a *capillary gate mechanism* that regulates blood flow resistance, which determines cardiac output, cardiac efficiency, tissue perfusion, and organ function.

Normally, the stress mechanism operates efficiently and unobtrusively, but unremitting combinations of excessive stresses can cause harmful stress mechanism hyperactivity that wastes energy, depletes the body's resources, and produces excessive quantities of its products that damage tissues and disrupt organ function. Such harmful hyperactivity manifests as *disease* that damages tissues and blood vessels, wastes energy and substrates, and causes fever, fatigue, fistulas, inflammation, edema (swelling), rashes, pus, exudates, atherosclerosis, thrombosis, infarction, scars, scabs, sclerosis, adhesions, and tumors. Severe stress mechanism hyperactivity threatens death by disrupting organ function, tissue perfusion, tissue oxygenation, tissue maintenance, and tissue repair.

Thus, *stress mechanism hyperactivity* explains the nature of disease.

THE THREE STRESS MECHANISM PRODUCTS

The stress mechanism produces three basic products that explain its effects:

1. *Thrombin* is the *Universal Enzyme of Extracellular Energy Transformation.* It converts ATP energy into cell and extracellular enzyme activities that repair tissues and regulate blood flow (see Appendix E on page 427 that reviews cell biology). The stress mechanism regulates thrombin generation in all tissues. It continuously generates thrombin in extravascular tissues to energize tissue maintenance. It continuously generates thrombin in blood to energize blood flow regulation. It elevates thrombin generation to energize coagulation and tissue repair, and restores thrombin to maintenance levels as the repair process resolves.
2. *Soluble fibrin* is the *Universal Substrate of Tissue Structure*. A substrate is a substance that is used to produce a product. Soluble fibrin is the substrate of collagen, scabs, scars, sclerosis, adhesions, mucus, and milk.
3. *Insoluble fibrin* is the *Universal Polymer of Hemostasis*. A polymer is a long-chain molecule consisting of a series of individual molecular elements called *monomers*. Insoluble fibrin monomers polymerize into strands that increase capillary flow resistance and bind blood cells into viscoelastic clots. Insoluble fibrin disintegrates into inert *fibrin split products* (FSB)

THE NERVOUS SYSTEM

All types of nervous activity, including consciousness and emotion, affect the autonomic nervous system. The autonomic (i.e. automatic) nervous system innervates the vascular endothelium, so that the vascular endothelium functions as a gigantic nervous gland.

The autonomic nervous system consists of *sympathetic* and *parasympathetic* subdivisions whose activities oppose each other. The net effect of this opposing activity is called *autonomic balance*.

Sympathetic nervous system (SNS) activity "closes" the capillary gate, which increases capillary flow resistance and decreases cardiac output, tissue perfusion, tissue oxygenation, and organ function.

Parasympathetic nervous system (PNS) activity "opens" the capillary gate, which decreases capillary flow resistance, and increases cardiac output, tissue perfusion, tissue oxygenation, and organ function.

THE THREE STRESS MECHANISM ACTIVATION PATHWAYS

Three synergistic, redundant pathways activate the stress mechanism and focus its activity to regulate tissue repair, blood flow, and organ function:

THE NOCICEPTION PATHWAY

The nociception pathway is a nervous pathway that activates the stress mechanism in accord with nervous stimuli that threaten or cause bodily harm. Specialized nerve endings called *nociceptors* detect tissue damage (cuts, bruises, burns, penetrations, surgery) in organs and peripheral tissues. They generate nervous activity called *nociception*. Peripheral sensory nerves communicate nociception to spinal cord nociception pathways. The spinal cord nociception pathways simultaneously communicate nociception to the brain *and* to sympathetic ganglia in the chest and abdomen. The sympathetic ganglia cause the vascular endothelium and the adrenal glands to release hormones that inhibit organ perfusion and oxygenation. Meanwhile, the awake brain generates consciousness, which perceives nociception as *pain*. The perception of pain causes emotional fear and anxiety but nociception cannot be perceived as pain in the absence of consciousness. Thus brain damage and drugs that inhibit consciousness can abolish the ability to perceive nociception as pain, while spinal cord nociception pathways continue to activate sympathetic ganglia in the chest and abdomen.

Analgesia inhibits the nociception pathway and thereby prevents nociception from reaching either the brain or sympathetic ganglia. Three types of analgesic agents are used to prevent surgical nociception:

1. Narcotics inhibit spinal cord nociception pathways.
2. Local analgesics such as lidocaine, procaine, and Marcaine block nociception signals in peripheral nerves and the spinal cord.
3. Non-Steroidal Anti-Inflammatory drugs (NSAIDs) prevent nociception by poisoning the nociceptors themselves.

THE COGNITIVE PATHWAY

The cognitive pathway is easily confused with the nociception pathway. The brain generates consciousness that continuously evaluates all forms of sensory information for signs of danger and opportunity. Consciousness interprets auditory sensation from the ears as sound, olfactory sensation from the nose as smell, visual sensation from the eyes as sight — and nociception as pain. When consciousness detects danger, it induces fear and anxiety.

Consciousness continuously inhibits spinal cord nociception pathways via descending pathways from the cerebral cortex to the spinal cord, and thereby regulates the activity of sympathetic ganglia in the chest and abdomen. When consciousness perceives impending danger, it pre-emptively generates anxiety, fear, and rage, all of which ordinarily exaggerate spinal cord nociception pathway activity. However, the modulation of spinal cord nociception pathways by consciousness depends on circumstances. The classic example is soldiers who remain unaware of pain and continue to fight for their lives during battle despite horrific combat wounds, but subsequently suffer severe pain when combat subsides.

To make the subject of pain even more confusing and complicated, emotional mechanisms work closely with consciousness to record vivid audiovisual memories of all experiences throughout life that occur when consciousness remains active. Subconscious emotional memories of painful events can sometimes mimic pain in the absence of actual danger or tissue damage. For example, a person who has experienced pain during dental procedures may perceive "imaginary" pain due to painful memories that are recalled by the sound of a dental drill.

Anesthesia inhibits the cognitive pathway. It *mimics* analgesia by inhibiting consciousness in a dose-related manner, and thereby abolishes anxiety, fear, and the perception of nociception as pain. The classic example is ether, which takes effect so slowly that patients exhibit loss of painful sensation before they lose the ability to move and speak coherently. This conveys the powerful — but misleading — impression that anesthetic agents possess analgesic properties, but anesthetics have negligible effect on spinal cord nociception pathways that elevate harmful sympathetic nervous activity despite the absence of consciousness. In fact, hypnotic agents *exaggerate* spinal cord nociception pathway activity by abolishing the descending inhibitory influence of consciousness. All hypnotic agents exhibit this dose-dependent ability to inhibit consciousness, including alcoholic beverages and sedatives such as Propofol, benzodiazepines (Valium, Versed, etc.), and barbiturates (sodium Pentothal, Brevital, and so forth). The only difference between such soporific agents and anesthetic agents is that inhalation agents can be conveniently administered via the lungs, be maintained at consistent levels during surgery, and rapidly be rid from the body.

THE TISSUE DISRUPTION PATHWAY

Trauma, surgery, radiation, toxic chemicals, and sepsis increase the permeability of the vascular endothelium and exaggerate the exposure of blood enzymes to *tissue factor* in extravascular tissues. This initiates tissue repair, and determines the location and magnitude of tissue repair activity. The resulting enzymatic interaction immediately initiates viscoelastic clot formation that substitutes for the vascular endothelium, re-isolates blood enzymes from tissue factor, and regulates contact between blood enzymes and exposed tissue factor in damaged tissues to govern tissue repair. Tissues rich in tissue factor, such as brain, nerves, lung, cervix, and placenta react more vigorously to blood enzymes than do tissues lacking in tissue factor. Such tissues are "targets" for stress mechanism hyperactivity, which explains why they exhibit malignancy and other manifestations of stress mechanism hyperactivity more readily than tissues lacking in tissue factor.

At present there is no direct means to inhibit the tissue disruption pathway. In theory, the pharmaceutical development of a safe and effective treatment that neutralizes tissue factor in blood circulation would revolutionize surgery and medicine. However, it is possible to inhibit harmful thrombin elevations, including tissue pathway activity, using simple chemical treatments that bind to Ca+, such as magnesium sulfate (used to control eclampsia), EDTA (used for chelation therapy), and trisodium citrate (most commonly used for blood preservation and renal dialysis).

THE SUB-COMPONENTS OF THE STRESS MECHANISM

Consistent with the concepts of the old stress researchers and theorists, the stress mechanism can best be understood as consisting of a *capillary gate component* and a *tissue repair component* that operate independently but share the same enzymes and substrates, so that the activity of each component exaggerates that of the other. Figure 6.1 on page 34 illustrates the relationships of its enzymes, their substrates, and its products:

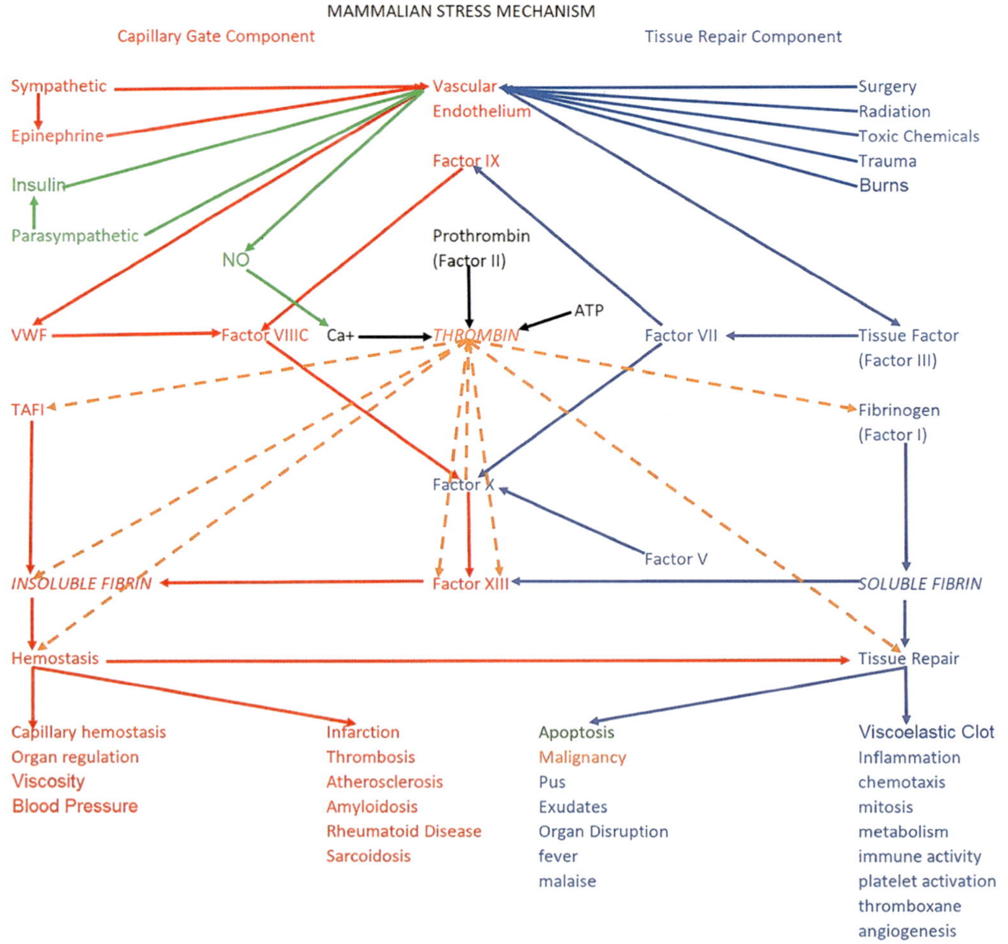

Figure 6.1: Mammalian stress mechanism.

A simplified diagram of a very complex stress mechanism, showing the main components of the system.

THE TISSUE REPAIR COMPONENT

The tissue disruption pathway activates the *tissue repair component* by exposing labile (variable or changeable) factor VII to tissue factor. This stabilizes labile factor VII and enables its enzymatic activity, which functions as a "trigger" for the interaction of factors VII, VIII, IX, and X, because factor VII activity is essential for the function of the other enzymes. Thus, factor VII activity determines the location and magnitude of the enzymatic interaction that enables tissue repair.

THE CAPILLARY GATE COMPONENT

The *capillary gate component* is more complex. It regulates blood viscosity (flow resistance), cardiac output, cardiac efficiency, tissue perfusion, tissue oxygenation, and organ function in accord with autonomic balance.

Both the cognitive and nociception pathways induce sympathetic nervous activity that "closes" the capillary gate. Sympathetic nervous activity generates strands of insoluble fibrin that increase flow resistance in capillaries — where surface area is vastly greater than that of all larger vessels combined, and pressure, flow, and turbulence are all minimal. This increases blood viscosity and reduces cardiac output, tissue perfusion, tissue oxygenation, and organ activity.

Parasympathetic nervous activity disintegrates insoluble fibrin in capillaries, and thereby reduces capillary flow resistance, which increases cardiac output, tissue perfusion, tissue oxygenation, and organ function.

THE TURBULENCE MECHANISM

The *turbulence sub-mechanism* functions in accord with the activity of the capillary gate component, so that their effects occur simultaneously. The turbulence mechanism maintains arterial patency to assure effective tissue perfusion and oxygenation. It generates cyclical turbulent pulse waves that propagate throughout the arterial tree during diastole, when the heart is in a state of relaxation. The turbulent waves scour particulate deposits from the inner walls of arteries that would otherwise incite atherosclerosis. They also disintegrate arterial thromboses (blood clots) that threaten to block arterial flow and disrupt perfusion of cells and tissues served by the artery. When arterial damage causes thrombus formation that obstructs the arterial lumen, pulsatile turbulence quickly tunnels through the obstructing thrombus (clot), re-canalizes (opens) the artery, and restores perfusion to downstream tissues.

Chronically increased capillary flow resistance, as in essential hypertension, and/or impaired cardiac contractility, as in congestive heart failure, undermines the efficiency of the turbulence mechanism, accelerates atherosclerosis, and promotes arterial thrombosis.

The turbulent pulsatile waves generate lateral forces that explain blood pressure and the palpable pulse. The fluctuating flow resistance caused by fluctuating autonomic balance affects the turbulent pulse wave, and thereby alters blood pressure measurement. Conventional blood pressure measurements are therefore counterintuitive and misleading, because sympathetic nervous activity closes the capillary gate, elevates blood pressure, and harmfully decreases tissue and organ perfusion and oxygenation. In contrast, parasympathetic nervous activity opens the capillary gate, lowers blood pressure, and helpfully increases tissue and organ perfusion and oxygenation.

POSITIVE FEEDBACK

The stress mechanism operates efficiently and unobtrusively within the normal confines of its capacities, but combinations of excessive and unremitting environmental stresses cause positive feedback to overwhelm negative feedback, causing confusing combinations of systemic and local stress mechanism hyperactivity that manifest as disease. Such disease activity consumes and wastes the body's resources (ATP energy, fibrinogen, and other substrates) and produces excessive and defective versions of its products. Such harmful hyperactivity disrupts organ function, and damages organs and tissues. Thus, controlling stress mechanism hyperactivity can restore homeostasis (normal physiologic function), reinstate effective organ function, promote tissue perfusion and oxygenation, enhance antibiotic effectiveness, and facilitate the body's ability to heal itself.

Specific forms of stress sometimes produce distinctive symptoms that identify specific causative stresses, particularly in the case of microbial infestations. For example, measles evokes Koplik spots; mumps causes painful salivary gland swelling; chickenpox produces skin pustules; diphtheria causes thick exudates that obstruct air flow; pneumonia and influenza cause alveolar exudates that disrupt gas exchange; and so forth. This conveys the misleading impression that diseases are distinctly different from one another and can be reliably diagnosed on the basis of their distinguishing symptoms and effectively managed with specialized treatments. This modern medical viewpoint overlooks the common manifestations of most diseases, including fever, fatigue, inflammation, rashes, edema (swelling), inflammation, exudates, cell proliferation, and increased blood viscosity and coagulability. It likewise ignores the close associations between and among seemingly unrelated diseases. For example, obesity, diabetes, hypertension, infarction, cancer, and congestive heart failure are all closely related. Such disease associations cannot be explained by conventional theory, but they are readily explained by stress mechanism hyperactivity, as this book endeavors to explain.

Because stress mechanism hyperactivity causes disease, the restoration of homeostasis (normal function) should be the primary objective of medical management. This, however, is easier said than done. The three stress mechanism activation pathways are synergistic and redundant, so that single treatments seldom suffice to suppress harmful stress mechanism hyperactivity. All three pathways must be simultaneously and synergistically inhibited. Furthermore, excessive suppression invites death, for stress mechanism activity is essential for life. Nevertheless, the realization that stress mechanism hyperactivity causes disease invites the prospect of a treatment regimen that cures everything from cancer to the common cold. Such a treatment strategy will be elaborated in this book.

It's no accident that the stress mechanism has escaped detection for so long. Its operation is complex, counterintuitive, confusing, and at odds with conventional beliefs and practices. Its fluctuating activity generates a bewildering barrage of constantly changing manifestations that obscure its relatively simple operation. Its enzymes and substrates are notoriously identified by Roman numerals instead of helpful names, a convention that was embraced during the heyday of stress research, when coagulation enzymes and substrates could be distinguished only by their locations in a *cascade* of interactions that produces clot formation (Appendix B: Hemostasis History, on page 407 describes the coagulation cascade, and Appendix I: Summary of Coagulation Enzymes and Substances, on page 449 provides a summary of enzymes and substrates that may prove informative).

Most of the substrates now have distinguishing names, but there is no escaping some use of the inscrutable Roman numerals to identify the enzymes. Furthermore, medical vocabulary is replete with terms unfamiliar to those lacking formal medical training. I will endeavor to provide pertinent definitions and the Internet should provide a convenient source for terms and more detailed definitions.

CONCLUSION

The mammalian stress mechanism is complex, confusing, counterintuitive, and at odds with orthodoxy. Hopefully, this abbreviated introduction will facilitate the understanding of subsequent chapters that add details and explain diseases in terms of stress mechanism hyperactivity.

NOTES

1 J. D. M. Loeser, in *Pain*, M. D. John D. Loeser, Ed. (International Association for the Study of Pain, 1994).

2 P. J. Rosch. (The American Institute of Stress).

Chapter 7

Products of the Stress Mechanism

"No great discovery was ever made without a bold guess."
— Isaac Newton

Explaining the stress mechanism is difficult because its various elements interact with one another in many ways. It is easiest to begin by explaining the effects of its three primary products, which are *thrombin, soluble fibrin,* and *insoluble fibrin.* The stress mechanism continuously governs the production of these products. Depending on circumstances, each of them produces multiple manifestations that seem unrelated.

THROMBIN

Thrombin is the most important of the three stress mechanism products. I call it the *universal enzyme of extracellular energy transformation.* It energizes the activities of both cells and extracellular enzymes.

Thrombin is analogous to ATPase enzymes that are found within cells. It is also analogous to auxin, which is found in plants. All three of these enzymes (thrombin, ATPase, and auxin) convert ATP (adenosine triphosphate) into cell and enzyme activities, and the activities of all three are dependent on calcium ion (Ca+) concentrations (the greater the Ca+ concentration, the greater their activities). This implies a common origin of animals and plants (see Appendix E, on page 427 that reviews cell biology). However, thrombin operates *outside* cells, while ATPase operates *inside* cells. Plants employ a different extracellular mechanism to alter auxin activity that is not pertinent to animals.

Cells govern their internal ATPase activity by regulating Ca+ levels within the cell. In tetrapods (amphibians, reptiles, and mammals), *extracellular* thrombin activity is optimized by the parathyroid glands, which maintain extracellular Ca+ concentrations within a narrow optimal range, so that thrombin activity is determined by thrombin concentrations, which in turn are regulated by the enzymatic interaction of factors VII, VIII, IX, and X. This explains why accidentally removing the parathyroid glands during surgery is lethal. The therapeutic effects of numerous drugs are explained by their ability to inhibit or elevate Ca+, and thereby alter thrombin activity. Plants appear to govern auxin activity by regulating access to its active site.

Little was known about the individual properties of blood enzymes during the era of stress research, and thrombin seemed indistinguishable from enzyme factors VII, VIII, IX, and X other than its location in the final common pathway of the sequence of enzymatic interactions in the coagulation cascade (see Appendix B: Hemostasis History, on page 407). However, it is now clear that thrombin is the *product* of the coagulation cascade. It energizes the activities of factors VIII, IX, and X and the production of soluble fibrin and insoluble fibrin, which are the other two stress mechanism products. Most thrombin is produced *after* coagulation is complete, which is consistent with its role in the tissue repair process that proceeds after clot formation.[1, 2] The complement cascade that attacks bacteria and foreign substances that invade circulating blood also generates thrombin that energizes the activities of its enzymes. Thrombin converts ATP energy into all of the following cell and enzyme activities:

- The conversion of fibrinogen to soluble fibrin.[3, 4]
- The conversion of soluble fibrin to insoluble fibrin.[5, 6, 7]
- Tissue repair activity, including embryogenesis, angiogenesis (capillary development), hemostasis, thrombogenesis, atherosclerosis, inflammation, edema, chemotaxis, cell differentiation, endothelial cell spreading, migration, and mitosis (cell division).[8, 9, 10, 11, 12, 13, 14, 15, 16, 17, 18, 19, 20, 21]
- Embryological development.[22]
- Cell hormone release (chemokines, cytokines, prostaglandins, etc.).[23]
- Tumor progression, cancer, coagulation, angiogenesis, mitosis, and metastasis.[24, 25, 26, 27, 28, 29, 30, 31, 32]
- Histamine release, and mast cell activation.[33]
- Caspase activation.[34]
- Smooth muscle cell mitosis.[35]
- Complement cascade enzyme activity in blood that attacks bacteria and foreign substances.
- Gelsolin enzyme activity that digests harmful actin released into blood by cell damage.
- Mitosis (eukaryotic cell division and multiplication).[36, 37]

- Cell differentiation, de-differentiation, and re-differentiation.[38]
- Prostaglandin synthesis.[39]
- Collagen production.[40]
- Cell metabolism that consumes glucose and oxygen and generates body heat and fever.
- Production of bone, muscle, nerves, and collagen connective tissue.
- Stabilization of insoluble fibrin via *Thrombin-Activated Fibrinolysis Inhibitor* (TAFI).[41, 42]
- Cellular mitosis and hypertrophy (increase in cell size, as occurs in muscle cells during weight training).[43, 44]
- Inflammation that loosens cell connections to facilitate chemotaxis.[45, 46]
- Chemotaxis (repair cell movement toward thrombin elevations in damaged tissues).
- Fibroblast proliferation and collagen production.
- Immune cell activity (phagocytosis and antibody production.[47, 48, 49, 50, 51]
- Capillary proliferation that perfuses tissues (angiogenesis).
- Cellular hormone release (chemokines, cytokines, prostaglandins, etc.) that facilitates cell-to-cell communication.[52, 53]
- Platelet activation and thromboxane synthesis.[54, 55, 56, 57]
- The release of von Willebrand Factor and nitric oxide by the vascular endothelium.
- Gap formation between the cells of the vascular endothelium that exaggerates the permeability of the vascular endothelium and increases contact between blood enzymes and tissue factor in extravascular tissues.[58]
- Thrombin-Activated Fibrinolysis Inhibitor (TAFI) prevents plasminogen from spontaneously disintegrating into plasmin that disintegrates the insoluble fibrin molecule.[59, 60, 61]
- Thrombin inhibits apoptosis.[62, 63, 64]
- Excessive thrombin causes cell death.[65, 66]

Thrombin energizes all stress mechanism activities, including embryological development, immune activity, tissue repair, and capillary gate operation. The stress mechanism continuously regulates thrombin generation in blood and extracellular tissues throughout the body to energize embryological development, tissue maintenance, tissue repair, and hemodynamic physiology:

- It generates small amounts of thrombin to energize tissue maintenance.
- It elevates thrombin generation to energize the enzyme and cellular activities that repair tissues.
- It restores normal thrombin levels as tissue repair nears completion.
- Thrombin starvation normally causes clot disintegration and apoptosis (cell suicide) during the conclusion of tissue repair, which shrinks tissues and enables wound closure.
- Apoptosis induced by thrombin starvation also plays an important role in the formation of anatomical structures during embryological development.
- Severe, unrelenting stress induces excessive thrombin generation that exaggerates repair cell proliferation. This invites a "vicious cycle" of invasive, self-sustaining tissue repair activity called *malignancy*.
- "Malignant" cells are normal cells that are hyper-stimulated by prolonged, excessive thrombin elevations. They cannot be distinguished from normal tissue repair cells engaged in peak tissue repair activity. Both exhibit broken chromosomes and high rates of cell division (mitosis) and metabolism.
- "Benign" tumors require unrelenting environmental stress to sustain their growth, as they are not self-sustaining.
- "Malignant" tumors are self-sustaining and capable of metastasis. They continuously invade and disrupt adjacent normal tissues, thereby releasing tissue factor into systemic circulation and stimulating nociception, both of which continuously stimulate both local and systemic stress mechanism hyperactivity that exaggerates thrombin generation.
- Tissue repair cells (fibroblasts) are *thrombin-dependent*. Thrombin starvation causes them to undergo apoptosis (harmless cell death). This normally occurs as thrombin levels decline to maintenance levels at the conclusion of ordinary tissue repair, whereupon apoptosis shrinks wound tissues to enable wound closure. Malignancy inhibits apoptosis and sustains systemic inflammation that promotes the survival and proliferation of *metastatic* cancer cells that are transported to distant locations by blood or lymph flow.

- In theory, cancer can be cured by inhibiting thrombin generation to disrupt the "vicious cycle" of repair cell proliferation, inflammation, and invasive tissue disruption that sustains the malignant state.

- Excessive thrombin elevations explain why both "benign" and "malignant" tumors consist of a variety of cells that replicate at various rates. The degree of thrombin elevation determines the difference between benign and malignant.

- Thrombin also energizes all extracellular enzyme activities, including gelsolin that disintegrates actin protein in blood circulation, the complement cascade enzymes that attack bacteria in flowing blood and Thrombin-Activated Fibrinolysis Inhibitor (TAFI) that stabilizes plasminogen and prevents the spontaneous self-destruction of the insoluble fibrin molecule.

THROMBIN AND IMMUNE ACTIVITY

Stress mechanism hyperactivity elevates thrombin generation, which energizes immune activity to combat infection, so that elevated immune activity is a natural consequence of both normal tissue repair and disease. However, elevated immune activity invites the mistaken assumption that immune activity causes cancer, diabetes, transplant rejection, and rheumatoid diseases (which are therefore called *autoimmune diseases*, if only for lack of a better explanation). In the case of cancer, some have embraced the opposite assumption — that inadequate immune activity causes cancer. There is no convincing evidence in support of either notion. Misguided treatments with steroids and toxic pharmaceuticals can suppress symptoms, but they fail to alleviate the underlying stress mechanism hyperactivity that causes disease.

THROMBIN AND PAR

Red blood cells, platelets, and endothelial cells that are directly exposed to blood enzymes are devoid of thrombin receptors, but all extravascular mammalian cells produce thrombin receptors on their outer surfaces.[67, 68] The research literature calls these *PAR* (protease activated receptors). Like sails on tiny ships, PAR determine how cells react to thrombin. Five types of PAR have been identified thus far. Specific types and numbers of PAR characterize mature cell types. PAR configurations normally remain stable in quiescent, mature cells, but cells can alter their PAR configurations to control their reactions to thrombin elevations to facilitate embryological development and tissue repair. PAR configurations are also altered during malignancy.[69, 70, 71, 72, 73, 74, 75, 76, 77, 78, 79, 80, 81, 82]

I hypothesize that chromosomal DNA governs PAR and tissue factor on cell surfaces via an as-yet undiscovered mechanism that energizes and orchestrates cell differentiation, de-differentiation, and specialization during embryological development, tissue repair, and malignancy.[83] (See Appendix E: Cell Biology, on page 427). This explains the mystery of embryological development, tissue repair, and malignancy, all of which is discussed in detail in Chapter 17: Unified Theory of Biology, on page 239.

Mammalian red cells and platelets are highly specialized cells that are unique to mammals. Both lack a nucleus, and neither produces PAR and tissue factor. Red cells expel their nucleus, mitochondria, and other non-essential cellular components during their development to shrink their size and enable their distinctive biconcave shape. Without a nucleus, they cannot differentiate, de-differentiate, and re-differentiate like the red cells of other classes of vertebrates and are therefore unable to participate in the tissue repair process. For this and other reasons, mammalian tissue repair, and especially bone repair, is inferior to that of lower vertebrates.

FIBRIN

A fibrillar (string-like) protein material called *fibrin* was identified as the end-product of the coagulation cascade during the era of stress research. It was clear that fibrin is essential for both coagulation (clot formation) and capillary hemostasis, but since it lacks enzymatic properties, its role remained unclear. Since then, it has become evident that *fibrin* consists of three closely related, chemically similar blood proteins: *fibrinogen, soluble fibrin,* and *insoluble fibrin*. While these three proteins are closely related and chemically similar, but they can now be distinguished from one another by chemical tests and physical tests such as Viscoelastograph and Thromboelastograph. The three proteins exist in a fluctuating equilibrium in blood that alters blood viscosity (flow resistance) and coagulability. Combinations of nervous activity and tissue disruption activate the enzymatic interaction of factors VII, VIII, IX, and X, which converts fibrinogen to soluble fibrin, and converts soluble fibrin to insoluble fibrin. *Of these three forms of fibrin, only insoluble fibrin affects blood viscosity and coagulability.*

FIBRINOGEN, THE "UNIVERSAL SUBSTRATE OF TISSUE PROTEIN" (AKA "FACTOR I")

The liver continuously manufactures and releases fibrinogen into flowing blood. Fibrinogen cannot escape from the intact vascular system because of its large molecular size. It is enzymatically inert, and has no direct effect on viscosity, coagulability, hemostasis, hemodynamic physiology, pathology, or tissue repair unless it becomes severely depleted.

Fibrinogen consists of fibrillar (string-like) alpha, beta, and gamma molecular sub-components that are bound together by disulfide bonds. See Figure 7.1. Thrombin disrupts the disulfide bonds and releases alpha, beta, and gamma fibrils of soluble fibrin. Like Lego toys, these fibrils of soluble fibrin are the substrates of fibronectin, collagen, insoluble fibrin, elastin, mucus, saliva, and other tissue proteins, depending on how they are re-arranged or combined with other substances.[84, 85, 86, 87, 88]

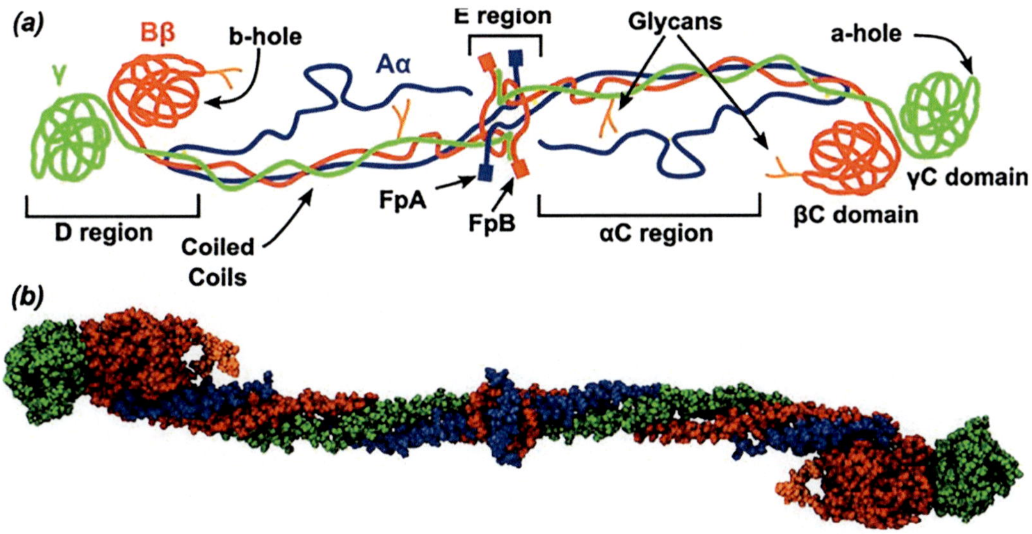

Figure 7.1: Fibrinogen.
Fibrinogen consists of alpha, beta, and gamma fibrillar subunits that are bound together by disulfide bonds. Thrombin disrupts the bonds and releases subunits of soluble fibrin.

Courtesy of Dr. Giovanni Settanni https://www.researchgate.net/profile/Giovanni_Settanni/info

SOLUBLE FIBRIN

I call soluble fibrin the *Universal Substrate of Tissue Structure*. Soluble fibrin is so ubiquitous as to be ignored, and its independent effects remain unappreciated in medical literature. Thrombin cleaves the disulfide bonds of fibrinogen to release fibrillar alpha, beta, and gamma fragments of soluble fibrin.[89] These protein fragments are customarily dismissed as intermediate substances that appear briefly during the conversion of fibrinogen to insoluble fibrin. However, soluble fibrin has significant pathophysiology of its own. It is the substrate of insoluble fibrin, collagen, elastin, saliva, scabs, sclerosis, adhesions, pus, exudates, milk, lipoproteins, and mucous.[90, 91, 92] It enables everything from digestion to sex. It is the substrate of insoluble fibrin that enables coagulation, capillary hemostasis, and blood flow regulation. It is also the substrate of collagen that binds cells together to form tissues and organs, and is the substance of ligaments, tendons, bone, adhesions, scars, scabs, and sclerosis.

Normally, soluble fibrin serves as a substrate of tissue repair. Fragments of soluble fibrin have no effect on blood viscosity or blood coagulation, but unlike fibrinogen and insoluble fibrin, they are small enough to escape from blood through thrombin-induced inflammatory gaps between the cells of the vascular endothelium and diffuse into inflamed or injured tissues, where fibroblasts convert them into collagen to create granulation tissue that binds damaged tissues together during tissue repair. Soluble fibrin also serves as a substrate for mucus, saliva, and exudates. See Figure 7.2, Figure 7.2, and Figure 7.3.

Stress mechanism hyperactivity causes excessive soluble fibrin generation that explains numerous disease manifestations. Soluble fibrin invades inflamed tissues and organs, causing tissue edema, organ disruption, adhesions, scars, scabs, and sclerosis. In the lung, soluble fibrin exudates perturb pulmonary gas exchange during pneumonia, asthma, and influenza. Soluble fibrin forms hyaline "casts" (protein plugs) in kidney ducts, causing renal failure. It disrupts liver, lung, kidney, bowel and brain function during eclampsia, sepsis, *Adult Respiratory Distress Syndrome* (ARDS), *Multi-Organ Failure Syndrome* (MOFS), *Systemic Inflammatory Response Syndrome* (SIRS), pneumonia, and other forms of critical illness. It deposits on arterial walls and causes characteristic "Rosette" formations during *Disseminated Intravascular Coagulation* (DIC).[93, 94] *Fibroblasts* (tissue repair cells) convert soluble fibrin to collagen that forms adhesions, granulation tissue, scars, scabs, and sclerosis.[95]

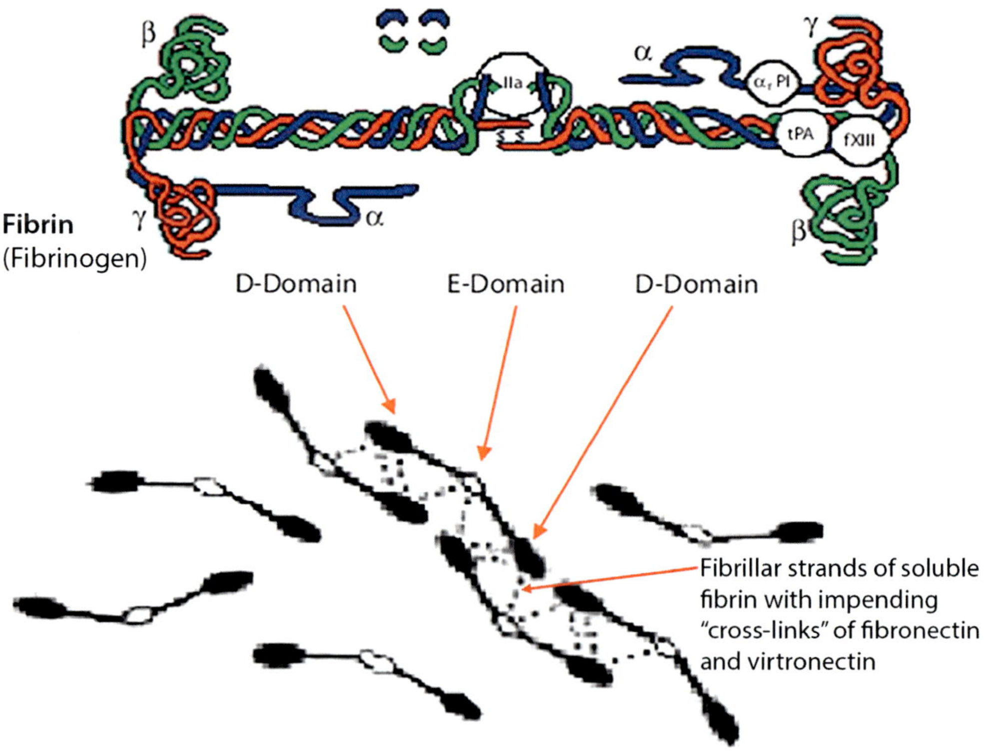

Figure 7.2: Fibrinogen and disulfide bonds.

This drawing incorrectly identifies fibrinogen as *fibrin*, which reflects lingering confusion in medical literature. Thrombin disrupts the disulfide bonds that connect the alpha, beta and gamma subunits of fibrinogen, and releases fibrillar fragments called *soluble fibrin* that escape from the vascular system via thrombin-induced inflammatory gaps in the vascular endothelium to enter extravascular tissues that are loosened by thrombin-energized inflammation as part of the normal tissue repair process. The addition of plasminogen, vitronectin, and fibronectin crosslinks converts soluble fibrin to insoluble fibrin.

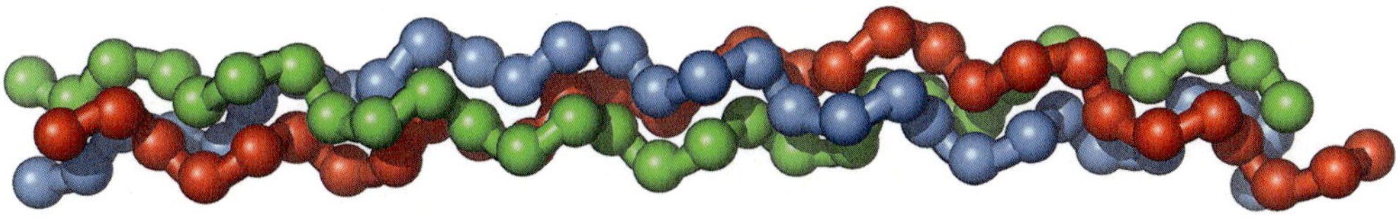

Figure 7.3: Collagen.
A cartoon of collagen. Like fibrinogen, collagen consists of three fibrillar protein subunits. I hypothesize that these derive from the alpha, beta and gamma fibrillar subunits of soluble fibrin. Collagen is the most common protein in the body. It is the connective tissue that binds cells together into tissues and forms bone, ligaments, tendons, adhesions, scars, sclerosis, and scabs.

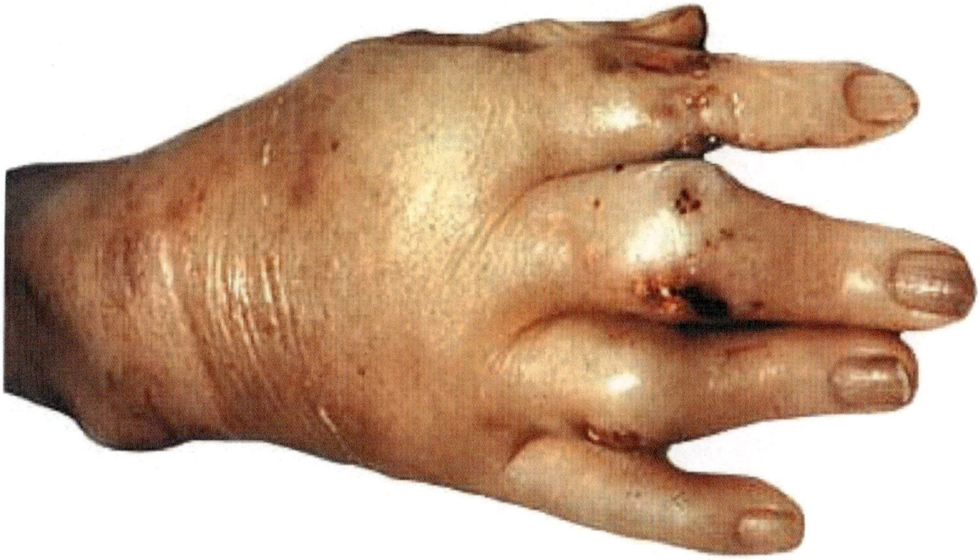

Figure 7.4: Inflammation.
Mammalian stress mechanism hyperactivity generates thrombin that causes inflammation. Thrombin energizes the cellular release of inflammatory substances that loosen cell connections and dissolve the basement membrane that holds cells in tight formation. Thrombin also energizes muscular fibers within the cells of the vascular endothelium, causing gaps to form between the cells, which increases the penetration of factor VII into extravascular tissues where it interacts with tissue factor, causing inflammation. The gaps also enable soluble fibrin to penetrate from escape from extravascular tissues into blood, causing tissue edema, and allow tissue factor to escape into blood, which activates blood enzymes that generate thrombin, causing systemic inflammation. Tissue damage causes localized inflammation that loosens cell connections to facilitate the movement of repair cells from adjacent undamaged tissues toward and into damaged tissues. The repair cells are attracted toward thrombin-rich damaged tissues via chemotaxis (the movement of cells towards a chemical stimulus). Systemic inflammation exaggerates metabolism, which causes fever and wastes body energy, causing fatigue and malaise. Severe stress mechanism hyperactivity causes systemic inflammation and soluble fibrin infiltration that manifests as edema, exudates, and organ disruption. This explains the nature of critical illnesses including sepsis, Adult Respiratory Distress Syndrome (ARDS), Multi-Organ Failure Syndrome (MOFS), eclampsia, and the surgical stress syndrome (SSS).

INSOLUBLE FIBRIN

I call insoluble fibrin *the Universal Polymer of Hemostasis*. Factor XIII adds cross-links of fibronectin, vitronectin, and plasminogen to fibrils of soluble fibrin to produce insoluble fibrin monomers that spontaneously polymerize into strands of insoluble fibrin.[96] (See Figure 7.5 and Figure 7.6.)

Insoluble fibrin is chemically unstable, because its plasminogen crosslinks spontaneously degenerate into plasmin, which enzymatically attacks and disintegrates insoluble fibrin into inert *fibrin split products* (also known as *d-Dimer*). Thrombin-Activated Fibrinolysis Inhibitor (TAFI) stabilizes plasminogen, which prevents the spontaneous disintegration of insoluble fibrin, and prolongs its effects.[97, 98, 99, 100, 101, 102]

Insoluble fibrin explains several seemingly unrelated effects that are elaborated in subsequent chapters. We've already discussed in overview the concept of a *capillary gate mechanism*. This is how insoluble fibrin acts both to operate the capillary gate, enable coagulation (clot formation), and regulate tissue repair:

- In capillaries, the generation and disintegration of insoluble fibrin in accord with autonomic balance regulates capillary flow resistance. This, in turn, affects blood pressure, pulse rate, cardiac work, cardiac efficiency, cardiac output, tissue perfusion, tissue oxygenation, organ function, and organ protection.

- Capillary damage directly exposes blood enzymes to tissue factor in extravascular tissues. This exposure exaggerates localized insoluble fibrin generation to enable hemostasis (coagulation and capillary hemostasis).[103]

- Insoluble fibrin binds blood cells into a viscoelastic clot that temporarily substitutes for the traumatized vascular endothelium. The selectively permeable viscoelastic clot governs tissue repair by regulating contact between blood enzymes and exposed tissue factor in the damaged tissues beneath its surface.[104]

- As tissue repair restores the integrity of the vascular endothelium, the barrier between blood enzymes and tissue factor is restored, thrombin generation declines, and the clot spontaneously disintegrates.

- Inadequate insoluble fibrin generation causes bleeding disorders such as hemophilia and von Willebrand Disease.

- Excessive insoluble fibrin generation elevates blood viscosity and blood coagulability, and promotes atherosclerosis, thrombosis, infarction, Disseminated Intravascular Coagulation (DIC), and pulmonary embolus.[105, 106, 107, 108, 109, 110, 111]

- Excessive insoluble fibrin generation causes abnormal insoluble fibrin digestion, which produces amyloid protein that damages tissues and causes rheumatoid diseases.[112, 113, 114, 115, 116, 117]

- Cell damage releases fibrillar actin, which becomes entangled with fibrillar insoluble fibrin, perturbs the permeability of the viscoelastic clot, and undermines tissue repair. Gelsolin enzyme, which is incorporated in the viscoelastic clot, attacks and dismembers actin into small segments that are digested by an enzymatic process that consumes vitamin D.[118, 119,]

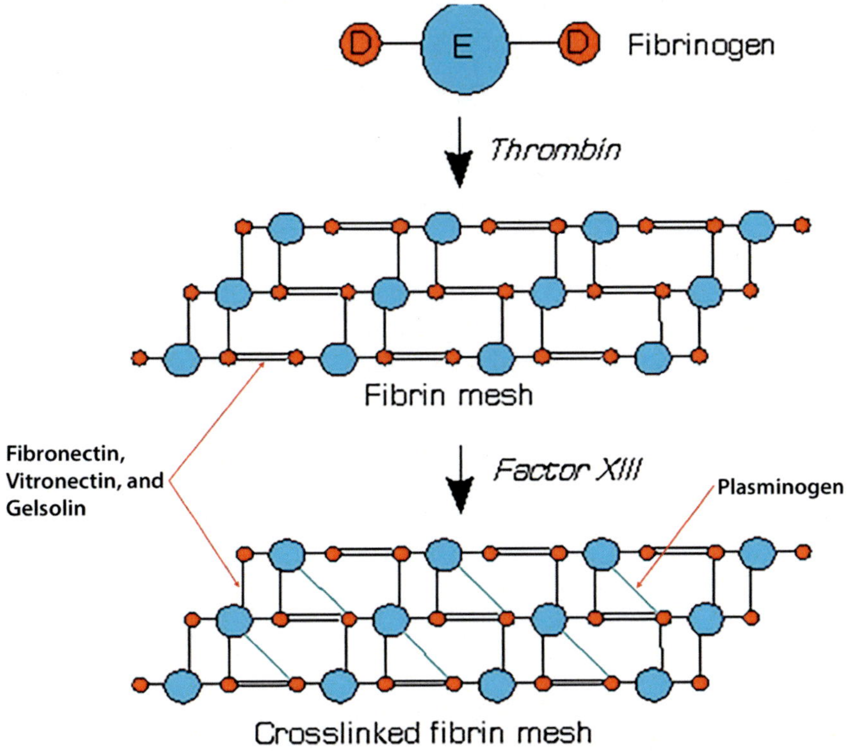

Figure 7.5: Conversion of fibrinogen to insoluble fibrin.

A diagram that portrays the conversion of fibrinogen to insoluble fibrin via the addition of fibronectin, vitronectin, and plasminogen crosslinks. Thrombin releases fibrillar strands of soluble fibrin from fibrinogen. Factor VIII accelerates thrombin generation to energize its interaction with factor X that produces factor XIII, which installs crosslinks of fibronectin and plasminogen to convert soluble fibrin to insoluble fibrin monomers. The monomers spontaneously polymerize into strands of insoluble fibrin (See Figure 7.6). Note that the diagram overlooks the involvement of factors VIII, IX, and X as well as the intermediate appearance of soluble fibrin. This reflects persisting confusion about the nature of coagulation. CC BY-SA 3.0, https://commons.wikimedia.org/w/index.php?curid=556303

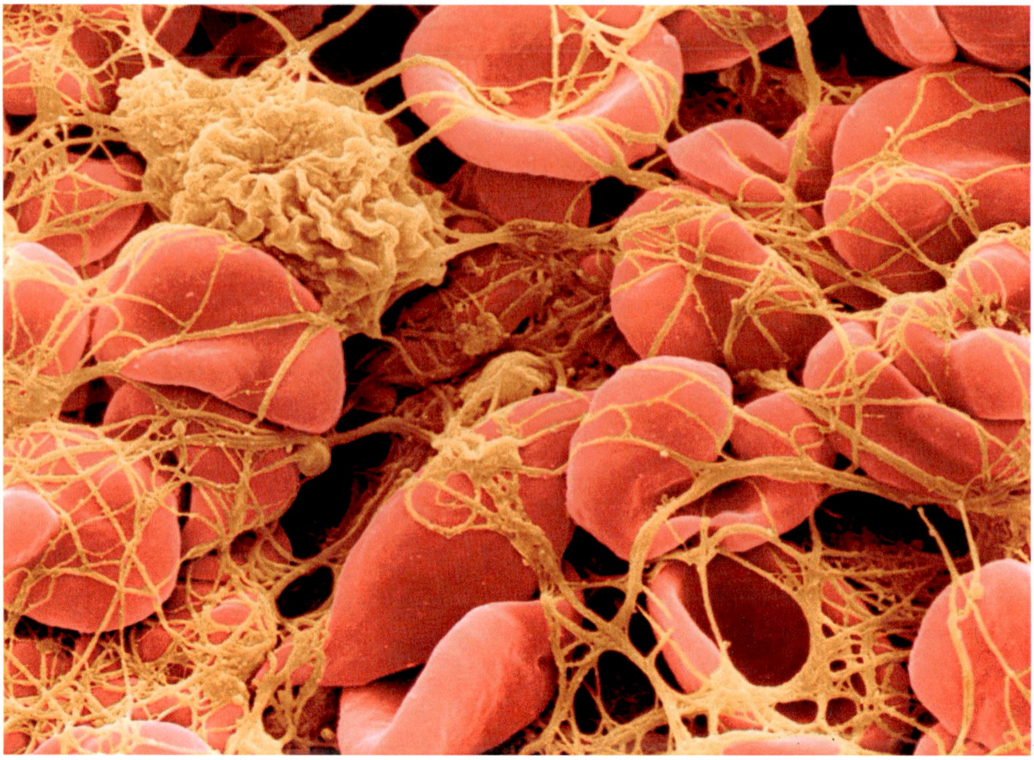

Figure 7.6: Insoluble fibrin.
Strands of insoluble fibrin bind blood cells into a selectively permeable viscoelastic clot that stems blood loss, isolates injured tissues from the enzymes in flowing blood, and regulates the tissue repair mechanism. Similar strands generated from binding sites in capillary walls induce capillary flow resistance and regulate the capillary gate mechanism that then regulates capillary flow resistance, tissue perfusion, and organ function. Courtesy David Gregory & Debbie Marshall Attribution 4.0 International (CC BY 4.0) https://www.nih.gov/news-events/nih-research-matters/microgel-particles-boost-blood-clotting

CONCLUSION

Modern research has clarified the relationships of fibrinogen, soluble fibrin, insoluble fibrin, collagen, elastin, milk, mucus, exudates, pus, fibrosis, sclerosis, amyloidosis, atherosclerosis, and infarction. This simplifies the understanding of embryological development, tissue maintenance, tissue repair, coagulation, capillary hemostasis, and hemodynamic physiology, which in turn lays the foundation for understanding how stress mechanism hyperactivity causes diseases, and how control of harmful stress mechanism hyperactivity can prevent and cure disease, which is elaborated in subsequent chapters.

NOTES

1 K. E. Brummel, S. G. Paradis, S. Butenas, K. G. Mann, Thrombin functions during tissue factor-induced blood coagulation. *Blood* 100, 148-152 (2002).

2 N. R. Goldsack, R. C. Chambers, K. Dabbagh, G. J. Laurent, Thrombin. *Int J Biochem Cell Biol* 30, 641-646 (1998).

3 M. E. Carr, Jr., D. A. Gabriel, J. McDonagh, Influence of factor XIII and fibronectin on fiber size and density in thrombin-induced fibrin gels. *J Lab Clin Med* 110, 747-752 (1987).

4 M. A. Gurrieri, [Thrombin-fibrinogen interaction: release of fibrinopeptides and the effects of ATP]. *Medicina (Firenze)* 10, 51-52 (1990).

5 K. E. Brummel, S. G. Paradis, S. Butenas, K. G. Mann, Thrombin functions during tissue factor-induced blood coagulation. *Blood* 100, 148-152 (2002).

6 Y. Dargaud *et al.*, Evaluation of thrombin generating capacity in plasma from patients with haemophilia A and B. *Thromb Haemost* 93, 475-480 (2005).

7 S. Haas, [Future potential indications for an oral thrombin inhibitor]. *Hamostaseologie* 22, 36-43 (2002).

8 S. Haas, [Future potential indications for an oral thrombin inhibitor]. *Hamostaseologie* 22, 36-43 (2002).

9 E. Archiniegas, C. Y. Neves, D. Candelle, J. E. Cardier, Thrombin and its protease-activated receptor-1 (PAR1) participate in the endothelial-mesenchymal transdifferentiation process. *DNA Cell Biol* 23, 815-825 (2004).

10 J. W. Fenton, 2nd, F. A. Ofosu, D. V. Brezniak, H. I. Hassouna, Thrombin and antithrombotics. *Semin Thromb Hemost* 24, 87-91 (1998).

11 Y. Gong, G. H. Xi, R. F. Keep, J. T. Hoff, Y. Hua, Complement inhibition attenuates brain edema and neurological deficits induced by thrombin. *Acta Neurochir Suppl* 95, 389-392 (2005).

12 R. Gruber *et al.*, Proliferation of dental pulp fibroblasts in response to thrombin involves mitogen-activated protein kinase signalling. *Int Endod J* 37, 145-150 (2004).

13 Y. Hua, R. F. Keep, T. Schallert, J. T. Hoff, G. Xi, A thrombin inhibitor reduces brain edema, glioma mass and neurological deficits in a rat glioma model. *Acta Neurochir Suppl* 86, 503-506 (2003).

14 Y. Q. Huang, J. J. Li, S. Karpatkin, Thrombin inhibits tumor cell growth in association with up-regulation of p21(waf/cip1) and caspases via a p53-independent, STAT-1-dependent pathway. *J Biol Chem* 275, 6462-6468 (2000).

15 J. M. Karp *et al.*, Thrombin mediated migration of osteogenic cells. *Bone* 37, 337-348 (2005).

16 E. Kipnis *et al.*, Massive alveolar thrombin activation in Pseudomonas aeruginosa-induced acute lung injury. *Shock* 21, 444-451 (2004).

17 V. B. Mahajan, K. S. Pai, A. Lau, D. D. Cunningham, Creatine kinase, an ATP-generating enzyme, is required for thrombin receptor signaling to the cytoskeleton. *Proceedings of the National Academy of Sciences of the United States of America* 97, 12062-12067 (2000).

18 K. G. Mann, S. Butenas, K. Brummel, The dynamics of thrombin formation. *Arterioscler Thromb Vasc Biol* 23, 17-25 (2003).

19 E. S. Mullins *et al.*, Genetic elimination of prothrombin in adult mice is not compatible with survival and results in spontaneous hemorrhagic events in both heart and brain. *Blood* 113, 696-704 (2009).
20 A. Naldini *et al.*, Thrombin inhibits IFN-gamma production in human peripheral blood mononuclear cells by promoting a Th2 profile. *J Interferon Cytokine Res* 26, 793-799 (2006).
21 J. O'Donnell, A. D. Mumford, R. A. Manning, M. A. Laffan, Marked elevation of thrombin generation in patients with elevated FVIII:C and venous thromboembolism. *British journal of haematology* 115, 687-691 (2001).
22 M. Moser, C. Patterson, Thrombin and vascular development: a sticky subject. *Arterioscler Thromb Vasc Biol* 23, 922-930 (2003).
23 A. Naldini, L. Aarden, A. Pucci, C. Bernini, F. Carraro, Inhibition of interleukin-12 expression by alpha-thrombin in human peripheral blood mononuclear cells: a potential mechanism for modulating Th1/Th2 responses. *Br J Pharmacol* 140, 980-986 (2003).
24 K. Asanuma *et al.*, Thrombin inhibitor, argatroban, prevents tumor cell migration and bone metastasis. *Oncology* 67, 166-173 (2004).
25 H. D. Bruhn, R. Bernsmeier, P. Luck, K. H. Zurborn, E. Christophers, Influences of thrombin, factor XIII and fibronectin on the growth of tumor cells and leukemic cells in vitro. *Klin Wochenschr* 61, 209-211 (1983).
26 D. Darmoul, V. Gratio, H. Devaud, T. Lehy, M. Laburthe, Aberrant expression and activation of the thrombin receptor protease-activated receptor-1 induces cell proliferation and motility in human colon cancer cells. *The American journal of pathology* 162, 1503-1513 (2003).
27 M. E. Maragoudakis, N. E. Tsopanoglou, P. Andriopoulou, M. M. Maragoudakis, Effects of thrombin/thrombosis in angiogenesis and tumour progression. *Matrix Biol* 19, 345-351 (2000).
28 M. L. Nierodzik, S. Karpatkin, Thrombin induces tumor growth, metastasis, and angiogenesis: Evidence for a thrombin-regulated dormant tumor phenotype. *Cancer cell* 10, 355-362 (2006).
29 F. R. Rickles, S. Patierno, P. M. Fernandez, Tissue factor, thrombin, and cancer. *Chest* 124, 58S-68S (2003).
30 H. Schiller *et al.*, Thrombin as a survival factor for cancer cells: thrombin activation in malignant effusions in vivo and inhibition of idarubicin-induced cell death in vitro. *Int J Clin Pharmacol Ther* 40, 329-335 (2002).
31 T. Tran, A. G. Stewart, Protease-activated receptor (PAR)-independent growth and pro-inflammatory actions of thrombin on human cultured airway smooth muscle. *Br J Pharmacol* 138, 865-875 (2003).
32 N. E. Tsopanoglou, M. E. Maragoudakis, Role of thrombin in angiogenesis and tumor progression. *Semin Thromb Hemost* 30, 63-69 (2004).
33 N. Ben Amor, J. A. Pariente, G. M. Salido, A. Bartegi, J. A. Rosado, Caspases 3 and 9 are translocated to the cytoskeleton and activated by thrombin in human platelets. Evidence for the involvement of PKC and the actin filament polymerization. *Cell Signal*, (2005).
34 R. Bobe *et al.*, Evidence for ERK1/2 activation by thrombin that is independent of EGFR transactivation. *Am J Physiol Heart Circ Physiol* 285, H745-754 (2003).

35 S. Haas, [Future potential indications for an oral thrombin inhibitor]. *Hamostaseologie* 22, 36-43 (2002).

36 A. Naldini, D. H. Carney, A. Pucci, F. Carraro, Human alpha-thrombin stimulates proliferation of interferon-gamma differentiated, growth-arrested U937 cells, overcoming differentiation-related changes in expression of p21CIP1/WAF1 and cyclin D1. *J Cell Physiol* 191, 290-297 (2002).

37 J. W. Fenton, 2nd, F. A. Ofosu, D. V. Brezniak, H. I. Hassouna, Thrombin and antithrombotics. *Semin Thromb Hemost* 24, 87-91 (1998).

38 S. L. Hong, L. Levine, Stimulation of prostaglandin synthesis by bradykinin and thrombin and their mechanisms of action on MC5-5 fibroblasts. *J Biol Chem* 251, 5814-5816 (1976).

39 D. C. Howell *et al.*, Direct thrombin inhibition reduces lung collagen, accumulation, and connective tissue growth factor mRNA levels in bleomycin-induced pulmonary fibrosis. *The American journal of pathology* 159, 1383-1395 (2001).

40 L. O. Mosnier, X. V. Yang, J. H. Griffin, Activated protein C mutant with minimal anticoagulant activity, normal cytoprotective activity, and preservation of thrombin activable fibrinolysis inhibitor-dependent cytoprotective functions. *J Biol Chem* 282, 33022-33033 (2007).

41 J. Stasko, J. Hudecek, P. Kubisz, [Thrombin activatable fibrinolysis inhibitor (TAFI) and its importance in the regulation of fibrinolysis]. *Vnitr Lek* 50, 36-44 (2004).

42 Y. Hua, R. F. Keep, T. Schallert, J. T. Hoff, G. Xi, A thrombin inhibitor reduces brain edema, glioma mass and neurological deficits in a rat glioma model. *Acta Neurochir Suppl* 86, 503-506 (2003).

43 M. A. Brostrom *et al.*, Ca2+ dynamics of thrombin-stimulated rat heart-derived embryonic myocytes: relationship to protein synthesis and cell growth. *Int J Biochem Cell Biol* 35, 1573-1587 (2003).

44 J. R. Day *et al.*, Aprotinin inhibits proinflammatory activation of endothelial cells by thrombin through the protease-activated receptor 1. *J Thorac Cardiovasc Surg* 131, 21-27 (2006).

45 Z. Suo, M. Wu, B. A. Citron, C. Gao, B. W. Festoff, Persistent protease-activated receptor 4 signaling mediates thrombin-induced microglial activation. *J Biol Chem* 278, 31177-31183 (2003).

46 G. L. Howells, M. Macey, M. A. Curtis, S. R. Stone, Peripheral blood lymphocytes express the platelet-type thrombin receptor. *British journal of haematology* 84, 156-160 (1993).

47 B. I. Kuznik, L. P. Malezhik, V. V. Al'fonov, Z. Mamedov Ia, [Effect of thrombin on macrophage and lymphocyte functional activity]. *Biull Eksp Biol Med* 99, 597-598 (1985).

48 A. Naldini, D. H. Carney, Thrombin modulation of natural killer activity in human peripheral lymphocytes. *Cell Immunol* 172, 35-42 (1996).

49 A. Naldini *et al.*, Thrombin enhances T cell proliferative responses and cytokine production. *Cell Immunol* 147, 367-377 (1993).

50 A. Tordai, J. W. Fenton, 2nd, T. Andersen, E. W. Gelfand, Functional thrombin receptors on human T lymphoblastoid cells. *J Immunol* 150, 4876-4886 (1993).

51 A. Naldini, D. H. Carney, A. Pucci, A. Pasquali, F. Carraro, Thrombin regulates the expression of proangiogenic cytokines via proteolytic activation of protease-activated receptor-1. *Gen Pharmacol* 35, 255-259 (2000).

52 A. Naldini *et al.*, The thrombin peptide, TP508, enhances cytokine release and activates signaling events. *Peptides* 25, 1917-1926 (2004).

53 K. E. Brummel, S. G. Paradis, S. Butenas, K. G. Mann, Thrombin functions during tissue factor-induced blood coagulation. *Blood* 100, 148-152 (2002).

54 R. A. Henriksen, G. P. Samokhin, P. B. Tracy, Thrombin-induced thromboxane synthesis by human platelets. Properties of anion binding exosite I-independent receptor. *Arterioscler Thromb Vasc Biol* 17, 3519-3526 (1997).

55 B. Klementsen, L. Jorgensen, Mechanisms involved in the early interaction between HeLa cells, platelets and endothelial cells in vitro under the influence of thrombin. Effects of acetylsalicylic acid and Na-salicylate. *Apmis* 105, 391-401 (1997).

56 L. Liu, J. Freedman, A. Hornstein, J. W. Fenton, 2nd, F. A. Ofosu, Thrombin binding to platelets and their activation in plasma. *British journal of haematology* 88, 592-600 (1994).

57 J. G. Garcia, F. M. Pavalko, C. E. Patterson, Vascular endothelial cell activation and permeability responses to thrombin. *Blood Coagul Fibrinolysis* 6, 609-626 (1995).

58 J. P. Antovic, A. Antovic, S. He, L. Tengborn, M. Blomback, Overall haemostatic potential can be used for estimation of thrombin-activatable fibrinolysis inhibitor-dependent fibrinolysis in vivo and for possible follow-up of recombinant factor VIIa treatment in patients with inhibitors to factor VIII. *Haemophilia* 8, 781-786 (2002).

59 B. N. Bouma, L. O. Mosnier, Thrombin activatable fibrinolysis inhibitor (TAFI)--how does thrombin regulate fibrinolysis? *Annals of medicine* 38, 378-388 (2006).

60 J. Malyszko, J. S. Malyszko, T. Hryszko, M. Mysliwiec, Thrombin activatable fibrinolysis inhibitor in hypertensive kidney transplant recipients. *Transplant Proc* 38, 105-107 (2006).

61 S. H. Choi *et al.*, Thrombin induces nigral dopaminergic neurodegeneration in vivo by altering expression of death-related proteins. *Neurobiol Dis* 14, 181-193 (2003).

62 C. N. Pagel *et al.*, Inhibition of osteoblast apoptosis by thrombin. Bone 33, 733-743 (2003).

63 C. J. Chalmers, K. Balmanno, K. Hadfield, R. Ley, S. J. Cook, Thrombin inhibits Bim (Bcl-2-interacting mediator of cell death) expression and prevents serum-withdrawal-induced apoptosis via protease-activated receptor 1. *Biochem* J 375, 99-109 (2003).

64 C. Chinni *et al.*, Thrombin, a survival factor for cultured myoblasts. *J Biol Chem* 274, 9169-9174 (1999).

65 V. L. Turgeon, E. D. Lloyd, S. Wang, B. W. Festoff, L. J. Houenou, Thrombin perturbs neurite outgrowth and induces apoptotic cell death in enriched chick spinal motoneuron cultures through caspase activation. *J Neurosci* 18, 6882-6891 (1998).

66 Y. Q. Huang, J. J. Li, S. Karpatkin, Thrombin inhibits tumor cell growth in association with up-regulation of p21(waf/cip1) and caspases via a p53-independent, STAT-1-dependent pathway. *J Biol Chem* 275, 6462-6468 (2000).

67 A. Naldini, C. Bernini, A. Pucci, F. Carraro, Thrombin-mediated IL-10 up-regulation involves protease-activated receptor (PAR)-1 expression in human mononuclear leukocytes. *J Leukoc Biol* 78, 736-744 (2005).

68 E. Archiniegas, C. Y. Neves, D. Candelle, J. E. Cardier, Thrombin and its protease-activated receptor-1 (PAR1) participate in the endothelial-mesenchymal transdifferentiation process. *DNA Cell Biol* 23, 815-825 (2004).

69 T. Tran, A. G. Stewart, Protease-activated receptor (PAR)-independent growth and pro-inflammatory actions of thrombin on human cultured airway smooth muscle. *Br J Pharmacol* 138, 865-875 (2003).

70 J. R. Day *et al.*, Aprotinin inhibits proinflammatory activation of endothelial cells by thrombin through the protease-activated receptor 1. *J Thorac Cardiovasc Surg* 131, 21-27 (2006).

71 Z. Suo, M. Wu, B. A. Citron, C. Gao, B. W. Festoff, Persistent protease-activated receptor 4 signaling mediates thrombin-induced microglial activation. *J Biol Chem* 278, 31177-31183 (2003).

72 A. Naldini, D. H. Carney, A. Pucci, A. Pasquali, F. Carraro, Thrombin regulates the expression of proangiogenic cytokines via proteolytic activation of protease-activated receptor-1. *Gen Pharmacol* 35, 255-259 (2000).

73 A. Naldini, C. Bernini, A. Pucci, F. Carraro, Thrombin-mediated IL-10 up-regulation involves protease-activated receptor (PAR)-1 expression in human mononuclear leukocytes. *J Leukoc Biol* 78, 736-744 (2005).

74 J. H. Jeng *et al.*, Protease-activated receptor-1-induced calcium signaling in gingival fibroblasts is mediated by sarcoplasmic reticulum calcium release and extracellular calcium influx. *Cell Signal* 16, 731-740 (2004).

75 E. Jin *et al.*, Protease-activated receptor (PAR)-1 and PAR-2 participate in the cell growth of alveolar capillary endothelium in primary lung adenocarcinomas. *Cancer* 97, 703-713 (2003).

76 W. Luo, Y. Wang, G. Reiser, Two types of protease-activated receptors (PAR-1 and PAR-2) mediate calcium signaling in rat retinal ganglion cells RGC-5. *Brain Res* 1047, 159-167 (2005).

77 A. Sabri *et al.*, Signaling properties and functions of two distinct cardiomyocyte protease-activated receptors. *Circ Res* 86, 1054-1061 (2000).

78 S. J. Song, C. N. Pagel, T. M. Campbell, R. N. Pike, E. J. Mackie, The role of protease-activated receptor-1 in bone healing. *The American journal of pathology* 166, 857-868 (2005).

79 H. Wang, G. Reiser, Thrombin signaling in the brain: the role of protease-activated receptors. *Biol Chem* 384, 193-202 (2003).

80 K. Y. Wang *et al.*, Expression of functional type 1 protease-activated thrombin receptors by mouse primary palatal mesenchymal cells in vitro. *Arch Oral Biol* 45, 819-825 (2000).

81 P. Zania, S. Kritikou, C. S. Flordellis, M. E. Maragoudakis, N. E. Tsopanoglou, Blockade of angiogenesis by small molecule antagonists to protease-activated receptor-1 (PAR-1): Association with endothelial cell growth suppression and induction of apoptosis. *J Pharmacol Exp Ther*, (2006).

82 K. Akassoglou, K. W. Kombrinck, J. L. Degen, S. Strickland, Tissue plasminogen activator-mediated fibrinolysis protects against axonal degeneration and demyelination after sciatic nerve injury. *J Cell Biol* 149, 1157-1166 (2000).

83 P. A. McKee, P. Mattock, R. L. Hill, Subunit structure of human fibrinogen, soluble fibrin, and cross-linked insoluble fibrin. *Proceedings of the National Academy of Sciences of the United States of America* 66, 738-744 (1970).

84 J. W. Weisel, Fibrinogen and fibrin. *Adv Protein Chem* 70, 247-299 (2005).

85 M. W. Mosesson, Fibrinogen and fibrin structure and functions. *J Thromb Haemost* 3, 1894-1904 (2005).

86 I. Juhan-Vague, M. Hans, [From fibrinogen to fibrin and its dissolution]. *Bull Acad Natl Med* 187, 69-82; discussion 83-64 (2003).

87 J. Halper, M. Kjaer, Basic components of connective tissues and extracellular matrix: elastin, fibrillin, fibulins, fibrinogen, fibronectin, laminin, tenascins and thrombospondins. *Adv Exp Med Biol* 802, 31-47 (2014).

88 P. A. McKee, P. Mattock, R. L. Hill, Subunit structure of human fibrinogen, soluble fibrin, and cross-linked insoluble fibrin. *Proceedings of the National Academy of Sciences of the United States of America* 66, 738-744 (1970).

89 I. Juhan-Vague, M. Hans, [From fibrinogen to fibrin and its dissolution]. *Bull Acad Natl Med* 187, 69-82; discussion 83-64 (2003).

90 G. Z. He, [Study on the fibrin fibrinogen degradation products levels and it's significance in the phlegm of the patients with chronic cor pulmonale through monoclonal antibodies]. *Zhonghua Jie He He Hu Xi Za Zhi* 13, 154-156, 191 (1990).

91 M. Rabaud *et al.*, Adduct formation between soluble fibrin monomers and elastin. *Biomaterials* 8, 217-222 (1987).

92 H. Wada *et al.*, Increased soluble fibrin in plasma of patients with disseminated intravascular coagulation. *Clin Appl Thromb Hemost* 9, 233-240 (2003).

93 H. Wada *et al.*, Increased plasma-soluble fibrin monomer levels in patients with disseminated intravascular coagulation. *Am J Hematol* 51, 255-260 (1996).

94 D. C. Howell *et al.*, Direct thrombin inhibition reduces lung collagen, accumulation, and connective tissue growth factor mRNA levels in bleomycin-induced pulmonary fibrosis. *The American journal of pathology* 159, 1383-1395 (2001).

95 P. A. McKee, P. Mattock, R. L. Hill, Subunit structure of human fibrinogen, soluble fibrin, and cross-linked insoluble fibrin. *Proceedings of the National Academy of Sciences of the United States of America* 66, 738-744 (1970).

96 J. Stasko, J. Hudecek, P. Kubisz, [Thrombin activatable fibrinolysis inhibitor (TAFI) and its importance in the regulation of fibrinolysis]. *Vnitr Lek* 50, 36-44 (2004).

97 B. N. Bouma, L. O. Mosnier, Thrombin activatable fibrinolysis inhibitor (TAFI)--how does thrombin regulate fibrinolysis? *Annals of medicine* 38, 378-388 (2006).

98 K. Akassoglou, K. W. Kombrinck, J. L. Degen, S. Strickland, Tissue plasminogen activator-mediated fibrinolysis protects against axonal

degeneration and demyelination after sciatic nerve injury. *J Cell Biol* 149, 1157-1166 (2000).

99 F. J. Castellino, J. R. Powell, Human plasminogen. *Methods in enzymology* 80 Pt C, 365-378 (1981).

100 Z. Hao *et al.*, Stimulated release of tissue plasminogen activator from artery wall sympathetic nerves: implications for stress-associated wall damage. *Stress* 8, 141-149 (2005).

101 D. Hevey, H. M. McGee, D. Fitzgerald, J. H. Horgan, Acute psychological stress decreases plasma tissue plasminogen activator (tPA) and tissue plasminogen activator/plasminogen activator inhibitor-1 (tPA/PAI-1) complexes in cardiac patients. *Eur J Appl Physiol* 83, 344-348 (2000).

102 L. S. Coleman, Insoluble fibrin may reduce turbulence and bind blood components into clots. *Med Hypotheses* 65, 820-821 (2005).

103 L. S. Coleman, Insoluble fibrin may reduce turbulence and bind blood components into clots. *Med Hypotheses* 65, 820-821 (2005).

104 L. S. Coleman, Insoluble fibrin may reduce turbulence and bind blood components into clots. *Med Hypotheses* 65, 820-821 (2005).

105 L. Coleman, A Hypothesis: Factor VII governs clot formation, tissue repair and apoptosis. *Medical Hypotheses* in press, (2007).

106 L. S. Coleman, An Improved Explanation of Atherosclerosis. *PLoS Medicine* 2(4):e98, (2005).

107 L. S. Coleman, A capillary hemostasis mechanism regulated by sympathetic tone and activity via factor VIII or von Willebrand's factor may function as a "capillary gate" and may explain angiodysplasia, angioneurotic edema, and variations in systemic vascular resistance. *Med Hypotheses*, (2005).

108 L. S. Coleman, To the Editor: Is von Willebrand Factor a Hormone that Regulates a Coagulation Mechanism? *World J Surg*, (2006).

109 L. S. Coleman, A capillary hemostasis mechanism regulated by sympathetic tone and activity via factor VIII or von Willebrand's factor may function as a "capillary gate" and may explain angiodysplasia, angioneurotic edema, and variations in systemic vascular resistance. *Med Hypotheses* 66, 773-775 (2006).

110 L. S. Coleman, A Stress Repair Mechanism that Maintains Vertebrate Structure during Stress. *Cardiovasc Hematol Disord Drug Targets*, (2010).

111 A. Bugeja, P. Blanco, E. G. Clark, M. M. Sood, Sickle cell disease: a case report of renal amyloidosis. *BMC Nephrol* 19, 256 (2018).

112 D. Foguel, J. L. Silva, New insights into the mechanisms of protein misfolding and aggregation in amyloidogenic diseases derived from pressure studies. *Biochemistry* 43, 11361-11370 (2004).

113 J. M. Frank, N. J. Palomino, Primary amyloidosis with diffuse splenic infiltration presenting as fulminant pneumococcal sepsis. *Am J Clin Pathol* 87, 405-407 (1987).

114 G. Gamba *et al.*, Clotting alterations in primary systemic amyloidosis. *Haematologica* 85, 289-292 (2000).

115 Y. A. Goffin, E. Gruys, G. D. Sorenson, F. Wellens, Amyloid deposits in bioprosthetic cardiac valves after long-term implantation in man. A new

localization of amyloidosis. *The American journal of pathology* 114, 431-442 (1984).

116 A. Jabor *et al.*, [Serum amyloid A as an effective marker for the assessment of surgical trauma and risk of post-operative complications]. *Ceska Gynekol* 71, 131-136 (2006).

117 P. A. Janmey *et al.*, Interactions of gelsolin and gelsolin-actin complexes with actin. Effects of calcium on actin nucleation, filament severing, and end blocking. *Biochemistry* 24, 3714-3723 (1985).

118 S. E. Lind, P. A. Janmey, Human plasma gelsolin binds to fibronectin. *J Biol Chem* 259, 13262-13266 (1984).

119 D. B. Smith, P. A. Janmey, T. J. Herbert, S. E. Lind, Quantitative measurement of plasma gelsolin and its incorporation into fibrin clots. *J Lab Clin Med* 110, 189-195 (1987).

Chapter 8

The Vascular Endothelium

"He who possesses art and science has religion; he who does not possess them, needs religion."
— Johann Wolfgang von Goethe

The *vascular endothelium* is the heart and soul of the stress mechanism, because it mediates all stress mechanism activities. It is a delicate layer of cells, only one cell thick, that lines the inner surface of the heart and all blood vessels (see Figure 8.1). The vascular endothelium is the sole constituent of capillaries, so it is ubiquitous throughout the body.

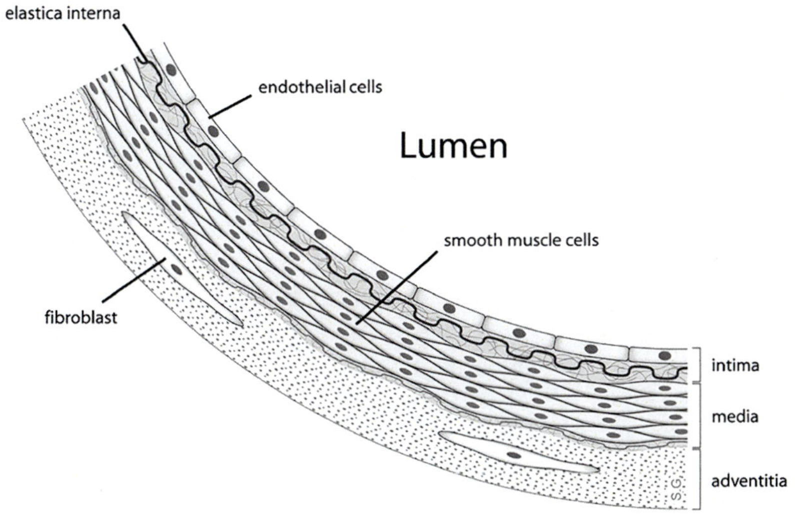

Figure 8.1: Endothelial cells.
This drawing portrays a section of arterial wall, showing how a single layer of endothelial cells lines the inner wall of the heart and all blood vessels. By DRosenbach at English Wikipedia, CC BY-SA 3.0, https://commons.wikimedia.org/w/index.php?curid=30308742

The vascular endothelium manufactures and releases several enzymes and substances that facilitate stress mechanism function, including von Willebrand Factor (VWF), nitric oxide (NO), Protein C, stoichiometric ATIII, fibronectin, vitronectin, and TFPI (Tissue Factor Pathway Inhibitor). It regulates the enzymatic interaction of factors VII, VIII, IX, and X, and thereby mediates both blood flow and tissue repair.

The vascular endothelium mediates tissue repair by isolating tissue factor in extravascular tissues from blood enzymes. It is *selectively permeable* so that it allows the slow penetration of factor VII from blood into extravascular tissues. This enables tissue maintenance, because tissue factor in extravascular tissues stabilizes factor VII, whereupon it generates small amounts of thrombin that energize slow fibroblast replication and collagen turnover that replenishes tissue strength.

Trauma disrupts the vascular endothelium, causing direct contact between exposed tissue factor in extravascular tissues with all the blood enzymes. The resulting simultaneous interaction of factors VII, VIII, IX, and X amplifies and accelerates thrombin generation to energize coagulation and tissue repair.

The vascular endothelium also allows the slow escape — or leakage — of tissue factor into flowing blood, where it stabilizes factor VII. This enables the continuous background activities of factors VIII, IX, and X in flowing blood, which otherwise remain inactive in the absence of factor VII activity.

CANNON'S CONVENTIONAL CONCEPT VERSUS STRESS THEORY

Medical literature presently embraces the conventional concept that blood flow to organs and tissues is mediated by *vasodilation* and *vasoconstriction* — muscular contraction and relaxation under direct nervous control. This idea originates from a persuasive book called "The Wisdom of the Body" by Walter B. Cannon, a prominent Harvard researcher in the WWI era. Cannon's concepts were so intuitively appealing that they became entrenched in medical literature but they cannot explain observations that have accumulated since his time.

Cannon viewed the muscular heart as the "Charles Atlas" of the body. He assumed that *cardiac reserve* enables the heart to increase cardiac output ten-fold during intense exercise. He further assumed that blood flow in arteries is the same as water flow in pipes, because water is incompressible, and blood consists of particles suspended and substances dissolved in water. However, we now know that water flow is profoundly different from blood flow. Water, steam, oil, and atmospheric gases are *Newtonian* fluids that exhibit exponentially *increasing* turbulent flow resistance when they are accelerated in pipes. In contrast, blood is a *non-Newtonian* fluid that exhibits exponentially *decreasing* turbulent flow resistance when it is accelerated in either pipes or arteries. The non-Newtonian nature of mammalian blood enables the heart to effortlessly expel its contents during systolic cardiac contraction.

Cannon's concept implies that *cardiac conditioning* increases cardiac size and strength to explain improved exercise capacity in trained athletes. This suggests that blood pressure should rise dramatically during intense exercise, and that the athletic heart should be much larger and heavier than a non-athletic heart, so as to generate the increased force and elevated blood pressure needed to propel oxygenated blood to skeletal muscles during intense exercise. It further indicates that blood pressure should rise identically and instantaneously throughout the arterial tree with the onset of cardiac contraction (systole) and fall to zero with the onset of cardiac relaxation (diastole). However, we now know that exercise has little effect on the size and strength of the heart, and that the blood pressure of trained athletes is abnormally low at rest and remains normal during intense exercise. Cannon's concept cannot explain this.

Cannon's concept implies that cardiac contraction works in opposition to muscular arterial contraction. We now know that both muscle contraction and muscle relaxation consume large quantities of ATP energy, which implies that muscular activity is inherently inefficient, which is generally at odds with the nature of life.

Lastly, Cannon's concept implies that blood pressure is directly related to the force of cardiac contraction, and that tissue perfusion is directly related to blood pressure. In contrast, modern studies have demonstrated that cardiac contractility varies little from beat to beat, and that cardiac output is mainly determined by flow resistance.

As early as 1925 such considerations inspired *capillary gate theory* — the idea that a submicroscopic, molecular-level mechanism governs cardiac output and tissue perfusion at the capillary level, where turbulence, flow, and pressures are minimal, and surface area is far greater than the sum total of all larger vessels combined.[1] This provides a superior explanation of cardiac output and blood flow regulation compared to Cannon's concept. Now, for the first time, the long-anticipated capillary gate mechanism can be described.

The vascular endothelium is directly and indirectly innervated by the autonomic (automatic, subconscious) nervous system, so that it functions as a gigantic nervous endocrine gland. In addition, the cells of the vascular endothelium react to local factors and communicate with one another to regulate microvascular tissue perfusion.[2]

SYMPATHETIC NERVOUS ACTIVITY CLOSES THE CAPILLARY GATE

Sympathetic nervous activity (shown in red in Figure 8.2) releases von Willebrand Factor (VWF) from the inner walls of capillaries.[3] The VWF production sites are next to binding sites for fibrinogen and fibronectin.[4] VWF binds to enzymatic factor VIIIC in flowing blood to form a gigantic molecular complex called *factor VIII* that converts fibrinogen, fibronectin, vitronectin, and plasminogen into insoluble fibrin monomers (a *monomer* is a molecule that can be bonded to other identical molecules to form a *polymer* — a chain of molecules). Insoluble fibrin monomers spontaneously polymerize into long molecular strands that increase capillary flow resistance (this increase in flow resistance is called *viscosity*). The increase in flow resistance inhibits cardiac output, tissue perfusion, tissue oxygenation, organ function, and glucose uptake by cells.

PARASYMPATHETIC NERVOUS ACTIVITY OPENS THE CAPILLARY GATE

Parasympathetic nervous activity (shown in green in Figure 8.2) releases nitric oxide (NO) from the vascular endothelium.[5] NO is a gas that diffuses into the capillary lumen, where it binds avidly to calcium ions (Ca+) that are essential for thrombin activity.[6] The binding of NO to free Ca+ inhibits thrombin activity, reduces insoluble fibrin generation, inhibits TAFI (Thrombin-Activated Fibrinolysis Inhibitor) and accelerates the spontaneous disintegration of insoluble fibrin by plasmin (plasminogen cross-links in the insoluble fibrin molecular structure spontaneously disintegrate into plasmin in the absence of TAFI).[7]

Thrombin is necessary to generate insoluble fibrin. It also energizes TAFI (Thrombin-Activated Fibrinolysis Inhibitor; *fibrinolysis* is the breakdown of insoluble fibrin). Fibrinolysis inhibitor stabilizes plasminogen and prevents it from breaking down into plasmin — and it is plasmin that attacks and dismembers the long strings of insoluble fibrin polymer.[8, 9]

By reducing fresh insoluble fibrin generation and reducing the inhibitor that prevents the breakdown of existing insoluble fibrin, nitric oxide (NO) disintegrates insoluble fibrin. By disintegrating insoluble fibrin and *decreasing* blood viscosity, microvascular flow resistance is decreased, cardiac output is enhanced, and tissue perfusion, tissue oxygenation, and organ function are all improved. Research literature calls this *nitrergic neurogenic vasodilation*.[10, 11]

The sympathetic nervous activity and parasympathetic nervous activity that increases and decreases blood viscosity can take place system-wide or at very specific sites in the body, depending on stress — body-wide stress (massive infection or radiological contamination) or something as minimal as a paper cut.

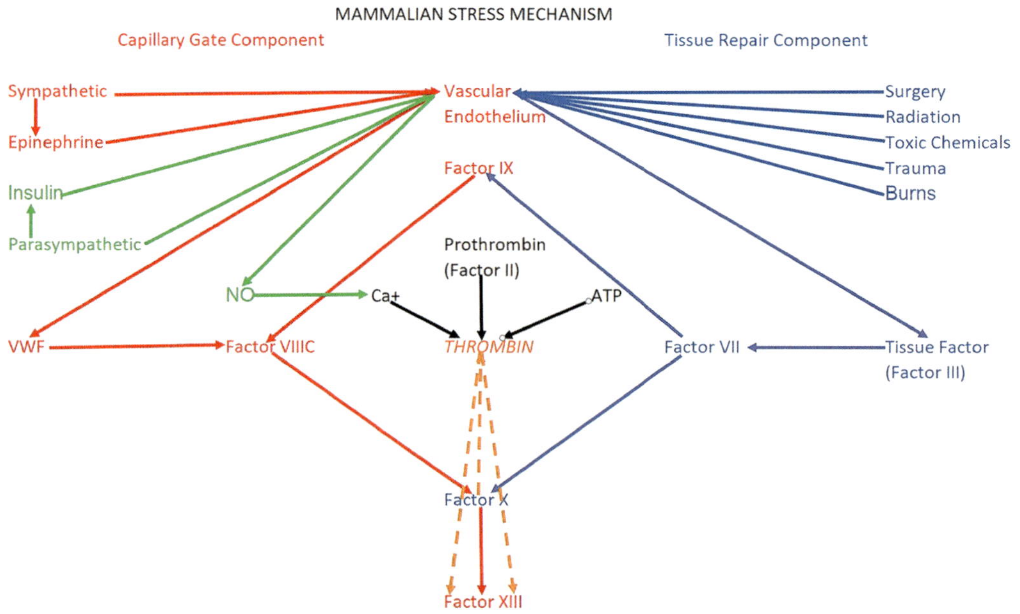

Figure 8.2: Mammalian stress mechanism.

The vascular endothelium governs the generation and disintegration of insoluble fibrin in accord with nervous activity to regulate capillary flow resistance (to open and close the capillary gate). Sympathetic nervous activity (shown in red) releases VWF from the vascular endothelium, which increases factor VIII activity that produces factor XIII and generates insoluble fibrin. Parasympathetic nervous activity (shown in green) releases NO from the vascular endothelium, which disintegrates insoluble fibrin. The vascular endothelium also regulates tissue repair by isolating tissue factor in extravascular tissues from blood enzymes, so that stressful disruption of the vascular endothelium exposes blood enzymes to tissue factor and initiates tissue repair. The tissue repair component is shown in blue.

THE GLYCOCALYX

The vascular endothelium (Figure 8.1) produces the glycocalyx, a delicate molecular matrix that lines the inner surface of all blood vessels. The purpose of the glycocalyx remains uncertain, but it is thought to regulate water absorption.[12, 13] Disease and intravenous fluids disrupt the glycocalyx.[14, 15, 16]

The glycocalyx incorporates *albumen protein*, so named because it congeals into a white layer when heated in a test tube. Albuminuria (albumen in the urine) is often misunderstood as a form of kidney disease, but it appears in urine when stress mechanism hyperactivity disturbs the vascular endothelium, disrupts the glycocalyx, and releases albumen into flowing blood. The kidneys filter albumen from blood, so that albuminuria serves as an early manifestation of diseases, including diabetes, hypertension, pre-eclampsia, Adult Respiratory Distress Syndrome, and Multi-Organ Failure Syndrome, all of which represent harmful stress mechanism hyperactivity.

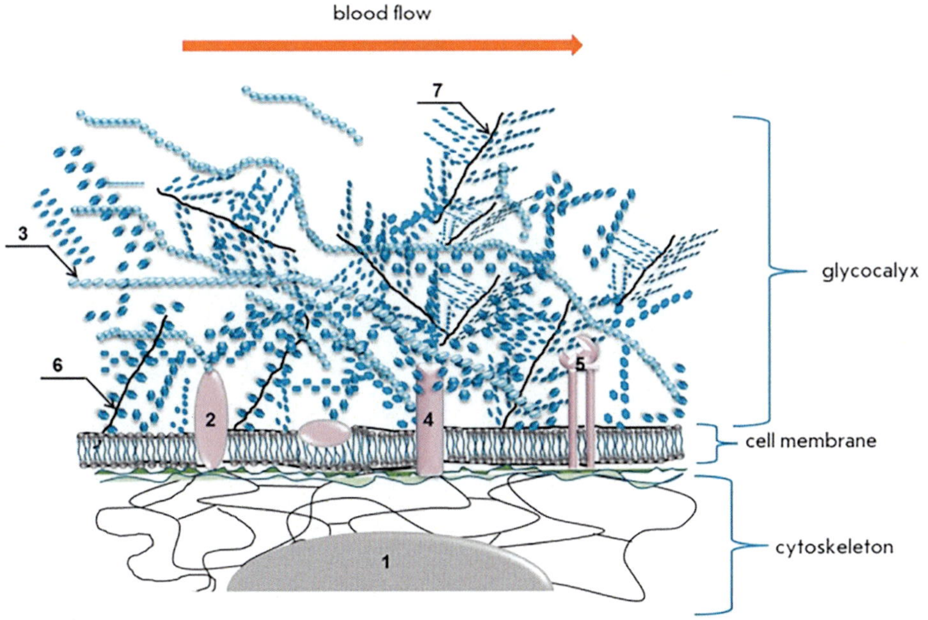

Figure 8.3: Glycocalyx.

The vascular endothelium produces the glycocalyx that lines the inner surface of all capillaries and blood vessels. Composition and structure of endothelial glycocalyx* 1 – nucleus;

2 – hyaladherin; 3 – hyaluronic acid (HA); 4 – glycoprotein; 5 – integrin; 6 – syndecan; 7 – small soluble proteoglycan.

*Glycocalyx is the external cell envelope associated with the plasma membrane that presents a polysaccharide gel. Glycocalyx executes receptor and protective functions, as well as the function of external signal transducer in circulating cells and cells in contact with biological fluids. Glycocalyx shedding leads to changes of cells' receptors.

SPECIALIZATION OF THE VASCULAR ENDOTHELIUM

The vascular endothelium is sub-specialized to serve the requirements of specialized organs and tissues.[17] For example, its cells are *continuous* (tightly joined) in brain tissue, which is rich in tissue factor (see Figure 8.4). This minimizes the permeability of the vascular endothelium in brain tissue and prevents harmful "leakage" of tissue factor from brain tissue into systemic circulation. It also minimizes harmful factor VII "penetration" from blood into brain tissue. In medical literature this is called the *blood-brain barrier* because it resists pharmaceutical penetration into brain tissue.

In other organs, such as intestines, muscles, kidneys, and endocrine glands, the vascular endothelium is *fenestrated* (it contains small windows). These small openings facilitate the function of the organs. In the liver, the vascular endothelium is sinusoid to maximize the exposure of liver cells to flowing blood. This enables liver cells to absorb large lipoproteins from portal circulation (blood flowing from the bowel to the liver) and re-process them into smaller lipoproteins that liver cells release into systemic circulation. Liver cells also de-toxify harmful substances including drugs and pollutants by attaching them to glucuronic acid, a process called *glucuronidation*.

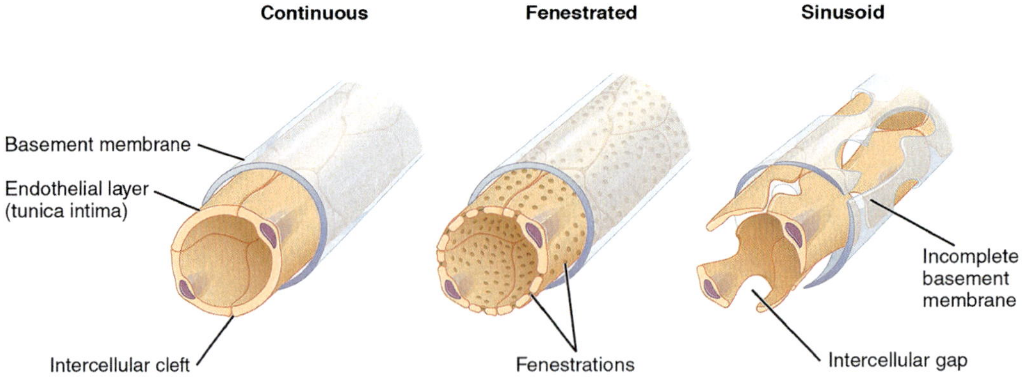

Figure 8.4: Vascular endothelium.

The vascular endothelium is the sole constituent of capillaries and lymph vessels. The capillary endothelium is specialized to facilitate organ function. In brain capillaries, the cells of the vascular endothelium bind tightly to one another to prevent the escape of tissue factor, which is present in high concentrations in brain tissue. This explains the *blood brain barrier* that prevents pharmaceuticals from entering brain tissue. Fenestrated vascular endothelium enhances molecular passage in bowel villi, kidneys, and endocrine glands. Sinusoid gaps in the vascular endothelium of liver tissue enhances the release of fibrinogen and blood enzymes into blood, and the absorption of lipoproteins produced by the bowel. By OpenStax College - Anatomy & Physiology, Connexions Web site. http://cnx.org/content/col11496/1.6/, Jun 19, 2013., CC BY 3.0, https://commons.wikimedia.org/w/index.php?curid=30148

ELECTRICAL SIGNALING IN THE VASCULAR ENDOTHELIUM

The cells of the vascular endothelium communicate with one another via electrical signals.[18] (See Figure 8.5). They also react to local circumstances. For example, exercise increases cellular metabolism, which elevates CO_2 concentrations, causing the vascular endothelium to release nitric oxide that disintegrates insoluble fibrin, reduces flow resistance, and enhances tissue perfusion and oxygen delivery.

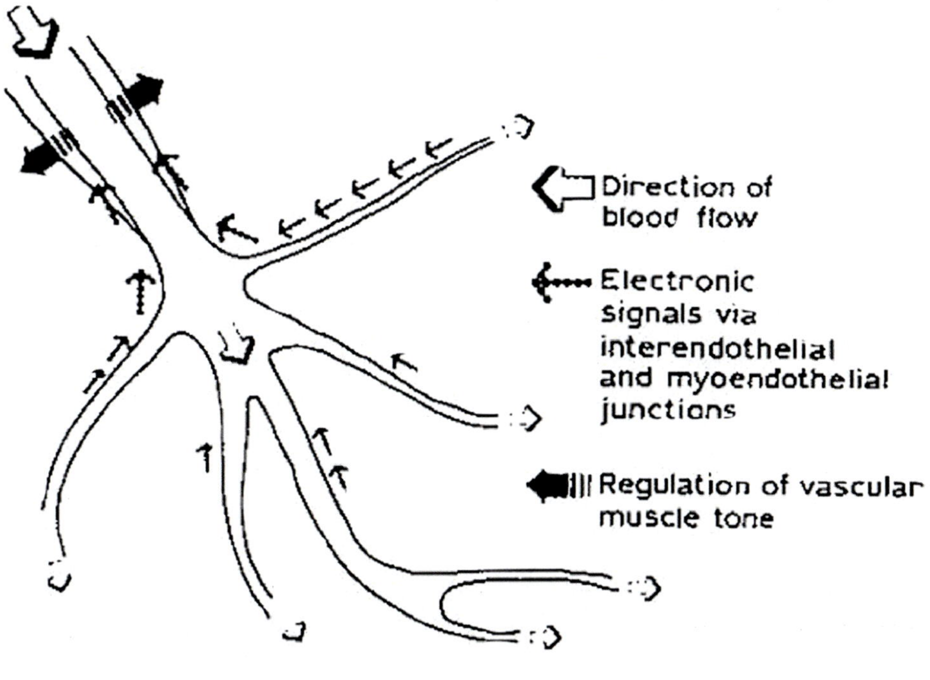

Figure 4 Electrotonic signal transmission in the microcirculation: schematic representation of a new control pathway for the regulation of coronary blood flow

Figure 8.5: Electronic signaling in microcirculation.

The cells of the vascular endothelium are sensitive to their surroundings and communicate with one another via electrical signals.[19] This figure shows electronic signal transmission in microcirculation: it is a schematic representation of a new control pathway for the regulation of coronary blood flow.

NOTES

1 J. Tannenberg, Uber die Kapillartatigkeit. *Verh dtsch. path.* Ges., 20:374, (1925).

2 S. Nees, Coronary flow increases induced by adenosine and adenine nucleotides are mediated by the coronary endothelium: a new principle of the regulation of coronary flow. *Eur Heart J* 10 Suppl F, 28-35 (1989).

3 L. S. Coleman, A capillary hemostasis mechanism regulated by sympathetic tone and activity via factor VIII or von Willebrand's factor may function as a "capillary gate" and may explain angiodysplasia, angioneurotic edema, and variations in systemic vascular resistance. *Med Hypotheses*, (2005).

4 S. Witte, The role of blood coagulation in capillary permeability Vitalmicroscopic contributions. *Biorheology* 21, 121-133 (1984).

5 M. L. Si, T. J. Lee, Alpha7-nicotinic acetylcholine receptors on cerebral perivascular sympathetic nerves mediate choline-induced nitrergic neurogenic vasodilation. *Circ Res* 91, 62-69 (2002).

6 V. B. Schini-Kerth, Vascular biosynthesis of nitric oxide: effect on hemostasis and fibrinolysis. *Transfus Clin Biol* 6, 355-363 (1999).

7 V. B. Schini-Kerth, Vascular biosynthesis of nitric oxide: effect on hemostasis and fibrinolysis. *Transfus Clin Biol* 6, 355-363 (1999).

8 B. N. Bouma, L. O. Mosnier, Thrombin activatable fibrinolysis inhibitor (TAFI)--how does thrombin regulate fibrinolysis? *Annals of medicine* 38, 378-388 (2006).

9 J. Stasko, J. Hudecek, P. Kubisz, [Thrombin activatable fibrinolysis inhibitor (TAFI) and its importance in the regulation of fibrinolysis]. *Vnitr Lek* 50, 36-44 (2004).

10 M. L. Si, T. J. Lee, Alpha7-nicotinic acetylcholine receptors on cerebral perivascular sympathetic nerves mediate choline-induced nitrergic neurogenic vasodilation. *Circ Res* 91, 62-69 (2002).

11 M. A. Vincent, M. Montagnani, M. J. Quon, Molecular and physiologic actions of insulin related to production of nitric oxide in vascular endothelium. *Current diabetes reports* 3, 279-288 (2003).

12 S. Reitsma, D. W. Slaaf, H. Vink, M. A. van Zandvoort, M. G. oude Egbrink, The endothelial glycocalyx: composition, functions, and visualization. *Pflugers Arch* 454, 345-359 (2007).

13 G. T. Sukhikh *et al.*, Differences of glycocalyx composition in the structural elements of placenta in preeclampsia. *Placenta* 43, 69-76 (2016).

14 S. Reitsma, D. W. Slaaf, H. Vink, M. A. van Zandvoort, M. G. oude Egbrink, The endothelial glycocalyx: composition, functions, and visualization. *Pflugers Arch* 454, 345-359 (2007).

15 G. T. Sukhikh *et al.*, Differences of glycocalyx composition in the structural elements of placenta in preeclampsia. *Placenta* 43, 69-76 (2016).

16 J. J. M. Zeron Ghazarian, Raminderjit Sekhon, Andrew Staron, Neil Parikh, Michael Hanna, M.A. Khan, in *American Thoracic Society 2015 International Conference*. (ATS Journals, 2015).

17 V. Stvrtinova, M. Ferencik, I. Hulin, E. Jahnova, [Vascular endothelium as a factor in information transfer between the cardiovascular and immune systems]. *Bratisl Lek Listy* 99, 5-19 (1998).
18 S. Nees, Coronary flow increases induced by adenosine and adenine nucleotides are mediated by the coronary endothelium: a new principle of the regulation of coronary flow. *Eur Heart J* 10 Suppl F, 28-35 (1989).
19 S. Nees, Coronary flow increases induced by adenosine and adenine nucleotides are mediated by the coronary endothelium: a new principle of the regulation of coronary flow. *Eur Heart J* 10 Suppl F, 28-35 (1989).

Chapter 9

The Enzymatic Engine of the Stress Mechanism

"Imagination is more important than knowledge. For knowledge is limited to all we now know and understand, while imagination embraces the entire world, and all there ever will be to know and understand."
— Albert Einstein

The elegant enzymatic interaction of blood enzyme factors VII, VIII, IX, and X is the engine of the stress mechanism, because it generates the three products of the stress mechanism, which are thrombin, soluble fibrin, and insoluble fibrin, that were discussed in the previous chapter.

Those with medical training will recognize factors VII, VIII, IX, and X as the enzymatic elements of the classical *coagulation cascade* that was first identified in the early 1960s, during the heyday of stress research.[1] When first described, the cascade was regarded as a sequence of enzymatic interactions that generates enzymatically inert fibrin to halt blood loss, but was otherwise unrelated to tissue repair. However, it was obvious that blood is a living substance that exhibits ongoing coagulation activity that is dependent on Ca+ even after it is removed from the body, and since fibrin is bereft of enzymatic properties, its ability to enable hemostasis remained unexplained.[2]

As a result, most medical literature assumes that platelet activity somehow orchestrates hemostasis, but the nature of platelet activation, and the means by which it enables coagulation, likewise remains unknown. Thus, the nature of hemostasis remains a medical mystery despite decades of intense medical research.

We now know that *fibrin* consists of a mixture of closely related fibrinogen, soluble fibrin, and insoluble fibrin; that soluble fibrin is derived from fibrinogen; and insoluble fibrin is derived from soluble fibrin. All three can now be distinguished from one another other by new physical and chemical tests. Nevertheless, their individual properties and functions, which were introduced in the previous chapter, are not generally appreciated by medical science.

The liver continuously releases factors VII, VIIIC (as distinguished from factor VIII), IX, and X into flowing blood.[3] Considerable information about their individual properties has been revealed since the era of stress research, and this enables a fresh interpretation of their functional relationships. As long suspected, these enzymes play a pivotal role in tissue repair as well as hemostasis. In fact, hemostasis is merely the first manifestation of a predictable sequence of tissue repair. To understand this, it is essential to study the individual properties of the cascade enzymes. The customary use of Roman numerals to identify these enzymes and their substrates is confusing, so this presentation will emphasize drawings and diagrams. A summary of the Roman numeral designations of the enzymes and substrates of the coagulation cascade can be found in Appendix I: Summary of Coagulation Enzymes and Substances, on page 449.

THE INTERACTION OF FACTORS VII AND X WITH TISSUE FACTOR

Figure 9.3 on page 73 illustrates the interaction of factors VII and X with tissue factor. Factor VII is *labile* (unstable) so that it cannot exert its enzymatic activity unless it is stabilized by tissue factor.[4, 5] Tissue factor is a glycoprotein (a sugar bound to a protein) that lacks enzymatic properties. It is produced by cells in extravascular tissues, as opposed to blood cells that are normally exposed to blood enzymes. In contrast, cells that are exposed to blood enzymes, including red cells, platelets, white cells, and the cells of the vascular endothelium are bereft of tissue factor.[6]

Factor VII in blood is insulated from tissue factor in extravascular tissues by the vascular endothelium, which is a delicate layer of specialized cells, one cell thick, that lines the inner surface of all blood vessels and is the sole constituent of capillaries. *Thus, there is no source of tissue factor within the circulatory system, and blood enzymes are exposed to tissue factor only when tissue factor leaks through the selectively permeable vascular endothelium into blood circulation or trauma disrupts the vascular endothelium and directly exposes blood enzymes to tissue factor in damaged tissues.*[7, 8] To this it may be added that tissue factor concentrations are greater in some tissues than others, and that tissues rich in tissue factor, such as the lung, brain, nerves, retina, perivascular tissues, gonads, cervix, placenta and amniotic fluid react to factor VII and/or thrombin elevations more readily than tissues lacking in tissue factor.[9] This explains why these tissues are targets for stress and malignancy.

Figure 9.3 illustrates how tissue factor binds to factor VII. Tissue factor stabilizes the factor VII enzyme, whereupon factor VII generates small amounts of thrombin that are adequate to energize tissue maintenance by energizing slow fibroblast turnover (replication via mitosis) and collagen generation that continuously replenishes tissue collagen. This explains how small amounts of factor VII energize tissue maintenance by penetrating from blood into extravascular tissues. However, factor VII by itself does not (under normal circumstances) generate enough thrombin to energize either the cellular activities that enable tissue repair or the enzymatic activities that enable coagulation.[10]

Figure 9.1 also illustrates how factor X binds to the complex of tissue factor and factor VII. The addition of factor X amplifies thrombin generation to levels that energize tissue repair and embryological development but are inadequate to energize coagulation.[11] Of course, thrombin generation is only one aspect of tissue repair and embryological development, as other elements of cellular communication are involved, including cellular hormones and subtle electromagnetic forces, most of which remain unclear.[12, 13, 14]

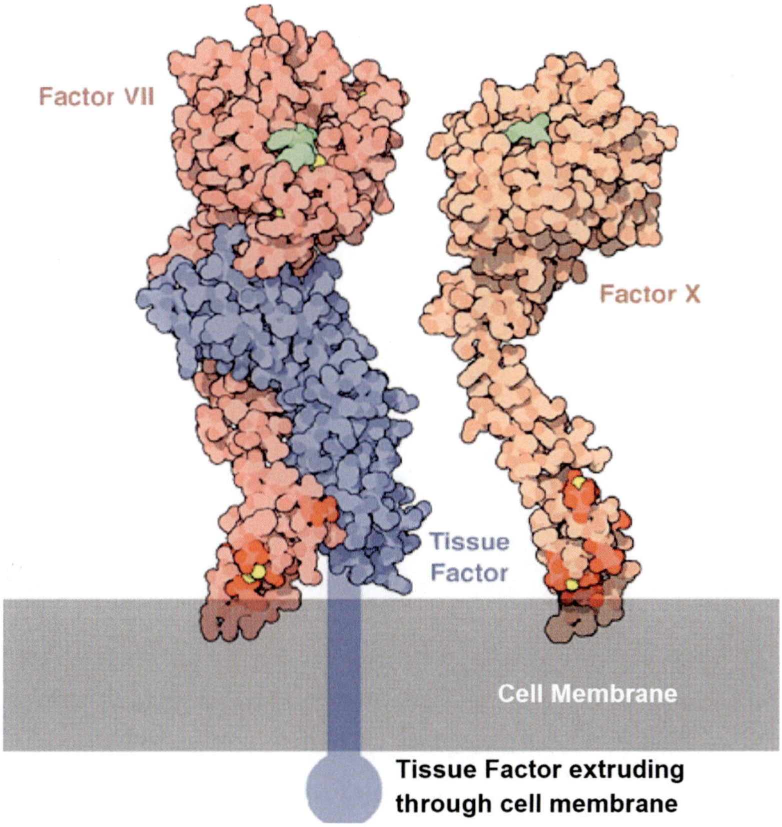

Figure 9.1: Tissue factor and factor VII.

This drawing illustrates how tissue factor and factor VII fit together like love and marriage, and how factor X fits to them like the North and South American continents originally fit together with Europe, Asia, and Africa before they were rifted apart by volcanic activity at the bottom of the Atlantic Ocean. Tissue factor is an enzymatically inert glycoprotein (a combination of protein and sugar) that is extruded to the surface of cells in extravascular tissues. The gray band portrays how the tail of tissue factor is tethered to the cell wall and penetrates to the cell interior. This suggests that the binding of tissue factor to factor VII somehow activates internal cell machinery, though such activity remains unconfirmed. (Image from the RCSB PDB March 2006 Molecule of the Month feature by David Goodsell (doi: http://dx.doi.org/10.2210/rcsb_pdb/mom_2006_3)

THE CHIMERIC NATURE OF FACTOR VIII

In Greek mythology, a *chimera* is a fire-breathing female monster with a lion's head, a goat's body, and a serpent's tail. The Minotaur (half man, half bull) and the

Centaur (half man, half horse) offer examples that are perhaps more "politically correct."

In genetics, a chimera is a rare organism or tissue that contains at least two different sets of DNA. True biological chimeras are exceedingly rare. Perhaps the most common example is *factor VIII* which consists of a chimeric combination of enzymatic factor VIIIC and enzymatically inert von Willebrand Factor (VWF), both of which have independent genetic origin and are produced by unrelated tissues. See Figure 9.2. VIIIC is continuously released into blood by the liver, while VWF is produced by the vascular endothelium and released into blood in accord with sympathetic nervous tone and activity. These two seemingly unrelated molecules bind together in blood circulation and exert their effects in concert in the form of factor VIII.[15, 16]

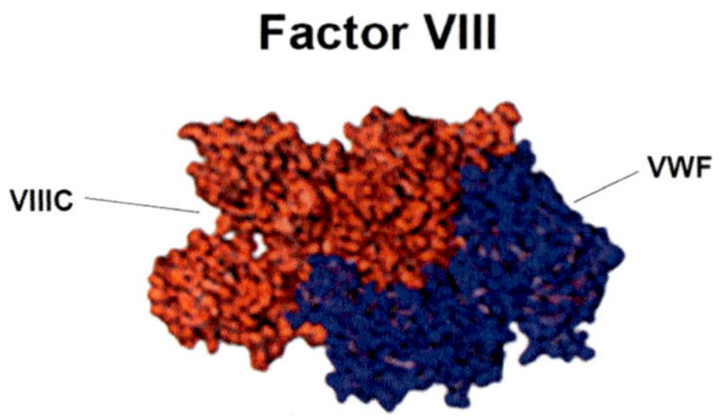

Figure 9.2: Factor VIIIC.

Factor VIIIC binds to von Willebrand Factor (VWF) to form the gigantic factor VIII molecular complex. VWF stabilizes the enzymatic activity of VIIIC, and the two molecules circulate together and exert their effects in concert in the form of chimeric factor VIII.[17] Factor VIII interacts with factors VII, IX and X to accelerate thrombin generation to produce factor XIII, which converts soluble fibrin to insoluble fibrin.[18]

Like factor VII, factor VIIIC is labile and cannot express its enzymatic activities unless it is stabilized. VWF stabilizes VIIIC and enables its enzymatic activities. *Therefore, factor VIII enzyme activity fluctuates in accord with sympathetic nervous activity, and links nervous activity to blood enzyme activity.*

The factor VIII chimera interacts with factors VII, X, and tissue factor to accelerate thrombin generation to energize its enzymatic generation of factor XIII, which converts soluble fibrin to insoluble fibrin to enable coagulation and capillary hemostasis. *The molecular factor VIII chimera is so huge that it cannot penetrate a viscoelastic clot of its own manufacture, which explains why coagulation is self-limiting.*

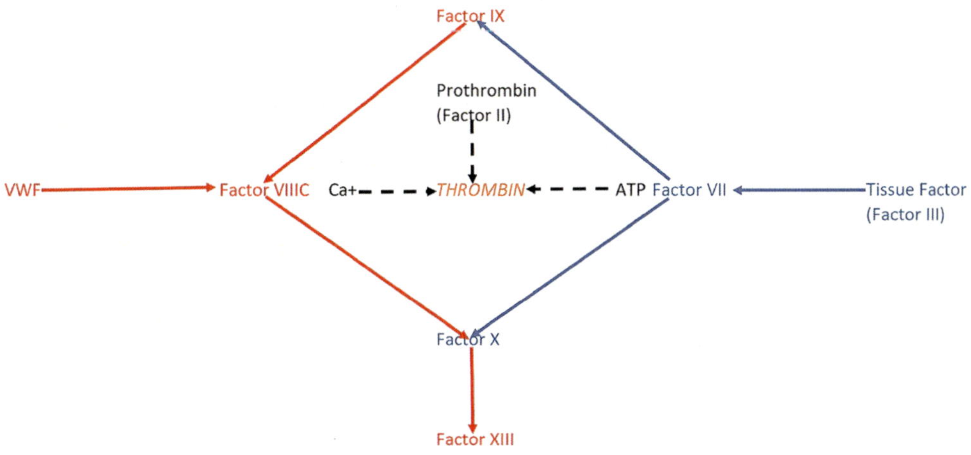

Figure 9.3: Factors VII, VIIIC, IX, and X.

This diagram illustrates the interactions of factors VII, VIIIC, IX and X with tissue factor (TF) and von Willebrand Factor (VWF). Tissue factor binds to, stabilizes, and activates factor VII, which generates small amounts of thrombin. This acts as a trigger for the enzymatic interaction, because the other enzymes remain inert in its absence. VWF binds to, stabilizes, and activates factor VIIIC to produce chimeric factor VIII that interacts with factor X to produce factor XIII. The fluctuating activities of factors VII and VIII determine the magnitude, location, and quantity of thrombin, soluble fibrin, and insoluble fibrin that is produced.

DEFECTS IN FACTORS VII, X, AND TISSUE FACTOR

Defects in tissue factor, and/or factors VII and X are incompatible with life, because they disrupt embryological development.[19] Medications such as Coumadin, Fondaparinux, and thalidomide that disrupt their interaction are incompatible with embryological development, tissue maintenance, and tissue repair.[20] Prolonged anticoagulant treatment with Coumadin disrupts tissue maintenance and causes sloughing of the skin and other forms of tissue deterioration. There is also evidence that Coumadin inhibits malignancy, which is elaborated later. In contrast, the use of factor VII concentrates to control bleeding may abnormally exaggerate tissue repair activity so as to cause malignancy.[21, 22] This may explain the mysterious increase in cancer in hemophilia patients in recent years.[23, 24] Meanwhile, factor VIII concentrates can be used safely to control bleeding in hemophilia, and factor VIII inhibitors such as heparin and its derivatives can be safely used as anticoagulants during pregnancy because they do not interfere with the interaction of tissue factor with factors VII and X.

AMYLOIDOSIS

Amyloidosis is a mysterious disease caused by the deposition of abnormal *amyloid protein* in various tissues. It is considered to be rare, but it may be far more common than presently believed, because its detection requires invasive tissue biopsy, so that its presence is overlooked unless it becomes so severe as to cause visible disease manifestations. The amyloid induces inflammatory tissue repair activity that causes inflammation and destructive tissue sclerosis. Abnormal amyloid concentrations are found in the afflicted tissues of all types of "Rheumatoid" diseases including atherosclerosis, diabetes, hypertension, rheumatoid arthritis, scleroderma, Parkinson's Disease, Alzheimer's Disease, Systemic Lupus Erythematosis, multiple sclerosis, cardiac myositis, cancer, and so forth. Such illnesses were rare before the industrial revolution introduced automobile exhaust, chlorinated water, pesticides, herbicides, harmful airborne particulates, and toxic industrial wastes into the environment. The simplest explanation is that environmental stress causes subclinical stress mechanism enzyme hyperactivity that produces excessive quantities of fibrin split products (FSP) that result from the disintegration of insoluble fibrin. The FSP somehow interacts with factor X to undergo a shape change (conformational change) that produces amyloid protein. The interaction consumes and depletes factor X, which sometimes causes bleeding. FSP occurs in many forms, and this is consistent with the observation that amyloid protein exists in at least 20 different forms. The successful use of artificial intelligence to decipher protein configurations may reveal the nature of amyloid formation, and confer fresh treatments to control and cure chronic illnesses.[25, 26, 27, 28, 29, 30, 31, 32, 33, 34, 35, 36, 37, 38, 39, 40, 41, 42, 43, 44, 45, 46, 47, 48, 49, 50, 51, 52, 53, 54, 55, 56, 57, 58, 59, 60, 61, 62, 63, 64, 65, 66, 67, 68, 69, 70, 71, 72, 73, 74]

HEMOPHILIA

Defects in the VIIIC enzyme cause hemophilia, which is an inherited sex-linked gene defect that cripples stress mechanism function, rather than being a true disease caused by harmful stress mechanism hyperactivity. It mostly afflicts males, and retards hemostasis, tissue repair, and blood flow regulation, so that it causes a peculiar form of exercise intolerance. However, it beneficially inhibits atherosclerosis ("hardening of the arteries") and does not disrupt embryological development. These effects provide valuable clues to the nature of stress mechanism operation that are elaborated in subsequent chapters.

VON WILLEBRAND DISEASE

Defects in the VWF molecule cause von Willebrand's Disease, which, like hemophilia, is not a genuine disease but rather an inherited somatic gene defect that undermines the quality and/or quantity of VWF production, and thereby undermines factor VIII function. It is sometimes mimicked by damage to the fragile

VWF molecule caused by heart valves and LVAD (Left Ventricle Assist Devices). Von Willebrand Disease causes a bleeding defect that is more common but usually less severe than hemophilia, though it can mimic hemophilia in severe cases. It afflicts males and females in equal numbers, and its severity fluctuates in accord with sympathetic nervous activity, which releases fresh reserves of VWF from the vascular endothelium. Defective or damaged VWF also causes flow-related capillary damage called angiodysplasia. These effects will also be elaborated in subsequent chapters.

CHRISTMAS DISEASE

Factor IX is a co-factor of factor VIII that enhances factor VIII activity but has no effect on factors VII and X.[75] Unlike labile factors VII and VIII, factor IX is stable, and has a prolonged half-life, so that its activity does not fluctuate. Defective factor IX causes a mild clotting disorder called *Christmas Disease*. Like hemophilia, it is not a true disease but rather an inherited genetic defect that undermines stress mechanism function.

FACTOR X DEFICIENCY

As noted, before, factor X is essential for embryological development, so that its inherited defects and deficiencies are incompatible with life. Like factor IX, factor X is stable and has a prolonged half-life, so that its activity does not fluctuate. However, it can sometimes become depleted by excessive coagulation activity, because it interacts with both factor VII and factor VIII. This usually occurs in the context of critical illnesses such as eclampsia and DIC (Disseminated Intravascular Coagulation). As previously noted, amyloidosis can also deplete factor X and cause bleeding.

TISSUE FACTOR IN ECLAMPSIA AND CRITICAL ILLNESSES

Pregnant women with pre-eclampsia exhibit elevated blood levels of tissue factor.[76] This is consistent with the hypothesis that tissue factor leaking into blood circulation from amniotic fluid, cervix, and placenta — all of which are rich in tissue factor — is the underlying cause of eclampsia. See Figure 9.4 and Figure 9.5. This is consistent with the hypothesis that all forms of critical illness, including ARDS, MOFS, SIRS, Ebola, COVID, SARS, cancer, and sepsis are due to excessive entry of tissue factor into systemic circulation that induces excessive stress mechanism enzyme activity.[77, 78, 79, 80, 81, 82, 83, 84, 85, 86, 87, 88, 89, 90, 91, 92, 93]

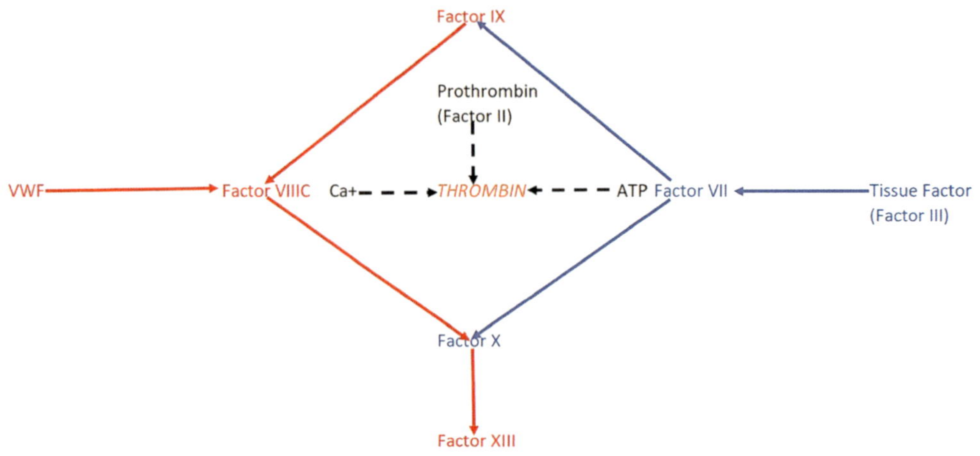

Figure 9.4: Factors IX and X.

Factors IX and X are stable and do not fluctuate. Factors VII and VIIIC are labile, and fluctuate in accord with tissue factor and VWF. Therefore, the fluctuating activity of factor VII alters factor VIII effects, and vice-versa. Their fluctuating activities determine the production of thrombin, soluble fibrin, and insoluble fibrin in accord with combinations of tissue disruption and nervous activity.

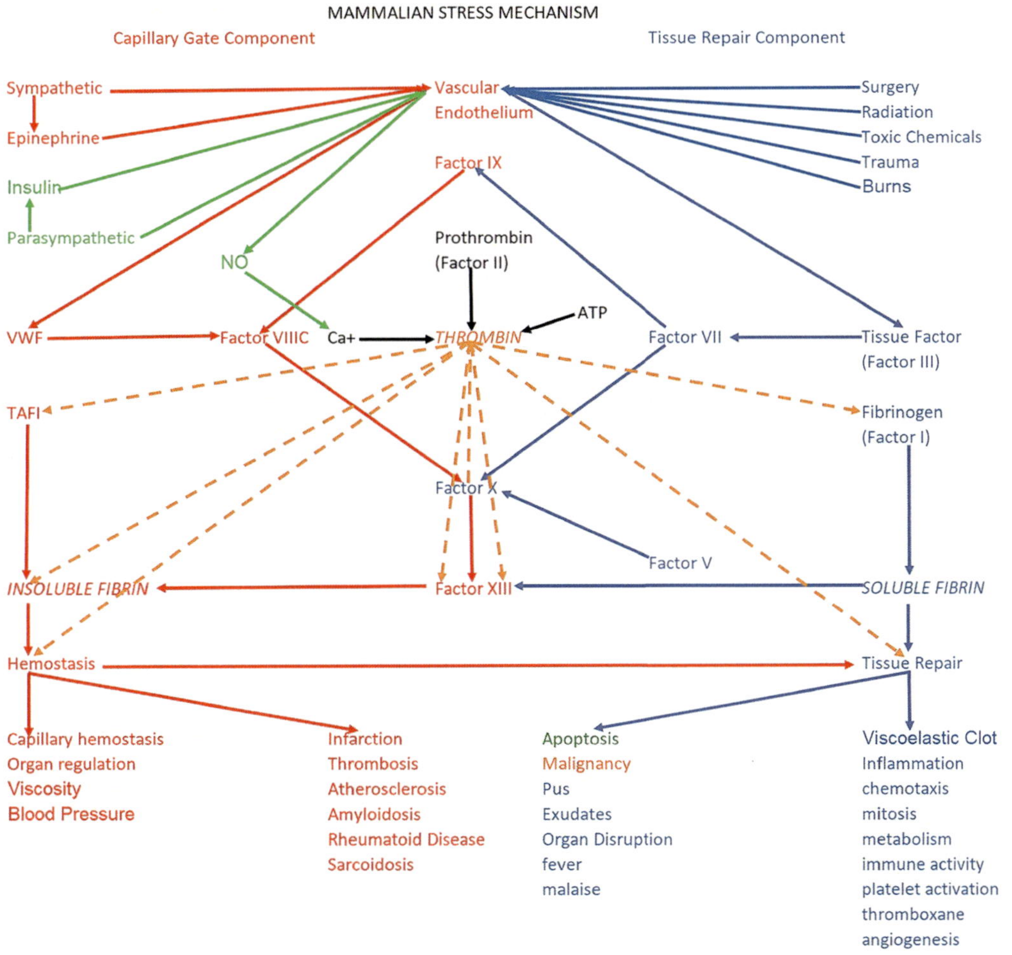

Figure 9.5: Mammalian stress mechanism (detail).

This more detailed diagram of the stress mechanism further illustrates how combinations of tissue disruption and nervous activity determine the production of thrombin, soluble fibrin, and insoluble fibrin. It clarifies and extends the explanatory power of the coagulation cascade. The capillary gate component, shown in red, is analogous to the intrinsic pathway of the coagulation cascade. The tissue repair component, shown in blue, is analogous to the extrinsic pathway of the coagulation cascade. Like the final common pathway of the coagulation cascade, the stress mechanism generates thrombin, soluble fibrin, and insoluble fibrin. The effects of these three products explain the manifestations of physiology, pathology, and stress.

Tissue damage exposes tissue factor to factor VII and activates factor VII, which interacts with factor X to amplify thrombin generation that converts fibrinogen to soluble fibrin to enable tissue repair.[94]

Nervous activity releases VWF that binds to enzyme factor VIIIC to form the gigantic molecular complex called factor VIII, which interacts with factor X to produce Factor XIII that converts soluble fibrin to insoluble fibrin to enable hemostasis and regulate capillary flow resistance.[95]

Stress mechanism enzyme hyperactivity induced by environmental stress is harmful. It exaggerates blood viscosity and coagulability, undermines tissue perfusion and oxygenation, and accelerates the gradual deterioration of capillary beds that occurs with advancing age.[96, 97, 98, 99, 100, 101, 102] Adding analgesia to general anesthesia during surgery inhibits harmful stress mechanism enzyme hyperactivity induced by surgical stress.[103, 104]

CONCLUSION

The elegant enzymatic engine of the stress mechanism generates thrombin, soluble fibrin, and insoluble fibrin to repair tissues and regulate blood flow in accord with combinations of tissue disruption and nervous activity induced by environmental stress. Hyperactivity of the enzymatic interaction produces harmful excesses of thrombin, soluble fibrin, and insoluble fibrin that manifests as disease. Treatments that mitigate enzyme hyperactivity can restore homeostasis and health.

NOTES

1 https://en.wikipedia.org/wiki/Coagulation#The_coagulation_cascade

2 K. S. Sakariassen, M. Ottenhof-Rovers, J. J. Sixma, Factor VIII-von Willebrand factor requires calcium for facilitation of platelet adherence. *Blood* 63, 996-103 (1984).

3 E. Rodriguez-Inigo *et al.*, Expression of factor VII in the liver of patients with liver disease: correlations with the disease severity and impairment in the hemostasis. *Blood Coagul Fibrinolysis* 12, 193-199 (2001).

4 P. W. Friederich *et al.*, Ability of recombinant factor VIIa to generate thrombin during inhibition of tissue factor in human subjects. *Circulation* 103, 2555-2559 (2001).

5 G. M. Hjortoe *et al.*, Tissue factor-factor VIIa-specific up-regulation of IL-8 expression in MDA-MB-231 cells is mediated by PAR-2 and results in increased cell migration. *Blood* 103, 3029-3037 (2004).

6 B. A. Bouchard, K. G. Mann, S. Butenas, No evidence for tissue factor on platelets. *Blood* 116, 854-855 (2010).

7 A. Siegbahn, Cellular consequences upon factor VIIa binding to tissue factor. *Haemostasis* 30 Suppl 2, 41-47 (2000).

8 A. Siegbahn *et al.*, Binding of factor VIIa to tissue factor on human fibroblasts leads to activation of phospholipase C and enhanced PDGF-BB-stimulated chemotaxis. *Blood* 96, 3452-3458 (2000).

9 R. A. Fleck, L. V. Rao, S. I. Rapaport, N. Varki, Localization of human tissue factor antigen by immunostaining with monospecific, polyclonal anti-human tissue factor antibody. *Thromb Res* 59, 421-437 (1990).

10 S. He, M. Blomback, G. Jacobsson Ekman, U. Hedner, The role of recombinant factor VIIa (FVIIa) in fibrin structure in the absence of FVIII/FIX. *J Thromb Haemost* 1, 1215-1219 (2003).

11 T. W. Geisbert *et al.*, Treatment of Ebola virus infection with a recombinant inhibitor of factor VIIa/tissue factor: a study in rhesus monkeys. *Lancet* 362, 1953-1958 (2003).

12 R. O. Becker, G. Selden, *The body electric : electromagnetism and the foundation of life*. (Quill, New York, ed. 1st Quill, 1985), pp. 364 p.

13 P. Cirillo *et al.*, Tissue factor binding of activated factor VII triggers smooth muscle cell proliferation via extracellular signal-regulated kinase activation. *Circulation* 109, 2911-2916 (2004).

14 P. J. Rosch, Bioelectromagnetic and subtle energy medicine: the interface between mind and matter. *Annals of the New York Academy of Sciences* 1172, 297-311 (2009).

15 P. Henke, T. W. Wakefield, Factor VIII and race: a new dimension in hypercoagulability. *Thromb Haemost* 90, 768-769 (2003).

16 J. C. Ngo, M. Huang, D. A. Roth, B. C. Furie, B. Furie, Crystal structure of human factor VIII: implications for the formation of the factor IXa-factor VIIIa complex. *Structure* 16, 597-606 (2008).

17 A. B. Federici, The factor VIII/von Willebrand factor complex: basic and clinical issues. *Haematologica* 88, EREP02 (2003).

18 G. A. Vehar, E. W. Davie, Formation of a serine enzyme in the presence of bovine factor VIII (antihemophilic factor) and thrombin. *Science* 197, 374-376 (1977).

19 K. Borensztajn *et al.*, Factor VII gene intronic mutation in a lethal factor VII deficiency: effects on splice-site selection. *Blood* 102, 561-563 (2003).

20 T. Lisman *et al.*, Recombinant factor VIIa reverses the in vitro and ex vivo anticoagulant and profibrinolytic effects of fondaparinux. *J Thromb Haemost* 1, 2368-2373 (2003).

21 V. Proulle *et al.*, Injection of recombinant activated factor VII can induce transient increase in circulating procoagulant microparticles. *Thromb Haemost* 91, 873-878 (2004).

22 S. Schulman *et al.*, Feasibility of using recombinant factor VIIa in continuous infusion. *Thromb Haemost* 75, 432-436 (1996).

23 A. Siegbahn, Cellular consequences upon factor VIIa binding to tissue factor. *Haemostasis* 30 Suppl 2, 41-47 (2000).

24 U. Martinowitz *et al.*, Recombinant activated factor VII for adjunctive hemorrhage control in trauma. *J Trauma* 51, 431-438; discussion 438-439 (2001).

25 F. V. Beardell, M. Varma, J. Martinez, Normalization of plasma factor X levels in amyloidosis after plasma exchange. *Am J Hematol* 54, 68-71 (1997).

26 A. Bohler, B. Lammle, [Decreased Quick percentage, acquired factor X deficiency, hemarthrosis and ecchymosis: amyloidosis]. *Ther Umsch* 56, 523-525 (1999).

27 E. B. Choufani *et al.*, Acquired factor X deficiency in patients with amyloid light-chain amyloidosis: incidence, bleeding manifestations, and response to high-dose chemotherapy. *Blood* 97, 1885-1887 (2001).

28 B. Furie, E. Greene, B. C. Furie, Syndrome of acquired factor X deficiency and systemic amyloidosis in vivo studies of the metabolic fate of factor X. *N Engl J Med* 297, 81-85 (1977).

29 M. Marcatti *et al.*, Unusual bleeding manifestations in a case of primary amyloidosis with factor X deficiency but elevations of in vivo markers of thrombin formation and activity. *Thromb Res* 80, 333-337 (1995).

30 J. Uprichard, D. J. Perry, Factor X deficiency. *Blood Rev* 16, 97-110 (2002).

31 A. Alwitry, E. T. Brackenbury, F. D. Beggs, I. Soomro, Vascular amyloidosis causing spontaneous mediastinal haemorrhage with haemothorax. *Eur J Cardiothorac Surg* 20, 871-873 (2001).

32 R. A. Armstrong, Classic beta-amyloid deposits cluster around large diameter blood vessels rather than capillaries in sporadic Alzheimer's disease. *Curr Neurovasc Res* 3, 289-294 (2006).

33 S. Beji *et al.*, [Renal amyloidosis complicating the outcome of chronic inflammatory colitis]. *Presse Med* 33, 862-865 (2004).

34 A. Bugeja, P. Blanco, E. G. Clark, M. M. Sood, Sickle cell disease: a case report of renal amyloidosis. *BMC Nephrol* 19, 256 (2018).

35 A. Chandra, Role of amyloid from a multiple sclerosis perspective: a literature review. *Neuroimmunomodulation* 22, 343-346 (2015).

36 A. Clark, M. R. Nilsson, Islet amyloid: a complication of islet dysfunction or an aetiological factor in Type 2 diabetes? *Diabetologia* 47, 157-169 (2004).

37 R. Deane, Z. Wu, B. V. Zlokovic, RAGE (yin) versus LRP (yang) balance regulates alzheimer amyloid beta-peptide clearance through transport across the blood-brain barrier. *Stroke* 35, 2628-2631 (2004).

38 W. EshKalback *et al.*, Atherosclerosis, vascular amyloidosis and brain hypoperfusion in the pathogenesis of sporadic Alzheimer's disease. *Neurol Res* 26, 525-539 (2004).

39 D. Foguel, J. L. Silva, New insights into the mechanisms of protein misfolding and aggregation in amyloidogenic diseases derived from pressure studies. *Biochemistry* 43, 11361-11370 (2004).

40 J. M. Frank, N. J. Palomino, Primary amyloidosis with diffuse splenic infiltration presenting as fulminant pneumococcal sepsis. *Am J Clin Pathol* 87, 405-407 (1987).

41 G. Gamba *et al.*, Clotting alterations in primary systemic amyloidosis. *Haematologica* 85, 289-292 (2000).

42 G. Gamba *et al.*, Abnormalities in thrombin-antithrombin pathway in AL amyloidosis. *Amyloid* 6, 273-277 (1999).

43 D. A. Gastineau, M. A. Gertz, T. M. Daniels, R. A. Kyle, E. J. Bowie, Inhibitor of the thrombin time in systemic amyloidosis: a common coagulation abnormality. *Blood* 77, 2637-2640 (1991).

44 Y. A. Goffin, E. Gruys, G. D. Sorenson, F. Wellens, Amyloid deposits in bioprosthetic cardiac valves after long-term implantation in man. A new localization of amyloidosis. *The American journal of pathology* 114, 431-442 (1984).

45 D. Hall, H. Edskes, Silent prions lying in wait: a two-hit model of prion/amyloid formation and infection. *J Mol Biol* 336, 775-786 (2004).

46 G. M. Hirschfield, P. N. Hawkins, Amyloidosis: new strategies for treatment. *Int J Biochem Cell Biol* 35, 1608-1613 (2003).

47 L. Ho *et al.*, Diet-induced insulin resistance promotes amyloidosis in a transgenic mouse model of Alzheimer's disease. *FASEB journal : official publication of the Federation of American Societies for Experimental Biology* 18, 902-904 (2004).

48 S. Iqbal, S. Reehana, D. Lawrence, Unique type of isolated cardiac valvular amyloidosis. *J Cardiothorac Surg* 1, 38 (2006).

49 A. Jabor *et al.*, [Serum amyloid A as an effective marker for the assessment of surgical trauma and risk of post-operative complications]. *Ceska Gynekol* 71, 131-136 (2006).

50 C. Jacquot, C. d'Auzac, P. Loirat, J. Bariety, [Dramatic aggravation of renal amyloidosis after surgery. Three cases (author's transl)]. *Nouv Presse Med* 10, 3389-3390, 3395 (1981).

51 E. Kaiserling, S. Krober, Massive intestinal hemorrhage associated with intestinal amyloidosis. An investigation of underlying pathologic processes. *Gen Diagn Pathol* 141, 147-154 (1995).

52 S. J. Kang *et al.*, Secondary amyloidosis associated with multiple sclerosis. *J Clin Neurol* 5, 146-148 (2009).

53 G. Leblebisatan, I. Sasmaz, B. Antmen, M. Ergin, Y. Kilinc, Concomitance of idiopathic myelofibrosis and amyloidosis. *Hematol Oncol Stem Cell Ther* 3, 206-207 (2010).

54 H. A. Liebman *et al.*, Excessive fibrinolysis in amyloidosis associated with elevated plasma single-chain urokinase. *Am J Clin Pathol* 98, 534-541 (1992).
55 U. Mousa, S. Aktas, H. Uner, Amyloid accumulation in the toxic nodule of the thyroid gland in a patient with end stage renal failure. *Case Rep Endocrinol* 2012, 741754 (2012).
56 M. R. Nilsson, Techniques to study amyloid fibril formation in vitro. *Methods* 34, 151-160 (2004).
57 R. Nortley *et al.*, Amyloid beta oligomers constrict human capillaries in Alzheimer's disease via signaling to pericytes. *Science* 365, (2019).
58 J. L. Palmero Marti *et al.*, [Secondary vesical amyloidosis]. *Actas Urol Esp* 28, 238-242 (2004).
59 D. S. Powell *et al.*, Apolipoprotein E genotype, islet amyloid deposition and severity of Type 2 diabetes. *Diabetes Res Clin Pract* 60, 105-110 (2003).
60 P. Schwartz, [Amyloidosis, expression and cause of presenile and senile mental and physical regression. A revision of the amyloid problem (author's transl)]. *Zentralbl Allg Pathol* 119, 533-548 (1975).
61 P. Schwartz, [Amyloidosis as a manifestation and origin of presenile and senile degeneration]. *Fortschr Med* 94, 890-896 (1976).
62 I. Schwarzinger *et al.*, Recurrent, isolated factor X deficiency in myeloma: repeated normalization of factor X levels after cytostatic chemotherapy followed by late treatment failure associated with the development of systemic amyloidosis. *Thromb Haemost* 68, 648-651 (1992).
63 B. V. Shekhonin, V. S. Rukosuev, [On the participation of fibrinogen (fibrin) in the formation of amyloid in experimental amyloidosis]. *Biull Eksp Biol Med* 65, 119-122 (1968).
64 G. R. Smith, K. E. Tymms, M. Falk, Etanercept treatment of renal amyloidosis complicating rheumatoid arthritis. *Intern Med J* 34, 570-572 (2004).
65 E. T. Sutton *et al.*, Amyloid-beta peptide induced inflammatory reaction is mediated by the cytokines tumor necrosis factor and interleukin-1. *J Submicrosc Cytol Pathol* 31, 313-323 (1999).
66 S. Takamori *et al.*, Amyloid tumor in the anterior mediastinum: report of a case. *Surg Today* 34, 518-520 (2004).
67 F. Tanaka *et al.*, Interleukin-18 induces serum amyloid A (SAA) protein production from rheumatoid synovial fibroblasts. *Life Sci* 74, 1671-1679 (2004).
68 J. Tian, J. Shi, K. Bailey, D. M. Mann, Relationships between arteriosclerosis, cerebral amyloid angiopathy and myelin loss from cerebral cortical white matter in Alzheimer's disease. *Neuropathol Appl Neurobiol* 30, 46-56 (2004).
69 M. P. Vitek *et al.*, Advanced glycation end products contribute to amyloidosis in Alzheimer disease. *Proceedings of the National Academy of Sciences of the United States of America* 91, 4766-4770 (1994).
70 Y. Wang, X. Feng, B. Shen, J. Ma, W. Zhao, Is Vascular Amyloidosis Intertwined with Arterial Aging, Hypertension and Atherosclerosis? *Front Genet* 8, 126 (2017).
71 D. C. Whitaker, M. F. Tungekar, J. E. Dussek, Angina with a normal coronary angiogram caused by amyloidosis. *Heart* 90, e54 (2004).

72 P. Wiland, R. Wojtala, J. Goodacre, J. Szechinski, The prevalence of subclinical amyloidosis in Polish patients with rheumatoid arthritis. *Clin Rheumatol* 23, 193-198 (2004).

73 Z. Xie *et al.*, The inhalation anesthetic isoflurane induces a vicious cycle of apoptosis and amyloid beta-protein accumulation. *J Neurosci* 27, 1247-1254 (2007).

74 R. A. Yood, M. Skinner, A. Rubinow, L. Talarico, A. S. Cohen, Bleeding manifestations in 100 patients with amyloidosis. *Jama* 249, 1322-1324 (1983).

75 J. H. McIntosh, D. Owens, C. A. Lee, S. Raut, T. W. Barrowcliffe, A modified thrombin generation test for the measurement of factor VIII concentrates. *J Thromb Haemost* 1, 1005-1011 (2003).

76 O. Erez *et al.*, Tissue factor activity in women with preeclampsia or SGA: a potential explanation for the excessive thrombin generation in these syndromes. *J Matern Fetal Neonatal Med* 31, 1568-1577 (2018).

77 A. Siegbahn, Cellular consequences upon factor VIIa binding to tissue factor. *Haemostasis* 30 Suppl 2, 41-47 (2000).

78 A. Siegbahn *et al.*, Binding of factor VIIa to tissue factor on human fibroblasts leads to activation of phospholipase C and enhanced PDGF-BB-stimulated chemotaxis. *Blood* 96, 3452-3458 (2000).

79 T. W. Geisbert *et al.*, Treatment of Ebola virus infection with a recombinant inhibitor of factor VIIa/tissue factor: a study in rhesus monkeys. *Lancet* 362, 1953-1958 (2003).

80 S. Gando, Tissue factor in trauma and organ dysfunction. *Semin Thromb Hemost* 32, 48-53 (2006).

81 S. Gando, S. Nanzaki, S. Sasaki, O. Kemmotsu, Significant correlations between tissue factor and thrombin markers in trauma and septic patients with disseminated intravascular coagulation. *Thromb Haemost* 79, 1111-1115 (1998).

82 J. Himber *et al.*, Inhibition of tissue factor limits the growth of venous thrombus in the rabbit. *J Thromb Haemost* 1, 889-895 (2003).

83 T. Imamura, [Tissue factor expression at the site of inflammation: a cross-talk between inflammation and the blood coagulation system]. *Rinsho Byori* 52, 342-349 (2004).

84 R. S. Kasthuri, M. B. Taubman, N. Mackman, Role of tissue factor in cancer. *J Clin Oncol* 27, 4834-4838 (2009).

85 A. A. Khorana *et al.*, Tissue factor expression, angiogenesis, and thrombosis in pancreatic cancer. *Clin Cancer Res* 13, 2870-2875 (2007).

86 M. Levi, T. van der Poll, H. Ten Cate, Tissue factor in infection and severe inflammation. *Semin Thromb Hemost* 32, 33-39 (2006).

87 C. J. Lockwood *et al.*, Amniotic fluid contains tissue factor, a potent initiator of coagulation. *Am J Obstet Gynecol* 165, 1335-1341 (1991).

88 J. H. McVey, Tissue factor pathway. *Baillieres Best Pract Res Clin Haematol* 12, 361-372 (1999).

89 B. Osterud, E. Bjorklid, The tissue factor pathway in disseminated intravascular coagulation. *Semin Thromb Hemost* 27, 605-617 (2001).

90 F. R. Rickles, S. Patierno, P. M. Fernandez, Tissue factor, thrombin, and cancer. *Chest* 124, 58S-68S (2003).

91 P. M. Sandset, B. J. Warn-Cramer, L. V. Rao, S. L. Maki, S. I. Rapaport, Depletion of extrinsic pathway inhibitor (EPI) sensitizes rabbits to disseminated intravascular coagulation induced with tissue factor: evidence supporting a physiologic role for EPI as a natural anticoagulant. *Proceedings of the National Academy of Sciences of the United States of America* 88, 708-712 (1991).

92 T. A. Warr, L. V. Rao, S. I. Rapaport, Disseminated intravascular coagulation in rabbits induced by administration of endotoxin or tissue factor: effect of anti-tissue factor antibodies and measurement of plasma extrinsic pathway inhibitor activity. *Blood* 75, 1481-1489 (1990).

93 J. L. Yao *et al.*, Tissue factor and VEGF expression in prostate carcinoma: a tissue microarray study. *Cancer Invest* 27, 430-434 (2009).

94 L. Coleman, A Hypothesis: Factor VII governs clot formation, tissue repair and apoptosis. *Medical Hypotheses* in press, (2007).

95 L. S. Coleman, A capillary hemostasis mechanism regulated by sympathetic tone and activity via factor VIII or von Willebrand's factor may function as a "capillary gate" and may explain angiodysplasia, angioneurotic edema, and variations in systemic vascular resistance. *Med Hypotheses*, (2005).

96 K. Kario *et al.*, Factor VII hyperactivity and endothelial cell damage are found in elderly hypertensives only when concomitant with microalbuminuria. *Arterioscler Thromb Vasc Biol* 16, 455-461 (1996).

97 T. Ogawa, Y. Suzuki, S. Sayama, H. Soma, Possible relationship between the ratio of Factor VIII complex and placental insufficiency in preeclampsia. *Biol Res Pregnancy Perinatol* 4, 155-157 (1983).

98 M. C. Scholtes, G. Gerretsen, H. L. Haak, The factor VIII ratio in normal and pathological pregnancies. *Eur J Obstet Gynecol Reprod Biol* 16, 89-95 (1983).

99 C. A. Thornton, J. Bonnar, Factor VIII-related antigen and factor VIII coagulant activity in normal and pre-eclamptic pregnancy. *Br J Obstet Gynaecol* 84, 919-923 (1977).

100 T. H. Tran, B. Lammle, F. Duckert, Factor VIII (procoagulant activity VIII:C, and antigen VIII:CAg, related antigen VIIIR:Ag and ristocetin cofactor VIIIR:Cof) in intensive care patients with clinically suspected disseminated intravascular coagulation (DIC). *Am J Clin Pathol* 82, 565-572 (1984).

101 R. von Kries, H. Jurgens, H. von Voss, U. Gobel, The clinical relevance of factor VIII: C and factor VII R: Ag determination in newborns. *Eur J Pediatr* 137, 189-194 (1981).

102 S. Weijer *et al.*, Inhibition of the tissue factor/factor VIIa pathway does not influence the inflammatory or antibacterial response to abdominal sepsis induced by Escherichia coli in mice. *J Infect Dis* 189, 2308-2317 (2004).

103 A. Kuitunen *et al.*, Anaesthesia affects plasma concentrations of vasopressin, von Willebrand factor and coagulation factor VIII in cardiac surgical patients. *Br J Anaesth* 70, 173-180 (1993).

104 L. W. Crile GW, *Anoci-association.* (Saunders, Philadelphia, 1914), pp. 223-225.

Chapter 10

The Autonomic Nervous System

"What is not fully understood is not possessed"
— Johann Wolfgang von Goethe

Stress theory suggests a fresh interpretation of the functional relationships of consciousness, sleep, dreaming, fight or flight, pain, PTSD (Post-Traumatic Stress Disorder), neurosis, and autonomic nervous activity (parasympathetic versus sympathetic nervous activity).

The autonomic (automatic) nervous system consists of two subdivisions: the sympathetic nervous system and the parasympathetic nervous system. Both operate subconsciously. In general, sympathetic nervous activity inhibits organ function, and parasympathetic nervous activity promotes it. Their opposing effects govern organ perfusion (blood flow) and thereby regulate organ function, including salivation, swallowing, digestion, urination, defecation, breathing, and cardiac output.

THE PARASYMPATHETIC NERVOUS SYSTEM (PNS)

Parasympathetic activity originates in the brain stem and distal portions of the spinal cord and communicates directly with the vascular endothelium of internal organs via cranial nerve that emanate from the brainstem and distal portion of the spinal cord. (See Figure 10.1 on page 88.) As discussed previously, parasympathetic activity releases nitric oxide (NO) from the vascular endothelium, which binds to Ca+ and inhibits thrombin activity that energizes the production and stabilization of insoluble fibrin. Parasympathetic nervous activity thus accelerates the spontaneous disintegration of insoluble fibrin by plasmin, which "opens" the capillary gate, reduces capillary flow resistance, enhances organ perfusion, and increases organ activity. In general, parasympathetic activity opposes sympathetic nervous activity and promotes organ function — but its effects are readily over-ridden by sympathetic nervous activity.

THE SYMPATHETIC NERVOUS SYSTEM (SNS)

You will recall that raising the tone of the sympathetic nervous system acts to reduce organ blood flow and thereby inhibit organ function. Two confusingly inter-related and synergistic pathways elevate sympathetic nervous activity: the *cognitive pathway* and the *nociception pathway*.

THE NOCICEPTION PATHWAY

Potentially harmful chemical, mechanical, and thermal stimuli activate tissue disruption sensors called *nociceptors* that are located in organs and peripheral tissues. Thus activated, they generate nervous signals called nociception. Peripheral sensory nerves communicate nociception to specialized spinal cord nociception pathways. Specialized *internuncial* (nerve cell-to-nerve cell) nerves communicate spinal cord nociception to sympathetic ganglia (nerve cell clusters) located in the chest and abdomen (see Figure 20). The sympathetic ganglia innervate the vascular endothelium of internal organs. Thus, nociception induces sympathetic nervous activity, which releases von Willebrand Factor from the vascular endothelium of internal organs, activates factor VIII, and generates insoluble fibrin that increases capillary flow resistance (closes the capillary gate). *The nociception pathway painlessly and subconsciously inhibits tissue perfusion and organ function.*

THE COGNITIVE PATHWAY

In addition to communicating nociception to sympathetic ganglia in the chest and abdomen, the spinal cord nociception pathways communicate nociception to the cerebral cortex via ascending nociception pathways in the brain stem. The cerebral cortex generates consciousness that interprets sensory information from the eyes as sight, from the ears as sound, from the nose as smell, from the tongue as taste, and from nociception as *pain*. When consciousness detects signs of danger it elevates sympathetic nervous activity via hypothalamic pathways that are independent of nociception pathways.

THE EMOTIONAL MECHANISM

The emotional mechanism_is situated in the amygdala, which is a "primitive" part of the brain that works closely with consciousness. It records and retains detailed audiovisual memories of all waking moments throughout life. These memories are vivid and lifelike, but they are usually suppressed so as to remain subconscious. However, under certain circumstances they can invade consciousness and mimic current reality. These memories were inadvertently discovered in the 1960s by Dr. Wilder Penfield, a neurosurgeon who stimulated various areas of the brain using tiny electrodes while searching for a surgical treatment for epilepsy. This caused his patients to suddenly recall distant memories with startling clarity. His startling observations were corroborated by a patient named Jill Price, whose unsuppressed memories impaired her ability to function.

Similar patients have been subsequently identified. This inability to suppress emotional memory is now called hyperthymesia and it explains photographic memory. For example, Marilu Henner, the television actress, is able to memorize her acting lines after a single reading. (https://abcnews.go.com/blogs/entertainment/2012/04/actress-marilu-henners-rare-super-memory-recalls-every-day-of-her-life)

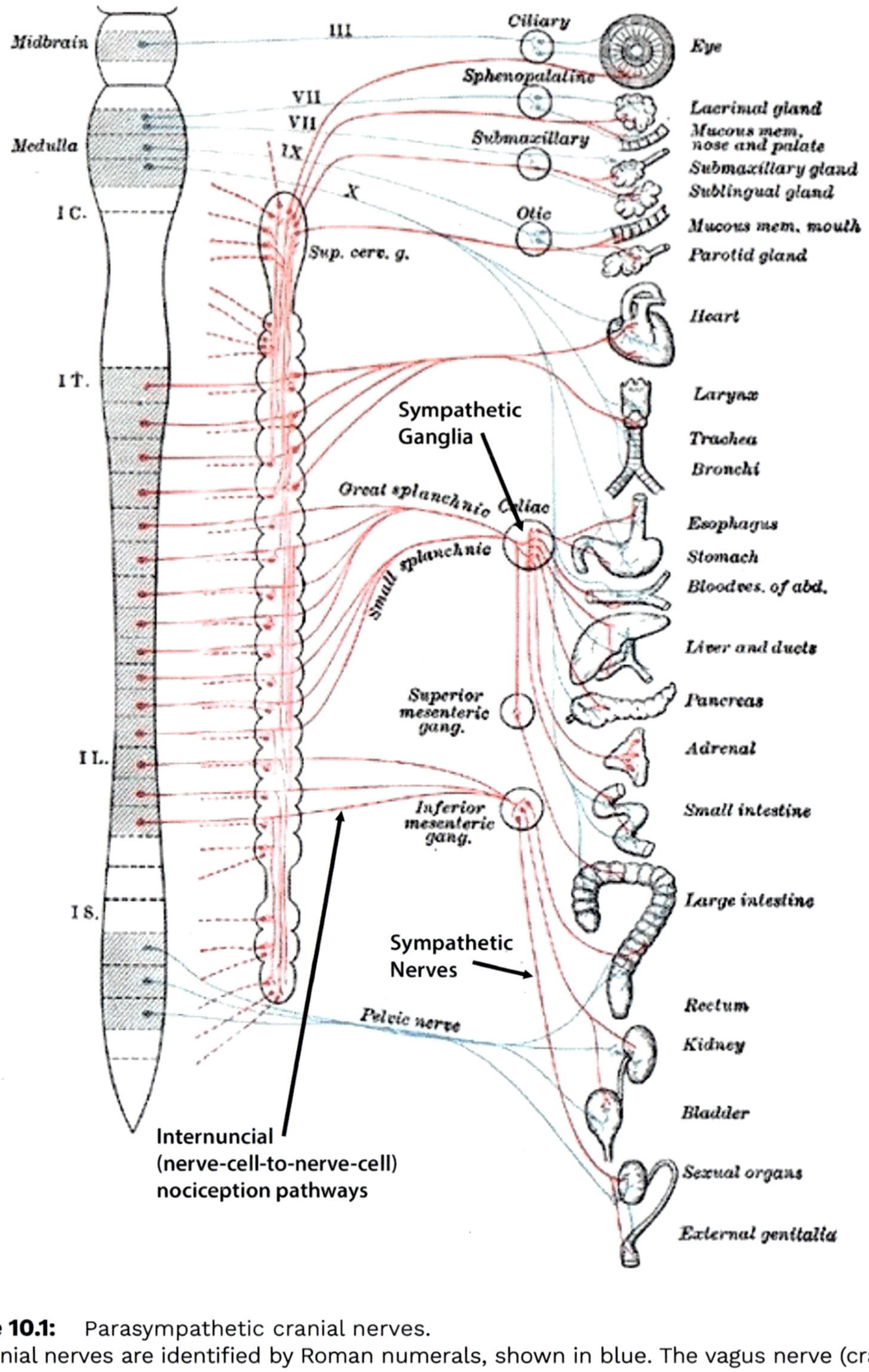

Figure 10.1: Parasympathetic cranial nerves.

Cranial nerves are identified by Roman numerals, shown in blue. The vagus nerve (cranial nerve X) provides parasympathetic innervation to the internal organs. Internuncial nerves communicate nociception to sympathetic ganglia in the chest and abdomen that innervate the internal organs (shown in red). The opposing effects of sympathetic and parasympathetic innervation enable autonomic balance that regulates the perfusion of internal organs, and thereby regulates their function. Note that direct autonomic innervation is lacking in peripheral muscles and tissues. Henry Vandyke Carter/Henry Gray (1918) Anatomy of the Human Body.

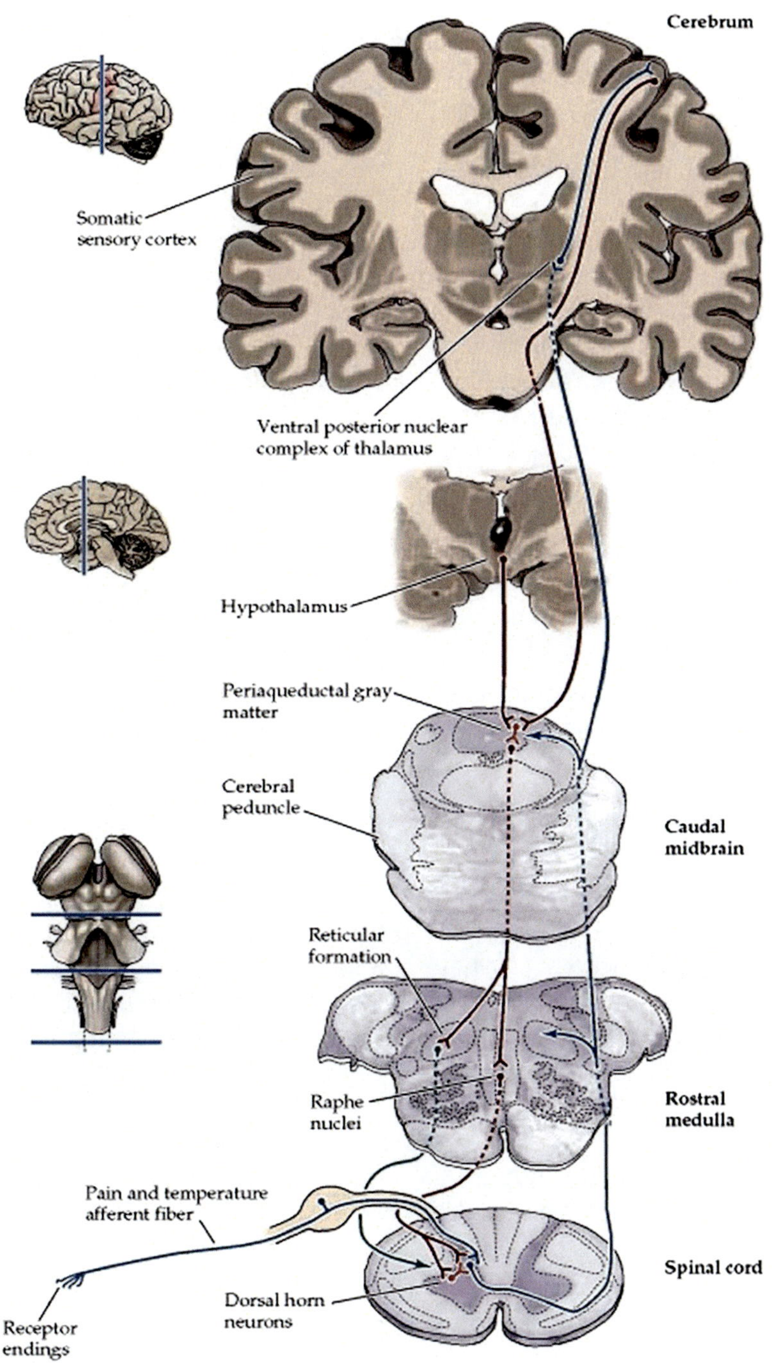

Figure 10.2: The *gate theory* of Melzack and Wall.[1]

Ascending spinal cord nociception pathways communicate with the cerebral cortex, where consciousness interprets nociception as *pain*. Consciousness works closely with emotional mechanisms to modulate spinal cord nociception via descending (*corticofugal*) pathways from the brain to spinal cord nociception pathways. General anesthesia abolishes consciousness, which eliminates the perception of nociception as pain in a dose-related manner, but it simultaneously eliminates its inhibition of spinal cord nociception pathways. This explains why general anesthesia harmfully exaggerates subconscious sympathetic tone and activity induced by surgical stimulation unless it is supplemented and complemented by analgesia that inhibits surgical nociception.[2, 3]

THE DREAMING MECHANISM

Sleep abolishes consciousness, whereupon the dreaming mechanism, which is part of the emotional mechanism, automatically evaluates emotional memories to identify dangerous environmental circumstances. The emotional mechanism subsequently monitors consciousness, and generates fear, anxiety, anger, and rage when it detects similar circumstances, all of which elevate sympathetic nervous system hyperactivity to facilitate fight or flight.[4, 5, 6, 7] The sympathetic nervous activity temporarily concentrates blood flow in critical organs and increases blood coagulability to minimize blood loss in the event of injury.

The dreaming mechanism continually re-evaluates the memory record to incorporate fresh memories, which gradually alters conscious behavior. This explains the phenomenon of *allostasis* — the ability of animals to alter their behavior in accord with changing environmental circumstances.[8, 9] The resulting improvement in survival may also explain the *Baldwin Effect* wherein a character or trait change induced by environmental circumstances gradually becomes assimilated into developmental genetic or epigenetic repertoire via natural selection.

THE CORTICOFUGAL MECHANISM

Consciousness continuously modulates spinal cord nociception pathway activity via corticofugal pathways (descending inhibitory pathways from the brain to the spinal cord nociception pathways). This enables consciousness to alter nociception in accord with circumstances. For example, soldiers sometimes ignore painful wounds suffered during the heat of battle but perceive excruciating pain in its aftermath. (See Figure 10.2 on page 89.) This descending cortical inhibition is abolished by:

1. General anesthesia and hypnotic drugs that abolish consciousness in a dose-related manner. This undermines the descending inhibition of spinal cord nociception pathways. The spinal cord nociception pathways resist the effects of general anesthesia and other hypnotic agents, so that their activity increases in the absence of corticofugal inhibition.
2. Spinal cord damage above the level of cranial nerve III disrupts communication between the brain and the spinal cord. The brain no longer receives ascending nociception from the spinal cord, and the spinal cord no longer receives corticofugal inhibition from the brain. The victim becomes paralyzed from the neck down and he loses all sensation below the neck.

The spinal cord nociception pathways remain intact, and become hyperactive in the absence of corticofugal inhibition. In this condition, nociceptor stimulation in the lower body may cause harmful sympathetic nervous hyperactivity that manifests as severe hypertension, undermines perfusion and oxygenation of organs and tissues, and limits longevity. This explains the limited life span of Christopher Reeve, the actor who portrayed Superman. He fell from a horse, broke his neck, became paraplegic, and died young.

HOW ANESTHESIA MIMICS ANALGESIA

Surgeons have long understood how to correct life-threatening surgical problems, but they usually avoided invasive surgical repairs before the discovery of anesthesia, because such procedures were invariably followed by a lethal stress reaction that culminated in agonizing pain, fever, delirium and death about 48 hours later. The original discovery of anesthesia was the invention of a simple glass device that provided a practical means to administer ether anesthesia, which revolutionized surgery.

Figure 10.3: Morton anesthesia inhaler.
Morton's early glass device for administering anesthesia. Before anesthesia, patients often succumbed to a lethal stress reaction in the aftermath of surgery.

The discovery of anesthesia enabled most patients to survive invasive surgical procedures, but anesthesia alone cannot control the surgical stress syndrome, which is harmful symptoms that occur distant from the location and time of surgery, including fever, nausea, vomiting, tachycardia, hypertension, dysrhythmias, strokes, heart attacks, pneumonia, atelectasis, cancer, and heart disease.[10]

Nociception is nervous sensory signals that consciousness interprets as pain.

Anesthesia inhibits consciousness, and thereby abolishes the ability of consciousness to interpret nociception as pain. Likewise, anesthesia inhibits fear and anxiety. However, anesthesia has negligible effect on nociception.

Analgesia inhibits nociception, and therefore prevents consciousness from receiving nociception signals.

General anesthesia can mimic analgesia because it inhibits consciousness in a dose-related manner. It progressively prevents fear, anxiety, and the perception of nociception as pain *before* it abolishes the ability hear and to speak coherently. This was readily observed during the Golden Era of Ether because ether takes effect very slowly compared to modern inhalation agents. Patients would continue to speak coherently even after surgery was underway, but they did not complain of pain. This conveyed the impression that anesthetic agents have analgesic properties. Anesthetic inhalation agents are but one form of hypnotic drugs. All hypnotic agents, including beverage alcohol and common sedatives, can similarly mimic analgesia by partially obtunding consciousness. However, hypnotic agents have no significant effect on harmful nociception. Only true analgesia inhibits nociception. For example, narcotic analgesia inhibits spinal cord nociception pathways, and local analgesics such as lidocaine and Marcaine disrupt nervous sensory and nociception pathways. This explains why the supplementation of general anesthesia with analgesia improves surgical outcome.[11, 12]

Anesthesia inhibits stress mechanism hyperactivity induced by surgery to the point that most (but not all) patients survive surgery. Unfortunately, things are not that simple. The abolition of consciousness also eliminates the ability to regulate spinal cord nociception pathways via inhibitory corticofugal pathways that descend from the cerebral cortex to spinal cord nociception pathways. Therefore, although general anesthesia beneficially eliminates fear and anxiety that harmfully exaggerates sympathetic nervous activity via hypothalamic pathways, it simultaneously abolishes the inhibition of spinal cord nociception pathways by consciousness. Such spinal cord nociception pathways activate sympathetic ganglia via internuncial nerve pathways that function independently of the hypothalamic pathways. General anesthesia alone cannot inhibit these spinal cord nociception pathways, so that harmful surgical nociception manifests as tachycardia, hypertension, untoward muscle tension, limb movements, and hyperventilation undermines cardiorespiratory function by depleting CO_2 tissue reserves. This has important implications for medical management that are nowadays overlooked.

One hundred years ago, George Washington Crile discovered that controlling harmful nociception using analgesia is a curative treatment unto itself that can be utilized in a variety of circumstances. For example, he supplemented general anesthesia with prilocaine tissue infiltration and intramuscular morphine analgesia (needle technology being too primitive to enable intravenous access in his time) to prevent harmful surgical stress (which was called “shock” in his time). These measures improved surgical outcome. He also cured life-threatening sepsis and peritonitis with massive intramuscular morphine dosage that rendered patients comatose for a week, without the need for intravenous fluids or food intake, during an era when sulfa drugs and antibiotics remained unknown. His book, “Anoci-Association,” is one of the great medical classics and should be *must reading* for all medical practitioners.

POST-TRAUMATIC STRESS DISORDER (PTSD), PHANTOM LIMB PAIN, AND CHRONIC PAIN SYNDROMES

Emotional memory explains *post-traumatic stress disorder* (PTSD), phantom limb pain, and chronic pain syndromes. Present circumstances can provoke emotional memories of past events that mimic reality, causing pain, fear, and panic that induce harmful sympathetic nervous system activity. For example, subconscious memories of painful dental procedures experienced during childhood can mimic acute pain during adulthood when provoked by the sight and sound of a dental drill.

INSULIN, EPINEPHRINE, AND AUTONOMIC BALANCE

Insulin and epinephrine are opposing hormones that extend the effects of autonomic balance to peripheral tissues such as skeletal muscle that lack direct autonomic innervation.

Sympathetic nervous activity releases epinephrine from the adrenal glands, and epinephrine releases von Willebrand Factor (VWF) from the vascular endothelium in peripheral tissues where direct autonomic innervation is lacking. The VWF binds to VIIIC and activates factor VIII, which generates insoluble fibrin that “closes” the capillary gate, increases microvascular flow resistance, and reduces tissue perfusion and glucose uptake.[13, 14, 15]

Parasympathetic nervous activity releases insulin from the pancreas, which releases nitric oxide (NO) from the vascular endothelium. The effect is pronounced in peripheral tissues, where direct autonomic innervation is lacking. The NO binds to Ca+, inhibits thrombin activity, and promotes the disintegration of insoluble fibrin. This reduces microvascular flow resistance and promotes tissue perfusion and glucose uptake.[16, 17, 18, 19, 20, 21, 22, 23, 24, 25, 26, 27]

CONCLUSION

The combined subconscious activities of emotional mechanisms and autonomic nervous mechanisms explain how animals alter their behavior to adapt to changing environmental circumstances. They also clarify the placebo effect, fight or flight, anesthesia, analgesia, allostasis, neurosis, PTSD (post-traumatic stress disorder), phantom limb pain, and chronic pain syndromes.

NOTES

1 R. Melzack, P. D. Wall, Pain mechanisms: a new theory. *Science* 150, 971-979 (1965).

2 L. W. Crile GW, *Anoci-association.* (Saunders, Philadelphia, 1914), pp. 223-225.

3 A. Kuitunen *et al.*, Anaesthesia affects plasma concentrations of vasopressin, von Willebrand factor and coagulation factor VIII in cardiac surgical patients. *Br J Anaesth* 70, 173-180 (1993).

4 W. B. Cannon, *The wisdom of the body.* (W.W. Norton & Company, New York,, 1939), pp. xviii, 19-333 p.

5 W. B. Cannon, The emergncy function of the adrenal medulla in pain and the major emotions. *The American journal of physiology* 33, 356-372 (1914).

6 W. B. Cannon, *Bodily Changes in Pain, Hunger, Fear and Rage: An Account of Recent Researches into the Function of Emotional excitement.* (Appleton, New York, 1915).

7 W. B. Cannon, Voodoo death. *Psychosom Med* 19, 182-190 (1957).

8 B. L. Ganzel, P. A. Morris, E. Wethington, Allostasis and the human brain: Integrating models of stress from the social and life sciences. *Psychol Rev* 117, 134-174 (2010).

9 S. M. Korte, J. M. Koolhaas, J. C. Wingfield, B. S. McEwen, The Darwinian concept of stress: benefits of allostasis and costs of allostatic load and the trade-offs in health and disease. *Neurosci Biobehav Rev* 29, 3-38 (2005).

10 T. G. Monk, V. Saini, B. C. Weldon, J. C. Sigl, Anesthetic management and one-year mortality after noncardiac surgery. *Anesth Analg* 100, 4-10 (2005).

11 L. W. Crile GW, *Anoci-association.* (Saunders, Philadelphia, 1914), pp. 223-225.

12 A. Kuitunen *et al.*, Anaesthesia affects plasma concentrations of vasopressin, von Willebrand factor and coagulation factor VIII in cardiac surgical patients. *Br J Anaesth* 70, 173-180 (1993).

13 F. R. Rickles, L. W. Hoyer, M. E. Rick, D. J. Ahr, The effects of epinephrine infusion in patients with von Willebrand's disease. *The Journal of clinical investigation* 57, 1618-1625 (1976).

14 M. Takeuchi, H. Nagura, T. Kaneda, DDAVP and epinephrine-induced changes in the localization of von Willebrand factor antigen in endothelial cells of human oral mucosa. *Blood* 72, 850-854 (1988).

15 U. M. Vischer, C. B. Wollheim, Epinephrine induces von Willebrand factor release from cultured endothelial cells: involvement of cyclic AMP-dependent signalling in exocytosis. *Thromb Haemost* 77, 1182-1188 (1997).

16 M. J. Beyak, P. I. Collman, S. Xue, D. T. Valdez, N. E. Diamant, Release of nitric oxide in the central nervous system mediates tonic and phasic contraction of the cat lower oesophageal sphincter. *Neurogastroenterol Motil* 15, 401-407 (2003).

17 M. J. Beyak, S. Xue, P. I. Collman, D. T. Valdez, N. E. Diamant, Central nervous system nitric oxide induces oropharyngeal swallowing and esophageal peristalsis in the cat. *Gastroenterology* 119, 377-385 (2000).

18 K. Ghimire, H. M. Altmann, A. C. Straub, J. S. Isenberg, Nitric oxide: what's new to NO? *Am J Physiol Cell Physiol* 312, C254-C262 (2017).

19 A. G. Hudetz, H. Shen, J. P. Kampine, Nitric oxide from neuronal NOS plays critical role in cerebral capillary flow response to hypoxia. *The American journal of physiology* 274, H982-989 (1998).

20 T. Kvasnicka, [NO (nitric oxide) and its significance in regulation of vascular homeostasis]. *Vnitr Lek* 49, 291-296 (2003).

21 M. A. Vincent, M. Montagnani, M. J. Quon, Molecular and physiologic actions of insulin related to production of nitric oxide in vascular endothelium. *Current diabetes reports* 3, 279-288 (2003).

22 C. C. Watkins *et al.*, Insulin restores neuronal nitric oxide synthase expression and function that is lost in diabetic gastropathy. *The Journal of clinical investigation* 106, 803 (2000).

23 A. D. Baron, Hemodynamic actions of insulin. *The American journal of physiology* 267, E187-202 (1994).

24 M. Coggins *et al.*, Physiologic hyperinsulinemia enhances human skeletal muscle perfusion by capillary recruitment. *Diabetes* 50, 2682-2690 (2001).

25 M. P. de Boer *et al.*, Insulin-induced changes in microvascular vasomotion and capillary recruitment are associated in humans. *Microcirculation*, (2014).

26 M. Kvist-Reimer, F. Sundler, B. Ahren, Effects of chemical sympathectomy by means of 6-hydroxydopamine on insulin secretion and islet morphology in alloxan-diabetic mice. *Cell Tissue Res* 307, 203-209 (2002).

27 Z. Zhao *et al.*, Advanced glycation end products inhibit glucose-stimulated insulin secretion through nitric oxide-dependent inhibition of cytochrome c oxidase and adenosine triphosphate synthesis. *Endocrinology* 150, 2569-2576 (2009).

Chapter 11

The Tissue Repair Mechanism

"The reasonable man adapts himself to the world. The unreasonable man persists in trying to adapt the world to himself. Therefore, all progress depends on the unreasonable man."
— George Bernard Shaw

Having described the basic elements of the stress mechanism, we can now combine them into a simplified diagram (see Figure 11.1 on page 98) that illustrates how the stress mechanism incorporates a *tissue repair component* and a *capillary gate component* within a single, cohesive mechanism. Both components activate independently, but because they share the same enzymes and produce the same products, the activity of each exaggerates that of the other. This enables them to generate positive feedback that focuses the powerful effects of the stress mechanism, and alters the individual locations, magnitudes, and speed of production of thrombin, soluble fibrin, and insoluble fibrin to simultaneously regulate tissue perfusion and tissue repair.

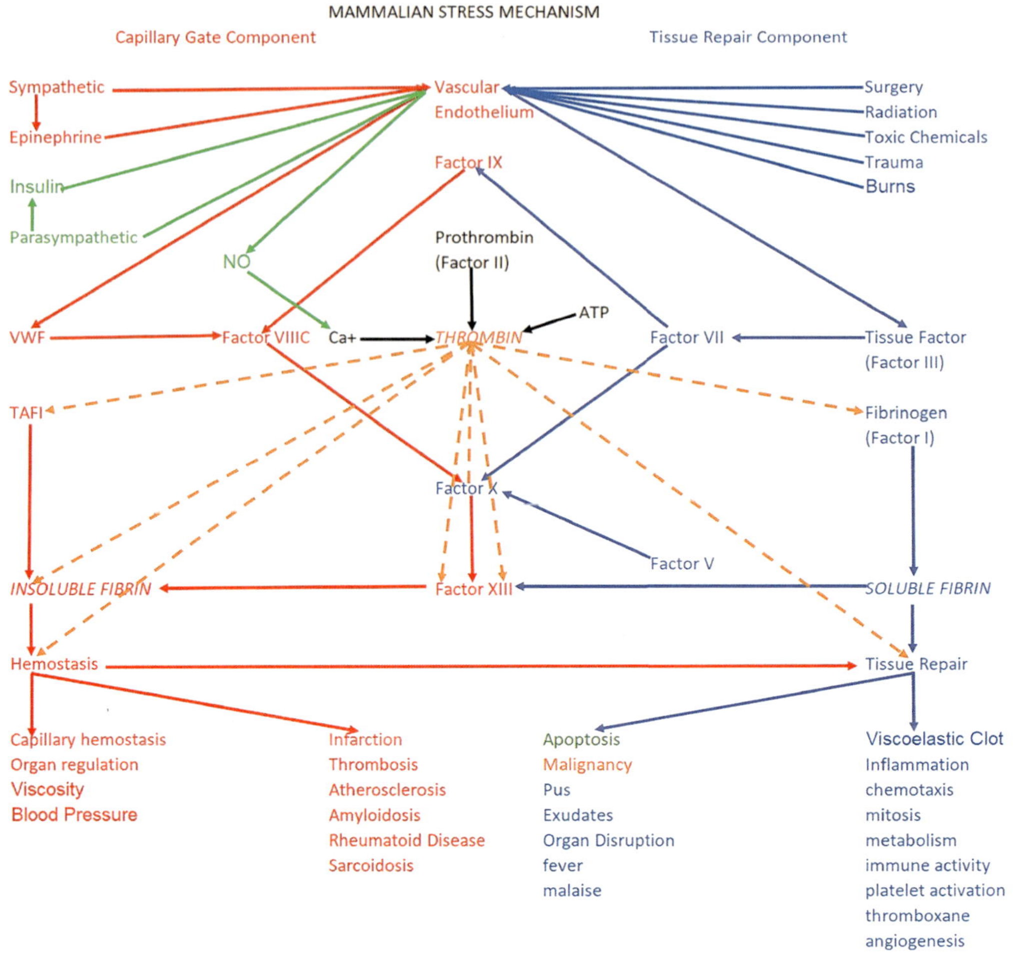

Figure 11.1: Mammalian stress mechanism (simplified).
A simplified diagram of the mammalian stress mechanism that illustrates its pathways, products, and effects. Capillary gate activation pathways portrayed in red, and its inhibitory pathways are portrayed in green. The tissue repair mechanism is portrayed in blue. Thrombin energy pathways are portrayed in orange. The diagram illustrates how hyperactivity of the mechanism causes disease.

The tissue repair component explains tissue maintenance:

- The vascular endothelium allows small amounts of factor VII to slowly penetrate from blood into extravascular tissues, where tissue factor activates it to generate small amounts of thrombin that energize slow fibroblast mitosis and collagen replenishment to enable tissue maintenance.
- The vascular endothelium allows small amounts of tissue factor to continuously escape from extravascular tissues into systemic circulation, where it binds to, stabilizes, and activates factor VII. This enables the background activity of factors VIIIC, IX and X, which otherwise remain inert in the absence of factor VII activity, and thereby enables coagulation (clot formation) and capillary hemostasis (control of capillary blood loss).

The tissue repair component explains the effects of radiation, burns, sepsis, and toxic chemicals:

- Sepsis, radiation, and toxic chemicals increase the permeability of the vascular endothelium, which exaggerates the exposure of tissue factor to factor VII and elevates thrombin generation in the affected tissues.

- Thrombin activates tiny intracellular muscle fibers *within* (not between) the cells of the vascular endothelium, thereby creating gaps between the cells.[1] This further exaggerates the permeability of the vascular endothelium, causing positive feedback so that maximum thrombin generation occurs hours later. This explains why sunburn symptoms appear hours after excessive radiation exposure.

- Increased permeability of the vascular endothelium does not cause clot formation so long as the vascular endothelium remains intact, because the gigantic molecular size of factor VIII prevents it from penetrating the intact vascular endothelium.

- The increased permeability of the vascular endothelium allows increased escape of tissue factor into systemic circulation, which induces systemic thrombin generation, causing systemic inflammation and ATP consumption that manifests as fever, fatigue, and malaise (general feeling of misery).[2, 3]

- Elevated thrombin generation causes cells in the damaged tissues to release cell hormones (chemokines, cytokines, etc.). These hormones induce inflammation that loosens cell connections to facilitate repair cell movement via chemotaxis.[4, 5]

- Soluble fibrin invades inflamed tissues via thrombin-induced gaps between the cells of the vascular endothelium, causing tissue edema (swelling).

- Thrombin causes cells to release bradykinins that increase nociceptor sensitivity, which exaggerates pain in damaged tissues.

The tissue repair component explains the effects of trauma:

Trauma (tissue damage, injury) disrupts the vascular endothelium and directly exposes blood enzymes to tissue factor in damaged tissues. This initiates coagulation (clot formation), which is the first step in a predictable sequence of tissue repair activities.

Traditional explanations of coagulation often assume that platelets orchestrate the coagulation process after they are *activated* by collagen exposure. The tissue repair mechanism clarifies the process of coagulation, which is the first step in the orderly sequence of tissue repair activities:

- Mammalian cells that are normally exposed to blood enzymes do not express tissue factor, so that they cannot activate factor VII. Mammalian cells not normally exposed to tissue factor include platelets, red cells, white cells, and the cells of the vascular endothelium.[6, 7]

- Trauma and surgery disrupt the vascular endothelium, which directly exposes tissue factor in damaged extravascular tissues to blood enzyme factors VII, VIII, IX, and X. This causes rapid generation of thrombin, soluble fibrin, and insoluble fibrin.[8, 9, 10, 11]

- Thrombin energizes all aspects of coagulation.[12]

- The stabilization of factor VII by tissue factor determines the location and magnitude of tissue repair because the other enzymes remain inert in the absence of factor VII activity.[13]

- Hemostasis is enhanced in lung, brain, retina, nerves, gonads, lung, cervix, perivascular tissues, placenta, amniotic fluid, and other tissues that are rich in tissue factor.[14] *These same tissues exhibit exaggerated risk of malignancy and metastasis.*[15, 16, 17, 18, 19]

- Platelets, which travel along the inner walls of blood vessels, are immediately thrust into damaged tissues by blood flow dynamics. Thrombin elevation in damaged tissues activates the platelets, whereupon they change shape and become “sticky.”[20, 21, 22]

- Rising levels of insoluble fibrin bind the activated platelets together to form a *white clot* that covers the surface of the damaged tissues.

- Factor VIII and factor XIII generate strands of insoluble fibrin that entangle red cells to reduce blood turbulence, increase blood viscosity, reduce blood flow, and increase blood coagulability in the vicinity of tissue damage to facilitate clot formation. See Figure 11.2. This explains blood flow stasis and hypercoagulability mandated by Virchow’s Triad.[23, 24, 25]

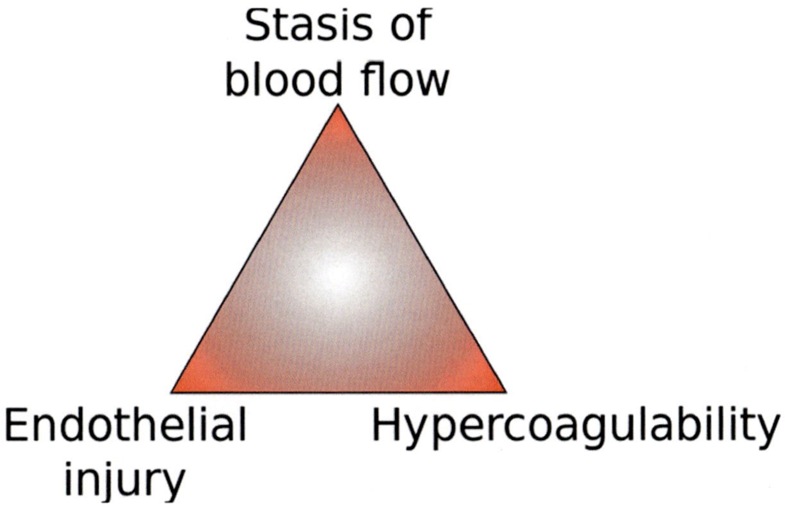

Figure 11.2: Virchow's Triad
https://en.wikipedia.org/wiki/Virchow%27s_triad (By Rudolf. Hellmuth - Own work, CC BY-SA 3.0, https://commons.wikimedia.org/w/index.php?curid=19153037)

- The rising levels of insoluble fibrin bind red blood cells into a viscoelastic *red clot* that supplants the *white clot* and restores the isolation of blood enzymes from extravascular tissues.

- Due to its massive molecular size, factor VIII cannot penetrate a clot of its own manufacture. This limits clot formation to the immediate vicinity of tissue damage.

The tissue repair mechanism clarifies the manifestations of inflammation: rubor (redness), tumor (swelling), calor (heat of inflammation), dolor (pain), and *functio laesa* (loss of function), and explains how inflammation enables tissue repair:

- The viscoelastic clot functions as a *hysteresis/synchronous mechanism* that regulates contact between factors VII, X, and tissue factor in damaged tissues to govern thrombin generation within an optimal range in the damaged tissues beneath the clot's protective surface. The contact between the factors energizes cellular tissue repair activities until the tissue repair process restores the integrity of the vascular endothelium.[26]

- The interaction of factors VII, X, and tissue factor is essential for embryological structural development, which explains why defects in any of them usually result in miscarriage. Their interaction is also essential for tissue repair. Warfarin and thalidomide disrupt the interaction of factors VII and X with tissue factor, which explains why those drugs disrupt both embryological development and tissue repair.[27]

- Factor VIII is not required for embryological development, which explains why heparin and similar drugs that disrupt factor VIII activity are safe for use as anticoagulants during pregnancy.

- Factor VIII generates factor XIII, which in turn generates insoluble fibrin that enables clot formation[28] — but factor VIII cannot penetrate a clot of its own manufacture on account of its gigantic size, so that it does not directly participate in tissue repair other than to enable viscoelastic clot formation.

- Factor IX facilitates factor VIII activity, but has no effect on factors VII and X, so it does not participate in either embryological development or tissue repair.

- The interaction of factors VII, X, and tissue factor generates thrombin. Thrombin energizes cellular tissue repair activities.

- During tissue repair, thrombin activates Thrombin-Activated Fibrinolysis Inhibitor (TAFI). TAFI stabilizes and strengthens insoluble fibrin, reduces clot permeability, inhibits the penetration of blood enzymes, and thereby limits excessive thrombin generation.[29, 30, 31] Excessive thrombin generation would otherwise threaten excessive cellular activity that might cause malignancy.[32, 33, 34]

- Thrombin elevations in damaged tissues induce cells to release inflammatory substances that dissolve the basement membrane and loosen cell connections that hold tissue cells in tight formation.

- Thrombin elevation opens gaps between the cells of the vascular endothelium that allow the penetration of fluids and soluble fibrin into damaged tissues, causing edema (*tumor*) that limits function (*functio laesa*).

- Thrombin-energized cells release bradykinins and other substances that exaggerate nociceptor function, which exaggerates pain in repairing tissues (*dolor*).

- Thrombin elevation in damaged tissues attracts fibroblasts (tissue repair cells) from adjacent undamaged tissues via chemotaxis, so that they move through the loosened tissues to enter the area of tissue damage. It is as though these cells move toward thrombin as toward a food source.[35]

- Thrombin energizes fibroblast proliferation and collagen production to form *granulation tissue* that fills empty spaces and binds damaged tissues together.

- The elevated metabolism of proliferating fibroblasts produces heat that elevates tissue temperature (*calor*).

- Cells communicate with one another and orient themselves using tiny electromagnetic forces to organize the cellular tissue repair process.[36, 37, 38]
- Thrombin energizes the cellular release of chemokines, cytokines, and other cell hormones that enable cell-to-cell communications.[39, 40, 41, 42, 43, 44, 45, 46]
- Thrombin elevations and cell-to-cell communications cause tissue repair cells to alter their thrombin receptor (PAR) configurations, which alter their responses to thrombin.[47, 48, 49, 50, 51, 52, 53, 54, 55]
- Thrombin energizes immune cell activity that removes foreign debris from damaged tissues and produces antibodies that inhibit microbial infection.
- Thrombin energizes cell differentiation, de-differentiation, and re-differentiation that replaces damaged cells and tissues.
- Thrombin energizes angiogenesis (capillary proliferation) that perfuses proliferating repair tissues, causing redness (*rubor*)
- Thrombin generation declines as tissue repair restores the vascular endothelium and isolates factor VII from tissue factor. The resulting thrombin starvation causes fibroblast apoptosis that shrinks wound tissues and pulls wound edges together to enable resolution of the repair process.
- Thrombin starvation undermines TAFI (Thrombin-Activated Fibrinolysis Inhibitor) causing the viscoelastic clot to disintegrate and disappear.

CONCLUSION

The improved understanding of individual blood enzyme properties produced by unrelated research now enables a testable mechanism that explains tissue maintenance and shows how the mechanism orchestrates the orderly sequence of tissue repair. It also provides powerful clues to the nature of the mechanism that enables embryological development, which will be elaborated in subsequent chapters.

NOTES

1 J. G. Garcia, F. M. Pavalko, C. E. Patterson, Vascular endothelial cell activation and permeability responses to thrombin. *Blood Coagul Fibrinolysis* 6, 609-626 (1995).

2 M. Levi, T. van der Poll, H. Ten Cate, Tissue factor in infection and severe inflammation. *Semin Thromb Hemost* 32, 33-39 (2006).

3 C. Savoia, E. L. Schiffrin, Inflammation in hypertension. *Curr Opin Nephrol Hypertens* 15, 152-158 (2006).

4 M. Chanson *et al.*, Gap junctional communication in tissue inflammation and repair. *Biochim Biophys Acta* 1711, 197-207 (2005).

5 L. S. Coleman, Stress repair mechanism activity explains inflammation and apoptosis. *Advances in Bioscience and Biotechnology* 3, 459-503 (2012).

6 B. A. Bouchard, K. G. Mann, S. Butenas, No evidence for tissue factor on platelets. *Blood* 116, 854-855 (2010).

7 R. A. Fleck, L. V. Rao, S. I. Rapaport, N. Varki, Localization of human tissue factor antigen by immunostaining with monospecific, polyclonal anti-human tissue factor antibody. *Thromb Res* 59, 421-437 (1990).

8 R. A. Fleck, L. V. Rao, S. I. Rapaport, N. Varki, Localization of human tissue factor antigen by immunostaining with monospecific, polyclonal anti-human tissue factor antibody. *Thromb Res* 59, 421-437 (1990).

9 S. Gando, Tissue factor in trauma and organ dysfunction. *Semin Thromb Hemost* 32, 48-53 (2006).

10 J. Himber *et al.*, Inhibition of tissue factor limits the growth of venous thrombus in the rabbit. *J Thromb Haemost* 1, 889-895 (2003).

11 H. J. Weiss, V. T. Turitto, H. R. Baumgartner, Y. Nemerson, T. Hoffmann, Evidence for the presence of tissue factor activity on subendothelium. *Blood* 73, 968-975 (1989).

12 K. G. Mann, S. Butenas, K. Brummel, The dynamics of thrombin formation. *Arterioscler Thromb Vasc Biol* 23, 17-25 (2003).

13 P. W. Friederich *et al.*, Ability of recombinant factor VIIa to generate thrombin during inhibition of tissue factor in human subjects. *Circulation* 103, 2555-2559 (2001).

14 R. A. Fleck, L. V. Rao, S. I. Rapaport, N. Varki, Localization of human tissue factor antigen by immunostaining with monospecific, polyclonal anti-human tissue factor antibody. *Thromb Res* 59, 421-437 (1990).

15 R. S. Kasthuri, M. B. Taubman, N. Mackman, Role of tissue factor in cancer. *J Clin Oncol* 27, 4834-4838 (2009).

16 A. A. Khorana *et al.*, Tissue factor expression, angiogenesis, and thrombosis in pancreatic cancer. *Clin Cancer Res* 13, 2870-2875 (2007).

17 F. R. Rickles, S. Patierno, P. M. Fernandez, Tissue factor, thrombin, and cancer. *Chest* 124, 58S-68S (2003).

18 J. L. Yao *et al.*, Tissue factor and VEGF expression in prostate carcinoma: a tissue microarray study. *Cancer Invest* 27, 430-434 (2009).

19 T. Ueno, M. Toi, M. Koike, S. Nakamura, T. Tominaga, Tissue factor expression in breast cancer tissues: its correlation with prognosis and plasma concentration. *Br J Cancer* 83, 164-170 (2000).

20 G. Hornstra, Platelet - vessel wall interaction: role of blood clotting. *Philos Trans R Soc Lond B Biol Sci* 294, 355-371 (1981).

21 C. Kanthou, G. Parry, E. Wijelath, V. V. Kakkar, C. Demoliou-Mason, Thrombin-induced proliferation and expression of platelet-derived growth factor-A chain gene in human vascular smooth muscle cells. *FEBS Lett* 314, 143-148 (1992).

22 M. Grunewald *et al.*, Absence of compensatory platelet activation in patients with severe haemophilia, but evidence for a platelet collagen-activation defect. *Platelets* 13, 451-458 (2002).

23 L. Coleman, A Hypothesis: Factor VII governs clot formation, tissue repair and apoptosis. *Medical Hypotheses* in press, (2007).

24 L. S. Coleman, Insoluble fibrin may reduce turbulence and bind blood components into clots. *Med Hypotheses* 65, 820-821 (2005).

25 L. S. Coleman, A capillary hemostasis mechanism regulated by sympathetic tone and activity via factor VIII or von Willebrand's factor may function as a "capillary gate" and may explain angiodysplasia, angioneurotic edema, and variations in systemic vascular resistance. *Med Hypotheses*, (2005).

26 L. Coleman, A Hypothesis: Factor VII governs clot formation, tissue repair and apoptosis. *Medical Hypotheses* in press, (2007).

27 A. J. Stewart, I. D. Penman, M. K. Cook, C. A. Ludlam, Warfarin-induced skin necrosis. *Postgrad Med J* 75, 233-235 (1999).

28 P. Wettstein *et al.*, Decreased factor XIII availability for thrombin and early loss of clot firmness in patients with unexplained intraoperative bleeding. *Anesth Analg* 99, 1564-1569; table of contents (2004).

29 J. Stasko, J. Hudecek, P. Kubisz, [Thrombin activatable fibrinolysis inhibitor (TAFI) and its importance in the regulation of fibrinolysis]. *Vnitr Lek* 50, 36-44 (2004).

30 T. Tran, A. G. Stewart, Protease-activated receptor (PAR)-independent growth and pro-inflammatory actions of thrombin on human cultured airway smooth muscle. *Br J Pharmacol* 138, 865-875 (2003).

31 N. E. Tsopanoglou, M. E. Maragoudakis, Role of thrombin in angiogenesis and tumor progression. *Semin Thromb Hemost* 30, 63-69 (2004).

32 F. R. Rickles, S. Patierno, P. M. Fernandez, Tissue factor, thrombin, and cancer. *Chest* 124, 58S-68S (2003).

33 M. L. Nierodzik, S. Karpatkin, Thrombin induces tumor growth, metastasis, and angiogenesis: Evidence for a thrombin-regulated dormant tumor phenotype. *Cancer cell* 10, 355-362 (2006).

34 H. Schiller *et al.*, Thrombin as a survival factor for cancer cells: thrombin activation in malignant effusions in vivo and inhibition of idarubicin-induced cell death in vitro. *Int J Clin Pharmacol Ther* 40, 329-335 (2002).

35 A. Siegbahn *et al.*, Binding of factor VIIa to tissue factor on human fibroblasts leads to activation of phospholipase C and enhanced PDGF-BB-stimulated chemotaxis. *Blood* 96, 3452-3458 (2000).

36 R. O. Becker, G. Selden, *The body electric : electromagnetism and the foundation of life*. (Quill, New York, ed. 1st Quill, 1985), pp. 364 p.

37 P. J. Rosch, Bioelectromagnetic and subtle energy medicine: the interface between mind and matter. *Annals of the New York Academy of Sciences* 1172, 297-311 (2009).
38 M. D. Bjorn E.W. Nordenstrom, *Exploring BCEC-Systems (Biologically Closed Electric Circuits).* (Nordic Medical Publications, Stockholm, Sweden, 1998).
39 A. Naldini, D. H. Carney, A. Pucci, A. Pasquali, F. Carraro, Thrombin regulates the expression of proangiogenic cytokines via proteolytic activation of protease-activated receptor-1. *Gen Pharmacol* 35, 255-259 (2000).
40 A. Naldini, L. Aarden, A. Pucci, C. Bernini, F. Carraro, Inhibition of interleukin-12 expression by alpha-thrombin in human peripheral blood mononuclear cells: a potential mechanism for modulating Th1/Th2 responses. *Br J Pharmacol* 140, 980-986 (2003).
41 A. Naldini, C. Bernini, A. Pucci, F. Carraro, Thrombin-mediated IL-10 up-regulation involves protease-activated receptor (PAR)-1 expression in human mononuclear leukocytes. *J Leukoc Biol* 78, 736-744 (2005).
42 A. Naldini *et al.*, Thrombin enhances T cell proliferative responses and cytokine production. *Cell Immunol* 147, 367-377 (1993).
43 A. Naldini *et al.*, The thrombin peptide, TP508, enhances cytokine release and activates signaling events. *Peptides* 25, 1917-1926 (2004).
44 A. Naldini *et al.*, Thrombin inhibits IFN-gamma production in human peripheral blood mononuclear cells by promoting a Th2 profile. *J Interferon Cytokine Res* 26, 793-799 (2006).
45 A. Naldini, A. Pucci, D. H. Carney, G. Fanetti, F. Carraro, Thrombin enhancement of interleukin-1 expression in mononuclear cells: involvement of proteinase-activated receptor-1. *Cytokine* 20, 191-199 (2002).
46 A. Naldini, L. Sower, V. Bocci, B. Meyers, D. H. Carney, Thrombin receptor expression and responsiveness of human monocytic cells to thrombin is linked to interferon-induced cellular differentiation. *J Cell Physiol* 177, 76-84 (1998).
47 T. Tran, A. G. Stewart, Protease-activated receptor (PAR)-independent growth and pro-inflammatory actions of thrombin on human cultured airway smooth muscle. *Br J Pharmacol* 138, 865-875 (2003).
48 K. Tantivejkul *et al.*, PAR1-mediated NFkappaB activation promotes survival of prostate cancer cells through a Bcl-xL-dependent mechanism. *J Cell Biochem* 96, 641-652 (2005).
49 E. Archiniegas, C. Y. Neves, D. Candelle, J. E. Cardier, Thrombin and its protease-activated receptor-1 (PAR1) participate in the endothelial-mesenchymal transdifferentiation process. *DNA Cell Biol* 23, 815-825 (2004).
50 L. F. Brass, M. Molino, Protease-activated G protein-coupled receptors on human platelets and endothelial cells. *Thromb Haemost* 78, 234-241 (1997).
51 D. Darmoul, V. Gratio, H. Devaud, T. Lehy, M. Laburthe, Aberrant expression and activation of the thrombin receptor protease-activated receptor-1 induces cell proliferation and motility in human colon cancer cells. *The American journal of pathology* 162, 1503-1513 (2003).
52 J. H. Jeng *et al.*, Protease-activated receptor-1-induced calcium signaling in gingival fibroblasts is mediated by sarcoplasmic reticulum calcium release and extracellular calcium influx. *Cell Signal* 16, 731-740 (2004).

53 E. Jin *et al.*, Protease-activated receptor (PAR)-1 and PAR-2 participate in the cell growth of alveolar capillary endothelium in primary lung adenocarcinomas. *Cancer* 97, 703-713 (2003).
54 W. Luo, Y. Wang, G. Reiser, Two types of protease-activated receptors (PAR-1 and PAR-2) mediate calcium signaling in rat retinal ganglion cells RGC-5. *Brain Res* 1047, 159-167 (2005).
55 S. J. Song, C. N. Pagel, T. M. Campbell, R. N. Pike, E. J. Mackie, The role of protease-activated receptor-1 in bone healing. *The American journal of pathology* 166, 857-868 (2005).

Chapter 12

The Capillary Gate Mechanism

"The tragedy of man is that he has developed an intelligence eager to uncover mysteries, but not strong enough to penetrate them. With minds but slightly evolved beyond those of our animal relations, we are tortured with precocious desires to pose questions which we are sometimes capable of asking, but rarely able to answer."
— Hans Zinsser

Marie Francois Xavier Bichat, the father of histology (see Appendix A: Medical History, on page 395), identified autonomic ganglia in the chest and abdomen, and Claude Bernard, the father of physiology, proved that organ perfusion determines organ function. They were contemporaries of Louis Pasteur, Rudolf Virchow, and Johannes Friedrich Miescher when medical advance accelerated late in the 19th century. Their efforts enabled the understanding that subconscious autonomic (automatic) nervous activity governs organ function. In 1925 Tannenberg proposed that a sub-microscopic molecular mechanism efficiently governs blood flow resistance in capillaries, where turbulence, pressures, and flows are minimal, and where surface area is vastly greater than all larger vessels combined.[1] His concept provided a powerful explanation of blood flow regulation, but his postulated capillary mechanism couldn't be confirmed, and after 1939 it was supplanted by Walter B. Cannon's intuitively appealing concept that opposing combinations of muscular cardiac and arterial contraction determine cardiac output, blood flow distribution, and organ function.[2] Stress researchers recognized the contradictions of Cannon's concept (see Appendix A: Medical History, on page 395) and sought to discover Tannenberg's *capillary gate mechanism*. They found compelling clues but were unable to identify any testable mechanism that could explain how it worked and confirm its existence. Since stress research was abandoned, fresh evidence from unrelated research has finally enabled a testable description of the capillary gate mechanism.[3, 4, 5]

AUTONOMIC BALANCE AND THE CAPILLARY GATE

The autonomic (automatic) nervous system directly innervates the vascular endothelium (the one-cell-thin lining of blood vessels and capillaries) of internal organs such as brain, lung, adrenal glands, pancreas, and digestive tract. This enables the vascular endothelium to function as a neuroendocrine gland that regulates capillary flow resistance — and therefore cardiac output, cardiac efficiency, tissue perfusion, tissue oxygenation, and organ function — in accord with autonomic balance.

CLOSING THE CAPILLARY GATE

The vascular endothelium manufactures and stores von Willebrand Factor (VWF), and extrudes it into the capillary lumen next to binding sites for fibrinogen and fibronectin in accord with sympathetic nervous activity.[6, 7, 8] VWF binds to hepatic (liver) factor VIIIC in flowing blood to form factor VIII. Factor VIII in turn generates factor XIII that converts fibrinogen and fibronectin into insoluble fibrin monomers. The insoluble fibrin monomers polymerize into strands of insoluble fibrin, and those strands of insoluble fibrin increase capillary flow resistance (viscosity) and decrease capillary flow. This "closes" the capillary gate. Capillary gate closure slows but does not halt tissue perfusion and tissue oxygenation, as evidenced by the absence of infarction in youth, when capillary beds remain intact.

OPENING THE CAPILLARY GATE

Parasympathetic nervous activity releases nitric oxide (NO) from the vascular endothelium. NO is a gaseous molecule that diffuses into the capillary lumen, where it binds avidly to Ca+, which inactivates thrombin and Thrombin-Activated Fibrinolysis Inhibitor (TAFI). See Figure 12.1 and Figure 12.2. This accelerates insoluble fibrin disintegration, which "opens" the capillary gate and increases organ function.[9, 10, 11, 12, 13, 14, 15, 16, 17, 18, 19, 20, 21, 22, 23, 24] In the research literature this is called *nitrergic neurogenic vasodilation*, which is a misnomer inspired by the conventional assumption that *vasoconstriction* regulates blood flow and tissue perfusion.[25, 26, 27, 28, 29, 30, 31, 32, 33, 34]

Brain astrocytes are specialized support cells that release NO to assure adequate capillary perfusion in brain tissue, which is exquisitely vulnerable to hypoxia. Brain astrocyte activity undermines coagulation and exaggerates brain vulnerability to bleeding, which explains why intracranial hematoma formation commonly occurs after head trauma. This, in turn, explains the high concentration of tissue factor in brain tissue that offsets the bleeding tendency caused by astrocytes. Delicate brain tissue is encased in bone to protect it from trauma. Brain astrocytes produce NO and TPA (Tissue Plasminogen Activator) that prevent ischemic (inadequate oxygenation) brain damage when large doses of epinephrine are administered during cardiopulmonary resuscitation.[35, 36, 37]

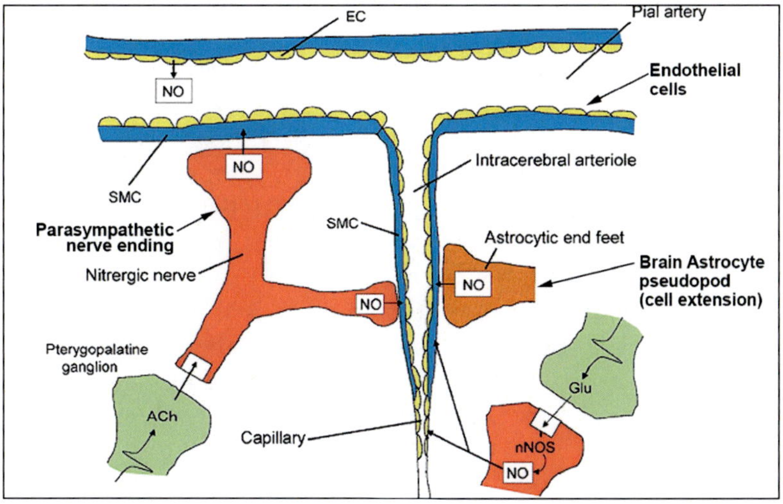

Figure 12.1: Releasing nitric oxide.
Parasympathetic nervous activity releases nitric oxide from endothelial cells.

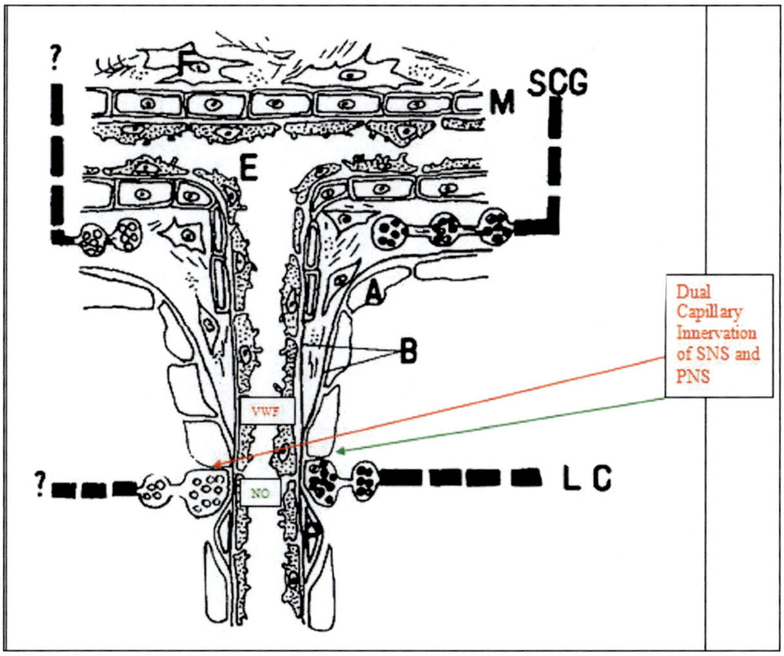

Figure 12.2: Dual innervation.
The "dual innervation" (both sympathetic and parasympathetic innervation) of the brain and internal organs enables autonomic balance to directly govern the capillary gate and regulate organ activity. Sympathetic nervous activity releases von Willebrand Factor (VWF) that generates insoluble fibrin to "close" the capillary gate. Parasympathetic nervous activity releases nitric oxide (NO) that disintegrates insoluble fibrin and "opens" the capillary gate.

CARBON DIOXIDE AND THE CAPILLARY GATE

Carbon dioxide is the primary regulator of the capillary gate, because it directly releases NO from endothelial cells.[38, 39, 40] Cellular metabolism generates CO_2 as a product of the Krebs cycle, which consumes oxygen and glucose and produces carbon dioxide and water. The Krebs cycle generates ATP that energizes cell activities, including muscle contraction and muscle relaxation.[41] CO_2 directly releases nitric oxide from the vascular endothelium to open the capillary gate, which increases tissue perfusion and tissue oxygen delivery.[42, 43] Carbon dioxide also releases oxygen from the hemoglobin molecule and stimulates breathing, so that the carbon dioxide produced by cellular metabolism promotes cellular oxygenation in multiple ways. Exercise exaggerates muscle metabolism that consumes glucose and oxygen to produce ATP (adenosine tri-phosphate) and carbon dioxide (see Appendix E: Cell Biology, on page 427). The ATP energizes muscle activity, while carbon dioxide enhances cellular oxygenation that sustains ATP generation.

EPINEPHRINE, INSULIN, AND THE CAPILLARY GATE

Epinephrine and insulin are opposing hormones that regulate the capillary gate in peripheral tissues such as skeletal muscle that are lacking in direct autonomic innervation:

- Sympathetic nervous activity releases epinephrine from the adrenal gland, which releases von Willebrand Factor (VWF) from the vascular endothelium to close the capillary gate.[44, 45, 46]

- Parasympathetic nervous activity releases insulin from the pancreas, which releases nitric oxide (NO) from the vascular endothelium to open the capillary gate, which enhances cardiac output, tissue perfusion, and glucose uptake.[47, 48, 49, 50, 51, 52]

CAPILLARY HEMOSTASIS AND THE CAPILLARY GATE

By itself, sympathetic nervous activity that "closes" the capillary gate does not explain capillary hemostasis, wherein capillary blood loss ceases within 15 minutes after capillary trauma. Sympathetic nervous activity produces insoluble fibrin that increases capillary flow resistance — but not to the point of causing tissue hypoxia that threatens infarction. This explains why infarction is rare in youth, when capillary beds remain intact, but occurs with increasing incidence as capillary beds deteriorate in accord with senescence. However, capillary trauma directly exposes blood enzymes to tissue factor in extravascular tissues, and this accelerates insoluble fibrin generation to enable capillary hemostasis.

CONCLUSION

The capillary gate mechanism is an efficient, submicroscopic, sub-molecular mechanism that regulates blood flow in capillaries, where pressures, flows, and turbulence are minimal and surface area is vastly greater than the sum total of all larger vessels combined. It provides a far more effective explanation of blood flow regulation than Cannon's inefficient concept of opposing muscular forces. The terms *vasoconstriction*, *vasodilation*, and *cardiac reserve* are all misnomers, and should be abolished from medical language. The nature of hemodynamic physiology is further elaborated in Chapter 13: The Turbulence Mechanism, on page 119, which explains the turbulent nature of blood pressure and the palpable pulse, and the non-Newtonian nature of blood flow.

NOTES

1 J. Tannenberg, Uber die Kapillartatigkeit. *Verh dtsch. path.* Ges., 20:374, (1925).

2 W. B. Cannon, *The wisdom of the body.* (W.W. Norton & Company, New York,, 1939), pp. xviii, 19-333 p.

3 L. Coleman, A Testable Stress Mechanism Explains Both Unified Stress Theory and Capillary Gate Theory. *FDA Anesthesia & Life Support Drugs Advisory Committee pp239-243 3/29/07,* (2007).

4 L. S. Coleman, A capillary hemostasis mechanism regulated by sympathetic tone and activity via factor VIII or von Willebrand's factor may function as a "capillary gate" and may explain angiodysplasia, angioneurotic edema, and variations in systemic vascular resistance. *Med Hypotheses* 66, 773-775 (2006).

5 L. S. Coleman, A Stress Repair Mechanism that Maintains Vertebrate Structure during Stress. *Cardiovasc Hematol Disord Drug Targets*, (2010).

6 J. A. Hovinga *et al.*, ADAMTS-13, von Willebrand factor and related parameters in severe sepsis and septic shock. *J Thromb Haemost*, (2007).

7 A. M. Muller, C. Skrzynski, G. Skipka, K. M. Muller, Expression of von Willebrand factor by human pulmonary endothelial cells in vivo. *Respiration* 69, 526-533 (2002).

8 J. E. Sadler, von Willebrand factor. *J Biol Chem* 266, 22777-22780 (1991).

9 M. J. Beyak, P. I. Collman, S. Xue, D. T. Valdez, N. E. Diamant, Release of nitric oxide in the central nervous system mediates tonic and phasic contraction of the cat lower oesophageal sphincter. *Neurogastroenterol Motil* 15, 401-407 (2003).

10 M. Amano, M. Takahashi, T. Kosaka, M. Kinoshita, Differential inhibition of platelet aggregation and calcium mobilization by nitroglycerin and stabilized nitric oxide. *J Cardiovasc Pharmacol* 24, 860-866 (1994).

11 M. J. Beyak, S. Xue, P. I. Collman, D. T. Valdez, N. E. Diamant, Central nervous system nitric oxide induces oropharyngeal swallowing and esophageal peristalsis in the cat. *Gastroenterology* 119, 377-385 (2000).

12 G. Dijkstra, H. van Goor, P. L. Jansen, H. Moshage, Targeting nitric oxide in the gastrointestinal tract. *Curr Opin Investig Drugs* 5, 529-536 (2004).

13 K. Ghimire, H. M. Altmann, A. C. Straub, J. S. Isenberg, Nitric oxide: what's new to NO? *Am J Physiol Cell Physiol* 312, C254-C262 (2017).

14 A. G. Hudetz, H. Shen, J. P. Kampine, Nitric oxide from neuronal NOS plays critical role in cerebral capillary flow response to hypoxia. *The American journal of physiology* 274, H982-989 (1998).

15 T. Kvasnicka, [NO (nitric oxide) and its significance in regulation of vascular homeostasis]. *Vnitr Lek* 49, 291-296 (2003).

16 M. J. Martin, M. D. Jimenez, V. Motilva, New issues about nitric oxide and its effects on the gastrointestinal tract. *Curr Pharm Des* 7, 881-908 (2001).

17 Y. Morita-Tsuzuki, J. E. Hardebo, E. Bouskela, Inhibition of nitric oxide synthase attenuates the cerebral blood flow response to stimulation of postganglionic parasympathetic nerves in the rat. *J Cereb Blood Flow Metab* 13, 993-997 (1993).

18 J. Saraiva, S. S. Marotta-Oliveira, S. A. Cicillini, O. Eloy Jde, J. M. Marchetti, Nanocarriers for nitric oxide delivery. *J Drug Deliv* 2011, 936438 (2011).

19 V. B. Schini-Kerth, Vascular biosynthesis of nitric oxide: effect on hemostasis and fibrinolysis. *Transfus Clin Biol* 6, 355-363 (1999).

20 J. J. Smith *et al.*, The role of nitric oxide in the cerebrovascular response to hypercapnia. *Anesth Analg* 84, 363-369 (1997).

21 N. Toda, K. Ayajiki, T. Okamura, Nitric oxide and penile erectile function. *Pharmacol Ther* 106, 233-266 (2005).

22 N. Toda, K. Ayajiki, T. Okamura, Interaction of endothelial nitric oxide and angiotensin in the circulation. *Pharmacol Rev* 59, 54-87 (2007).

23 D. J. Webb *et al.*, Sildenafil citrate potentiates the hypotensive effects of nitric oxide donor drugs in male patients with stable angina. *J Am Coll Cardiol* 36, 25-31 (2000).

24 O. Yalcin, P. Ulker, U. Yavuzer, H. J. Meiselman, O. K. Baskurt, Nitric oxide generation by endothelial cells exposed to shear stress in glass tubes perfused with red blood cell suspensions: role of aggregation. *Am J Physiol Heart Circ Physiol* 294, H2098-2105 (2008).

25 Y. Morita-Tsuzuki, J. E. Hardebo, E. Bouskela, Inhibition of nitric oxide synthase attenuates the cerebral blood flow response to stimulation of postganglionic parasympathetic nerves in the rat. *J Cereb Blood Flow Metab* 13, 993-997 (1993).

26 L. S. Coleman, The Mammalian Stress Mechanism (MSM) Explains Radial Artery Spasm (RAS). *J Anes Surg* 5, 81-94 (2018).

27 M. L. Si, T. J. Lee, Alpha7-nicotinic acetylcholine receptors on cerebral perivascular sympathetic nerves mediate choline-induced nitrergic neurogenic vasodilation. *Circ Res* 91, 62-69 (2002).

28 L. C. Anderson *et al.*, The influence of gender on parasympathetic vasodilatation in the submandibular gland of the rat. *Exp Physiol* 91, 435-444 (2006).

29 K. Ayajiki, H. Hayashida, M. Tawa, T. Okamura, N. Toda, Characterization of nitrergic function in monkey penile erection in vivo and in vitro. *Hypertens Res* 32, 685-689 (2009).

30 K. Ayajiki, N. Toda, T. Okamura, [Nitroxidergic (nitrergic) nerve and erectile dysfunction]. *Nihon Yakurigaku Zasshi* 119, 21-28 (2002).

31 T. Okamura *et al.*, Effects of endothelial impairment by saponin on the responses to vasodilators and nitrergic nerve stimulation in isolated canine corpus cavernosum. *Br J Pharmacol* 127, 802-808 (1999).

32 N. Toda, [Nitrergic cerebrovascular regulation as affected by donepezil]. *Nihon Yakurigaku Zasshi* 141, 150-154 (2013).

33 N. Toda, A. G. Herman, Gastrointestinal function regulation by nitrergic efferent nerves. *Pharmacol Rev* 57, 315-338 (2005).

34 N. Toda, T. Okamura, Recent advances in research on nitrergic nerve-mediated vasodilatation. *Pflugers Arch* 467, 1165-1178 (2015).

35 J. A. dKim, N. D. Tran, S. J. Wang, M. J. Fisher, Astrocyte regulation of human brain capillary endothelial fibrinolysis. *Thromb Res* 112, 159-165 (2003).

36 H. Wang, G. Reiser, The role of the Ca2+-sensitive tyrosine kinase Pyk2 and Src in thrombin signalling in rat astrocytes. *J Neurochem* 84, 1349-1357 (2003).
37 H. Xin *et al.*, Increasing tPA activity in astrocytes induced by multipotent mesenchymal stromal cells facilitate neurite outgrowth after stroke in the mouse. *PloS one* 5, e9027 (2010).
38 G. Curley, J. G. Laffey, B. P. Kavanagh, Bench-to-bedside review: carbon dioxide. *Crit Care* 14, 220 (2010).
39 O. J. Kirkeby, S. Kutzsche, C. Risoe, I. R. Rise, Cerebral nitric oxide concentration and microcirculation during hypercapnia, hypoxia, and high intracranial pressure in pigs. *J Clin Neurosci* 7, 531-538 (2000).
40 N. Toda, K. Ayajiki, T. Okamura, Cerebral blood flow regulation by nitric oxide: recent advances. *Pharmacol Rev* 61, 62-97 (2009).
41 L. S. Coleman, The Mammalian Stress Mechanism (MSM) Explains Radial Artery Spasm (RAS). *J Anes Surg* 5, 81-94 (2018).
42 J. J. Smith *et al.*, The role of nitric oxide in the cerebrovascular response to hypercapnia. *Anesth Analg* 84, 363-369 (1997).
43 G. B. Stefano *et al.*, Presence of the mu3 opiate receptor in endothelial cells. Coupling to nitric oxide production and vasodilation. *J Biol Chem* 270, 30290-30293 (1995).
44 L. S. Coleman, A capillary hemostasis mechanism regulated by sympathetic tone and activity via factor VIII or von Willebrand's factor may function as a "capillary gate" and may explain angiodysplasia, angioneurotic edema, and variations in systemic vascular resistance. *Med Hypotheses*, (2005).
45 F. R. Rickles, L. W. Hoyer, M. E. Rick, D. J. Ahr, The effects of epinephrine infusion in patients with von Willebrand's disease. *The Journal of clinical investigation* 57, 1618-1625 (1976).
46 U. M. Vischer, C. B. Wollheim, Epinephrine induces von Willebrand factor release from cultured endothelial cells: involvement of cyclic AMP-dependent signalling in exocytosis. *Thromb Haemost* 77, 1182-1188 (1997).
47 M. J. Beyak, P. I. Collman, S. Xue, D. T. Valdez, N. E. Diamant, Release of nitric oxide in the central nervous system mediates tonic and phasic contraction of the cat lower oesophageal sphincter. *Neurogastroenterol Motil* 15, 401-407 (2003).
48 M. Amano, M. Takahashi, T. Kosaka, M. Kinoshita, Differential inhibition of platelet aggregation and calcium mobilization by nitroglycerin and stabilized nitric oxide. *J Cardiovasc Pharmacol* 24, 860-866 (1994).
49 M. H. Ahmed, in *Med Hypotheses*. (2006), vol. 66, pp. 208-209.
50 M. A. Vincent, M. Montagnani, M. J. Quon, Molecular and physiologic actions of insulin related to production of nitric oxide in vascular endothelium. *Current diabetes reports* 3, 279-288 (2003).
51 C. C. Watkins *et al.*, Insulin restores neuronal nitric oxide synthase expression and function that is lost in diabetic gastropathy. *The Journal of clinical investigation* 106, 803 (2000).
52 Z. Zhao *et al.*, Advanced glycation end products inhibit glucose-stimulated insulin secretion through nitric oxide-dependent inhibition of cytochrome c

oxidase and adenosine triphosphate synthesis. *Endocrinology* 150, 2569-2576 (2009).

Chapter 13

The Turbulence Mechanism

"Big whirls have little whirls,
That feed on their velocity;
And little whirls have lesser whirls,
And so on to viscosity."
— Lewis Fry Richardson 1881-1953

"When I meet God, I am going to ask him two questions: Why relativity? And why turbulence? I really believe he will have an answer for the first."
—Werner Heisenberg 1901-1976

"Turbulence is the most important unsolved problem of classical physics."
—Richard P. Feynman 1918-1988

Walter B. Cannon, the WWI Harvard professor whose ideas have dominated medical physiology for the past century,[1, 2, 3, 4, 5, 6, 7, 8, 9] can be excused for his crippled concept of hemodynamic physiology because turbulence remained the greatest mystery of physics, and its medical implications were unimagined in his time. However, the recently emerged science of Chaos Theory reveals that turbulence has underlying order that can be described by a class of mathematical equations that were previously regarded as useless.[10] Based on a combination of Chaos Theory, stress theory, the non-Newtonian nature of blood, plus the recent studies of Hof *et al* that confirm the nature of Newtonian pipe flow turbulence,[11] this chapter will explain the mammalian *turbulence mechanism*. The mechanism generates bursts of diastolic pulsatile turbulence that propagate throughout the arterial tree after each heartbeat to maintain arterial patency by inhibiting thrombosis and atherosclerosis.[12, 13, 14] Capillary gate operation regulates flow resistance, which alters turbulence mechanism effects, so that the two mechanisms function in concert. Their combined effects explain the confusing characteristics of hemodynamic physiology, including cardiac output, cardiac efficiency, tissue perfusion, tissue oxygenation, organ function, blood pressure, the palpable pulse, blood viscosity, blood coagulability, atherosclerosis, and their relationships.

THE NATURE OF TURBULENCE

Turbulence is any motion characterized by chaotic (erratic) changes in pressure and velocity. In the context of fluid flow in pipes and arteries, turbulence contrasts with *laminar flow* that occurs when a fluid (a gas or a liquid) flows in parallel layers, with no disruption between those layers (see Figure 13.4 on page 123).

Laminar flow is rare in nature, but turbulence is ubiquitous, and can be observed in rising smoke, flowing rivers, irregular coastlines, weather systems, storm clouds, cardiac rhythms, lung structure, plant anatomy — and pipe flow.[15] Turbulence generates powerful forces, as in the Bernoulli Principle that propels sailing ships and levitates airplanes. It was regarded as the greatest mystery of physics until the 1980s, when three scientists working in unrelated fields simultaneously realized that turbulence has underlying order. Their discovery created the fledgling science of *Chaos Theory* that may one day be appreciated as the greatest triumph of 20th century science.[16] Turbulence provides a simple explanation of numerous medical phenomena, including blood pressure, the palpable pulse, bruits, atherosclerosis, coagulation, and shear stress.[17, 18, 19, 20, 21]

Turbulent interactions are exponential, so that small changes can produce drastic results. This is known as the *Butterfly Effect*: a butterfly flapping its wings in China can cause a tornado in North America. This explains why gun design is a triumph of trial and error: small changes in cartridge design, bullet density, barrel length, barrel diameter, and gunpowder chemistry, quality, and quantity can exponentially alter the explosion that propels the bullet, and thereby optimize performance, render the gun useless, or cause it to explode. Similar considerations apply to rocket motor design and piston engine design. This chapter will endeavor to provide practical examples that illustrate how turbulence clarifies medical phenomena.

NEWTONIAN PIPE FLOW

Familiar Newtonian fluids such as water, oil, steam, and atmospheric gases exhibit exponentially increasing viscosity (i.e. flow resistance) when they accelerate in pipes. They also generate exponentially increasing turbulent forces that press and pull on the inner walls of pipes. Numerous factors affect these forces, including pipe materials, length, diameter, bifurcations, curvatures, and surface irregularities. Such forces cause banging noises and vibrations, and they can unpredictably damage pipes at any point along their path. Plumbers call this *water hammer*. See Figure 13.1.

Bathtubs and swimming pools fill slowly because pipe flow resistance rises exponentially with flow velocity. Suddenly opening a water faucet in an old house causes exponential flow acceleration. The resulting turbulence produces pipe "chatter" that is accompanied by decreased water flow filled with rust particles dislodged from the inner walls of the pipe. Such turbulence can usefully inhibit the accumulation of particulate debris on the inner walls of pipes. For example, sludge accumulation accelerated corrosion and necessitated expensive repairs when oil flow rates were reduced in the Alaska oil pipeline. This ability of pipe turbulence to efficiently mobilize particles and cleanse the inner walls of pipes is pertinent to pulsatile turbulence in mammalian arteries, which maintains arterial patency by inhibiting atherosclerosis and thrombosis.[22, 23]

Figure 13.1: Newtonian pipe flow and water hammer.
Newtonian pipe flow generates turbulence that produces lateral forces that can unpredictably appear anywhere in a pipe system. Such forces can cause banging sounds or destroy the pipe. Numerous factors affect such forces, including temperature and pipe length, diameter, curvature, and junctions. In this case turbulence explodes a pipe junction. https://www.youtube.com/watch?v=JragfgkoVSI

NON-NEWTONIAN BLOOD FLOW

There is a second category of less familiar *non-Newtonian* fluids that exhibit exponential *reductions* in turbulent flow resistance when they are accelerated in pipes. These typically consist of particulates suspended in what would otherwise be Newtonian fluids. Ketchup, corn starch, and mammalian blood are common examples, but there are numerous others. In the case of blood, the primary particulates are red blood cells that are suspended in the water medium of blood. See Figure 13.2. Their unique combination of neutral buoyancy, small size, and bi-concave shape causes them to spontaneously form stacks called *aggregates* (aka *Rouleaux*) that block the formation of turbulent flow patterns when blood accelerates in arteries during systole (the period of contraction of the heart, when blood flows into the aorta and pulmonary artery).

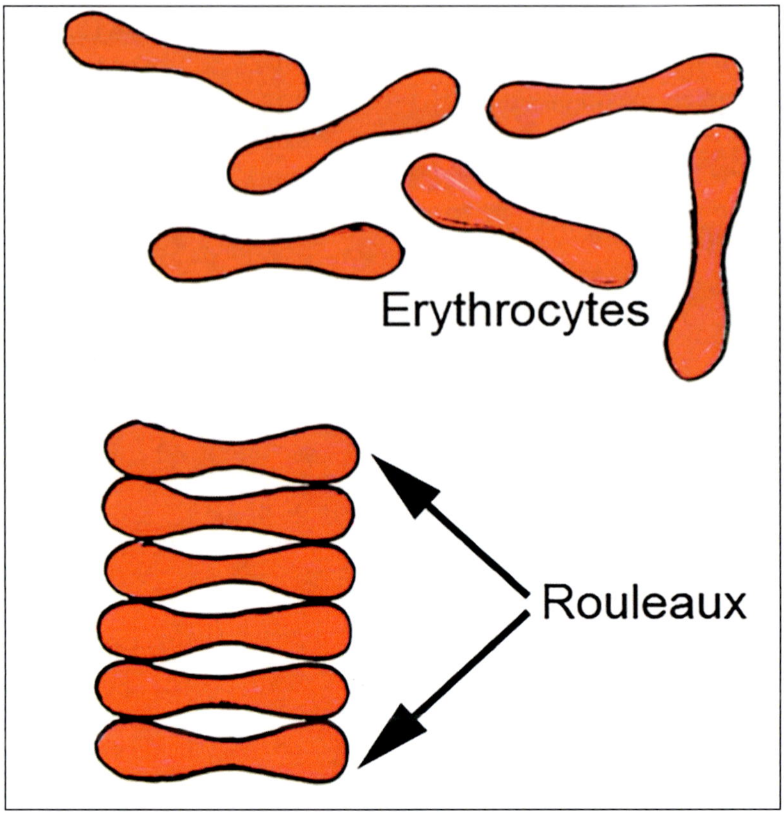

Figure 13.2: Rouleaux.
Mammalian red cells spontaneously form *Rouleaux*, aka *aggregate formations* that inhibit turbulent flow patterns during systolic blood acceleration, and reduce turbulent flow resistance during systolic ejection. As a result, blood exhibits laminar (i.e. turbulence-free) flow when it is accelerated in arteries. https://www.medicine.mcgill.ca/physio/vlab/bloodlab/ESR.htm

THE HAGEN-POISEUILLE EQUATION

Pipe flow dynamics are of great interest to plumbers and engineers who design pipe systems that carry hot and toxic fluids, and this has inspired considerable effort to devise some means to guide the design of practical pipe systems. In the mid-1800s Jean Leonard Marie Poiseuille (a French physician) and Gotthlif Heinrich Ludwig Hagen (a German engineer) simultaneously and independently postulated the classical Hagen-Poiseuille equation in an effort to predict pipe flow characteristics (Figure 13.3):

$$\Delta P = \frac{8\mu LQ}{\pi R^4}$$

Where:
ΔP = pressure differential between two ends of the pipe
L = Length of pipe
μ = Dynamic viscosity
Q = Volumetric flow rate
π = Ratio of a circle's circumference to its radius
R = Pipe radius

Figure 13.3: The Hagen-Poiseuille Equation.

In all walks of science and philosophy, false assumptions destroy deductive reasoning. Both Hagen and Poiseuille incorrectly assumed that fluid moving within a tube forms *lamina* (layers) that move at different speeds. They imagined that the layers closest to the inner walls of the tube are slowed by *shear stress* (i.e. fluid flow friction), while those near the center move faster, thus forming a parabolic laminar flow profile (Figure 13.4 on page 123). Unfortunately, the assumption is invalid, and the equation is useless, because Newtonian fluid flow is inherently turbulent rather than laminar. As a result, the equation agrees with experimental evidence only when flow rates are so low that turbulence is negligible.

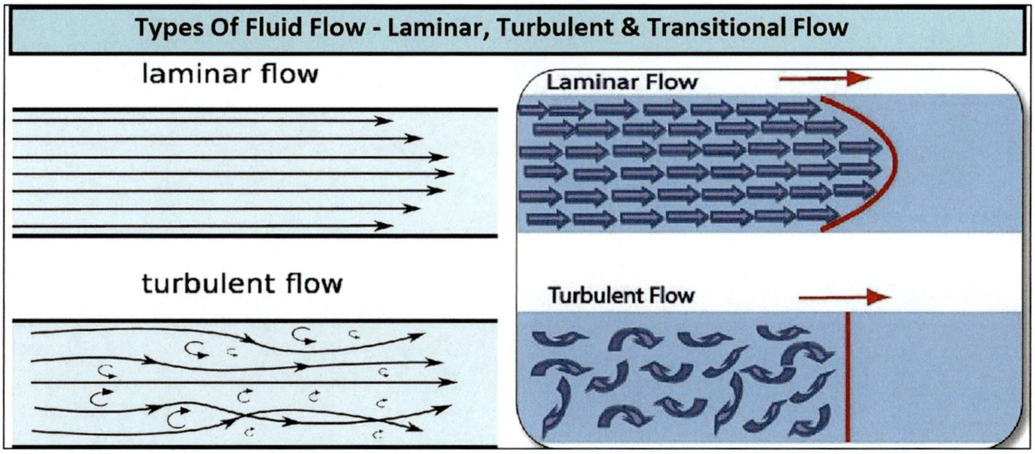

Figure 13.4: Laminar flow.
Hagen and Poiseuille independently and simultaneously derived the same mathematical equation in an effort to predict pipe flow. Both assumed that pipe flow is laminar (free of turbulence), but particulate-free fluids, including water, oil, and atmospheric gases, exhibit turbulent (aka *Newtonian*) flow in pipes. Because of this flawed assumption, their equation is useless. Blood differs from familiar Newtonian liquids and gases such as water, oil, steam, and atmospheric gases. It is a *non-Newtonian* fluid that exhibits laminar (turbulence-free) flow when it is accelerated in pipes and arteries. https://engineeringinsider.org/types-fluid-flow/

THE REVOLUTIONARY HOF EXPERIMENT

The dynamics of Newtonian pipe flow defied description until 2004, when Belgian mechanical engineers used laser photography to confirm mathematical models of water pipe turbulence.[24] They proved that turbulence is the sole source of pipe flow resistance, and refuted earlier concepts including the Hagen-Poiseuille equation, shear stress, Navier-Stokes equations, and Reynolds' numbers.

Pipe flow turbulence consists of small *jet streams* of forward-moving fluid that appear along the inner walls of pipes (see Figure 13.5). The jet streams force slower-moving fluid to the center of the pipe, where it moves *backwards*. This explains how turbulence induces flow resistance. The forward-moving jet streams exert lateral forces (Figure 13.6 on page 125) along the inner walls of the pipe. The Butterfly Effect unpredictably exaggerates these forces, which explains water hammer. Fluctuating combinations of turbulent flow resistance and turbulent lateral forces explain why suddenly opening the faucet in an old house causes pipe vibration that simultaneously slows water flow and dislodges rust from the inner walls of the pipe. Conventional concepts cannot explain this.

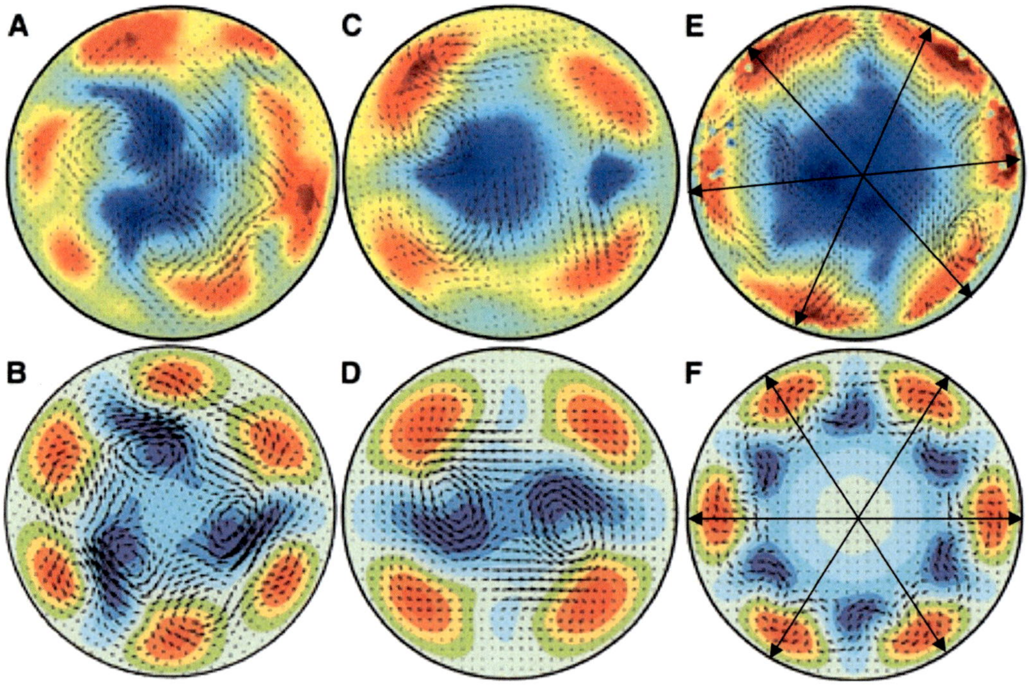

Figure 13.5: Newtonian water flow in a pipe.

Turbulent Newtonian water flow in a pipe. A, C and E are laser photographs that demonstrate changing patterns of water pipe flow turbulence induced by increasing acceleration. "A" is fast; "C" is faster; and "E" is fastest. They confirm computer generated mathematical simulations shown in B, D and F.[25] Accelerating water forms small jet streams (shown in red) that flow forward along the inner walls of the pipe. These induce lateral forces that are analogous to Bernoulli forces (illustrated by black arrows). Slower-moving water (shown in blue) is forced to the center of the pipe, where it flows backwards, so that increasing acceleration causes exponential increases in flow resistance and lateral forces. Such turbulence explains why suddenly opening the water faucet in an old house dislodges rust from the inner walls of the pipe. Turbulence explains why the Alaska oil pipeline was damaged by "sludge" accumulation when its flow rates were reduced, and why pulsatile turbulence prevents atherosclerosis in arteries. The turbulence generates lateral forces on the inner walls of the pipe, depicted by the black arrows, that can collapse or explode a pipe. Plumbers call this *water hammer*. Likewise, lateral forces generated by diastolic pulsatile turbulence explain the mysterious phenomenon of blood pressure, which appears during diastole, when the heart is in a state of relaxation.

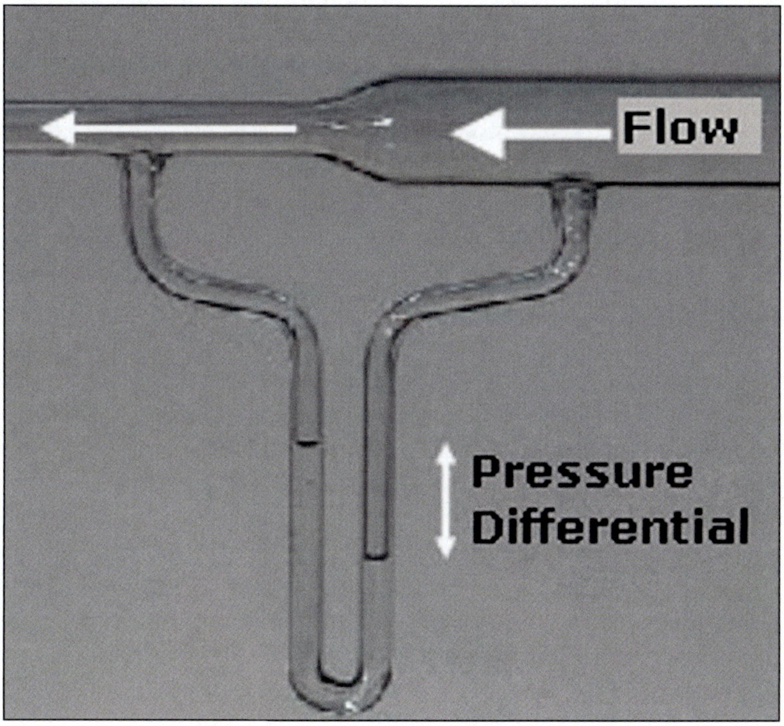

Figure 13.6: Lateral forces in pipe flow.
This picture illustrates the lateral forces induced by pipe flow turbulence, which vary in accord with flow velocity and direction. Fluid flow in pipes exerts turbulent lateral forces that press and pull on the inner walls of pipes. This phenomenon explains water hammer in Newtonian water pipe flow but is absent in non-Newtonian blood flow. By user: Computer Geezer and Geof - Modified from: de: Bild: Venturirohr.jpg, CC BY-SA 3.0, https://commons.wikimedia.org/w/index.php?curid=8957957

NON-NEWTONIAN MAMMALIAN BLOOD

Unlike familiar Newtonian fluids such as water, oil, and atmospheric gases, mammalian blood is a non-Newtonian fluid that exhibits exponential *reductions* of flow resistance — as opposed to increases — when it accelerates in arteries or pipes. This is because the small size, biconcave shape, and neutral buoyancy of mammalian red cells causes them to spontaneously form aggregates (i.e. stacks, or Rouleaux) during systolic blood acceleration. These aggregates inhibit the formation of turbulent flow patterns that exponentially exaggerate flow resistance. Thus, the faster red cells accelerate, the better they form Rouleaux stacks which enable friction-free laminar blood flow that is free of turbulent flow resistance. This non-Newtonian property of mammalian blood perfectly matches the capabilities of the mammalian heart, which is commonly regarded as the "Charles Atlas" of the body, but more closely resembles a "90-pound weakling." The heart is replete with energy-generating mitochondria and perfused by a dense network of capillaries, so that it can contract very quickly and reliably, but is so weak that it can barely propel blood to the brain, which explains why most mammalian heads are located only a short distance above the heart. The giraffe requires a disproportionately large heart to propel blood to its elevated head. Thus, the non-Newtonian nature of mammalian blood is the secret of superior mammalian exercise capacity. It enables mammals to rapidly transport oxygen to the cells of distant muscles and tissues, which enables them to migrate great distances for food, evade enemies, and capture prey. The human heart can reliably pump 5 liters of blood per minute for a lifetime using the energy of a 40-watt light bulb, elevate its output ten-fold during intense exercise, and sustain the enlarged human brain, which requires exaggerated perfusion, nutrition, and uninterrupted oxygen delivery.

The benefits of mammalian blood flow efficiency do not come cheap. Mammalian red cells must abandon their nucleus, mitochondria, and other non-essential organelles during their development to enable their small size and biconcave shape. Without mitochondria, red cells must rely on inefficient anaerobic metabolism, which limits their longevity. Without a nucleus, red cells are unable to de-differentiate and re-differentiate to facilitate tissue repair.[26] This largely explains why wound healing, and especially bone healing, is inferior in mammals compared to "lower" vertebrates such as amphibians and reptiles, whose red cells retain both nuclei and mitochondria. Worse yet, mammals must maintain their body temperature at 100 degrees Fahrenheit in order to liquefy lipoproteins that would otherwise exaggerate blood viscosity and undermine exercise tolerance.[27] This, in turn, exaggerates mammalian caloric requirements, and necessitates a large, burdensome brain to find and outsmart their food sources. The elevated sentience of the mammalian brain exaggerates susceptibility to malignancy, which is discussed in Chapter 17: Unified Theory of Biology, on page 239.

A FRESH INTERPRETATION OF MAMMALIAN HEMODYNAMIC PHYSIOLOGY

Cardiac contraction is called *systole*, and cardiac relaxation is called *diastole*. In the absence of turbulent flow resistance, systolic cardiac contraction effortlessly accelerates aortic blood velocity from zero to 120 cm/sec as it expels its contents into the arterial tree in less than a tenth of a second (see Figure 13.7 on page 130). The arterial tree expands to accommodate the increased volume, for blood, like water, is incompressible. The expanded arterial tree slowly contracts to restore its resting volume after each traveling turbulent pulse wave. Thus, the arterial tree functions as a *secondary heart* that gently and efficiently completes the propulsion of laminar blood flow to distal (distant) capillary beds, so that capillary flow slows but continues between heartbeats. Arterial smooth muscle is specialized for this role. It contains elastic protein and lacks the energy-generating mitochondria and perfusing capillary network of cardiac and skeletal muscle, so that it contracts slowly.[28]

After the heart has discharged its contents, its flexible cartilage skeleton re-expands in a spring-like fashion to facilitate re-filling of the ventricular chambers with blood. This causes a brief reversal of aortic blood flow that closes the aortic valve. This momentary flow reversal is exaggerated by the decreased (it is sharply constricted) diameter of the distal aorta, which disrupts the red cell aggregates and converts the kinetic energy of speeding aortic blood into a burst of pulsatile turbulence that propagates rapidly throughout the distal arterial tree, momentarily halting the forward blood flow as it travels. This cyclical laminar acceleration and turbulent halting of pulsatile blood can be observed when a surgeon accidentally slices an artery in half (which is fortunately rare), but the ejected blood travels only a short distance from the severed artery before being suddenly halted by pulsatile turbulence. The heart cannot match the propulsive power of a cardiac bypass pump, which causes greater ejection volume and ejection force from a severed artery, so that dramatic blood loss occurs when arteries are severed during cardiac bypass surgery.

The traveling waves of pulsatile turbulence maintain arterial patency by mobilizing particulate deposits from the inner walls of the arteries to prevent atherosclerosis, by preventing thrombus (arterial blood clot) formation, and by tunneling through thrombi when they occur. Turbulent intensity is exponentially exaggerated by declining arterial diameter with distance from the heart, which explains why atherosclerosis is absent in fingers and toes, and appears instead at the curvatures and bifurcations of large proximal arteries where turbulent intensity is reduced.[29, 30, 31, 32]

THE PATHOPHYSIOLOGY OF HYPERTENSION, DIABETES, AND CONGESTIVE HEART FAILURE

So-called cardiac conditioning exercise such as running does not significantly increase the size, weight, or work capacity of the heart itself. Instead, it induces *angiogenesis* (capillary proliferation) in oxygen-starved muscle tissues. This reduces flow resistance, enhances cardiac efficiency, and increases exercise tolerance.[33, 34, 35, 36, 37, 38, 39]

In contrast, chronic illness and environmental stress induce microvascular inflammation that accelerates the natural deterioration of capillary beds that occurs with age. This undermines cellular glucose uptake, causing Type II diabetes, and increases flow resistance that alters the turbulent pulse wave, accelerates atherosclerosis, and elevates blood pressure. The sustained microvascular flow resistance exaggerates cardiac work, which causes cardiac cellular hypoxia that induces inflammatory tissue repair activity that deposits collagen between cardiac muscle cells that thickens heart walls, undermines heart function, and causes congestive heart failure.[40, 41, 42, 43, 44, 45, 46, 47, 48, 49, 50, 51, 52, 53, 54, 55]

BLOOD PRESSURE AND THE TURBULENCE MECHANISM

"Each time a change can be observed in the height of a manometer applied to an artery, the experimenter must ask himself which of the two factors of the arterial pressure has varied, or whether the two factors, power and resistance, have been modified at the same time. In the absence of a criterion which would allow one to decide with complete certainty this litigious question, often enough physiologists have chosen the hypothesis which best accorded with their preconceived ideas."
— Etienne Jules Marey 1881

"The development of our knowledge of the circulation has been bedeviled by the fact that the measurement of blood flow is so complicated, whereas that of blood pressure is so easy: hence the blood pressure manometer has exerted an almost hypnotic influence, though bodily organs don't need pressure but flow."
— Adolf Jarisch 1928

Blood pressure is not the driving force of blood flow, as almost universally assumed, because conventional instruments cannot measure the tiny force caused by the contraction of vascular muscle that propels turbulence-free arterial laminar flow. Instead, conventional blood pressure instruments measure the lateral forces exerted by traveling turbulent pulse waves, which are only distantly, deceptively, and indirectly related to blood flow and tissue perfusion. The lateral forces generated by traveling pulsatile turbulence are determined by a combination of cardiac contractility, declining arterial diameter, temperature,[56] arterial bifurcations and curvatures, turbulent wave reflections, and fluctuating capillary flow resistance, so that blood pressure measurement varies from point-to-point throughout the arterial tree. In general, elevated blood pressure reflects exaggerated microvascular flow resistance that undermines cardiac output, while low blood pressure reflects beneficial reductions of microvascular flow resistance that reflects good health, low stress, athletic conditioning, and enhanced cardiac output. However, life-threatening diseases and toxins that impair cardiac contractility can also lower blood pressure. Therefore, blood pressure is a misleading indication of medical condition, and the significance of blood pressure readings must be judged in context.

The capillary gate mechanism and the turbulence mechanism function in concert, as if they were the same mechanism. Fluctuating autonomic balance regulates microvascular flow resistance, which explains why blood viscosity (i.e. flow resistance), blood coagulability, cardiac output, cardiac efficiency, tissue perfusion, tissue oxygenation, organ function, blood pressure, pulse rate, factor VIII activity,

VWF levels, and insoluble fibrin levels all fluctuate in concert with autonomic balance.

BLOOD PRESSURE MEASUREMENT VIA THE RIVA-ROCCI TECHNIQUE

The familiar Riva-Rocci blood pressure measurement technique applies cuff pressure to a limb until it halts blood flow in the arteries beneath the cuff. As cuff pressure is gradually released, arterial flow is restored, but the reduced arterial diameter accelerates flow velocity. The speeding blood suddenly decelerates as it emerges from beneath the distal edge of the cuff, where arterial diameter is normal, and this disrupts red cell aggregates and converts the kinetic energy of speeding blood into audible turbulence called *Korotkoff sounds* that can be detected by a stethoscope. The measured cuff pressure when the Korotkoff sounds first become audible is arbitrarily called *systolic pressure*. As the cuff pressure continues to decline, and the arterial diameter is restored, the turbulence subsides, and the Korotkoff sounds become inaudible. The measured cuff pressure when the Korotkoff sounds can no longer be heard is arbitrarily called *diastolic pressure*. These terms are misnomers, because the waves of pulsatile turbulence appear only during diastole, when the cardiac muscle is in a state of relaxation. Thus, conventional blood pressure measurement is only distantly and indirectly related to cardiac function, and blood pressure is inversely related to cardiac output, tissue perfusion, and tissue oxygenation.

BLOOD PRESSURE MEASUREMENT VIA DIRECT ARTERIAL CANNULATION

The standard Riva-Rocci blood pressure measurement technique cannot measure blood pressure lower than 40 torr (about 0.77 psi). For this reason, invasive arterial cannulas were introduced to measure blood pressure during cardiac bypass operations. The cannulas are installed in arteries and connected to transducers that measure blood pressure. However, the transducers must be calibrated using blood pressure cuff measurements, which in turn are affected by cuff size, limb location, arterial diameter, bifurcations, curvatures, and wave reflections, so that blood pressure measurement varies from point-to-point throughout the arterial tree. Meanwhile, blood pressure fluctuates in accord with stress mechanism activity. Thus the "accuracy" of blood pressure measurement is mythical. *Flow resistance*, as opposed to blood pressure, is the primary determinant of cardiac output, cardiac efficiency, tissue perfusion, and tissue oxygenation. Low blood pressure typifies good health, normal sleep, and athletic conditioning, while hypertension invariably reflects pathology. However, low blood pressure can also be caused by congestive heart failure and toxic effects that undermine cardiac contractility, such as anesthetic inhalation agent overdose. Calculations derived from blood pressure measurement, including Systemic Vascular Resistance (SVR), Pulmonary Vascular Resistance (PVR), and Mean Arterial Pressure (MAP) are as confusing and counterproductive as blood pressure itself.

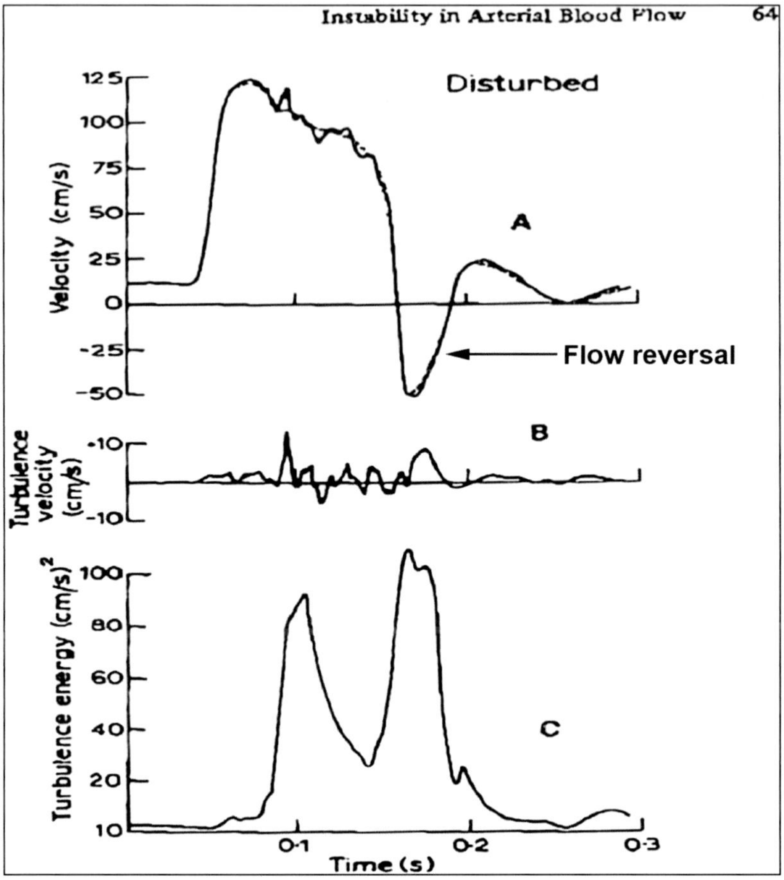

Figure 13.7: Blood viscosity vs. turbulence.

Blood velocity versus blood turbulence in the aorta of a dog. Aortic blood flow accelerates from 0 to 125cm/sec as the heart efficiently ejects its contents within a tenth of a second. Within 0.2 sec there is a brief flow reversal that closes the aortic valves and initiates a wave of pulsatile turbulence that propagates throughout the arterial tree during diastole, halting forward flow as it travels.(108) Forward flow resumes in the aftermath of the turbulent pulse wave as the muscular arterial tree slowly contracts to restore its resting volume.

BRUITS AND THE TURBULENCE MECHANISM

Bruits are analogous to Korotkoff sounds. They are produced by atherosclerotic plaque that forms *stenotic lesions* that decrease arterial diameter, typically at the carotid bifurcation in the neck. The decreased diameter of the diseased segment accelerates laminar arterial flow, but the speeding blood slows abruptly as it emerges from the distal end of the stenotic segment. This disrupts red cell aggregates, causing turbulence that is audible with the aid of a stethoscope.

LVAD AND THE TURBULENCE MECHANISM

The LVAD (Left Ventricle Assist Device) is a turbine device that is inserted in the left ventricle to enable patients to "rest and recover." However, it was soon discovered that the device not only sustains life but also enables moderate exercise tolerance despite the complete absence of heart function. This caused consternation because the device produces neither a palpable pulse nor measurable blood pressure. *This proves that blood pressure is not the driving force of blood flow, and that neither blood pressure nor the palpable pulse is essential to sustain life.*

The intermittent waves of pulsatile turbulence produced by cardiac contraction are the source of both the palpable pulse and blood pressure, but the LVAD turbine pump operates continuously. As it accelerates blood from the left ventricle into the aorta, the red blood cells spontaneously form aggregates that abolish turbulence, so that the pump propels blood throughout the arterial tree with extreme efficiency. Thus, the LVAD does not generate turbulence, so that it does not generate a palpable pulse, and conventional instruments cannot detect or measure blood pressure in its presence. However, the pump damages the fragile von Willebrand Factor protein, so that it causes angiodysplasia. In addition, it traumatizes red cells and releases their contents, including harmful actin protein, into systemic blood circulation. This causes amyloidosis, which reflects occult (hidden) stress mechanism hyperactivity induced by the pump. The subject of amyloidosis is discussed in Chapter 16: The Unified Theory of Medicine, on page 175. Since the LVAD does not generate turbulence, it should theoretically accelerate atherosclerosis if it is employed for prolonged periods of time.

ANEMIA VERSUS POLYCYTHEMIA

Conventional medical theory assumes that the sole purpose of red cells is to transport oxygen to cells, but red cells also regulate blood turbulence, which explains the manifestations of anemia and polycythemia:

- Inadequate red cell mass (anemia) allows excessive turbulence that prevents arterial coagulation,[57] exaggerates arterial turbulent flow resistance (viscosity), inhibits atherosclerosis,[58] and undermines exercise tolerance.[59]

- Excessive red cell mass (polycythemia) reduces turbulence to the point that spontaneous coagulation sometimes occurs in flowing blood, causing potentially lethal thrombus formation.[60] It also invites *Disseminated Intravascular Coagulation,* (DIC) which is spontaneous coagulation throughout the vascular system that is usually fatal.[61, 62]

In the past, athletes trained at high altitudes to induce natural erythropoietin (red blood cell) production that increases red cell mass and enhances exercise capacity. Since the development of synthetic erythropoietin (Epogen), athletes have embraced erythropoietin treatments (aka "blood doping"), which artificially elevates red cell mass to enhance athletic performance. This practice is dangerous as well as unethical and illegal because excessive red cell mass can precipitate sudden, systemic arterial thromboembolism known as Disseminated Intravascular Coagulation (DIC).

DIC most often occurs in pregnant women. Pregnancy elevates red cell mass, which enhances oxygen transport and delivery. Pregnancy thus reduces arterial flow resistance, enhances cardiac output and optimizes oxygen transport and delivery to sustain fetal development, but it also exaggerates the susceptibility of pregnant patients to thromboembolism and DIC. This most often happens during eclampsia, when labor releases tissue factor from amniotic fluid, placenta, and cervix into systemic circulation and precipitates sudden systemic coagulation.

CONCLUSION

The Turbulence Mechanism is my favorite element of the stress mechanism because it clarifies the nature of hemodynamic physiology and refutes false medical concepts that confuse conventional medical management.

NOTES

1 W. B. Cannon, The emergncy function of the adrenal medulla in pain and the major emotions. *The American journal of physiology* 33, 356-372 (1914).

2 W. B. Cannon, *Bodily Changes in Pain, Hunger, Fear and Rage: An Account of Recent Researches into the Function of Emotional excitement.* (Appleton, New York, 1915).

3 W. B. Cannon, Physiological Regulation of Normal States Some tentative Postulates Concerning Biological Homeostasis. *Jubilee Volume for Charles Richet*, 91 (1927).

4 W. B. Cannon, Organization for physiological homeostasis. *Physiological reviews* 9, 399-431 (1929).

5 W. B. Cannon, The Sympathetic Division of the Autonomic System in Relation To Homeostasis. *Arch Neur Psych* 22, 282-294 (1929).

6 W. B. Cannon, Chemical Mediators of Autonomic Nerve Impulses. *Science* 78, 43-48 (1933).

7 W. B. Cannon, The Stress and Strains of Homeostasis. *Am J Med Sci* 189, 1-14 (1935).

8 W. B. Cannon, *The wisdom of the body.* (W.W. Norton & Company, New York,, 1939), pp. xviii, 19-333 p.

9 W. B. Cannon, Voodoo death. *Psychosom Med* 19, 182-190 (1957).

10 J. Gleick, *Chaos: Making a New Science.* (1988).

11 B. Hof *et al.*, Experimental observation of nonlinear traveling waves in turbulent pipe flow. *Science* 305, 1594-1598 (2004).

12 L. Coleman, A Hypothesis: Factor VII governs clot formation, tissue repair and apoptosis. *Medical Hypotheses* in press, (2007).

13 L. S. Coleman, An Improved Explanation of Atherosclerosis. *PLoS Medicine* 2(4):e98, (2005).

14 L. S. Coleman, Insoluble fibrin may reduce turbulence and bind blood components into clots. *Med Hypotheses* 65, 820-821 (2005).

15 B. Hof *et al.*, Experimental observation of nonlinear traveling waves in turbulent pipe flow. *Science* 305, 1594-1598 (2004).

16 J. Gleick, *Chaos: Making a New Science.* (1988).

17 L. S. Coleman, Insoluble fibrin may reduce turbulence and bind blood components into clots. *Med Hypotheses* 65, 820-821 (2005).

18 D. Bluestein, S. Einav, Transition to turbulence in pulsatile flow through heart valves--a modified stability approach. *J Biomech Eng* 116, 477-487 (1994).

19 F. H. Busse, Physics. Visualizing the dynamics of the onset of turbulence. *Science* 305, 1574-1575 (2004).

20 L. S. Coleman, Atherosclerosis may be caused by inadequate levels of turbulence and mixing. *World J Surg* 30, 638-639 (2006).

21 F. Ghalichi *et al.*, Low Reynolds number turbulence modeling of blood flow in arterial stenoses. *Biorheology* 35, 281-294 (1998).

22 L. S. Coleman, An Improved Explanation of Atherosclerosis. *PLoS Medicine* 2(4):e98, (2005).

23 L. S. Coleman, Insoluble fibrin may reduce turbulence and bind blood components into clots. *Med Hypotheses* 65, 820-821 (2005).

24 B. Hof *et al.*, Experimental observation of nonlinear traveling waves in turbulent pipe flow. *Science* 305, 1594-1598 (2004).
25 B. Hof *et al.*, Experimental observation of nonlinear traveling waves in turbulent pipe flow. *Science* 305, 1594-1598 (2004).
26 R. O. Becker, G. Selden, *The body electric : electromagnetism and the foundation of life.* (Quill, New York, ed. 1st Quill, 1985), pp. 364 p.
27 L. Dintenfass, Some Observations on the Viscosity of Pathological Human Blood Plasma. *Thrombosis et diathesis haemorrhagica* 13, 492-499 (1965).
28 A. Scarpa, J. Vallieres, B. Sloane, A. P. Somlyo, Smooth muscle mitochondria. *Methods in enzymology* 55, 60-65 (1979).
29 L. S. Coleman, An Improved Explanation of Atherosclerosis. *PLoS Medicine* 2(4):e98, (2005).
30 L. S. Coleman, Insoluble fibrin may reduce turbulence and bind blood components into clots. *Med Hypotheses* 65, 820-821 (2005).
31 L. S. Coleman, Atherosclerosis may be caused by inadequate levels of turbulence and mixing. *World J Surg* 30, 638-639 (2006).
32 L. S. Coleman, The Mammalian Stress Mechanism (MSM) Explains Radial Artery Spasm (RAS). *J Anes Surg* 5, 81-94 (2018).
33 P. A. Ades *et al.*, Skeletal muscle and cardiovascular adaptations to exercise conditioning in older coronary patients. *Circulation* 94, 323-330 (1996).
34 E. R. Eichner, Exercise and heart disease. Epidemiology of the "exercise hypothesis". *Am J Med* 75, 1008-1023 (1983).
35 F. L. Gobel, L. A. Norstrom, R. R. Nelson, C. R. Jorgensen, Y. Wang, The rate-pressure product as an index of myocardial oxygen consumption during exercise in patients with angina pectoris. *Circulation* 57, 549-556 (1978).
36 B. M. Prior, P. G. Lloyd, H. T. Yang, R. L. Terjung, Exercise-induced vascular remodeling. *Exerc Sport Sci Rev* 31, 26-33 (2003).
37 A. Schmermund, [Cardiorespiratory fitness. Importance of exercise for healthiness]. *Herz* 29, 365-372 (2004).
38 J. A. Wijnen *et al.*, Effect of exercise training on ambulatory blood pressure. *Int J Sports Med* 15, 10-15 (1994).
39 A. Zanesco, E. Antunes, Effects of exercise training on the cardiovascular system: Pharmacological approaches. *Pharmacol Ther* 114, 307-317 (2007).
40 A. Genovese *et al.*, Inhibition of hypoxia-induced cardiac hypertrophy by verapamil in rats. *Basic Res Cardiol* 75, 757-763 (1980).
41 A. Genovese, M. Chiariello, S. Latte, W. DeAlfieri, M. Condorelli, Bilateral ventricular hypertrophy in rats exposed to acute or chronic hypobaric hypoxia. *Respiration* 44, 289-293 (1983).
42 A. Genovese, W. De Alfieri, S. Latte, M. Chiariello, M. Condorelli, Regression of myocardial hypertrophy in the rat following removal of acute or chronic hypobaric hypoxia. *Eur Heart J* 3 Suppl A, 161-164 (1982).
43 A. Genovese, S. Latte, M. Bozzaotre, M. Chiariello, Response of the left ventricular connective tissue to hypoxia. *Res Exp Med (Berl)* 183, 111-115 (1983).
44 S. Archer, S. Rich, Primary pulmonary hypertension: a vascular biology and translational research "Work in progress". *Circulation* 102, 2781-2791 (2000).
45 L. Bogar, Hemorheology and hypertension: not "chicken or egg" but two chickens from similar eggs. *Clin Hemorheol Microcirc* 26, 81-83 (2002).

46 H. A. Boudier, Arteriolar and capillary remodelling in hypertension. *Drugs* 58 Spec No 1, 37-40 (1999).
47 G. Cicco, P. Vicenti, G. D. Stingi, Tarallo, A. Pirrelli, Hemorheology in complicated hypertension. *Clin Hemorheol Microcirc* 21, 315-319 (1999).
48 C. Crump, J. Sundquist, M. A. Winkleby, K. Sundquist, Interactive Effects of Physical Fitness and Body Mass Index on the Risk of Hypertension. *JAMA Intern Med* 176, 210-216 (2016).
49 B. Dahlof, H. Herlitz, M. Aurell, L. Hansson, Reversal of cardiovascular structural changes when treating essential hypertension. The importance of the renin-angiotensin-aldosterone system. *Am J Hypertens* 5, 900-911 (1992).
50 K. Farkas *et al.*, Impairment of skin microvascular reactivity in hypertension and uraemia. *Nephrology, dialysis, transplantation : official publication of the European Dialysis and Transplant Association - European Renal Association* 20, 1821-1827 (2005).
51 Z. Hlubocka *et al.*, [Is mild essential hypertension without obvious organ complications and risk factors associated with increased levels of circulating markers of endothelial dysfunction? Effect of ACE inhibitor therapy]. *Vnitr Lek* 48, 718-723 (2002).
52 F. H. Messerli, B. Williams, E. Ritz, Essential hypertension. *Lancet* 370, 591-603 (2007).
53 C. Savoia, E. L. Schiffrin, Inflammation in hypertension. *Curr Opin Nephrol Hypertens* 15, 152-158 (2006).
54 E. H. Serne *et al.*, Capillary recruitment is impaired in essential hypertension and relates to insulin's metabolic and vascular actions. *Cardiovasc Res* 49, 161-168 (2001).
55 E. Vicaut, Microcirculation and arterial hypertension. *Drugs* 59 Spec No, 1-10 (1999).
56 L. Dintenfass, Some Observations on the Viscosity of Pathological Human Blood Plasma. *Thrombosis et diathesis haemorrhagica* 13, 492-499 (1965).
57 L. S. Coleman, Insoluble fibrin may reduce turbulence and bind blood components into clots. *Med Hypotheses* 65, 820-821 (2005).
58 L. S. Coleman, An Improved Explanation of Atherosclerosis. *PLoS Medicine* 2(4):e98, (2005).
59 J. R. Dunne, D. Malone, J. K. Tracy, C. Gannon, L. M. Napolitano, Perioperative anemia: an independent risk factor for infection, mortality, and resource utilization in surgery. *J Surg Res* 102, 237-244 (2002).
60 H. C. Kwaan, J. Wang, Hyperviscosity in polycythemia vera and other red cell abnormalities. *Semin Thromb Hemost* 29, 451-458 (2003).
61 J. W. Fisher, Erythropoietin: physiology and pharmacology update. *Exp Biol Med (Maywood)* 228, 1-14 (2003).
62 https://www.washingtonpost.com/business/economy/anemia-drug-made-billions-but-at-what-cost/2012/07/19/gJQAX5yqwW_story.html

Chapter 14

Oxygen Transport and Delivery

"Over the oxygen supply of the body carbon dioxide spreads its protecting wings."
— Johannes Friedrich Miescher

The mechanism of oxygen transport and delivery is the primary regulator of the capillary gate mechanism, and therefore an important element of the stress mechanism. Oxygen transport is essential to mammalian life, because all eukaryotic cells require a constant supply of oxygen, and the transport mechanism captures oxygen from the atmosphere and delivers it to cells deep within the body that are isolated from atmospheric oxygen.

The mechanism of oxygen transport and delivery is obviously fundamental to vertebrate physiology, and it was widely understood 100 years ago, but it is disturbingly absent from current medical literature, training, and application. A possible explanation of this bizarre oversight is presented in Chapter 18: The Great Medical Hoax of the 20th Century, on page 277. In addition, I recommend my previously published review of the confusing chemistry, origin, pathophysiology, and therapeutic properties of carbon dioxide and its relationship to narcotics that is entitled "Four Forgotten Giants of Anesthesia History." It can be downloaded free of charge from the publisher's website:[1]

https://www.ommegaonline.org/article-details/Four-Forgotten-Giants-of-Anesthesia-History/468

Atmospheric gases are odorless, tasteless, clear, colorless, and invisible, and their chemistry is subtle and confusing. Atmospheric air consists of 78% nitrogen, 21% oxygen, and some 15 trace gases such as argon and xenon. Nitrogen and most of the trace gases are benign and inert. Oxygen is essential for life, but it is chemically reactive and toxic, so that breathing pure oxygen for as little as 24 hours at sea level pressure causes lethal lung inflammation that cannot be distinguished from ARDS (adult respiratory distress syndrome).

Carbon dioxide (CO_2) is a trace gas that constitutes only 0.04% of the atmosphere. Unlike other trace gases, carbon dioxide possesses a unique combination of chemical characteristics that explains its confusing effects and its essential role in oxygen transport and delivery:

- CO_2 *inhibits the binding of oxygen to the hemoglobin molecule.* The hemoglobin molecule binds avidly to oxygen. The act of breathing lowers the partial pressure of CO_2 in the lung to 40 mm Hg (as opposed to 45 mm Hg elsewhere in the body). This optimizes the saturation of all four hemoglobin receptor sites with oxygen as blood traverses the lung (see Figure 17.1 on page 252). Because CO_2 inhibits binding of oxygen to hemoglobin, the extracellular CO_2 levels of 45 mm Hg in organs and tissues release oxygen from hemoglobin into extracellular fluids to enable cellular oxygenation.

- CO_2 *activates respiratory chemoreceptors* located in brain ventricles and extracellular tissues that stimulate breathing. Breathing replenishes oxygen in the lung and lowers that CO_2 partial pressure in the lung to 40 mm Hg. This optimizes the binding of hemoglobin to oxygen in the lung.

- CO_2 *directly releases nitric oxide (NO) from the vascular endothelium* in the manner of *nitrergic neurogenic vasodilation* (see Chapter 12: The Capillary Gate Mechanism, on page 109). This opens the capillary gate, reduces microvascular flow resistance, increases cardiac output, enhances tissue perfusion, and accelerates the transport of oxygenated arterial blood to tissues and organs, where higher levels of CO_2 release oxygen from hemoglobin into extravascular tissues.[2, 3]

- CO_2 *causes a harmless — but painful and fearsome — sense of suffocation when it is inhaled in concentrations of 1% and above.* This provides warning of its presence. Other atmospheric gases lack this effect. Because it is so necessary to oxygen transport, mammals must have evolved to notice an improper concentration of carbon dioxide.

- CO_2 *has refrigerant properties.* Compared to other atmospheric gases, CO_2 is easily isolated from atmospheric air by compression and cooling that converts it to liquid form. When liquid CO_2 is released from pressurized canisters it vaporizes into a cool, invisible, dense cloud that is temporarily affected by gravity, so that it pours like water and pools in low-lying places. Chemists call this a *transition state*. Such transition states can persist for hours if air movement is minimal, until CO_2 gradually absorbs enough heat to blend with the surrounding atmosphere. Unwary victims who blunder into such invisible CO_2 clouds can suffer lethal asphyxiation.[4]

Figure 14.1: Carbon dioxide flows like water.
As carbon dioxide transitions from a pressurized liquid to a gas, it can temporarily form a cool, invisible cloud that is temporarily denser than atmospheric air, so that it pours like water and dangerously accumulates in low-lying areas.

The Krebs cycle in living eukaryotic cells continuously consumes oxygen and produces carbon dioxide (see Appendix E: Cell Biology, on page 427). *Carbon dioxide plays an indispensable role in oxygen transport and delivery, and is therefore as essential to life as oxygen.*

CO_2 dissolves in water and saturates body fluids and tissues, so that the adult human body contains about 20 Liters of CO_2, as compared to 1 Liter of oxygen and 1 Liter of nitrogen. The CO_2 leaks continuously from the skin to establish an equilibrium partial pressure of 45 mm Hg within the body as compared to 0.3 mm Hg in ambient air. Mosquitoes home in on this plume of carbon dioxide. Breathing mechanisms in mammals adapt to this equilibrium and seek to maintain it. As we have stated, breathing replenishes oxygen in the lung and lowers the partial pressure of CO_2 in the lung to 40 mm Hg. This optimizes the binding of oxygen to hemoglobin in red blood cells as they transit the lung.[5] The heart propels the oxygenated red cells to organs and tissues, where the slightly higher partial pressure of CO_2 (45 mm Hg) releases oxygen from hemoglobin in the red cells to enable cellular oxygen uptake.

The mechanism of oxygen transport and delivery automatically matches oxygen delivery to oxygen consumption. For example, exercise elevates muscle cell metabolism, which increases cellular oxygen consumption and CO_2 production. The CO_2 elevation in exercising muscle opens the capillary gate (see Chapter 12: The Capillary Gate Mechanism, on page 109), which lowers flow resistance, increases cardiac output, and accelerates the delivery of oxygenated blood to the exercising muscle cells. In the blood near the muscle cells, the elevated CO_2 levels release oxygen from the hemoglobin molecule, which enables neighboring cells to absorb it. Rising CO_2 levels in venous blood stimulate breathing, which replenishes oxygen in the lung, and maintains the partial pressure of CO_2 in the lung to optimize hemoglobin binding to oxygen. This explains the brief delay between the onset of exercise and the increased respiratory rate and volume induced by exercise.

THE BENIGN NATURE OF CARBON DIOXIDE

Science has long since proved that carbon dioxide is biologically inert (it doesn't chemically react with body tissues, and therefore lacks toxicity). It is benign, beneficial, and — indeed — essential for life. It might be compared to water. Both water and CO_2 can exist as liquid, solid, or transparent gas. Both are essential for life, but both can cause drowning (asphyxiation) in rare circumstances. *Unlike water vapor (water as gas), CO_2 gas causes a fearsome sense of suffocation that warns of its presence when it is inhaled in elevated concentrations.* While the sensation of suffocation is *not* associated with toxicity (toxicity causes tissue damage), carbon dioxide does cause an emotional reaction that *mimics the natural reaction to toxicity* by elevating blood pressure and pulse rate: hence, "fearsome."

Common sense should negate the notion that carbon dioxide has either toxic or narcotic properties. The human body is saturated with carbon dioxide. Were it toxic, we would be dead. Were it narcotic, we would be drunk. Carbonated water and beverages are ingested harmlessly. CO_2 is used for stage show cloud effects, food processing, food preservation, fire extinguishing, spray can propellant, and numerous commercial and industrial purposes, all with remarkable safety. CO_2 is harmlessly injected into the peritoneal cavity to enable laparoscopic surgery and injected into arteries to enable angiography. Supersaturation studies in hyperbaric chambers as well as numerous documented reports of accidental or inadvertent hypercarbia illustrate the absence of CO_2 toxicity — *as long as brain oxygenation is sustained.*[6, 7, 8, 9, 10, 11]

My favorite example of the resilience of oxygen transport and delivery is that of a 16-year-old Canadian boy who accidentally fell into a grain transport wagon, was buried in grain for 5–7 minutes, and was unresponsive and cyanotic (indicating severe hypoxemia) when he was extricated from the grain.[12] Oxygen was administered by mask but there was no obvious gas exchange during the subsequent 40-minute ambulance ride to the nearest hospital. On arrival he was unresponsive, but his pupils were equal and reacted slowly to light, which indicated absence of hypoxic brain damage. He was making "shallow respiratory efforts" at a rate of 25–30 breaths/minute. After his mouth and pharynx were cleared of grain, he was intubated using a 7.5mm endotracheal tube, but the cuff could not be extended beyond the vocal cords on account of impacted grain in his trachea, and "no effective tidal volumes could be detected by auscultation or inspection during positive pressure ventilation." His arterial blood gases documented the following:

pH 6.62 (normal 7.38-7.42)
$PaCO_2$ 501 mm Hg (normal 38-42)
PaO_2 76 mm Hg (normal 75-100)
Bicarbonate 51 (normal 23-30)

It is remarkable that the partial pressure of oxygen in his blood remained adequate despite the elevated CO_2 partial pressure that might have inhibited the binding of oxygen to hemoglobin as blood traversed his lungs. However, available evidence suggests that carbon dioxide sustains tissue oxygenation primarily by opening the capillary gate and exaggerating cardiac output, which speeds oxygen transport from the lung to oxygen-starved tissues.

The boy was treated with Sufentanil 200mcg and midazolam 4mg in preparation for emergency ECMO (extra corporeal membrane oxygenation) but when he was placed in Trendelenberg position (tilted head down) for installation of a central venous catheter it was discovered that the grain in his trachea and bronchi could be gradually extricated from the proximal end of the endotracheal tube after each exhalation. Some 200 ml of grain was removed with the help of chest percussion in various positions, and it became possible to extricate the remainder using rigid bronchoscopy.

After the grain was removed, he remained unresponsive with bilaterally constricted pupils but his level of consciousness gradually improved over the next six hours and by the following morning he was alert, oriented, and had no residual neurologic defect. This incident reflects not only the absence of CO_2 toxicity but also the remarkable resiliency of the mechanism of oxygen transport and delivery.

CO_2 THERAPEUTIC BENEFITS

The mechanism of oxygen transport and delivery explains the therapeutic benefits of carbon dioxide supplementation. Short of disrupting oxygen binding to hemoglobin (as described above), adding CO_2 to ordinary air improves cardiac output and tissue oxygenation, with or without oxygen enrichment.[13, 14] The improvement is primarily enabled by increased cardiac output that accelerates oxygen transport, and secondarily by the release of oxygen from hemoglobin.[15] In contrast, soda lime (a useless but ubiquitous feature of anesthesia machines) harmfully removes carbon dioxide from inhaled gas mixtures and undermines tissue oxygenation.[16, 17, 18]

Carbogen, a mixture of carbon dioxide and oxygen in a pressurized tank (usually 5% CO_2 and 95% oxygen), is safer and more effective than either oxygen or CO_2 alone. *Carbogen is perhaps the most potent, practical, affordable, simple, and safe medical treatment ever discovered.* Its oxygen supplementation assures hemoglobin saturation with oxygen in the lung. Its carbon dioxide supplementation opens the capillary gate throughout the body, increases cardiac output, accelerates oxygen transport, and exaggerates the release of oxygen from hemoglobin to optimize tissue oxygenation. Carbogen is thus a perfect emergency treatment for heart attacks and strokes because it immediately restores oxygen to ischemic tissues without inviting the hazards caused by toxic medications, anticoagulants, and angioplasty. It is also the best emergency treatment for asthma, atelectasis, pneumonia, smoke inhalation, carbon monoxide poisoning, alcohol inebriation, and breathing problems in newborn babies.

Carbogen was commonplace on fire trucks and in hospitals and doctors' offices 100 years ago. It served as an emergency treatment for asthma, atelectasis, influenza, pneumonia, smoke inhalation, drowning, narcotic respiratory depression, carbon monoxide poisoning, alcohol inebriation, and breathing problems in newborn babies.[19, 20, 21, 22] Anesthetists used CO_2 supplementation to prevent CO_2 depletion, accelerate ether induction and emergence, prevent ether explosions and fires, optimize cardiorespiratory performance, counteract narcotic respiratory depression, and prevent postoperative respiratory depression, nausea, atelectasis, pneumonia, asthma, laryngospasm, heart attacks, and strokes.[23, 24, 25, 26]

Science waits patiently while medicine discovers, abandons, and re-discovers its secrets. Carbogen and CO_2 supplementation were abandoned and forgotten about 70 years ago due to untoward fear of mythical CO_2 narcosis and toxicity[27, 28] (see also Chapter 18: The Great Medical Hoax of the 20th Century, on page 277). The therapeutic benefits of carbon dioxide supplementation were re-discovered during the 1990s by critical care specialists who were seeking better treatments for ARDS (Adult Respiratory Distress Syndrome).[29, 30] Similarly, the therapeutic benefits of opioid supplementation were discovered 100 years ago but abandoned along with carbon dioxide, and then re-discovered during the 1960s, and re-re-discovered during the 1980s.[31, 32, 33, 34] Unfortunately, the synergistic benefits of combining oxygen supplementation, CO_2 supplementation, anesthesia, and narcotic analgesia — which was understood 100 years ago — is now unappreciated.[35] This therapeutic synergy, which promotes tissue repair, tissue perfusion, and tissue oxygenation, would be even more useful than in the forgotten past when combined with modern machines, medications, and monitors. For example, combinations of oxygen, hypercarbia, antibiotics, anesthesia, and analgesia could revolutionize the treatment of heart attacks, strokes, bone fractures, stasis ulcers, diabetes, malignancy, burn injuries, severe trauma, sepsis, gas gangrene, MRSA, Lyme Disease, and tuberculosis (see Chapter 16: The Unified Theory of Medicine, on page 175).

THE MYTH OF CO_2 NARCOSIS

The dictionary defines *narcosis* as "a state of stupor, drowsiness, or unconsciousness produced by drugs." It defines *anesthesia* as "insensitivity to pain, especially as artificially induced by the administration of gases or the injection of drugs before surgical operations." Both terms are confusing and imprecise. The terms anesthesia and analgesia as explained in Chapter 15: Three Stress Mechanism Activation Pathways, on page 155 are more precise, but they are often misused and confused with each other.

Analgesia inhibits nociception. Nociception is perceived by consciousness as pain. Nociception increases harmful sympathetic nervous activity.

Anesthesia extinguishes consciousness in a dose-related fashion. It inhibits the ability of consciousness to perceive nociception as pain before it abolishes the ability to speak coherently. In other words, *anesthesia mimics analgesia under special circumstances*. Anesthesia indirectly exaggerates nociception by extinguishing inhibitory signals descending from the cortex to nociception pathways in the spinal cord.

CO_2 asphyxiation mimics anesthesia under special circumstances. It disrupts oxygen transport and delivery, which causes brain hypoxia that extinguishes consciousness. This explains so-called "CO_2 narcosis."

Adding 10% carbon dioxide to room air produces a gas mixture that disrupts oxygen transport and delivery and causes moderate brain hypoxia that extinguishes consciousness. This mimics general anesthesia (so-called *narcosis*) because it can be quickly reversed by breathing pure atmospheric air. However, general anesthesia and CO_2 asphyxiation are profoundly different. Anesthesia causes a gradual loss of consciousness, without reducing brain oxygen. CO_2 asphyxiation causes sudden loss of consciousness by reducing brain oxygen below a critical threshold. Anesthesia is pleasant and addictive. Inhaling CO_2 at asphyxiating concentrations causes an alarming sense of suffocation that is not addictive. Anesthesia inhibits surgical stress, but brain hypoxia that eliminates consciousness is extremely dangerous, because it occurs at the threshold of brain damage and death.

CARBON DIOXIDE AND RESPIRATORY DRIVE

The mechanism of oxygen transport and delivery is robust, but respiratory drive is readily disrupted by circumstances that are seldom appreciated. Hypercarbia is the main source of respiratory drive because respiratory chemoreceptors are only weakly sensitive to hypoxia in the absence of hypercarbia. CO_2 stimulates respiratory drive via two different types of chemoreceptors, as follows:

1. pH receptors located in brain ventricles stimulate breathing in accord with acid levels in cerebrospinal fluid (CSF). Hypoventilation causes CO_2 accumulation, which elevates carbonic acid levels in CSF that activate the pH receptors in brain ventricles and stimulate breathing. This primary type of respiratory drive sustains breathing despite CO_2 depletion, but it is extinguished by loss of consciousness (as during normal sleep), heavy sedation, or general anesthesia, in which case breathing becomes dependent on the secondary form of respiratory drive.
2. Secondary respiratory drive is enabled by chemoreceptors that are activated by CO_2 elevations in extracellular fluids. This secondary form of respiratory drive remains active and sustains breathing during normal sleep and general anesthesia. This explains why sleep is normally accompanied by mild hypoventilation, hypercarbia, hypoxemia, and acidosis, but the hypercarbia opens the capillary gate, lowers blood pressure, increases cardiac efficiency, and sustains effective tissue oxygenation. However, these chemoreceptors are inhibited by synergistic combinations of the following:
    - Hyperventilation that depletes CO_2 tissue reserves.
    - Narcotics.
    - Hypnotic agents.

There is yet a third type of respiratory drive: nociception. Nociception stimulates breathing with or without consciousness. For example, surgical nociception stimulates hyperventilation despite general anesthesia, and thereby depletes CO_2 tissue reserves and paralyzes respiratory chemoreceptors. This can lead to unexpected respiratory arrest during or after surgery.[36, 37]

Mechanical hyperventilation during anesthesia depletes CO_2 tissue reserves, obtunds respiratory chemoreceptors, and unpredictably undermines postoperative respiratory drive. Hyperventilated patients regain consciousness, which restores primary respiratory drive during anesthetic emergence despite CO_2 depletion, so that breathing appears to be normal. However, chemoreceptor function remains paralyzed due to CO_2 depletion, so that if CO_2 depleted patients fall asleep or lose consciousness for any reason, they may unexpectedly stop breathing and die. This typically happens when elderly patients are treated with combinations of narcotics and sedatives after uneventful surgery.[38, 39]

HARMFUL HYPERVENTILATION

Everyone knows that breathing is essential for life, and this invites the alluring assumption that more breathing is better, so that hyperventilation beneficially enhances oxygen reserves. However, this intuitive assumption is ignorant and false. Tissue oxygen reserves cannot be improved by hyperventilation because breathing room air normally saturates hemoglobin 100%. Therefore, *hyperventilation cannot improve oxygen transport and delivery.* Instead, it dangerously depletes CO_2 tissue reserves,[40, 41, 42, 43, 44, 45] which undermines respiratory drive, elevates microvascular flow resistance, reduces cardiac output, inhibits the release of oxygen from hemoglobin, and undermines tissue perfusion and tissue oxygenation, as illustrated by the following examples:

- Every year healthy swimmers accidentally drown themselves by voluntarily hyperventilating themselves in preparation for swimming under water in the naïve assumption that this will prolong their ability to swim under water without breathing. Unfortunately, oxygen reserves cannot be increased, while CO_2 tissue reserves are rapidly depleted by such voluntary hyperventilation. As a result, the hyperventilated swimmer suffers painless brain hypoxia that manifests as euphoria, suffers no urge to breathe, and thus falls asleep at the bottom of the pool and drowns. This is called "shallow water blackout syndrome."

- Anesthesiologists routinely employ mechanical hyperventilation to "rid the body" of CO_2 so as to avoid its supposedly narcotic and toxic effects.[46] This perilous practice obtunds respiratory chemoreceptors and exaggerates the respiratory depression caused by narcotics, so that patients recovering from surgery may unexpectedly stop breathing if they fall asleep and lose consciousness for any reason. Most victims are elderly patients recovering from otherwise uneventful surgery who are treated with combinations of narcotics and sedatives after surgery. The problem is incorrectly attributed to *re-narcotization syndrome* or *narcotic hypersensitivity*.[47, 48]

- Mechanical hyperventilation during general anesthesia for cesarean section depletes CO_2 tissue reserves and obtunds respiratory chemoreceptors in newborn babies as well as their mothers. The mothers usually survive because inadequately controlled postoperative pain keeps them awake. The babies are awakened by stimulation during the delivery, but they subsequently fall asleep before they replenish their CO_2 tissue reserves and thus unexpectedly stop breathing and die soon afterwards.[49] Spinal and epidural analgesia avoids the problem.

- More than 400 Darwin Award winners reside in the "dead zone" of Mt. Everest, where ambient oxygen levels are very low and carbon dioxide is virtually absent. The low oxygen levels at high altitude cause hypoxic hyperventilation that depletes CO_2 and undermines brain oxygenation, causing "mountain sickness" and unexpected loss of consciousness that is often lethal under such circumstances.

- Professional guides who repeatedly climb Mt. Everest suffer hypoxic brain damage.[50]

- Airplane pilots flying at high altitudes in unpressurized airplanes suffer the same unexpected loss of conscious as mountain climbers. In both cases, breathing carbon dioxide helps to sustain brain oxygenation and prevent loss of consciousness.[51, 52, 53, 54, 55, 56, 57, 58, 59, 60, 61, 62, 63, 64, 65, 66, 67, 68, 69, 70, 71]

- Polio victims hyperventilated in iron lungs suffered high mortality until Hewlett-Packard engineers, who were developing early capnography technology, discovered that these patients were being hyperventilated and corrected the ventilator settings to restore normal CO_2 levels, whereupon the death rate fell in half.[72]

- Extra-Corporeal Membrane Oxygenation (ECMO) treatments and cardiac bypass machines can severely deplete CO_2 tissue reserves, causing increased morbidity and mortality.[73]

CO_2 AND ACID/BASE BALANCE

Water readily absorbs carbon dioxide, and small amounts of water and carbon dioxide molecules reversibly react with each other to form harmless carbonic acid (H_2CO_3). *Hypo*ventilation elevates CO_2 levels in the body and thereby elevates carbonic acid levels in blood and tissues. This is called *respiratory acidosis*. *Hyper*ventilation lowers CO_2 levels in the body and thereby lowers carbonic acid levels in the blood and tissues. This is called *respiratory alkalosis*. Respiratory *acidosis* is benign and harmless, but respiratory *alkalosis* undermines respiratory chemoreceptor activity, and is therefore inherently dangerous.

Respiratory acidosis is commonly equated with metabolic acidosis, but they are profoundly different. Respiratory acidosis is essentially harmless and usually beneficial. Metabolic acidosis is also harmless *per se*, but some types of metabolic acidosis reflect harmful or dangerous conditions. As a result, both metabolic acidosis and respiratory acidosis are often regarded with dread as harbingers of death.

Here are some examples of metabolic acidosis:

Lactic acidosis is a form of metabolic acidosis that occurs when cellular oxygen starvation causes cells to revert to anaerobic respiration (aka glycolysis), which generates less ATP than the Krebs cycle but does not require oxygen (see Appendix E that explains cell biology). Anaerobic pathways produce lactic acid that painfully accumulates in tissues. This commonly causes harmless muscle pain and soreness during intense exercise. The Krebs cycle consumes lactic acid to produces ATP if oxygen is restored to cells in a timely fashion, whereupon the muscle pain subsides. Anaerobic metabolism temporarily sustains cell survival in the presence of oxygen starvation, but prolonged and severe oxygen starvation cripples cell function, and ultimately causes cell death.

Ketoacidosis is a form of metabolic acidosis that occurs when cellular glucose starvation causes cells to digest tissue protein to generate ATP. This abnormal form of cellular metabolism occurs in prolonged starvation and uncontrolled diabetes. It undermines muscle and tissue structure and generates acidic ketone bodies that reduce pH in blood and tissues. Cells metabolize the harmless ketone bodies, but prolonged starvation and uncontrolled diabetes cause progressive tissue damage that is potentially lethal.

Intoxication acidosis is a form of metabolic acidosis that occurs when toxic drugs and chemicals harmfully disrupt cellular metabolism. Such drugs and chemicals include salicylates, ethanol, methanol, formaldehyde, sulfates, and metformin.

CO_2 ASPHYXIATION

All atmospheric gases can cause asphyxiation (suffocation, oxygen deprivation, inadequate oxygen in tissues) by diluting the concentration of oxygen below 10% in the inhaled gas mixture, *but CO_2 causes asphyxiation at much lower concentrations than other atmospheric gases because of its ability to inhibit the binding of oxygen to hemoglobin in the lung.* This property can sometimes mimic either toxicity or general anesthesia. It is not a *manifestation* of toxicity, because it does not cause tissue damage in the manner of truly toxic gases such as military mustard gas.

Oxygen constitutes 21% of atmospheric air. Hemoglobin binds avidly to oxygen, so that hemoglobin is effectively saturated with oxygen in the lung when inhaled oxygen concentrations are 10% or higher. This enables hemoglobin to efficiently extract oxygen from the atmosphere.

All atmospheric gases can dangerously disrupt oxygen binding to hemoglobin in the lung if they dilute oxygen in the inhaled gas mixture below 10%. For any single atmospheric gas (other than carbon dioxide) to cause asphyxiation, it must be present in concentrations high enough to reduce oxygen below 10% in the inhaled gas mixture. The resulting hypoxemia (inadequate oxygen in the blood) causes hypoxia (inadequate oxygen in tissues). The brain is the organ most vulnerable because it requires more oxygen than any other organ. Mild hypoxia impairs judgment and causes muscle weakness. Moderate hypoxia abolishes consciousness. Severe hypoxia causes brain seizures and brain cell death.

In humans, hypoxia is painless and only weakly stimulates respiratory drive in the absence of hypercarbia, so it can kill without warning. For example, methane sometimes accumulates in agricultural silos and pits, and dilutes oxygen below 10%. Victims entering such pits and silos quickly lose consciousness and die without warning. Another example is welding masks that are accidentally connected to nitrogen instead of compressed air. In either case, the victims quickly die without warning.

Although carbon dioxide causes a warning sense of suffocation, it can cause asphyxiation at much lower concentrations than other atmospheric gases because of its unique ability to inhibit the binding of oxygen to hemoglobin in the lung. Breathing CO_2 in concentrations as low as 10% in atmospheric air can cause loss of consciousness that is reversible by simply breathing pure room air. This behavior mimics general anesthesia. For example, a common trick was performed in circus side shows during the early days of anesthesia in which a "shill" in the audience would "volunteer" to undergo *CO2 anesthesia* consisting of 10% CO_2 in atmospheric air, *but the margin of safety for this stunt is small.* Permanent brain damage and/or death will ensue if the oxygen starvation is prolonged, or if the CO_2 concentration climbs higher. For example, the professional guides who assist amateur fools who climb to the top of Mt. Everest eventually suffer a subtle form of hypoxic brain damage that undermines memory.[74]

If oxygen is the only other gas present, as during general anesthesia, then CO_2 concentrations as high as 30% can be tolerated, but if the CO_2 concentration exceeds 30% the danger of brain hypoxia suddenly becomes extreme despite the high concentration of oxygen.[75]

CO_2 ASPHYXIATION DURING ANESTHESIA

General anesthesia abolishes consciousness, and therefore eliminates the painful sense of suffocation caused by breathing elevated concentrations of CO_2. This facilitates beneficial hypercarbia during anesthesia that enhances cardiac output, tissue perfusion, and tissue oxygenation, especially in the presence of oxygen supplementation that offsets the inhibition of hemoglobin affinity for oxygen that is caused by carbon dioxide. Under these circumstances, CO_2 concentrations of up to 30% (228 mm Hg at sea level) can be safely tolerated.

Anesthetists routinely used CO_2 supplementation during the Golden Era of Ether 100 years ago to optimize cardiac output and tissue perfusion, accelerate ether induction and emergence, maximize morphine dosage, prevent ether fires and explosions, and prevent postoperative asthma, atelectasis, pneumonia, nausea, and vomiting. Unfortunately, the early "closed circuit" anesthesia machines that dominated anesthesia practice in that era were defective and dangerous. They were designed to conserve expensive anesthetic gases, so that overenthusiastic CO_2 supplementation sometimes elevated CO_2 concentrations to asphyxiating levels within the breathing circuit.[76, 77] This caused unexpected hypoxic brain seizures that were called *ether fits*. Most victims were children because of their elevated metabolism compared to adults. At first these problems were attributed to fever or occult medical conditions, but their association with CO_2 supplementation was soon recognized. Dr. Ralph Waters, the founder of the anesthesiology profession, condemned CO_2 supplementation and characterized CO_2 as "toxic waste, like urine, that must be rid from the body." As a result, physicians and anesthetists abandoned the use of CO_2, forgot its therapeutic benefits, and embraced harmful hyperventilation.[78, 79, 80, 81, 82, 83] Waters thus inadvertently reversed anesthesia progress and derailed medical progress (see Chapter 18: The Great Medical Hoax of the 20th Century, on page 277).[84]

CARBON MONOXIDE POISONING

Like carbon dioxide, carbon monoxide (CO) inhibits the binding of oxygen to the hemoglobin molecule. In fact, hemoglobin binds to carbon monoxide far more avidly than it binds to oxygen. Furthermore, carbon monoxide is odorless, colorless, tasteless, and invisible, and it does not cause an alarming sense of suffocation in the manner of carbon dioxide. Because of this, it causes lethal asphyxiation painlessly and without warning in very small concentrations. It is not a true poison, because lacks tissue toxicity — it does not damage tissues in the manner of toxic mustard gas. It is produced by incomplete combustion, as when fires burn in the absence of adequate oxygen. For example, car exhaust invariably contains carbon monoxide, so that operating a car engine in a closed garage can cause lethal carbon monoxide accumulations within the garage. Fires consume oxygen and produce carbon monoxide, so that fire fighters sometimes succumb to combinations of carbon monoxide and hypoxia.

Oxygen alone is ineffective as a treatment for CO asphyxiation, because the CO binds to hemoglobin more avidly than oxygen. The best-known antidote for CO poisoning is Carbogen, because carbon dioxide not only displaces CO from the hemoglobin molecule, but also stimulates breathing that removes CO from the body, promotes the saturation of hemoglobin with oxygen, and accelerates the transport and delivery of oxygen to hypoxic tissues and organs.

CONCLUSION

Carbon dioxide pathophysiology is fundamental to medical science, and its absence from medical literature, teaching, and practice represents a shocking failure of professional medicine and a triumph of commercial interests that profit at the price of public health (see Chapter 18: The Great Medical Hoax of the 20th Century, on page 277). A correct understanding of oxygen transport and delivery, combined with modern machines, monitors, and medications, could enable practical, powerful, proven treatments that save lives and preserve health and longevity.

NOTES

1 L. S. Coleman, Four Forgotten Giants of Anesthesia History. *Journal of Anesthesia and Surgery* 3, 1-17 (2015).

2 T. C. Harvey *et al.*, Effect of carbon dioxide in acute mountain sickness: a rediscovery. *Lancet* 2, 639-641 (1988).

3 O. Akca *et al.*, Hypercapnia improves tissue oxygenation. *Anesthesiology* 97, 801-806 (2002).

4 https://www.osha.gov/dts/hib/hib_data/hib19960605.html

5 A. Campbell, E. P. Poulton, *Oxygen and carbon dioxide therapy*. Oxford medical publications (Oxford University Press, H. Milford, London,, 1934), pp. xv, 179 p.

6 L. Litt *et al.*, Cerebral intracellular changes during supercarbia: an in vivo 31P nuclear magnetic resonance study in rats. *J Cereb Blood Flow Metab* 5, 537-544 (1985).

7 Y. Cohen *et al.*, Stability of brain intracellular lactate and 31P-metabolite levels at reduced intracellular pH during prolonged hypercapnia in rats. *J Cereb Blood Flow Metab* 10, 277-284 (1990).

8 Y. Xu, Y. Cohen, L. Litt, L. H. Chang, T. L. James, Tolerance of low cerebral intracellular pH in rats during hyperbaric hypercapnia. *Stroke* 22, 1303-1308 (1991).

9 P. Slinger, P. E. Blundell, I. R. Metcalf, Management of massive grain aspiration. *Anesthesiology* 87, 993-995 (1997).

10 A. Ohmura, M. Sha, J. Katagiri, How far can we go with permissive hypercapnia? A case presentation and some biased comments with emphasis on maintaining normal haemoglobin level. *Acta Anaesthesiol Scand Suppl* 107, 209-213 (1995).

11 C. H. Imray *et al.*, Carbon dioxide contributes to the beneficial effect of pressurization in a portable hyperbaric chamber at high altitude. *Clin Sci (Lond)* 100, 151-157 (2001).

12 P. Slinger, P. E. Blundell, I. R. Metcalf, Management of massive grain aspiration. *Anesthesiology* 87, 993-995 (1997).

13 T. C. Harvey *et al.*, Effect of carbon dioxide in acute mountain sickness: a rediscovery. *Lancet* 2, 639-641 (1988).

14 O. Akca *et al.*, Hypercapnia improves tissue oxygenation. *Anesthesiology* 97, 801-806 (2002).

15 O. Akca *et al.*, Tissue oxygenation response to mild hypercapnia during cardiopulmonary bypass with constant pump output. *Br J Anaesth* 96, 708-714 (2006).

16 C. H. Imray *et al.*, Carbon dioxide contributes to the beneficial effect of pressurization in a portable hyperbaric chamber at high altitude. *Clin Sci (Lond)* 100, 151-157 (2001).

17 L. S. Coleman, Should soda lime be abolished? *Anesth Analg* 102, 1290-1291 (2006).

18 L. S. Coleman, Soda lime is a dangerous relic of a forgotten past. *Anaesthesia* 70, (2016).

19 Y. Henderson, Physiological Observations on Pike's Peak, Colorado, made in the Summer of 1911. *Trans Am Climatol Assoc* 28, 11-15 (1912).
20 Y. Henderson, A Lecture ON RESPIRATION IN ANAESTHESIA: CONTROL BY CARBON DIOXIDE. *Br Med J* 2, 1170-1175 (1925).
21 Y. Henderson, THE PREVENTION AND TREATMENT OF ASPHYXIA IN THE NEW-BORN. *JAMA* JAMA. 1928;90(8):583-586. doi:10.1001/jama.1928.02690350001001, 583-586 (1928).
22 Y. Henderson, in *Cyclopedia of Medicine*. (1940).
23 Y. Henderson, A Lecture ON RESPIRATION IN ANAESTHESIA: CONTROL BY CARBON DIOXIDE. *Br Med J* 2, 1170-1175 (1925).
24 Y. Henderson, in *Cyclopedia of Medicine*. (1940).
25 Y. Henderson, The Wesley M. Carpenter Lecture: Atelectasis, Massive Collapse and Related Post-Operative Conditions. *Bull N Y Acad Med* 11, 639-656 (1935).
26 Y. Henderson, Resuscitation with Carbon Dioxide. *Science* 83, 399-402 (1936).
27 R. M. Waters, Toxic Effects of Carbon Dioxide. *J.A.M.A* 100:519, 1933, 219-224 (1933).
28 R. M. Waters, Carbon Dioxide. *Can Med Assoc J* 38, 240-243 (1938).
29 A. Ohmura, M. Sha, J. Katagiri, How far can we go with permissive hypercapnia? A case presentation and some biased comments with emphasis on maintaining normal haemoglobin level. *Acta Anaesthesiol Scand Suppl* 107, 209-213 (1995).
30 K. G. Hickling, S. J. Henderson, R. Jackson, Low mortality associated with low volume pressure limited ventilation with permissive hypercapnia in severe adult respiratory distress syndrome. *Intensive Care Med* 16, 372-377 (1990).
31 T. H. Stanley, *Anesthesia for the New Millennium*. (Springer-Science + Business Media, 1999), vol. 34, pp. 288.
32 L. W. Crile GW, *Anoci-association*. (Saunders, Philadelphia, 1914), pp. 223-225.
33 E. Lowenstein *et al.*, Cardiovascular response to large doses of intravenous morphine in man. *N Engl J Med* 281, 1389-1393 (1969).
34 S. N. Raja, E. Lowenstein, The birth of opioid anesthesia. *Anesthesiology* 100, 1013-1015 (2004).
35 L. S. Coleman, 30 Years Lost in Anesthesia Theory. *Cardiovasc Hematol Agents Med Chem* 10, 31-49 (2012).
36 Y. Henderson, in *Cyclopedia of Medicine*. (1940).
37 Y. Henderson, Resuscitation with Carbon Dioxide. *Science* 83, 399-402 (1936).
38 L. S. Coleman, in *apsf Newsletter*. (Anesthesia Patient Safety Foundation, Administrator, Deanna Walker Anesthesia Patient Safety Foundation Building One, Suite Two 8007 South Meridian Street Indianapolis, IN 46217-2922 e-mail address: walker@apsf.org FAX: (317) 888-1482, 2010), vol. Winter 2009-2020.
39 L. S. Coleman, A call for standards on perioperative CO(2) regulation. *Can J Anaesth*, (2011).

40 Y. Henderson, THE PREVENTION AND TREATMENT OF ASPHYXIA IN THE NEW-BORN. *JAMA* JAMA. 1928;90(8):583-586. doi:10.1001/jama.1928.02690350001001, 583-586 (1928).

41 Y. Henderson, in *Cyclopedia of Medicine*. (1940).

42 Y. Henderson, The Wesley M. Carpenter Lecture: Atelectasis, Massive Collapse and Related Post-Operative Conditions. *Bull N Y Acad Med* 11, 639-656 (1935).

43 Y. Henderson, Resuscitation with Carbon Dioxide. *Science* 83, 399-402 (1936).

44 Y. Henderson, Observations on RESUSCITATION FROM ASPHYXIA AND PREVENTION AND TREATMENT OF SECONDARY PNEUMONIA BY INHALATION OF CO(2). *Br Med J* 2, 687-689 (1931).

45 Y. Henderson, How Breathing Begins at Birth. *Science* 85, 89-91 (1937).

46 M. M. S. a. J. Wang, Central Nervous System Toxicity during the Induction of General Anesthesia After Intravenous Bolus of Lidocaine. *Journal of Clinical Research and Anesthesia* 2(1)-10004 www.auctoresonline.org, 4 (2019).

47 L. S. Coleman, in *apsf Newsletter*. (Anesthesia Patient Safety Foundation, Administrator, Deanna Walker Anesthesia Patient Safety Foundation Building One, Suite Two 8007 South Meridian Street Indianapolis, IN 46217-2922 e-mail address: walker@apsf.org FAX: (317) 888-1482, 2010), vol. Winter 2009-2020.

48 L. S. Coleman, A call for standards on perioperative CO(2) regulation. *Can J Anaesth*, (2011).

49 R. Gale, I. Zalkinder-Luboshitz, P. E. Slater, Increased neonatal risk from the use of general anesthesia in emergency cesarean section. A retrospective analysis of 374 cases. *J Reprod Med* 27, 715-719 (1982).

50 N. Fayed, P. J. Modrego, H. Morales, Evidence of brain damage after high-altitude climbing by means of magnetic resonance imaging. *Am J Med* 119, 168 e161-166 (2006).

51 T. C. Harvey *et al.*, Effect of carbon dioxide in acute mountain sickness: a rediscovery. *Lancet* 2, 639-641 (1988).

52 O. Akca *et al.*, Hypercapnia improves tissue oxygenation. *Anesthesiology* 97, 801-806 (2002).

53 O. Akca *et al.*, Tissue oxygenation response to mild hypercapnia during cardiopulmonary bypass with constant pump output. *Br J Anaesth* 96, 708-714 (2006).

54 M. V. Singh *et al.*, Blood gases, hematology, and renal blood flow during prolonged mountain sojourns at 3500 and 5800 m. *Aviat Space Environ Med* 74, 533-536 (2003).

55 M. E. Raichle, F. Plum, Hyperventilation and cerebral blood flow. *Stroke* 3, 566-575 (1972).

56 P. Bartsch, A. Haeberli, M. Franciolli, E. K. Kruithof, P. W. Straub, Coagulation and fibrinolysis in acute mountain sickness and beginning pulmonary edema. *J Appl Physiol* 66, 2136-2144 (1989).

57 J. Ramirez *et al.*, Oxygen-carrying capacity during 10 hours of hypercapnia in ventilated dogs. *Critical care medicine* 28, 1918-1923 (2000).

58 K. G. Hickling, Permissive hypercapnia. *Respir Care Clin N Am* 8, 155-169, v (2002).

59 B. Kavanagh, Normocapnia vs hypercapnia. *Minerva Anestesiol* 68, 346-350 (2002).
60 U. H. Thome, W. A. Carlo, Permissive hypercapnia. *Semin Neonatol* 7, 409-419 (2002).
61 G. M. Hare *et al.*, Hypercapnia increases cerebral tissue oxygen tension in anesthetized rats. *Can J Anaesth* 50, 1061-1068 (2003).
62 J. G. Laffey, D. O'Croinin, P. McLoughlin, B. P. Kavanagh, Permissive hypercapnia--role in protective lung ventilatory strategies. *Intensive Care Med* 30, 347-356 (2004).
63 A. T. Mazzeo, A. Spada, C. Pratico, T. Lucanto, L. B. Santamaria, Hypercapnia: what is the limit in paediatric patients? A case of near-fatal asthma successfully treated by multipharmacological approach. *Paediatric anaesthesia* 14, 596-603 (2004).
64 O. Akca, Hypercapnia: what is the upper limit? *Paediatric anaesthesia* 15, 80-81 (2005).
65 H. Hager *et al.*, Hypercapnia improves tissue oxygenation in morbidly obese surgical patients. *Anesth Analg* 103, 677-681 (2006).
66 M. Komori *et al.*, Permissive range of hypercapnia for improved peripheral microcirculation and cardiac output in rabbits. *Critical care medicine* 35, 2171-2175 (2007).
67 O. Akca, Carbon dioxide and tissue oxygenation: is there sufficient evidence to support application of hypercapnia for hemodynamic stability and better tissue perfusion in sepsis? *Intensive Care Med* 34, 1752-1754 (2008).
68 A. M. Mukhtar, G. M. Obayah, A. Elmasry, N. M. Dessouky, The therapeutic potential of intraoperative hypercapnia during video-assisted thoracoscopy in pediatric patients. *Anesth Analg* 106, 84-88, table of contents (2008).
69 Z. Wang, F. Su, A. Bruhn, X. Yang, J. L. Vincent, Acute hypercapnia improves indices of tissue oxygenation more than dobutamine in septic shock. *American journal of respiratory and critical care medicine* 177, 178-183 (2008).
70 Q. Zhou *et al.*, Effects of permissive hypercapnia on transient global cerebral ischemia-reperfusion injury in rats. *Anesthesiology* 112, 288-297 (2010).
71 C. C. Stubs *et al.*, Acute, short-term hypercapnia improves microvascular oxygenation of the colon in an animal model of sepsis. *Microvasc Res* 90, 180-186 (2013).
72 T. H. Stanley, T. D. Egan, H. Van Aken, A tribute to Dr. Paul A. J. Janssen: entrepreneur extraordinaire, innovative scientist, and significant contributor to anesthesiology. *Anesth Analg* 106, 451-462, table of contents (2008).
73 J. Hanawald, Rapid Gas Change During ECMO Linked to Mortality Risk. *Anesthesiology News* 37:8, (2011).
74 N. Fayed, P. J. Modrego, H. Morales, Evidence of brain damage after high-altitude climbing by means of magnetic resonance imaging. *Am J Med* 119, 168 e161-166 (2006).
75 J. H. Eisele, E. I. Eger, 2nd, M. Muallem, Narcotic properties of carbon dioxide in the dog. *Anesthesiology* 28, 856-865 (1967).
76 L. S. Coleman, Four Forgotten Giants of Anesthesia History. *Journal of Anesthesia and Surgery* 3, 1-17 (2015).

77 J. H. Eisele, E. I. Eger, 2nd, M. Muallem, Narcotic properties of carbon dioxide in the dog. *Anesthesiology* 28, 856-865 (1967).

78 L. S. Coleman, Four Forgotten Giants of Anesthesia History. *Journal of Anesthesia and Surgery* 3, 1-17 (2015).

79 L. S. Coleman, Should soda lime be abolished? *Anesth Analg* 102, 1290-1291 (2006).

80 L. S. Coleman, Soda lime is a dangerous relic of a forgotten past. *Anaesthesia* 70, (2016).

81 L. S. Coleman, 30 Years Lost in Anesthesia Theory. *Cardiovasc Hematol Agents Med Chem* 10, 31-49 (2012).

82 L. S. Coleman, in *apsf Newsletter*. (Anesthesia Patient Safety Foundation, Administrator, Deanna Walker Anesthesia Patient Safety Foundation Building One, Suite Two 8007 South Meridian Street Indianapolis, IN 46217-2922 e-mail address: walker@apsf.org FAX: (317) 888-1482, 2010), vol. Winter 2009-2020.

83 L. S. Coleman, A call for standards on perioperative CO(2) regulation. *Can J Anaesth*, (2011).

84 L. S. Coleman, Four Forgotten Giants of Anesthesia History. *Journal of Anesthesia and Surgery* 3, 1-17 (2015).

Chapter 15

Three Stress Mechanism Activation Pathways

"When humans have once acquired a superstition, nothing short of death is ever likely to remove it."
— Mark Twain

Three fluctuating, synergistic pathways activate the mammalian stress mechanism and focus its powerful effects to repair tissues and regulate blood flow and organ function. Stress mechanism activity is normally unobtrusive, efficient, and effective, but because the three pathways are synergistic, and because many forms of stress activate two or all three of the pathways in varying degrees, the sum of their activities can induce positive feedback that overwhelms negative feedback and produces harmful stress mechanism hyperactivity that manifests as disease. Such disease activity consumes and wastes energy, resources, and substrates; disrupts organ function; damages tissues; and generates a bewildering blizzard of disease symptoms and manifestations that vary in accord with the fluctuating activities of the three pathways and obscure the otherwise relatively straightforward operation of the mammalian stress mechanism. Since hyperactivity of the stress mechanism equates to disease, it follows that all three activation pathways must be controlled simultaneously to restore homeostasis and cure disease.

The *Tissue Factor Pathway* activates the tissue repair mechanism. The *Nociception Pathway* and the *Cognitive Pathway* are closely related nervous pathways that activate the capillary gate mechanism and the turbulence mechanism. The three pathways are redundant and synergistic; the activity of each pathway exaggerates that of the other two, so that stress mechanism activity causes fluctuating effects that are difficult to control by any single measure. All three pathways must be controlled simultaneously to restore effective stress mechanism function and cure disease, but excessive suppression is potentially lethal, for stress mechanism activity is essential to life. The three pathways are relevant to all types of medical care, and are particularly pertinent to critical care specialists, to cancer specialists, and to anesthetists who seek to optimize surgical outcome. The three pathways appear in detail Figure 15.1.

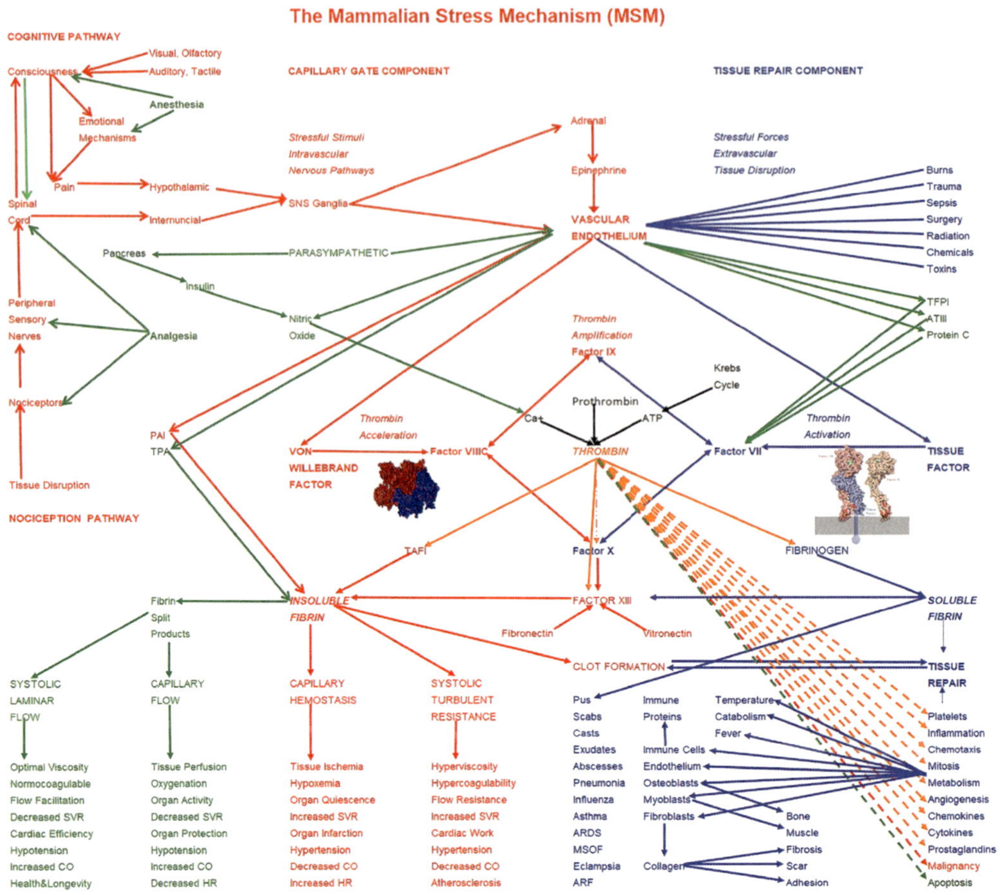

Figure 15.1: Mammalian stress mechanism (detailed).

A detailed diagram of the mammalian stress mechanism demonstrates the Nociception Pathway, the Cognitive Pathway, and the Tissue Disruption Pathway. The Capillary Gate Component and its cognitive and nociception activation pathways are shown in red. The Tissue Repair component and its tissue factor pathway are shown in blue. Thrombin energy pathways are shown in orange. Inhibitory pathways are shown in green. The vascular endothelium releases TFPI, ATIII and Protein C inhibit the Tissue Repair Component, and it releases nitric oxide to open the capillary gate. Anesthesia inhibits the cognitive pathway, and analgesia inhibits the nociception pathway.

Some definitions from the diagram:
PAI = Plasminogen Activator Inhibitor
PA = Tissue Plasminogen Activator
SVR = Systemic Vascular Resistance
CO = Cardiac Output
HR = Heart Rate
TFPI = Tissue Factor Pathway Inhibitor
ATIII = Stoichiometric Anti-Thrombin III
TFPI = Tissue Factor Pathway Inhibitor

THE TISSUE FACTOR PATHWAY

The Tissue Factor Pathway consists of the vascular endothelium, tissue factor, and factor VII. It activates the tissue repair component in accord with the magnitude and location of tissue damage that disrupts the vascular endothelium and exposes tissue factor to factor VII. This sounds simple, but the tissue disruption pathway is more nuanced than it first appears because of the *selective permeability* of the vascular endothelium and the disparate distribution of tissue factor in organs and tissues.

Tissue factor affects these issues:

- Tissue factor is concentrated lung, brain, bowel, nervous tissue, cervix, gonads, placenta, amniotic fluid, and peri-arterial tissues.[1] These tissues are more sensitive to stress than those lacking tissue factor. For example, the lung and the brain are "target organs" because they are rich in tissue factor. Thus, severe stress mechanism hyperactivity typically manifests as lung inflammation called ARDS (*Adult Respiratory Distress Syndrome*) that is soon followed by dementia (brain effects) that progresses to MOFS (*Multi-Organ Failure Syndrome*).

- Tissue factor activates factor VII, and factor VII activation is essential for the operation of factors VIII, IX, and X, so that the presence and concentration of tissue factor determines the location, rate, and magnitude of the blood enzyme interaction.[2]

- The intact vascular endothelium isolates blood enzymes from tissue factor in extravascular tissues. It is normally impermeable to factors VIII, IX, and X, but it allows the slow and continuous penetration of factor VII that enables tissue maintenance and the slow and continuous escape of tissue factor that enables background blood enzyme activity that sustains continuous capillary gate function.

- Nervous activity increases the permeability of the vascular endothelium.[3]

- Complement cascade enzymes (see Figure 15.3 on page 160) generate thrombin to energize their attack on bacteria and foreign substances that invade the bloodstream, so that complement cascade activity exaggerates stress mechanism activity, and vice-versa.[4, 5, 6]

- Thrombin induces *gap formation* between the cells of the vascular endothelium (see Figure 15.2 on page 159), which increases its permeability and allows greater penetration of soluble fibrin and factor X as well as the escape of tissue factor into systemic circulation.[7, 8, 9]

- Radiation, sepsis, and toxic chemicals damage the vascular endothelium, increase its permeability, and exaggerate the leakage of tissue factor and penetration of factors VII and X, causing systemic and local positive feedback that reaches peak effect in about 48 hours. This explains why sunburn pain occurs hours after excessive sunlight exposure, and why it causes systemic symptoms such as fever and fatigue.[10]

- Tissue damage releases tissue factor into systemic circulation, which activates blood enzymes, causes systemic inflammation, and increases blood viscosity and coagulability.

- Trauma and surgery directly expose tissue factor to blood enzymes, causing an intense enzymatic interaction that rapidly generates thrombin, soluble fibrin, and insoluble fibrin to enable coagulation (clot formation).

- The viscoelastic clot substitutes for the vascular endothelium and restores the barrier between blood enzymes and damaged tissues.

- Like the vascular endothelium, the viscoelastic clot exhibits *selective permeability.* It allows penetration of factors VII and X, which interact with exposed tissue factor in damaged tissues beneath the protective surface of the clot.

- Factor VIII cannot penetrate a clot of its own manufacture on account of its gigantic size, so it does not participate in tissue repair other than by enabling clot formation. This limits clot formation to the site of tissue damage.

- Factor IX is a co-factor of factor VIII and exerts no activity in the absence of factor VIII, so it also does not participate in tissue repair.

- Amniotic fluid is rich in tissue factor that is shed from the gut of the fetus. It is normally isolated within the amniotic sac, but it drastically exaggerates blood enzyme activity if it enters maternal circulation via *amniotic fluid embolus*.

- The placenta is likewise rich in tissue factor that can harmfully activate blood enzymes if it enters maternal circulation.

- During pregnancy, placental tissue factor leakage, amniotic fluid embolus, and cervical trauma can release abnormal quantities of tissue factor into systemic circulation that precipitate eclampsia, which is analogous to Multi-Organ Failure Syndrome (MOFS).

- Trauma disrupts the vascular endothelium and exposes tissue factor to blood enzymes, triggering hemostasis, followed by tissue repair activity.

- Major trauma releases excessive tissue factor into systemic circulation, causing systemic thrombin generation, inflammation, exaggerated blood viscosity and coagulability, infarction, tissue edema, and organ failure.

- Brain trauma exaggerates morbidity and mortality because the brain is rich in tissue factor, so that it releases more tissue factor into systemic circulation than other tissues.

- Surgery releases tissue factor into systemic circulation, causing symptoms and manifestations distant from the site of tissue disruption that reach peak intensity about 48 hours later. This is known as the surgical stress syndrome (SSS).[11]

- At present there is no clinically available treatment that can directly inhibit the interaction of factor VII with tissue factor, and inhibit the tissue disruption pathway, but experimental evidence suggests that safe tissue factor antidotes are possible.[12]

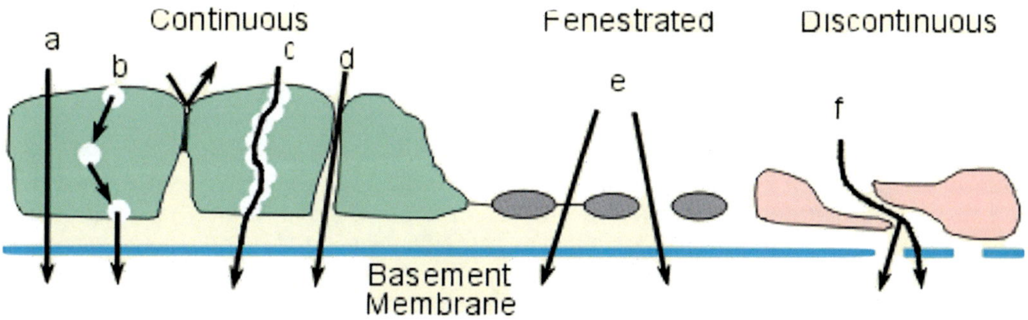

Figure 15.2: Openings in the vascular endothelium.

The vascular endothelium governs contact between blood enzymes and tissue factor in extravascular tissues. Continuous (intact) vascular endothelium allows slow penetration of factor VII (and escape of tissue factor). Toxic chemicals and radiation damage the vascular endothelium, causing it to become fenestrated so that it allows passage of soluble fibrin and factor X. This explains sunburn. Trauma and surgery cause discontinuous (disrupted) vascular endothelium that directly exposes tissue factor to all blood enzymes.

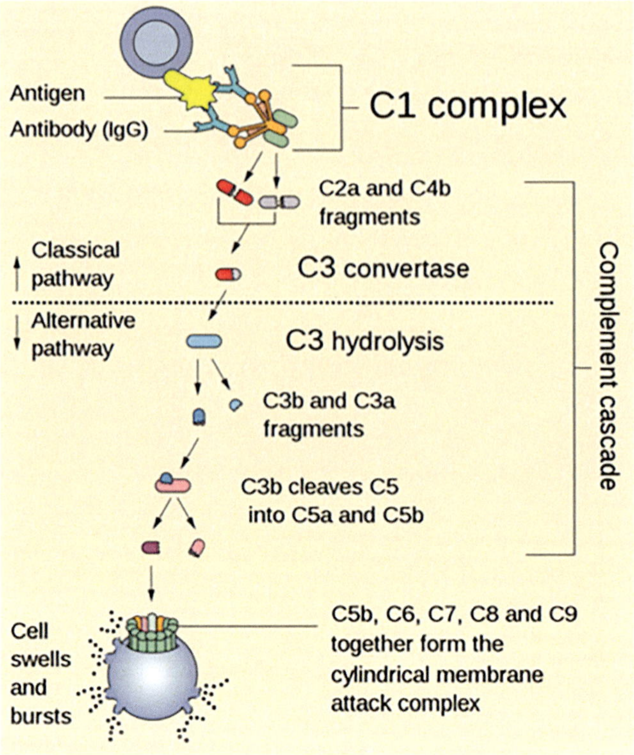

Figure 15.3: Complement cascade.

Both the complement cascade and the stress mechanism generate thrombin, so that the activity of each exaggerates that of the other to generate positive feedback. The complement cascade consists of blood enzymes that react to bacterial invasion of blood. The cascade enzymes generate thrombin and energizes immune activity to attack bacteria and foreign substances that invade the bloodstream. This explains the critical illness called *sepsis*. By English text of 'Image: Complement pathway.png' by DO11.10German translation of 'Image: Complement pathway.png' by Hduman Galician translation Miguel ferig Catalan translation Leptictidium SVG by Perhelion - Own work based on: http://www.niaid.nih.gov/publications/immune/the_immune_system.pdf, Public Domain, https://commons.wikimedia.org/w/index.php?curid=45352985

THE NOCICEPTION PATHWAY

The nociception pathway consists of *nociceptors* in the skin and internal organs and peripheral sensory nerves that convey nociceptor signals to spinal cord nociception pathways. Nociceptor signals are called *nociception*. They detect chemical, thermal, and mechanical stresses that threaten or cause tissue damage. Inflammation exaggerates their sensitivity, but they cannot directly detect harmful radiation, sepsis, and many toxic chemicals.

The spinal cord nociception pathways communicate with the cerebral cortex, where consciousness interprets nociception as *pain*. Simultaneously, the spinal cord nociception pathways convey nociception to sympathetic ganglia in the chest and abdomen that innervate the vascular endothelium of internal organs, causing the release of von Willebrand Factor that closes the capillary gate and inhibits organ function. *The spinal cord nociception pathway continues to activate sympathetic ganglia even in the absence of brain function, whether caused by brain damage or anesthesia.*

Analgesia inhibits nociception but has little effect on the cognitive pathway (see "The Melzack and Wall Gate Theory of Pain"). There are three types of analgesic agents:

1. Non-Steroidal Anti-Inflammatory Drugs (NSAIDs) such as aspirin, ibuprofen (Advil), and acetaminophen (Tylenol) that neutralize nociceptors.
2. Narcotics that inhibit spinal cord nociception pathways.
3. Local analgesics such as lidocaine, bupivacaine, and tetracaine that inhibit nociceptors, peripheral sensory nerve pathways, and spinal cord pathways.

The term *local anesthetic* is often applied to analgesic agents such as lidocaine, Marcaine, and tetracaine that inhibit nociception, but the term is a misnomer. These agents should be called *local analgesics*. Analgesics are not anesthetics. They prevent the communication of nociception to the brain, which prevents consciousness from interpreting nociception as pain, but they do not inhibit consciousness or brain function, and therefore lack anesthetic effects.

THE MELZACK AND WALL GATE THEORY OF PAIN

In 1955 Ronald Melzack and Patrick Wall hypothesized that consciousness inhibits spinal cord nociception pathways via descending inhibitory nervous pathways from the brain to the spinal cord.[13] This enables the brain to modulate pain perception. The classic example is that of a soldier who is severely wounded during combat but perceives no pain until the danger dissipates, whereupon his pain becomes agonizing. This has numerous implications.

Spinal cord damage around the level of T5 disrupts ascending nociception pathways that communicate nociception to the cerebral cortex, and simultaneously disrupts descending pathways from the cortex that inhibit spinal cord nociception pathways. This results in a condition called *autonomic dysreflexia* wherein analgesia is necessary to prevent exaggerated nociception that dangerously elevates sympathetic nervous activity during surgery, even though the patient cannot perceive surgical pain.[14]

Analgesia minimizes surgical morbidity and mortality more effectively than anesthesia, because analgesia directly inhibits nociception, while anesthesia inhibits only the conscious perception of nociception as pain (see "The Cognitive Pathway" below), while simultaneously paralyzing the descending pathways from brain that inhibit spinal cord nociception. This explains why supplementing general anesthesia with analgesic opioids and analgesic blocks improves surgical outcome.[15, 16, 17, 18, 19]

Fetuses and newborn babies require intense analgesia to prevent potentially lethal surgical nociception, even though they cannot understand language and danger, and therefore lack fear and apprehension.[20, 21, 22, 23]

General anesthesia abolishes consciousness, which paralyzes the descending cortical pathways that inhibit spinal cord nociception, so that harmful spinal cord nociception is exaggerated during surgery. Such spinal cord hyperactivity persists in the aftermath of surgery and is called *windup syndrome*.[24, 25, 26] This explains why general anesthesia must be supplemented with analgesia in order to optimize surgical outcome.

THE COGNITIVE PATHWAY

Alcohol is the anesthesia by which we endure the operation of life.
— George Bernard Shaw

The cognitive pathway consists of the following elements:

- Corticofugal mechanisms in the cerebral cortex that generate consciousness.
- Nervous pathways that convey sensory information to cortical consciousness, which interprets auditory signals as sound, visual signals as sight, olfactory signals as smell, and nociception signals as pain.
- Consciousness evaluates sensory information for signs of danger.
- Consciousness elevates respiratory drive and releases HPA hormones in response to peripheral nociception.
- Consciousness regulates spinal cord nociception pathways via descending nervous pathways from the brain to spinal cord. Depending on circumstances, consciousness can either inhibit or exaggerate spinal cord nociception activity.
- Emotional mechanisms in the amygdala work closely with consciousness to refine the ability to detect and evaluate danger.
- The emotional mechanism maintains a complete audiovisual memory of all waking moments.
- A dreaming mechanism reviews and updates the audiovisual memory record during sleep to identify dangerous circumstances.
- Emotional memory and dreaming mechanisms work closely with consciousness to pre-emptively detect dangerous circumstances and automatically induce anxiety, fear, rage that release HPA hormones and elevate sympathetic nervous activity to facilitate fight or flight.

CONSCIOUSNESS

Consciousness is presently described as "awareness of internal or external existence" but its nature remains one of the great mysteries of medicine and biology despite centuries of philosophical debate and modern scientific scrutiny. I hypothesize that consciousness is a simulation of environmental surroundings created by the brain using combinations of sensory input and stored memory.

MEMORY

The memory mechanism of the brain was discovered in the 1950s by Dr. Wilder Penfield, a neurosurgeon who was searching for a surgical solution for epilepsy.[27] At that time epilepsy was a frightening and mysterious phenomenon, but today it is understood to be caused in most cases by brain scar formation that disturbs brain electrical activity. Using only local anesthesia (the brain itself is devoid of sensation) he stimulated various parts of the brain with tiny electrodes and elicited distant memories along with sensations and motor activity. This enabled him to produce a so-called *cortical homunculus* — a map of brain functions. However, the ability of the brain to retain distant audiovisual memory was not fully appreciated until a patient named Jill Price consulted neurobiologists at U.C. Irvine. She complained that her ability to function was distracted by distant memories.[28] The researchers confirmed her condition and called it *Hyperthymestic Syndrome*.[29] It soon became apparent that the condition is fairly common. It appears to be caused by inadequate suppression of distant memories. Mild cases probably account for "photographic memory," as in the case of Marilu Henner, an actress who memorizes her scripts with ease. By implication, all vertebrates possess this memory mechanism, but its nature remains mysterious.

THE DREAMING MECHANISM

I hypothesize that during sleep — which quenches consciousness and halts memory accumulation — a *dreaming mechanism* automatically reviews retained memory records to identify dangerous environmental circumstances. During this dreaming process motor activity is normally disconnected from brain activity to prevent untoward movements. Sometimes such untoward movements are inadequately suppressed, causing sleepwalking, restless leg syndrome, periodic limb movement disorder, sleep bruxism (tooth-grinding), and sleep talking (somniloquy). Sleep paralysis can also manifest when falling asleep or waking up.

THE EMOTIONAL MECHANISM

Animals lack the reasoning capability of humans, so they rely primarily on their emotional mechanism, which enhances survival by automatically detecting danger, increasing sympathetic nervous activity, and releasing HPA hormones, and inducing fear, anxiety, and rage to facilitate fight or flight.

I hypothesize that the emotional mechanism consists of the memory mechanism and the dreaming mechanism. The dreaming mechanism continually re-reviews and updates its assessment of memory records during sleep. This enables animals to adapt their behavior to changing environmental circumstances over time. During wakeful periods, the emotional mechanism works closely with consciousness to automatically detect dangerous circumstances previously identified by the dreaming mechanism, whereupon it pre-emptively induces anxiety, fear, and rage that elevate sympathetic nervous activity and release HPA hormones to facilitate impending fight or flight. VWF and epinephrine close the capillary gate and concentrate perfusion in critical organs and tissues. They also increase blood coagulability to reduce blood loss in the event of injury. Glucagon releases glucose from glycogen stores in the liver to provide cellular energy. Consciousness simultaneously inhibits spinal cord nociception via descending pathways from the brain to prevent distracting pain that might hinder survival.[30] These effects facilitate fight or flight.[31, 32, 33]

The re-introduction of wolves into Yellowstone Park after many years of absence illustrates how the emotional mechanism facilitates survival. The caribou population was familiar with harmless coyotes, and initially ignored the wolves, but wolf attacks quickly altered their behavior so that they ceased to linger near rivers and vegetation where they could be ambushed, and preferred open spaces where they could easily detect impending attacks.

The cognitive pathway synergizes nociception pathway activity, and vice versa, but the two pathways function independently, synergystically, and redundantly. The cognitive pathway elevates sympathetic nervous system (SNS) activity despite the absence of spinal cord nociception, and the spinal cord nociception pathway elevates SNS activity in the absence of consciousness.

Emotional memories can be so vivid as to mimic reality. Emotional memories normally remain subconscious, but inadequate suppression of emotional memory causes Hyperthymestic Syndrome, wherein distracting memories can undermine the ability to function.[34] Mild cases explain photographic memory.

ANESTHESIA

Anesthesia inhibits the cognitive pathway. Inhalation anesthetics and sedatives are hypnotic agents that inhibit consciousness in a dose-related manner, and thereby prevent sympathetic nervous system activity due to anxiety and fear.[35] However, hypnotics lack the ability to inhibit spinal cord nociception pathways that induce sympathetic nervous system hyperactivity in the absence of consciousness. For example, surgical tissue disruption activates the nociception pathway despite general anesthesia, causing tachycardia, hypertension, dysrhythmias, and spontaneous hyperventilation.[36] Furthermore, anesthesia paralyzes descending pathways from the brain to the spinal cord that normally enable consciousness to inhibit the nociception pathway.

Volatile inhalation anesthetics are more practical than intravenous and oral sedatives such as alcohol or benzodiazepines because they can be introduced and eliminated quickly and maintained at precise levels. Their effects are otherwise equivalent to oral, intramuscular, and intravenous hypnotic pharmaceuticals.

The following examples illustrate cognitive pathway activity:

- Painful memories that are retained and exaggerated by the emotional mechanism can mimic reality, causing the perception of non-existent pain and danger, as in PTSD (Post-Traumatic Stress Disorder).

- We know that the cognitive pathway resists analgesia in clinically practical doses because the perception of sight, smell, vibration, and sound remains intact despite the absence of nociception. Spinal and epidural analgesia, analgesic block techniques, and high-dose narcotic analgesia for cardiac surgery often require supplementation with hypnotic agents to inhibit consciousness and prevent sharp increases in blood pressure, pulse rate and muscle activity caused by frightening sounds and sensations, even though nociception and pain are absent.[37]

- Anesthesia increases surgical safety by abolishing fear, apprehension, and the perception of nociception as pain, but it abolishes descending cortical inhibition of spinal cord nociception pathways, and thereby indirectly exaggerates harmful spinal cord nociception pathway activity.[38, 39]

- The intense emotional trauma of earthquake exposure activates the cognitive pathway, which sharply elevates blood viscosity and coagulability that invites myocardial infarction, stroke, and sudden death. Residual stress mechanism hyperactivity reduces life span in the distant aftermath of such emotional trauma, despite the absence of injury, pain, or nociception.

- Chronic emotional stress such as job difficulties elevates VWF and Factor VIII activity, accelerates atherosclerosis, and undermines life span.[40]

- Sedatives inhibit emotional stress.[41]

- The addition of analgesia to general anesthesia during surgery prevents *emergence infarction* by inhibiting surgical nociception that is perceived as intense pain with the restoration of consciousness.[42, 43]

POSITIVE FEEDBACK AND THE THREE ACTIVATION PATHWAYS

The three stress mechanism activation pathways share the enzymatic interaction of factors VII, VIII, IX, and X so that the activity of each pathway exaggerates that of the others, and their fluctuating activities cause chaotic manifestations that belie the relative simplicity of the mammalian stress mechanism. Normally their synergism focuses stress mechanism activity to repair tissues and regulate hemodynamic physiology, but acute and/or unremitting stress produces potentially lethal positive feedback that overwhelms negative feedback and causes stress mechanism hyperactivity that manifests as disease.

Emotional memory and spinal cord wind-up prolong factor VIII elevations and cause harmful stress mechanism hyperactivity to linger long after stressors subside. For example, unscathed earthquake survivors exhibit elevated VWF, factor VIII activity, blood viscosity, blood coagulability and exaggerated morbidity and mortality in the distant aftermath of earthquake exposure.[44, 45, 46, 47, 48, 49, 50]

If stressors subside, negative feedback restores stress mechanism activity to maintenance levels. Tissue repair restores the vascular endothelium and reduces exposure of tissue factor to blood enzymes to normal levels. Stoichiometric ATIII, Tissue Factor Pathway Inhibitor (TFPI), Tissue Plasminogen Activator (TPA), and protein C are released by the vascular endothelium to inhibit the interaction of factor VII with tissue factor and factor X.[51, 52, 53, 54, 55, 56, 57, 58, 59, 60, 61, 62] Elevated immune activity eliminates the stress of microbial invasion. Parasympathetic nervous activity releases NO, accelerates the disintegration of insoluble fibrin, opens the capillary gate, and restores tissue perfusion.[63, 64, 65, 66, 67, 68] Thrombin generation declines, inflammation subsides, and homeostasis is restored.

TREATMENTS THAT INHIBIT THE STRESS MECHANISM ACTIVATION PATHWAYS

Anesthesia (hypnosis) inhibits the cognitive pathway. *Analgesia* inhibits the nociception pathway. Synergistic combinations of anesthesia and analgesia control hyperactivity of the mammalian stress mechanism more effectively than either anesthesia or analgesia alone, but neither anesthesia nor analgesia inhibits the tissue factor pathway.

The vascular endothelium produces Tissue Factor Pathway Inhibitor (TFPI), stoichiometric ATIII, and Protein C that inhibit the interaction of factor VII with tissue factor. This normally inhibits tissue repair pathway activity, but at this time there is no medical treatment available that can safely inhibit the tissue factor pathway. However, in situations where hemostasis is assured, simple treatments with magnesium sulfate ($MgSO_4$),[69, 70, 71] EDTA (Ethylenediaminetetraacetic acid),[72, 73, 74, 75] or trisodium citrate[76, 77] can temporarily inhibit thrombin activity and synergistically complement the effects of anesthesia and analgesia to control harmful stress mechanism hyperactivity. All these chemicals have been used to prevent coagulation in stored blood, and all three have therapeutic properties and are relatively safe for human exposure. $MgSO_4$ is routinely used to treat eclampsia. EDTA is widely used for chelation therapy. Trisodium citrate has been used to prevent coagulation during dialysis. However, their anticoagulant properties obviously invite bleeding, so they are incompatible with surgery.

Permissive hypercarbia added to anesthesia, analgesia, and thrombin suppression can offset the respiratory depression caused by narcotics, and optimize cardiopulmonary function, tissue oxygenation, and antibiotic potency.[78, 79, 80, 81, 82, 83, 84, 85, 86, 87, 88, 89] Such combinations, used with elective endotracheal intubation and modern monitoring, can theoretically revolutionize the treatment of critical illnesses, including cancer, asthma, pneumonia, influenza, Legionnaire's Disease, SIRS, SARS, Ebola, DIC, eclampsia, ARDS, necrotizing fasciitis, Lyme Disease, Tuberculosis, MRSA, and other recalcitrant infections.

RESPIRATORY DRIVE

Mammalian respiratory drive is surprisingly tenuous. Specialized pH nociceptors in brain ventricles stimulate breathing in accord with cerebrospinal fluid acid levels, which fluctuate in accord with CO_2 concentrations in blood. This form of respiratory drive resists hyperventilation and hypocarbia but is dependent on consciousness and is thus inactivated by either general anesthesia or normal sleep.

Specialized nociceptors called *respiratory chemoreceptors* detect hypoxia and hypercarbia during sleep and general anesthesia, when consciousness is absent, but they are readily paralyzed by hypocarbia, react weakly to hypoxia, and are inhibited by narcotics.

Nociception stimulates respiratory drive, and surgical nociception can cause hyperventilation that depletes CO_2 tissue reserves during general anesthesia, causing unpredictable postoperative respiratory arrest. Mechanical hyperventilation during surgery causes the same phenomenon.

Combinations of narcotics and sedatives are extremely dangerous because they simultaneously inhibit both the respiratory drive of consciousness and the respiratory drive of chemoreceptors. This probably causes most recreational drug deaths.

NOTES

1 R. A. Fleck, L. V. Rao, S. I. Rapaport, N. Varki, Localization of human tissue factor antigen by immunostaining with monospecific, polyclonal anti-human tissue factor antibody. *Thromb Res* 59, 421-437 (1990).

2 H. J. Weiss, V. T. Turitto, H. R. Baumgartner, Y. Nemerson, T. Hoffmann, Evidence for the presence of tissue factor activity on subendothelium. *Blood* 73, 968-975 (1989).

3 G. Burnstock, V. Ralevic, New insights into the local regulation of blood flow by perivascular nerves and endothelium. *Br J Plast Surg* 47, 527-543 (1994).

4 Y. Gong, G. H. Xi, R. F. Keep, J. T. Hoff, Y. Hua, Complement inhibition attenuates brain edema and neurological deficits induced by thrombin. *Acta Neurochir Suppl* 95, 389-392 (2005).

5 A. P. Kaplan, M. Silverberg, J. T. Dunn, B. Ghebrehiwet, Interaction of the clotting, kinin-forming, complement, and fibrinolytic pathways in inflammation. *Annals of the New York Academy of Sciences* 389, 25-38 (1982).

6 B. Ghebrehiwet, B. P. Randazzo, J. T. Dunn, M. Silverberg, A. P. Kaplan, Mechanisms of activation of the classical pathway of complement by Hageman factor fragment. *The Journal of clinical investigation* 71, 1450-1456 (1983).

7 J. G. Garcia, F. M. Pavalko, C. E. Patterson, Vascular endothelial cell activation and permeability responses to thrombin. *Blood Coagul Fibrinolysis* 6, 609-626 (1995).

8 J. G. Garcia *et al.*, Regulation of endothelial cell myosin light chain kinase by Rho, cortactin, and p60(src). *The American journal of physiology* 276, L989-998 (1999).

9 V. Stvrtinova, M. Ferencik, I. Hulin, E. Jahnova, [Vascular endothelium as a factor in information transfer between the cardiovascular and immune systems]. *Bratisl Lek Listy* 99, 5-19 (1998).

10 M. Greaves, The Pharmacology of sunburn erythema. *Journal of Dermatological Treatment* 2:1, 35-36 (1991).

11 C. C. Finnerty, N. T. Mabvuure, A. Ali, R. A. Kozar, D. N. Herndon, The surgically induced stress response. *JPEN J Parenter Enteral Nutr* 37, 21S-29S (2013).

12 K. E. Welty-Wolf *et al.*, Coagulation blockade prevents sepsis-induced respiratory and renal failure in baboons. *American journal of respiratory and critical care medicine* 164, 1988-1996 (2001).

13 R. Melzack, P. D. Wall, Pain mechanisms: a new theory. *Science* 150, 971-979 (1965).

14 C. P. Ho, A. V. Krassioukov, Autonomic dysreflexia and myocardial ischemia. *Spinal Cord* 48, 714-715 (2010).

15 L. W. Crile GW, *Anoci-association*. (Saunders, Philadelphia, 1914), pp. 223-225.

16 K. Giesecke, B. Hamberger, P. O. Jarnberg, C. Klingstedt, B. Persson, High- and low-dose fentanyl anaesthesia: hormonal and metabolic responses during cholecystectomy. *Br J Anaesth* 61, 575-582 (1988).

17 A. I. McEwan *et al.*, Isoflurane minimum alveolar concentration reduction by fentanyl. *Anesthesiology* 78, 864-869 (1993).

18 P. S. Myles *et al.*, Remifentanil, fentanyl, and cardiac surgery: a double-blinded, randomized, controlled trial of costs and outcomes. *Anesth Analg* 95, 805-812, table of contents (2002).

19 https://en.wikipedia.org/wiki/Autonomic_dysreflexia

20 K. J. Anand, D. D. Hansen, P. R. Hickey, Hormonal-metabolic stress responses in neonates undergoing cardiac surgery. *Anesthesiology* 73, 661-670 (1990).

21 K. J. Anand, P. R. Hickey, Halothane-morphine compared with high-dose sufentanil for anesthesia and postoperative analgesia in neonatal cardiac surgery. *N Engl J Med* 326, 1-9 (1992).

22 K. J. Anand, M. Maze, Fetuses, fentanyl, and the stress response: signals from the beginnings of pain? *Anesthesiology* 95, 823-825 (2001).

23 K. J. Anand, W. G. Sippell, A. Aynsley-Green, Randomised trial of fentanyl anaesthesia in preterm babies undergoing surgery: effects on the stress response. *Lancet* 1, 243-248 (1987).

24 J. F. Herrero, J. M. Laird, J. A. Lopez-Garcia, Wind-up of spinal cord neurones and pain sensation: much ado about something? *Prog Neurobiol* 61, 169-203 (2000).

25 R. D'Mello, A. H. Dickenson, Spinal cord mechanisms of pain. *Br J Anaesth* 101, 8-16 (2008).

26 https://www.vin.com/apputil/content/defaultadv1.aspx?id=5709881&pid=11372&print=1

27 W. Penfield, Gordon Wilson Lecture, The Mechanism of Memory. *Trans Am Clin Climatol Assoc* 62, 165-169 (1950).

28 J. Price, *The Woman Who Can't Forget: The Extraordinary Story of Living with the Most Remarkable Memory Known to Science — A Memoir.* (Free Press, a division of Simon and Schuster, 2008), pp. 263.

29 Wickipedia, Hyperthymestic Syndrome.

30 R. Melzack, P. D. Wall, Pain mechanisms: a new theory. *Science* 150, 971-979 (1965).

31 W. B. Cannon, The emergncy function of the adrenal medulla in pain and the major emotions. *The American journal of physiology* 33, 356-372 (1914).

32 W. B. Cannon, *Bodily Changes in Pain, Hunger, Fear and Rage: An Account of Recent Researches into the Function of Emotional excitement.* (Appleton, New York, 1915).

33 W. B. Cannon, *The wisdom of the body.* (W.W. Norton & Company, New York,, 1939), pp. xviii, 19-333 p.

34 J. Price, *The Woman Who Can't Forget: The Extraordinary Story of Living with the Most Remarkable Memory Known to Science — A Memoir.* (Free Press, a division of Simon and Schuster, 2008), pp. 263.

35 B. F. Rusy, H. Komai, Anesthetic depression of myocardial contractility: a review of possible mechanisms. *Anesthesiology* 67, 745-766 (1987).

36 L. S. Coleman, Four Forgotten Giants of Anesthesia History. *Journal of Anesthesia and Surgery* 3, 1-17 (2015).

37 K. C. Wong, Narcotics are not expected to produce unconsciousness and amnesia. *Anesth Analg* 62, 625-626 (1983).

38 R. Melzack, P. D. Wall, Pain mechanisms: a new theory. *Science* 150, 971-979 (1965).

39 T. J. Coderre, R. Melzack, Cutaneous hyperalgesia: contributions of the peripheral and central nervous systems to the increase in pain sensitivity after injury. *Brain Res* 404, 95-106 (1987).
40 S. J. Chang, S. B. Koh, B. S. Cha, J. K. Park, Job characteristics and blood coagulation factors in Korean male workers. *J Occup Environ Med* 44, 997-1002 (2002).
41 V. Dorges *et al.*, The effect of midazolam on stress levels during simulated emergency medical service transport: a placebo-controlled, dose-response study. *Anesth Analg* 95, 417-422, table of contents (2002).
42 W. S. Beattie, N. H. Badner, P. Choi, Epidural analgesia reduces postoperative myocardial infarction: a meta-analysis. *Anesth Analg* 93, 853-858 (2001).
43 J. P. Cravero, M. Beach, B. Thyr, K. Whalen, The effect of small dose fentanyl on the emergence characteristics of pediatric patients after sevoflurane anesthesia without surgery. *Anesth Analg* 97, 364-367, table of contents (2003).
44 K. Kario, T. Matsuo, Increased incidence of cardiovascular attacks in the epicenter just after the Hanshin-Awaji earthquake. *Thromb Haemost* 74, 1207 (1995).
45 K. Kario *et al.*, Earthquake-induced cardiovascular disease and related risk factors in focusing on the Great Hanshin-Awaji Earthquake. *J Epidemiol* 8, 131-139 (1998).
46 K. Kario *et al.*, Factor VII hyperactivity and endothelial cell damage are found in elderly hypertensives only when concomitant with microalbuminuria. *Arterioscler Thromb Vasc Biol* 16, 455-461 (1996).
47 K. Kario, T. Matsuo, H. Kobayashi, K. Yamamoto, K. Shimada, Earthquake-induced potentiation of acute risk factors in hypertensive elderly patients: possible triggering of cardiovascular events after a major earthquake. *J Am Coll Cardiol* 29, 926-933 (1997).
48 K. Kario, B. S. McEwen, T. G. Pickering, Disasters and the heart: a review of the effects of earthquake-induced stress on cardiovascular disease. *Hypertens Res* 26, 355-367 (2003).
49 T. Matsuo, S. Suzuki, K. Kario, H. Kobayashi, [Acute myocardial infarction in the 1995 Hanshin-Awaji Earthquake]. *Rinsho Byori* Suppl 104, 133-141 (1997).
50 T. Matsuo, S. Suzuki, K. Kodama, K. Kario, Hemostatic activation and cardiac events after the 1995 Hanshin-Awaji earthquake. *Int J Hematol* 67, 123-129 (1998).
51 J. H. Griffin, J. A. Fernandez, A. J. Gale, L. O. Mosnier, Activated protein C. *J Thromb Haemost* 5 Suppl 1, 73-80 (2007).
52 L. O. Mosnier, X. V. Yang, J. H. Griffin, Activated protein C mutant with minimal anticoagulant activity, normal cytoprotective activity, and preservation of thrombin activable fibrinolysis inhibitor-dependent cytoprotective functions. *J Biol Chem* 282, 33022-33033 (2007).
53 L. Robriquet *et al.*, Intravenous administration of activated protein C in Pseudomonas-induced lung injury: impact on lung fluid balance and the inflammatory response. *Respir Res* 7, 41 (2006).

54 F. B. Taylor, Jr. *et al.*, Protein C prevents the coagulopathic and lethal effects of Escherichia coli infusion in the baboon. *The Journal of clinical investigation* 79, 918-925 (1987).

55 P. Audu *et al.*, The impact of tissue factor pathway inhibitor on coagulation kinetics determined by thrombelastography. *Anesth Analg* 103, 841-845 (2006).

56 M. S. Bajaj, J. J. Birktoft, S. A. Steer, S. P. Bajaj, Structure and biology of tissue factor pathway inhibitor. *Thromb Haemost* 86, 959-972 (2001).

57 M. S. Bajaj, U. Pendurthi, K. Koenig, S. Pueblitz, S. Idell, Tissue factor pathway inhibitor expression by human pleural mesothelial and mesothelioma cells. *The European respiratory journal* 15, 1069-1078 (2000).

58 S. Sajadi, M. D. Ezekowitz, A. Dhond, P. Netrebko, Tissue factor pathway inhibitors as a novel approach to antithrombotic therapy. *Drug News Perspect* 16, 363-369 (2003).

59 K. Akassoglou, K. W. Kombrinck, J. L. Degen, S. Strickland, Tissue plasminogen activator-mediated fibrinolysis protects against axonal degeneration and demyelination after sciatic nerve injury. *J Cell Biol* 149, 1157-1166 (2000).

60 Z. Hao *et al.*, Stimulated release of tissue plasminogen activator from artery wall sympathetic nerves: implications for stress-associated wall damage. *Stress* 8, 141-149 (2005).

61 D. Hevey, H. M. McGee, D. Fitzgerald, J. H. Horgan, Acute psychological stress decreases plasma tissue plasminogen activator (tPA) and tissue plasminogen activator/plasminogen activator inhibitor-1 (tPA/PAI-1) complexes in cardiac patients. *Eur J Appl Physiol* 83, 344-348 (2000).

62 B. V. Zlokovic *et al.*, Expression of tissue plasminogen activator in cerebral capillaries: possible fibrinolytic function of the blood-brain barrier. *Neurosurgery* 37, 955-961 (1995).

63 O. D. Altland, D. Dalecki, V. N. Suchkova, C. W. Francis, Low-intensity ultrasound increases endothelial cell nitric oxide synthase activity and nitric oxide synthesis. *J Thromb Haemost* 2, 637-643 (2004).

64 J. W. Coleman, Nitric oxide in immunity and inflammation. *Int Immunopharmacol* 1, 1397-1406 (2001).

65 K. Ghimire, H. M. Altmann, A. C. Straub, J. S. Isenberg, Nitric oxide: what's new to NO? *Am J Physiol Cell Physiol* 312, C254-C262 (2017).

66 A. G. Hudetz, H. Shen, J. P. Kampine, Nitric oxide from neuronal NOS plays critical role in cerebral capillary flow response to hypoxia. *The American journal of physiology* 274, H982-989 (1998).

67 O. J. Kirkeby, S. Kutzsche, C. Risoe, I. R. Rise, Cerebral nitric oxide concentration and microcirculation during hypercapnia, hypoxia, and high intracranial pressure in pigs. *J Clin Neurosci* 7, 531-538 (2000).

68 M. J. Martin, M. D. Jimenez, V. Motilva, New issues about nitric oxide and its effects on the gastrointestinal tract. *Curr Pharm Des* 7, 881-908 (2001).

69 N. M. Elsharnouby, M. M. Elsharnouby, Magnesium sulphate as a technique of hypotensive anaesthesia. *Br J Anaesth* 96, 727-731 (2006).

70 H. Kondo, E. Kobayashi, T. Itani, N. Tatsumi, I. Tsuda, Hematology tests of blood anticoagulated with magnesium sulphate. *Southeast Asian J Trop Med Public Health* 33 Suppl 2, 6-9 (2002).
71 D. Memis, A. Turan, B. Karamanlioglu, N. Oguzhan, Z. Pamukcu, Comparison of sufentanil with sufentanil plus magnesium sulphate for sedation in the intensive care unit using bispectral index. *Crit Care* 7, R123-128 (2003).
72 E. Escolar *et al.*, The Effect of an EDTA-based Chelation Regimen on Patients With Diabetes Mellitus and Prior Myocardial Infarction in the Trial to Assess Chelation Therapy (TACT). *Circulation. Cardiovascular quality and outcomes*, (2013).
73 S. Finnegan, S. L. Percival, EDTA: An Antimicrobial and Antibiofilm Agent for Use in Wound Care. *Adv Wound Care (New Rochelle)* 4, 415-421 (2015).
74 L. J. Leipzig, A. J. Boyle, D. S. McCann, Case histories of rheumatoid arthritis treated with sodium or magnesium EDTA. *J Chronic Dis* 22, 553-563 (1970).
75 R. E. Wooley, M. S. Jones, Action of EDTA-Tris and antimicrobial agent combinations on selected pathogenic bacteria. *Veterinary microbiology* 8, 271-280 (1983).
76 D. A. Pierce, M. V. Rocco, Trisodium citrate: an alternative to unfractionated heparin for hemodialysis catheter dwells. *Pharmacotherapy* 30, 1150-1158 (2010).
77 J. H. von Brecht, M. J. Flanigan, R. M. Freeman, V. S. Lim, Regional anticoagulation: hemodialysis with hypertonic trisodium citrate. *American journal of kidney diseases : the official journal of the National Kidney Foundation* 8, 196-201 (1986).
78 O. Akca, Hypercapnia: what is the upper limit? *Paediatric anaesthesia* 15, 80-81 (2005).
79 O. Akca, Carbon dioxide and tissue oxygenation: is there sufficient evidence to support application of hypercapnia for hemodynamic stability and better tissue perfusion in sepsis? *Intensive Care Med* 34, 1752-1754 (2008).
80 O. Akca *et al.*, Hypercapnia improves tissue oxygenation. *Anesthesiology* 97, 801-806 (2002).
81 O. Akca *et al.*, Tissue oxygenation response to mild hypercapnia during cardiopulmonary bypass with constant pump output. *Br J Anaesth* 96, 708-714 (2006).
82 R. Arellano *et al.*, Acute graded hypercapnia increases collateral coronary blood flow in a swine model of chronic coronary artery obstruction. *Critical care medicine* 27, 2729-2734 (1999).
83 C. R. Carvalho *et al.*, Temporal hemodynamic effects of permissive hypercapnia associated with ideal PEEP in ARDS. *American journal of respiratory and critical care medicine* 156, 1458-1466 (1997).
84 H. Hager *et al.*, Hypercapnia improves tissue oxygenation in morbidly obese surgical patients. *Anesth Analg* 103, 677-681 (2006).
85 G. M. Hare *et al.*, Hypercapnia increases cerebral tissue oxygen tension in anesthetized rats. *Can J Anaesth* 50, 1061-1068 (2003).
86 K. G. Hickling, Permissive hypercapnia. *Respir Care Clin N Am* 8, 155-169, v (2002).

87 B. Kavanagh, Normocapnia vs hypercapnia. *Minerva Anestesiol* 68, 346-350 (2002).

88 M. Komori *et al.*, Permissive range of hypercapnia for improved peripheral microcirculation and cardiac output in rabbits. *Critical care medicine* 35, 2171-2175 (2007).

89 J. G. Laffey, D. O'Croinin, P. McLoughlin, B. P. Kavanagh, Permissive hypercapnia — role in protective lung ventilatory strategies. *Intensive Care Med* 30, 347-356 (2004).

Chapter 16

The Unified Theory of Medicine

"Disease is not an entity, but a fluctuating condition of the patient's body, a battle between the substance of disease and the natural self-healing tendency of the body."
— Hippocrates

"Illnesses do not come upon us out of the blue. They are developed from small daily sins against Nature. When enough sins have accumulated, illnesses will suddenly appear."
— Hippocrates

"There is only one health, but diseases are many. Likewise, there appears to be one fundamental force that heals, although the myriad schools of medicine all have their favorite ways of cajoling it into action."
— Robert O. Becker

Now empowered by its explanatory stress mechanism, the *Unified Theory of Medicine* provides a nuclear weapon in the war against disease. It explains the nature of known diseases and confers universal treatments that reinstate homeostasis and health by halting harmful stress mechanism hyperactivity that depletes resources and damages tissues; by restoring tissue perfusion, tissue oxygenation, and organ function; and by facilitating surgery, antibiotic therapy, and other measures that promote homeostasis and health.

THE CATEGORIES OF ILLNESS

For the purposes of this presentation, *disease* will be defined as harmful stress mechanism hyperactivity induced by combinations of environmental stresses including microbial infestation, surgery, trauma, fear, anxiety, and toxic substances such as pesticides, herbicides, automobile exhaust, chlorinated water, and food additives. The terms *illness* and *sickness* are generally synonymous with the term *disease* but they encompass conditions that do not involve stress mechanism hyperactivity.

For the sake of discussion, this presentation will be divided into the following categories:

1. Genetic Defects.
2. Dietary Deficiencies.
3. Chronic Illness.
4. Critical Illness.
5. Heart Disease.
6. Malignancy (Cancer).
7. Trauma and Surgery.
8. Universal Treatments.

GENETIC DEFECTS OVERVIEW

If you lack knowledge of cell biology, please review Appendix E: Cell Biology, on page 427. The category of Genetic Defects encompasses both gene defects and chromosome defects but these are very different from each other. Each gene determines an individual factor. Chromosomes consist of numerous genes. Both gene defects and chromosome defects cause a wide variety of problems, *but neither are true* diseases, *because they are not caused by harmful stress mechanism hyperactivity.* Anyone can suffer disease, but only those with defective genes or chromosomes can suffer genetic illnesses. At present there are no effective treatments or cures for these conditions. CRISPR technology will hopefully provide the means to repair defective genes in the future, but at present there is no theoretical prospect of a means to correct chromosome defects.

Some gene and chromosome defects do not affect the stress mechanism at all, but most affect stress mechanism activity either directly or indirectly. In these latter cases an understanding of stress theory can clarify their manifestations.

CHROMOSOME DEFECTS

In the nucleus of eukaryotic plant and animal cells, multiple strands of DNA are entwined in the form of chromosomes, each of which retains vast numbers of genes that determine anatomy, intelligence, behavior, and other characteristics. Each gene codes for an individual characteristic. At least, that is the presently prevailing concept. However, much remains to be understood, and combinations of genes may determine what appears to be an individual characteristic.

The chromosomes exist in matched pairs, one from the mother, and the other from the father; therefore, genes also exist in matched pairs (except for sex chromosomes). During eukaryotic cell division, which is called *mitosis*, chromosome pairs are separated and duplicated, so that both resulting cells retain chromosome pairs that are identical to the original chromosome pairs.

Mitosis is a complex process that sometimes goes awry and causes chromosome breakage, abnormal duplication of broken fragments, or even duplication of entire chromosomes. Accelerated cell division exaggerates the incidence of chromosomal damage during both peak wound healing and malignancy, so that malignant tissues cannot be reliably distinguished from healthy tissue repair. Mitotic mayhem also occurs during embryological development, when cell division proceeds faster than any other time of life. Chromosome damage usually results in miscarriage, but certain types of survivable chromosome damage are common enough to have names. For example, the accidental duplication of chromosome #21 causes *Down's Syndrome* (otherwise known as *trisomy 21*). The resulting offspring usually suffer impaired intelligence, anatomy, life span, and fertility. Such problems occur at random and are not inheritable.

Since there are matching sets of chromosomes, the effects of harmful or deficient genes are usually negated by the matching healthy genes on the matching chromosome unless both parents contribute the same defective gene, whereupon its harmful or deficient consequences prevail. Sickle cell anemia is a prototypical example of such a homozygous recessive trait. In classical genetics homozygous means inherited from both parents, and recessive means a dysfunctional gene on one chromosome that can be compensated for by the matching functional gene on the chromosome contributed by the other parent. There are also autosomal-dominant genes that dominate the normal gene on their matching chromosome, such as the defective gene that causes Huntington's Chorea (an inherited form of progressive brain deterioration). Last, but not least, there are sex-linked genes such as the defective hemophilia gene that mostly afflicts males.

The following conditions exemplify chromosome damage:

Trisomy 21 (Down's Syndrome) is caused by having three copies of chromosome 21 instead of the usual two copies. It causes intellectual disability, a small nose, slanted eyes, and low muscle tone in infancy. The degree of intellectual disability varies from mild to moderate. People with Down's Syndrome may be born with heart defects, digestive abnormalities, short stature, and a single deep crease across the center of the palm. They sometimes suffer gastrointestinal reflux, celiac disease, hypothyroidism, hearing and vision problems, leukemia, and senile dementia.

Victims of Down's Syndrome enjoy enhanced resistance to atherosclerosis. Stress theory explains this. Down's Syndrome victims lack the ability to perceive danger, so they suffer less anxiety and fear than normal people and are thereby protected from the sympathetic nervous system hyperactivity that accelerates atherosclerosis.

Trisomy 13 causes severe intellectual disability and physical abnormalities including congenital heart, brain, and spinal cord defects, small or poorly developed eyes, extra fingers and toes, and cleft lip and palate. Most do not survive for more than a few weeks.

Klinefelter's syndrome (XXY) occurs in males when they have an extra X (female) chromosome. Some have no obvious symptoms, and others have varying degrees of cognitive, social, behavioral, and learning difficulties. Adults may suffer decreased testosterone production, small penis, undescended testicles, enlarged breasts, hypospadias, and infertility.

Turner's syndrome (XO) produces defective females. It results when a person is born with only one normal X chromosome and the other sex chromosome is either missing or damaged. Symptoms include short stature, "webbed neck," skeletal abnormalities, kidney problems, infertility, and congenital heart defects. Most have normal intelligence, but some have learning disabilities and behavior problems.

GENE DEFECTS

A *gene* is a set of DNA information that codes for individual characteristic. A *mutation* is a permanent alteration in the DNA code that changes an individual gene. Such mutations can be beneficial, harmful, or meaningless. They occur at random, and are not influenced by environmental stress.[1] Once mutations occur, they predictably re-appear unchanged in subsequent generations.

Sickle Cell Anemia is a prototypical example of a homozygous recessive trait caused by the contribution of the same defective gene from both the mother and the father. The defective gene undermines the strength of red blood cells, so that environmental stress causes them to collapse into "sickle" shapes that release toxic cell contents such as actin[2, 3] into the bloodstream. The "sickled" cells form *microthrombi* that obstruct capillaries and cause tissue hypoxia.[4, 5] The resulting cellular hypoxia causes cells to revert to inefficient anaerobic metabolism that produces painful lactic acidosis.[6] The pain exaggerates sympathetic nervous activity that promotes amyloidosis (which will be discussed shortly) and atherosclerosis.[7, 8, 9, 10, 11, 12] The condition can be ameliorated with anticoagulants and narcotics, but there is no cure.[13, 14]

Hemophilia was mentioned in the chapter describing blood enzyme interactions. It is caused by a defective sex-linked gene that mothers convey to their sons. It causes severe bleeding by crippling the ability of factor VIII to generate thrombin that is essential to energize platelet chemotaxis, platelet activation, and insoluble fibrin generation to enable coagulation and capillary hemostasis.[15] The inability to form a viscoelastic clot impairs tissue repair[16] and causes a peculiar form of exercise intolerance by undermining capillary gate regulation of blood flow to organs and tissues.[17] On the other hand, the inability to generate insoluble fibrin inhibits atherosclerosis.[18, 19, 20, 21, 22, 23]

In recent years it has become popular to treat hemophilia with factor VII instead of factor VIII. This controls bleeding, but it promotes cancer and heart attacks.[24, 25, 26, 27] Stress theory explains why: excessive factor VII activity promotes harmful cellular tissue repair activity that damages capillaries, undermines tissue oxygenation, invites infarction, and promotes malignancy. Stress theory thus indicates that hemophiliacs should be treated with purified preparations of factor VIIIC instead of factor VII.

Von Willebrand Coagulopathy was mentioned in the chapter describing blood enzymes. It is caused by a defective autosomal-dominant (non sex-linked) gene, so that it afflicts both sexes equally. The gene defect undermines the quality or quantity of von Willebrand Factor (VWF), which is produced by the vascular endothelium and released into blood by sympathetic nervous activity.[28, 29, 30]

The defect undermines the ability of VWF to activate VIIIC, and thereby inhibits atherosclerosis in the manner of hemophilia.[31, 32, 33, 34, 35]

Von Willebrand Coagulopathy affects 1–3% of people. It is usually mild and remains undiagnosed in most patients. It exaggerates nosebleeds, menstrual bleeding, and blood loss during surgery and dental procedures.[36] In severe cases it mimics hemophilia.

VWF and Angiodysplasia

The large VWF molecule is extruded from the inner walls of capillaries, and it protects capillary structure by controlling capillary blood flow, so that von Willebrand coagulopathy exaggerates the incidence of flow-related capillary damage called angiodysplasia.[37, 38, 39, 40] The gigantic VWF molecule is fragile and readily damaged by excessive blood turbulence caused by severe anemia, prosthetic heart valves, LVAD machines, and so forth. Such VWF destruction causes angiodysplasia and mimics von Willebrand coagulopathy.[41]

VWF and Nervous Activity

The severity of the von Willebrand clotting defect fluctuates with nervous activity, which releases VWF reserves from the vascular endothelium that activate factor VIII.[42] This explains why various types of stress, including sepsis, increase blood viscosity and coagulability by exaggerating nervous activity, which elevates factor VIII activity.[43, 44, 45, 46, 47, 48] This, in turn, explains the therapeutic properties of narcotics. For example, long before the development of antibiotics, George Washington Crile cured sepsis and peritonitis using massive doses of morphine to mitigate harmful sympathetic nervous hyperactivity.[49, 50]

VWF and Angioneurotic Edema

Transfusions, bee stings, and foreign substances that invade the bloodstream can precipitate complement enzyme activity that attacks the VWF molecule, causing anaphylaxis.[51] Thrombin generation and VWF destruction by complement cascade hyperactivity causes sudden onset systemic inflammation and tissue edema that threatens airway patency, and sometimes causes death by suffocation. The VWF destruction paralyzes capillary gate function, causing sudden, reduction of systemic blood flow resistance that manifests as low blood pressure and inadequate brain perfusion, causing sudden loss of consciousness (fainting). However, cardiac output remains intact, so that supine position restores brain perfusion. The best treatment is epinephrine, which releases VWF reserves from the vascular endothelium, restores microvascular flow regulation, and relieves tissue edema.[52, 53]

VWF and Pharmacology

Streptokinase restores perfusion and oxygenation to infarcted tissues by attacking the VWF molecule, which opens the capillary gate. This explains why its powerful therapeutic effects cause hypotension.[54] Similarly, ACE inhibitors relieve hypertension by attacking the VWF molecule and reducing microvascular flow resistance. This accounts not only for their therapeutic effects, but also explains why these medications sometimes cause life-threatening angioneurotic edema.[55, 56]

Defects of factors VII, X, and tissue factor, as mentioned previously, are incompatible with life, because their interaction is essential for embryological tissue development. Such defects invariably cause miscarriage. Likewise, drugs such as Coumadin that disrupt this enzymatic interaction are incompatible with embryological development.

Factor IX deficiency (aka "Christmas Disease") is a mild clotting deficiency caused by an inherited defect in the gene that produces non-enzymatic factor IX. The defect inhibits the addition of crosslinks to soluble fibrin to generate insoluble fibrin.[57]

Factor XIII deficiency is another clotting deficiency caused by an inheritable genetic defect. It is rare, but it causes multiple problems. For example, it promotes cancer by exaggerating viscoelastic clot permeability to blood enzymes.[58]

Porphyria is caused by a group of inheritable gene defects that disrupt hemoglobin production and metabolism. It causes a wide variety of manifestations, including photosensitivity and digestive disorders.

Cystic Fibrosis is caused by a defective gene that produces abnormally thick mucus that disrupts the function of internal organs including the lung, pancreas, liver, kidneys, and intestine. It does not directly affect stress mechanism activity.

Familial Dysautonomia is caused by a gene defect that disrupts the development of the autonomic nervous system, causing chronic autonomic instability that disrupts organ function.

Ehlers-Danlos Syndromes are rare inherited genetic connective tissue disorders.

DIETARY DEFICIENCIES

Like genetic illnesses, dietary deficiencies should be understood as illnesses caused by malnourishment rather than true diseases caused by environmental stresses, even though they produce symptoms similar to those caused by disease. Dietary deficiencies often cause anorexia (inadequate appetite), cachexia, weight loss, short stature, and failure to thrive. These illnesses can be treated and cured by providing adequate nutrition in the diet. The following are examples of dietary deficiencies:

Marasmus is caused by inadequate energy intake. It is usually due to inadequate carbohydrates and fats, although proteins also provide energy. It usually occurs in children below the age of 1 year, and causes muscle wasting, because metabolic pathways digest muscle to obtain energy in the presence of inadequate energy intake.

Kwashiorkor is a form of malnutrition caused by inadequate protein in the diet. It is characterized by edema, enlarged liver, and ascites. It typically occurs after children are weaned from breast milk, which contains essential amino acids needed to manufacture protein.

Fat-Soluble Vitamin deficiencies

Vitamin A deficiency (Bitot's spots)
Vitamin D deficiency (Ricketts)
Vitamin E
Vitamin K

Water-Soluble Vitamin deficiencies

- Vitamin C deficiency (scurvy)
- Vitamin B1 (thiamine) deficiency (Beriberi)
- Vitamin B2 (riboflavin) deficiency
- Vitamin B12 deficiency
- Vitamin K deficiency

Mineral Deficiencies

- Calcium
- Sodium
- Potassium
- Iron
- Zinc
- Manganese
- Copper
- Iodine
- Chromium
- Molybdenum
- Selenium

CHRONIC DISEASE (CHRONIC ILLNESS)

Unlike genetic derangements and dietary deficiencies, true disease is caused by stress mechanism hyperactivity that is induced by environmental stresses — including surgery, trauma, microbial infestation, toxic substances, excessive radiation, fear, anxiety, and pain. Low levels of environmental stress cause subclinical stress mechanism hyperactivity that progressively damages tissues, accelerates capillary senescence, and undermines organ function, even though it doesn't initially produce overt disease symptoms. Individual stresses, such as microbial infestations, can provoke specific stress mechanism reactions, which sometimes, but not always, produce distinctive symptoms that enable diagnosis. However, all forms of stress mechanism hyperactivity (i.e., disease activity), if severe enough, produce common symptoms (fever, fatigue, misery, inflammation, edema, and so forth), which reflects the universal nature of disease and disease relationships. Thus, in most cases, it is impossible to identify individual causative sources of stress, because disease results from the sum total of various environmental stresses that induce harmful stress mechanism hyperactivity. As we shall see, chronic illnesses exacerbate the incidence and severity of critical illnesses by accelerating capillary senescence, damaging tissues, and undermining organ function.

AMYLOIDOSIS

I hypothesize that chronic, subacute stress mechanism hyperactivity accelerates amyloid protein production that causes chronic illnesses.

Amyloidosis is the abnormal accumulation of fibrillar amyloid protein in blood vessels, organs, and tissues. It occurs in all vertebrates, including fish, amphibians, mammals, reptiles, and birds.

Although amyloidosis is commonplace, it is usually overlooked because its detection requires tissue biopsy unless it becomes so severe as to be noticed by the naked eye.[59]

The coagulation/capillary gate mechanism, which is common to all vertebrates, continually generates and disintegrates insoluble fibrin into multiple forms of fibrin split products (FSP).[60, 61, 62, 63, 64, 65, 66, 67, 68, 69, 70, 71, 72] The simplest explanation of amyloidosis is that persistent exposure to environmental toxins and stresses causes subclinical stress mechanism hyperactivity that exaggerates FSP turnover, causing FSP to interact with factor X and undergo *conformational changes* (mis-folding) that produce harmful amyloid protein.[73, 74] This is consistent with the observation that amyloid exists in multiple forms.[75] This interaction between FSP and factor X depletes factor X and undermines insoluble fibrin generation. The combination of tissue damage and factor X depletion explains the bleeding problems caused by amyloidosis.[76, 77, 78, 79, 80, 81, 82, 83, 84, 85, 86, 87, 88]

Numerous forms of stress accelerate amyloidosis, including infectious diseases, dialysis, cardiac bypass, left-ventricular assist devices (LVAD), prosthetic heart valves, surgery, starvation, and sickle cell disease. This indicates that subclinical stress mechanism hyperactivity causes amyloidosis.[89, 90, 91, 92, 93, 94, 95, 96] These illnesses were unknown before the Industrial Revolution in North America, but since then they have risen to epidemic proportions. This suggests that the industrial revolution has introduced toxic environmental stresses that induce subclinical stress mechanism hyperactivity, including herbicides, pesticides, automobile exhaust, chlorinated water, air pollution, and food additives.

Amyloid protein progressively accumulates in organs and tissues with age.[97] Depending on the type of amyloid protein, and where and how quickly it accumulates, amyloid causes chronic illnesses called “Rheumatoid” diseases, including diabetes, essential hypertension, atherosclerosis, congestive heart failure, heart disease, rheumatoid arthritis, scleroderma, Sjögren’s Disease, psoriasis, multiple sclerosis, Parkinson’s disease, Alzheimer’s disease, Autism, and Systemic Lupus Erythematosis (SLE).[98, 99, 100, 101, 102, 103, 104, 105]

Amyloid deposition induces subclinical tissue repair activity including inflammation, fibroblast proliferation, and sclerosis that damages tissues, destroys capillaries, disrupts organ function, and causes bleeding.[106, 107, 108, 109, 110, 111, 112] Such stress mechanism hyperactivity automatically elevates immune activity, which is often blamed for these harmful effects, if only because the immune activity can be detected, while stress mechanism activity has yet to be recognized, but there is no evidence that immune activity *per se* causes harmful effects.

Amyloidosis accelerates capillary senescence (age-related capillary deterioration). I call this *accelerated capillary senescence*. Capillary senescence explains why diabetes, essential hypertension, atherosclerosis, congestive heart failure, and infarction (hypoxic tissue damage) are rare in youth but commonplace in old age. Capillary senescence increases flow resistance (causing hypertension), alters pulsatile blood turbulence (causing atherosclerosis), inhibits glucose uptake (causing diabetes), and undermines cellular oxygenation that provokes inflammatory tissue repair activity, causing tissue sclerosis (collagen deposition) that progressively stiffens tissues and undermines cell, tissue, and organ function (which explains the manifestations of rheumatoid diseases).[113, 114, 115, 116, 117, 118] These destructive effects explain why chronic illnesses exaggerate the incidence and severity of one another, and why pre-existing chronic illness exaggerates the incidence and severity of critical illnesses (see "Critical Diseases (Critical Illnesses)" on page 186). For example, obesity, diabetes, hypertension, cancer, and heart disease are closely related, and patients with these conditions exaggerate the incidence, morbidity, and mortality of critical diseases (critical illnesses) such as eclampsia, Adult Respiratory Distress Syndrome, and Multi-Organ Failure Syndrome.[119, 120, 121, 122, 123, 124, 125, 126]

The following examples illustrate how amyloidosis causes the variety of rheumatoid disease:

1. Amyloid accumulation damages capillaries, which exaggerates blood flow resistance that manifests as *essential hypertension*.[127]
2. Amyloidosis causes *excessive fibrinolysis*, which undermines hemostasis.[128]
3. Amyloid invades the spleen and undermines its function.[129]
4. Amyloid accumulates on the inner walls of arteries and induces tissue repair activity that causes *atherosclerosis*.[130]
5. Amyloidosis causes *atherosclerosis* by accelerating capillary senescence, which undermines the pulsatile turbulence that mobilizes particulates from the inner walls of arteries. The particulate deposits incite inflammatory tissue repair activity than produces atherosclerotic plaque.[131]
6. Amyloid accumulation in bone marrow causes *multiple myeloma*.[132, 133]
7. Amyloid deposits in lymph nodes causes *Hodgkin's disease*.[134]
8. Amyloid deposits in lung tissue and causes *lung cancer*.[135]
9. Amyloid deposits in bowel tissues, causing *Crohn's Disease*.[136]
10. Amyloid accumulates in pancreatic tissues and prevents the release of insulin from pancreatic islet cells, causing *Type I diabetes*.[137] This explains why Type I diabetics require precise insulin management to prevent ketoacidosis.
11. Amyloid causes *Type II diabetes* by accelerating capillary senescence, which inhibits cellular glucose uptake.
12. Amyloid undermines cellular oxygen delivery, which induces pathological collagen generation that causes *tissue sclerosis*.[138, 139, 140, 141] This explains the tissue deterioration and poor wound healing in *Type II diabetes*.
13. Amyloid causes *congestive heart failure* by increasing static systemic microvascular flow resistance, which exaggerates cardiac work, causing tissue repair activity that deposits collagen in cardiac tissues. The collagen thickens heart walls and undermines cardiac function.[142, 143, 144, 145]
14. Amyloid accelerates *atherosclerosis* by increasing static flow resistance that alters pulsatile turbulence and undermines arterial patency.[146]
15. Amyloid invites *infarction* by undermining cardiac tissue perfusion and tissue oxygenation and by accelerating atherosclerosis.
16. Amyloid causes *Sjögren's Syndrome* by damaging salivary gland capillaries, which undermines salivary gland perfusion and inhibits saliva production.

17. Amyloid causes *rheumatoid arthritis* by accumulating in joint tissues, where it damages capillary perfusion and induces painful inflammation and sclerosis that inhibits joint lubrication. In severe cases the amyloid accumulation forms *rheumatoid nodules* that consist of dense collections of amyloid protein.
18. Amyloid accumulates in bone marrow and causes *idiopathic myelofibrosis*.[147]
19. Amyloid causes *Systemic Lupus Erythematosis* (SLE) by accumulating in skin, organs, and brain, causing dementia and organ dysfunction.
20. Amyloid deposits in skin and causes *scleroderma* by inducing sclerosis and capillary deterioration that undermines skin elasticity.[148, 149]
21. Amyloid accumulation in skin pre-disposes to *skin cancer and psoriasis.*
22. Amyloidosis causes *multiple sclerosis* (which damages nerves) by accumulating in nerve tissues, inducing inflammation, destroying capillaries, causing sclerosis, and undermining nerve perfusion, oxygenation, and function.[150, 151]
23. Amyloidosis causes *Parkinson's Disease* (cerebellar dysfunction) by accumulating in the cerebellum, causing inflammation and sclerosis.
24. Amyloidosis causes *autism* and *Asperger's Syndrome* (a mild form of autism) by accumulating in fetal brain tissue during gestation. These illnesses primarily afflict males, because the male brain is more highly specialized than the female brain.
25. Amyloidosis causes *Alzheimer's Disease* by accumulating in the cerebral cortex, causing inflammation, accelerated capillary senescence, decreased capillary flow and cellular oxygenation, and sclerosis that undermines cognitive function.[152, 153, 154, 155]
26. Renal amyloidosis disrupts kidney function.[156]
27. *Obesity* is a stressful condition that is closely associated with amyloidosis, Type II diabetes, hypertension, heart disease, and cancer.[157, 158, 159, 160, 161, 162, 163, 164] The close associations of these seemingly disparate conditions reflect subclinical stress mechanism hyperactivity.
28. Methotrexate and other drugs used to treat rheumatoid diseases by inhibiting immune activity can exaggerate stress mechanism hyperactivity and increase the risk of *cancer*.

FAMILIAL AMYLOIDOSIS

Familial amyloidosis is not true amyloidosis. It is an inherited genetic defect rather than a true disease induced by environmental stress.[165] The defect produces an aberrant form of gelsolin, which normally functions as part of an actin scavenger system.[166, 167, 168, 169]

Actin is present within every eukaryotic cell, and is the most abundant protein in the body. It is a fibrillar protein that becomes entangled with insoluble fibrin in the viscoelastic clot.[170, 171, 172, 173] I hypothesize that actin disturbs the permeability of the viscoelastic clot and undermines its role in tissue repair.

Gelsolin binds to fibronectin in the viscoelastic clot, and it cleaves fibrillar actin protein into pieces that are subsequently digested by a poorly understood enzymatic interaction that consumes vitamin D.[174, 175, 176, 177, 178, 179, 180] Vitamin D also appears to be involved in digesting amyloid protein.[181] Gelsolin activity is calcium-dependent, which suggests that it may be energized by thrombin, so that stress mechanism hyperactivity automatically exaggerates actin digestion.[182, 183, 184, 185, 186] Thus, gelsolin and vitamin D may be elements of a generalized scavenging mechanism that removes harmful protein debris from body tissues.

SARCOIDOSIS

Sarcoidosis is similar in principle to amyloidosis, but less common. I hypothesize that subclinical stress mechanism hyperactivity causes sarcoidosis by releasing albumen protein from the glycocalyx that lines all blood vessels.[187, 188] The kidneys remove albumen protein from blood, which explains why its appearance in urine serves as an early indication of sub-clinical stress mechanism hyperactivity. It is called *albumen* because if a test tube of urine is boiled using a Bunsen burner, the albumen protein will coagulate into a layer of denatured (insoluble) protein that resembles cooked egg white, which is also called albumen. When released from the vascular glycocalyx, the albumen preferentially deposits lung tissue, where it induces harmful inflammatory tissue repair activity that causes sclerosis. This proffers the simplest explanation of sarcoidosis.[189, 190, 191, 192]

TREATMENTS FOR CHRONIC ILLNESSES

Humans, animals, and plants are no better adapted to modern environmental pollutants than they are adapted to conditions at the bottom of the ocean, the top of Mount Everest, or Mars. At present there are no acknowledged cures for chronic illnesses, and conventional treatments such as prednisone and methotrexate suppress harmless immune activity and cause counterproductive side effects. Therefore, the most effective measure is prevention. Patients should avoid environmental stresses to the extent possible, including smoking, vaping, chlorinated water, polluted air, contaminated food, unnecessary drug treatments, and emotional stress. Such is easier said than done. Mitigating these problems will necessitate constructive politics.

Stress theory suggests that the following measures may be helpful. Diet and exercise have been known to be therapeutic since the days of Hippocrates. Daily fat-soluble Vitamin D3 supplementation is more effective than water-soluble vitamin D1-D2 food supplements found in dairy products. D3 helps to digest harmful actin, and there is evidence that it prevents or mitigates cancer, viral infections, and other maladies. In the distant past it was a staple of medical care. Sunlight converts D2 to D3, so that sunbathing is healthful. Otherwise, the best D3 source is fresh cod liver oil taken on a daily basis.[193, 194, 195, 196, 197, 198, 199, 200]

Chelation therapy inhibits stress mechanism hyperactivity, reverses atherosclerosis, reduces the risk of infarction, mitigates rheumatoid inflammation and sclerosis, and enhances the potency and penetration of antibiotics.[201, 202, 203] Sildenafil mimics parasympathetic nervous activity and promotes tissue perfusion. It has been found helpful in treating achalasia (swallowing difficulty), diabetes, high-altitude pulmonary edema, and other chronic illnesses.[204, 205, 206, 207, 208, 209, 210, 211, 212, 213, 214, 215]

CRITICAL DISEASES (CRITICAL ILLNESSES)

I hypothesize that excessive tissue factor entry into systemic circulation causes critical illnesses.

Critical illnesses are life-threatening conditions that cannot be explained by conventional medical theory. They are typically categorized as *syndromes* (groups of symptoms that consistently occur together in specific circumstances). The term *syndrome* is synonymous with “We don’t know.”

Critical illnesses are often confused with one another because they are all caused by the same thing: severe systemic stress mechanism hyperactivity induced by the abnormal entry of tissue factor into systemic circulation. Such syndromes include Adult Respiratory Distress Syndrome (ARDS), Multi-Organ Failure Syndrome (MOFS), Systemic Inflammatory Response Syndrome (SIRS), viral pneumonia (influenza, SARS, MERS, COVID), bacterial pneumonia, vaping syndrome (caused by inhaling toxic electronic cigarette vapor), multiple sclerosis (MS), bacterial Meningitis, brain trauma, viral encephalitis, Severe Burn Syndrome, Disseminated Intravascular Coagulation (DIC), eclampsia, sepsis, asthma, and the surgical stress syndrome (SSS).

Organs and tissues rich in tissue factor are targets for critical illnesses. These include lung, brain, retina, nerves, spinal cord, cervix, gonads, vascular adventitia, renal and gastrointestinal mucosa, placenta, and amniotic fluid.[216, 217, 218, 219, 220] Severe systemic inflammation preferentially disrupts these tissues compared to tissues lacking in tissue factor, such as bone, ligaments, skin, muscle, and so forth. The lung and the brain are especially rich in tissue factor, so that they are typically the first organs to manifest stress mechanism hyperactivity. This also explains why the brain and the lung exhibit exaggerated vulnerability to both primary and secondary (metastatic) cancer, which will be elaborated presently. Eclampsia is a unique form of stress mechanism hyperactivity that occurs in pregnant women, because the placenta, cervix, and amniotic fluid are all rich in tissue factor.

CRITICAL DISEASE AND THE LUNG

The lung is typically the first organ to manifest systemic stress mechanism hyperactivity. Being specialized to optimize gas exchange with atmospheric air, it consists of a delicate network of air sacs called *alveoli* that are covered with a dense capillary network over 70% of their surface (see Figure 16.1 on page 188). The interface between alveolar capillaries and the atmospheric air inside them is as little as 0.2 micrometers, and the total surface area of the alveoli is 750 sq ft, which is comparable to the floor space of a room.

Because of its fragility, lung tissue is vulnerable to damage and bleeding, and is thus rich in tissue factor that promotes both hemostasis (control of blood loss) and tissue repair. Being thus richly endowed with tissue factor and directly exposed to atmospheric air, the lung reacts violently to airborne toxins, pollutants, viruses, and microbes that irritate or attack delicate lung tissue and expose tissue factor to blood enzymes. The resulting enzymatic reaction produces *exudates* (fluids, cells, soluble fibrin, and other proteins) that flood the alveoli and disrupt gas exchange. This explains why viral pneumonia, bacterial pneumonia, mustard gas, and vaping[221] are lethal: inhaled toxic substances and contagious microbes attack delicate lung tissues and expose tissue factor to blood enzymes, which induces inflammation and soluble fibrin exudates that disrupt gas exchange and cause asphyxiation (suffocation).[222]

The lung reacts the same way to systemic thrombin elevations. For example, bacterial invasion of flowing blood provokes complement cascade blood enzyme activity that elevates thrombin generation, causing systemic inflammation that exaggerates the permeability of the vascular endothelium and increases the penetration of factor VII into extravascular tissues. Lung tissue, being rich in tissue factor, reacts by producing copious glycoprotein exudates that disrupt gas exchange. All organs and tissues are affected by such systemic inflammation, but the lung is typically the first organ to manifest disturbances on account of its rich concentration of tissue factor, and depending on the severity of the systemic inflammation, this begins as atelectasis and progresses to ARDS (Adult Respiratory Distress Syndrome) and culminates in MOFS (Multi-Organ Failure Syndrome).

Another example is surgical tissue disruption that releases tissue factor into blood circulation, which exaggerates blood thrombin generation that affects lung tissue before it affects other organs with lesser tissue factor. Thrombin elevations increase the permeability of lung tissue, and release additional tissue factor into flowing blood. This causes *positive feedback* that further exacerbates thrombin generation. The resulting lung inflammation manifests as atelectasis that sometimes progresses to ARDS (Adult Respiratory Distress Syndrome).[223] Pulmonary exudates contain soluble fibrin, which induces collagen generation that causes sclerosis and permanent lung damage in ARDS survivors.[224]

ARDS is merely the first manifestation of Multi-Organ Failure Syndrome (MOFS) because the pulmonary problems are soon followed by similar disturbances in other organs, along with fever, dementia, albuminuria, abnormal liver function, acute renal failure, tissue edema, fatigue, malaise, and other manifestations of stress mechanism hyperactivity. In the context of surgery, these symptoms are called the surgical stress syndrome (SSS).

As might be suspected, severe lung contusion (bruising) due to chest injury is closely associated with ARDS, MOFS, and death. Like other forms of severe trauma and stress, it is often asymptomatic at the time of injury, but grows progressively worse, reaching peak severity about 48 hours later. However, since lung injury is usually accompanied by injuries to the head, heart, and bones it is difficult to determine the exact contribution of lung injury to outcome.

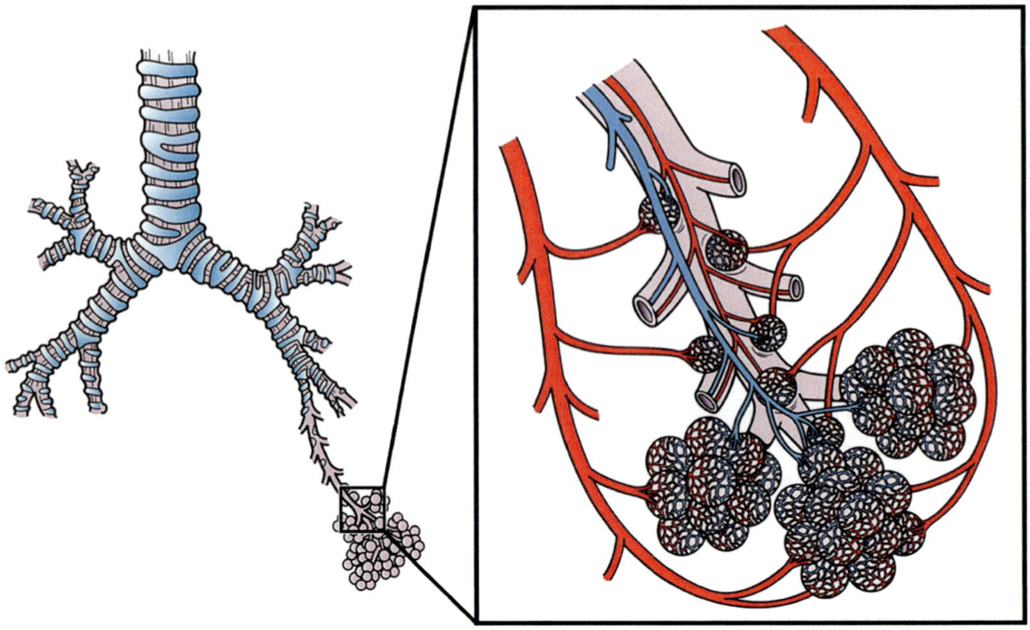

Figure 16.1: Alveoli.

A drawing that portrays the interface between blood and atmospheric air in the alveoli of the lung. The large surface area of lung tissue that is directly exposed to atmospheric air, plus the thin interface between air and blood enzymes, plus the high concentration of tissue factor in lung tissues, renders the lung uniquely vulnerable to destructive stress mechanism hyperactivity.

https://commons.wikimedia.org/wiki/File:Pulmonary_Blood_Circulation.png#/media/File:Pulmonary_Blood_Circulation.png

The following examples further illustrate how excessive contact between tissue factor and blood enzymes causes stress mechanism hyperactivity that produces critical illnesses:

Multi-Organ Failure Syndrome (MOFS) follows ARDS and dementia as worsening stress mechanism hyperactivity affects additional organs with lesser tissue factor than the lung and brain. Dementia typically appears soon after the onset of ARDS because brain tissue is also rich in tissue factor, so that the brain is typically the second organ to exhibit dysfunction in the form of dementia (disordered thinking). As brain inflammation releases still more tissue factor into systemic circulation, the stress mechanism hyperactivity affects additional organs and tissues with lesser concentrations of tissue factor. Meanwhile exaggerated blood viscosity and coagulability increases the risk of infarction, thrombosis, and embolus.

The Surgical Stress Syndrome (SSS) is analogous to MOFS. Like trauma, surgical tissue disruption releases tissue factor into blood circulation, causing systemic inflammation. Surgery simultaneously induces harmful sympathetic nervous activity that releases VWF, activates factor VIII, exaggerates blood viscosity and coagulability, and increases the risk of infarction. The SSS thus manifests as symptoms and manifestations distant from the time and location of surgical tissue disruption, such as exaggerated pain (allodynia), dementia, edema, fever, organ dysfunction, infarction, and death.

Before the discovery of anesthesia, most surgeons avoided invasive body cavity surgery because patients died in agony about two days after such procedures. General anesthesia mitigated the surgical stress syndrome and enabled patients to survive life-saving surgery, and was thus regarded as miraculous. Supplementation of general anesthesia with analgesia such as morphine further mitigates but does not eliminate the surgical stress syndrome, which persists in the guise of postoperative fever, nausea, vomiting, bowel ileus, acute renal failure, atelectasis, pneumonia, and exaggerated pain. These problems reach peak intensity about 48 hours after surgery, and then subside, but lingering increases in morbidity and mortality persist for months, largely due to cancer and heart disease.[225, 226, 227, 228]

SIRS (Systemic Inflammatory Response Syndrome) is a special example of the surgical stress syndrome that is induced by cardiac bypass machines. It is most simply explained by red blood cell damage that releases toxic cell contents such as actin into systemic circulation. SIRS caused severe morbidity and mortality in the early days of cardiac bypass surgery until the 1960s when Lowenstein re-discovered Crile's principle that supplementing general anesthesia with morphine analgesia enhances surgical outcome.[229, 230, 231] The SIRS phenomenon re-appeared in the 1990s after medical insurance companies altered reimbursements to promote "fast tracking" (early discharge), whereupon cardiac anesthesiologists promptly abandoned narcotic supplementation to accommodate the insurance company dictates.[232, 233, 234] There are numerous similar examples of medical knowledge being forgotten, re-discovered, and then forgotten all over again.

Viral Pneumonia most often causes devastating epidemics during wartime when abnormal crowding exaggerates viral virulence. For example, the influenza virus caused the disastrous viral pneumonia epidemic that devastated Athenians and killed Pericles during their war with Sparta. The entire Athenian population had crowded into a small island offshore from Athens to evade the marauding Spartan army. Similar crowding caused the worldwide WWI influenza epidemic that killed more people than the war.[235]

Around 2010, researchers in Europe and the United States discovered the means to "weaponize" the normally benign coronavirus:

- https://www.scientificamerican.com/article/a-man-made-contagion/
- https://www.scientificamerican.com/article/contagion-controversy-erupts/

Soon thereafter, the coronavirus suddenly caused unprecedented avian, mammalian, and human epidemics. The first human epidemic appeared in China and was called SARS (Severe Acute Respiratory Syndrome). The second was called MERS (Middle Eastern Respiratory Syndrome). The third so-called COVID version is less lethal but has caused a worldwide epidemic. These circumstances are most simply explained by "weaponized" versions of the coronavirus that do not occur naturally.[236, 237, 238] As we shall see, stress theory confers powerful treatments for all forms of critical disease that can salvage the victims of these epidemics by preventing lethal stress mechanism hyperactivity induced by the virus.

Bacterial Pneumonia causes an inflammatory lung reaction that is analogous to viral pneumonia. Pneumococcal pneumonia spreads from person to person via coughing and sneezing. *Legionnaire's Disease* is caused by normally innocuous *Legionella* microbes that thrive in commercial air conditioning systems. It can be prevented by chlorinating the water in these air conditioning systems. Like viral pneumonia, bacterial pneumonia causes intense lung inflammation and lethal exudate production that disrupts pulmonary gas exchange. However, bacterial virulence is not exaggerated by crowding as in the case of viral pneumonia.

Fungal Pneumonia such as coccidiomycosis and histoplasmosis is caused by fungal spores launched into the air from ground-dwelling fungi by farming and construction activities. As in the case of viral and bacterial pneumonia, fungal pneumonia causes pulmonary inflammation and exudate production that threatens gas exchange in the lung.

Vaping syndrome is caused by inhaling toxic electronic cigarette fumes, which cause chemical inflammation of lung tissues that produces potentially lethal exudate production, as in the case of viral, bacterial, and fungal pneumonia. All three of these conditions are analogous to ARDS.

Mustard Gas is toxic war gas that exaggerates the exposure of tissue factor to blood enzymes by chemically denaturing (cooking) lung protein and exposing lung tissue factor to blood enzymes, and thereby causing ARDS.

Asthma is stress mechanism hyperactivity that exaggerates glycoprotein mucus exudates in lung airway passages. This decreases airway diameter and increases flow resistance. The effect is most pronounced during exhalation, which causes further reductions in airway diameter. This causes a distressing sense of "air trapping" that exaggerates sympathetic nervous activity and aggravates the stress mechanism hyperactivity. Most victims are children, whose airway diameters are smaller than those of adults, and most children "outgrow" asthma. However, if asthma isn't adequately treated and controlled, its exudates can stimulate sclerosis of lung tissue, and cause permanent lung damage that limits exercise tolerance.

100 years ago, asthma was quickly and easily controlled using Carbogen (a mixture of carbon dioxide and oxygen in pressurized tanks), but Carbogen was abandoned for the sake of unfounded fear of CO_2 toxicity.[239]

Eclampsia is MOFS that occurs in pregnant women. Pregnancy exaggerates the risk of MOFS for these reasons:

- The placenta, cervix, and amniotic fluid are rich in tissue factor.
- Sustaining fetal growth and development tests the limits of hemodynamic physiology, even if health is intact.
- Accelerated capillary senescence is easily overlooked in breeding age women.
- The placenta produces progesterone, which inhibits ureteral peristalsis and invites ascending urinary tract infections that are easily overlooked amid the discomforts of pregnancy.

As in the case of other critical illnesses, the risk and severity of eclampsia is exaggerated by smoking, alcohol abuse, and chronic illnesses that accelerate capillary senescence. Albuminuria is the first warning of its danger.

As the demands of fetal development inexorably increase, pre-existing capillary senescence manifests as gestational hypertension and diabetes. This is called *pre-eclampsia*. With the onset of labor, uterine contractions release tissue factor into circulation from the dilating cervix and the compressed placenta, and combinations of pain, fear, increased cardiac output, capillary senescence, and capillary gate closure cause severe hypertension. The resulting systemic inflammation and edema disrupts organ function, causing brain dementia, ARDS, and bursting of the liver.[240] This is called *eclampsia*. To make matters worse, amniotic fluid, which is also rich in tissue factor, sometimes escapes into systemic circulation. This is called *amniotic fluid embolus* and it can precipitate lethal Disseminated Intravascular Coagulation (DIC) that disrupts oxygen transport and delivery.[241]

Fortunately, obstetricians understand the therapeutic benefits of magnesium sulfate, which competes with Ca+, inhibits thrombin generation, and mitigates stress mechanism hyperactivity. Narcotics, spinal analgesia, and epidural analgesia can helpfully control nociception that causes harmful sympathetic nervous activity.

Bacterial Meningitis and Viral Encephalitis are life-threatening conditions because of the rich concentration of tissue factor in brain tissue. Viral and bacterial infections of the brain thus provoke intense inflammation and edema that causes brain swelling, which can cause lethal brain herniation.

Severe Burn Syndrome releases tissue factor into systemic circulation, causing potentially lethal stress mechanism hyperactivity.

Sepsis (blood poisoning) is caused by bacterial invasion of systemic blood circulation. It is dangerous because it activates complement enzymes that generate thrombin, causing systemic inflammation. The systemic inflammation exaggerates the leakage of tissue factor into circulation, which causes even more thrombin generation. This positive feedback causes painless systemic inflammation that exaggerates blood viscosity and coagulability, which threatens death.

Spinal Cord Shock Syndrome The spinal cord is confined within a narrow protective bony canal, so that spinal cord trauma causes edema and swelling that compromises perfusion in the bruised tissue. This explains why spinal cord function remains intact for a few hours after injury, and then deteriorates as increasing edema compromises spinal cord function and culminates in permanent spinal cord damage, so that the patient becomes hemiplegic or paraplegic. Stress theory suggests that such damage can be prevented or minimized using measures that inhibit stress mechanism hyperactivity and maintain microvascular perfusion, such as EDTA and Carbogen.

Surgical Stress Syndrome (SSS) Surgery simultaneously stimulates all three pathways of stress mechanism activation and causes stress mechanism hyperactivity known as the Surgical Stress Syndrome (SSS) that is analogous to ARDS and MOFS. The resulting stress mechanism hyperactivity can be substantially muted using general endotracheal anesthesia supplemented by hypercarbia and narcotics, but these measures have no effect on the tissue repair pathway, and cannot eliminate the SSS. The ideal solution would be a benign and safe antidote to tissue factor in blood circulation that could be administered at the conclusion of surgery to mitigate the dangerous positive feedback reaction that reaches peak effect about 48 hours after surgery. In the absence of such an antidote, it may be possible to mute the positive feedback reaction using intravenous $MgSO_4$ infusions to mitigate harmful thrombin generation within a few hours after the conclusion of surgery, after hemostasis is assured.

Trauma is analogous to surgery. It activates all three stress mechanism pathways and can cause stress mechanism hyperactivity that manifests as ARDS and MOFS. It should be managed with the same combination of anesthesia, analgesia, permissive hypercarbia, normothermia, and thrombin suppression used to prevent the SSS.

Disseminated Intravascular Coagulation (DIC) is sudden, unexpected, life-threatening systemic arterial coagulation that disrupts oxygen transport and delivery to cells, consumes and depletes coagulation enzymes and substrates, produces ineffective versions of insoluble fibrin, and causes severe bleeding. It is clinically confusing because coagulation involves a fluctuating balance of multiple factors that promote and prevent coagulation.

Arterial coagulation (thrombosis) normally occurs when tissue damage exposes tissue factor to blood enzymes. This generates thrombin and insoluble fibrin in the exposed tissues. Blood flow dynamics thrust platelets into the damaged tissues, where thrombin activates them, causing them to adhere to one another, and insoluble fibrin binds them into a white clot. Soon thereafter rising levels of insoluble fibrin bind red cells into a stronger viscoelastic red clot that isolates the damaged tissues from flowing blood, halts blood loss, and regulates the exposure of blood enzymes to tissue factor in the damaged tissues beneath its surface. Clot formation is normally limited to the immediate area of tissue damage because factor VIII cannot penetrate the clot of its own manufacture. Thus, clot formation normally involves a balance of blood turbulence, tissue damage, and blood enzyme activity.

Red cells regulate blood turbulence. Excessive red cell mass lowers blood turbulence to the point that spontaneous coagulation takes place. Inadequate red cell mass exaggerates blood turbulence to the point that coagulation cannot take place.

Excessive red cell mass caused by *polycythemia vera*, excessive transfusion with packed red blood cells, or overenthusiastic treatment with erythropoietin (Epogen) can elevate red cell mass to dangerous levels that reduce blood turbulence below a threshold where DIC begins spontaneously. Professional athletes sometimes treat themselves with Epogen to artificially elevate red cell mass because this reduces turbulent blood flow resistance and optimizes exercise capacity. However, the practice is inherently dangerous because it invites DIC.

Inadequate red cell mass allows excessive turbulence that inhibits coagulation. This can cause unexpected bleeding that mimics DIC, but it doesn't halt oxygen transport and delivery, and is thus less lethal. It is most often caused by chronic renal failure, which inhibits erythropoietin production by the kidneys that causes severe anemia. It can also be caused by excessive intravenous fluid administration that dilutes red cell mass and coagulation substrates. Such bleeding can be controlled by transfusion with packed red blood cells.

Pre-existing disease exaggerates the risk of DIC, because stress mechanism hyperactivity generates insoluble fibrin that exaggerates blood viscosity and coagulability. A common example is cancer, which exaggerates blood viscosity and coagulability by shedding tissue factor into systemic circulation.[242, 243] This explains why most cancer victims succumb to strokes, heart attacks, and pulmonary embolus as opposed to catastrophic bleeding or invasive disruption of organ tissues as commonly supposed. All forms of illness caused by stress mechanism hyperactivity exaggerate blood viscosity and coagulability in a similar manner.

DIC most often occurs in the following situations:

1. *Eclampsia*, where amniotic fluid embolus suddenly introduces large quantities of tissue factor into systemic circulation.
2. *Severe trauma*, such as automobile accidents, that release excessive quantities of tissue factor into systemic circulation.
3. *Sepsis*, which increases the permeability of the vascular endothelium and exaggerates the leakage of tissue factor into systemic circulation.
4. *Prolonged surgical procedures* with extensive tissue disruption and inadequate supplemental analgesia.
5. *Multi-Organ Failure Syndrome*, especially when it is complicated by sepsis.

Once DIC begins, there is no effective treatment. The best strategy is prevention, as follows:

1. Prevent hypothermia, which induces stress mechanism hyperactivity.
2. Prevent nociception using analgesic blocks and narcotics.
3. Prevent fear and anxiety using hypnotic agents or general anesthesia.
4. Prevent sepsis using effective antibiotic treatments.[244]
5. Perform only essential surgical repairs in stages at least two days apart to avoid excessive tissue factor release.
6. Maintain red cell mass within normal limits using transfusions where necessary.
7. Maintain permissive hypercarbia to minimize peripheral flow resistance and optimize blood turbulence.

HEART DISEASE

Like rheumatoid disease, heart disease largely escaped medical notice until after the Industrial Revolution. In 1768 William Heberden, an English physician, coined the term *angina* (heart pain that accompanies myocardial ischemia). In 1880 William Osler, who was one of the four founding professors at Johns Hopkins medical school, attributed myocardial infarction to thrombus (arterial blood clot) formation that blocks coronary artery flow. In 1912 James B. Herrick recommended bed rest for several weeks following myocardial infarction, and in 1928 Parkinson and Bedford recommended morphine. Aspirin, sublingual nitroglycerin, and oxygen supplementation were recommended in the early 1950s.[245]

CONVENTIONAL EXPLANATIONS OF HEART DISEASE

Presently prevailing perspective has changed little since the days of Osler. It attributes angina (heart pain) to *stenotic atherosclerotic lesions* that obstruct blood flow. It attributes infarction to the release of debris from atherosclerotic lesions, which induces clot formation that obstructs blood flow to distal tissues. This conventional concept has numerous shortcomings when viewed in the light of accumulating evidence:

- It alleges that thrombus formation causes infarction but it fails to explain why thrombus formation occurs *after* infarction damage becomes permanent.
- It fails to explain why angina (infarction pain), autonomic imbalance, and microvascular flow disturbances occur *before* thrombus formation.
- It fails to explain why powerful anticoagulants such as streptokinase, urokinase, and TPA (Tissue Plasminogen Activator) must be administered *before* thrombus formation to prevent permanent myocardial damage.[246, 247, 248]
- It fails to explain why powerful anticoagulants such as streptokinase sharply lower blood pressure even as they restore perfusion to infarcted tissues, and despite the anticoagulants' lack of effect on muscle or nervous activity.[249]
- It fails to explain how stress induces infarction.[250, 251]
- It fails to explain *emergence infarction* where patients suffer infarction as they emerge (regain consciousness) after the completion of surgery despite the absence of atherosclerosis.[252]
- It fails to explain why analgesia prevents infarction during anesthetic emergence.[253]
- It fails to explain *sudden death syndrome* where extreme fright or pain precipitates fatal infarction despite the absence of atherosclerotic debris or clot formation.
- It fails to explain the increased von Willebrand Factor levels, Factor VIII activity, blood viscosity, and blood hypercoagulability that occurs in sudden death syndrome, where victims are exposed to extreme fright such as earthquakes and gunfire.[254, 255]
- It fails to explain why factor VII treatment for hemophilia and von Willebrand disease invites infarction and cancer, while factor VIII treatment does not.[256]

- It fails to explain the *no reflow phenomenon*, where stent installation restores arterial flow but fails to restore microvascular perfusion.[257, 258, 259, 260, 261]
- It fails to explain the nature of atherosclerosis.
- It fails to explain why atherosclerosis occurs in arteries but not veins.
- It fails to explain how advancing age accelerates atherosclerosis.
- It fails to explain why atherosclerosis first appears at the bifurcations and greater curvatures of large proximal arteries but is rare in small distal arteries and arterioles.
- It fails to explain why blood pressure measures higher in small distal arteries than in large proximal arteries.
- It ignores the role of capillary senescence.
- It ignores the non-Newtonian laminar (turbulence-free) nature of arterial blood flow that causes blood flow to exponentially accelerate in the presence of reduced arterial diameter.
- It ignores the close association of amyloidosis and atherosclerosis.

STRESS THEORY AND HEART DISEASE

Stress theory addresses the above-mentioned shortcomings of conventional medical theory. *I hypothesize that amyloidosis causes heart disease by accelerating capillary senescence.*

Senescence is a condition or process of deterioration with age. It occurs in all cells and manifests as disability and disease. Diabetes, hypertension, infarction, and congestive heart failure are rare in youth, because abundant youthful capillaries minimize microvascular flow resistance, which enables cardiac efficiency, tissue perfusion, tissue oxygenation, and glucose uptake. This youthful *capillary reserve* enables capillary gate activity to optimize exercise tolerance and regulate organ function without compromising physiological function. I call the fluctuating microvascular flow resistance that regulates organ function and optimizes exercise tolerance *dynamic flow resistance.*

Capillary flow resistance inexorably increases and capillary reserve declines as capillaries deteriorate and disappear with age. I call this *static flow resistance* (fixed, unchanging flow resistance due to limited microvascular passageways).[262, 263] Static flow resistance increases with age because capillary cells, like all cells, suffer declining cellular mitosis and metabolism in accord with the *Hayflick limit* that determines maximum human life span (normal human cells can replicate and divide no more than forty to sixty times).

Persistent stress mechanism hyperactivity is the fundamental cause of heart disease because it accelerates amyloidosis, which deteriorates capillaries, exaggerates static flow resistance, undermines cleansing pulsatile turbulence, increases cardiac work, damages cardiac tissues, and inhibits cardiac efficiency, tissue perfusion, glucose uptake, and tissue oxygenation. This is the fundamental cause of diabetes, hypertension, congestive heart failure and infarction. Such stress-related microvascular deterioration can be counteracted by exercise, which induces angiogenesis (capillary proliferation) that improves cardiac efficiency, reduces blood pressure, reverses atherosclerosis, and enables trained athletes to increase their exercise performance ten-fold without increasing normal blood pressure. Exercise can thus optimize or even restore health, but it cannot alter the Hayflick limit on maximum life span. Nobody has yet managed to outrun the reaper, not even David Rockefeller, who died at 101 despite five heart transplants.

ATHEROSCLEROSIS

Pulsatile turbulence efficiently mobilizes molecular debris from the inner walls of arteries to inhibit atherosclerosis, as exemplified by brain-damaged, bedridden patients who survive for years without developing atherosclerosis because they are immune to harmful fear and anxiety. However, increasing static flow resistance alters pulsatile turbulence, which exaggerates turbulent lateral forces that elevate blood pressure, and undermines turbulent cleansing that mobilizes molecular debris from the inner walls of arteries. The accumulating debris induces inflammatory tissue repair activity that causes atherosclerosis. This explains why atherosclerosis is closely associated with hypertension, and why it begins at the curvatures and bifurcations of large arteries close to the heart, where turbulence is reduced, and why atherosclerosis is rare in arteries and arterioles distant from the heart, where declining arterial diameter exaggerates pulsatile turbulence.

Atherosclerosis causes stenotic lesions that reduce arterial diameter. This does not obstruct arterial flow as commonly assumed. Instead, friction-free laminar arterial blood flow accelerates exponentially in accord with decreasing diameter as it moves through the stenotic segment. This maintains flow volume in the stenotic segment. However, as the accelerated blood emerges into the greater diameter distal to the narrowed segment, its kinetic energy converts into turbulence that inhibits arterial flow. This undermines perfusion and oxygenation in the downstream tissues served by the artery. The turbulence sometimes produces audible noise that explains why *bruits* occur distal to atherosclerotic lesions. Thus, tissue perfusion and oxygenation are compromised by combinations of static (capillary) flow resistance and atherosclerotic lesions that induce turbulent flow resistance. The resulting cellular oxygen starvation renders tissues vulnerable to infarction.

ESSENTIAL HYPERTENSION AND TYPE II DIABETES

Essential hypertension and Type II diabetes are the same disease. Accelerated capillary senescence simultaneously increases microvascular flow resistance, which causes essential hypertension, and inhibits microvascular perfusion, which undermines cellular glucose uptake, and causes Type II diabetes. The first warning of the disease is albuminuria caused by harmful systemic stress mechanism hyperactivity that disrupts the glycocalyx, releases albumen protein into blood circulation and causes albuminuria, accelerates capillary senescence, compromises tissue perfusion, and exaggerates static flow resistance. At first, blood pressure remains normal when the patient is resting, but exercise and emotional excitement cause abnormal blood pressure elevations. As capillary compromise continues, blood pressure and blood glucose levels remain continuously elevated, and insulin requirements increase.

CONGESTIVE HEART FAILURE

Mammalian hemodynamic physiology is a marvel of efficiency, as opposed to power and force. The heart is a weak muscle that generates the same force with each contraction, and cannot significantly increase its contractile force for any length of time without suffering damage. It can barely propel blood as high as the brain, which explains why most mammalian heads are located immediately above the heart. The giraffe requires an outsized heart to propel blood to its elevated head. Worse yet, muscle cells suffer exaggerated energy requirements, because ATP is needed to energize both contraction and relaxation.[264]

Capillary senescence exaggerates microvascular flow resistance, causes hypertension and diabetes, increases cardiac work, elevates cardiac oxygen requirements, and simultaneously undermines oxygen delivery to heart tissues, causing chronic cardiac oxygen starvation. The cardiac tissue hypoxia induces inflammatory fibroblast repair activity that deposits collagen between cardiac muscle cells.[265, 266, 267, 268] As collagen thickens cardiac walls, the heart assumes a distinctive pear shape that characterizes early heart failure. The collagen compromises cardiac contractility, which limits the ability of the heart to accelerate blood during systolic ejection, undermines the ability of the turbulent pulse wave to maintain arterial patency, and accelerates atherosclerosis.[269] This explains the close relationships of diabetes, hypertension, atherosclerosis, and congestive heart failure.

The heart is enclosed within a pericardial sac consisting of two layers of tough connective tissue that are separated from each other by lubricating pericardial fluid. The inflexibility of this pericardial sac tissue explains why bleeding into the pericardial sac causes potentially lethal *cardiac tamponade* by pressing on the exterior of the heart and preventing diastolic filling of the ventricles. The purpose of the pericardium is easily overlooked: it prevents overdistension of the heart due to excessive venous filling pressure that can stretch and damage the cardiac muscle.

Congestive heart failure is caused by the inability of the heart to completely expel its contents during systolic contraction. This increases the filling pressure of venous return to the heart, which eventually causes the pericardium to stretch, whereupon the failing heart "balloons" as its ventricular volume increases. This, in turn, harmfully stretches cardiac muscle cells to an extreme, causing damage that further undermines cardiac contractility. This final stage of congestive heart failure is a prelude to death. However, the heart can be "rescued" using ventricular restraint therapy and/or LVAD (Left Ventricular Assist Device) that relieves the strain on myocardial cells and allows them to recover.[270, 271, 272] Ventricular restraint therapy is surgery that encloses the heart with mesh material that substitutes for the stretched pericardium, and relieves the overstretched heart muscle. The LVAD device propels blood from the left ventricle into the aorta. These measures often allow the heart muscle to recover.

The unappreciated role of the pericardium has confused research reasoning. When Yandell Henderson performed dog experiments in 1936 to investigate the effect of carbon dioxide on blood circulation, he devised a crude but clever cardiometer by enclosing the dog's heart within an airtight child's rubber ball that was connected to a tambour by a large rubber tube. With this makeshift apparatus he was able to record a volume curve of the heart. In so doing he enclosed the heart in such a way as to prevent excessive expansion of the ventricles. In other words, his apparatus unwittingly substituted for the pericardium, and prevented damaging distention of the heart. As a result, he correctly concluded that hyperventilation depletes CO_2 and causes cardiovascular collapse, while CO_2 supplementation beneficially enhances cardiac output and venous return to the heart.[273] About 20 years later Kenneth J. Boniface and John M. Brown performed a similar experiment, except that they opened the pericardium and sewed a strain gauge apparatus to the exposed dog heart muscle, and thus unwittingly left the heart directly exposed to the atmosphere without pericardial protection. When they exposed the dog to CO_2 supplementation, they observed a sharp decrease in cardiac contractility that was accompanied by exaggerated venous return that caused the heart to balloon so severely as to disrupt the function of their strain gauge apparatus. As a result, they mistakenly concluded that carbon dioxide toxicity had poisoned the heart. In reality, they were observing a beneficial decrease in microvascular flow resistance that reduced cardiac work, and thus decreased cardiac contractility, and exaggerated venous return that distended the exposed and unprotected ventricles.[274] This lingering confusion lurks in the lore of current medicine.

INFARCTION

From the perspective of stress theory, *infarction* represents the culmination of accelerated capillary senescence caused by perpetual pollution. The capillary deterioration undermines tissue perfusion and oxygenation, accelerates atherosclerosis, impairs heart function, and sets the stage for infarction in tissues with the greatest oxygen requirements, namely the brain, bowel, and heart.

Infarction is a microvascular event caused by cellular oxygen starvation that triggers cyclical, spasmodic opening and closing of the capillary gate. It culminates in cell death and permanent tissue damage unless microvascular perfusion is promptly restored, so that the oxygen starvation is relieved prior to cell death. The cause of the cyclical opening and closing of the capillary gate remains in question, but it appears to be a combination of autonomic instability, anaerobic cellular respiration which produces a painful lactic acidosis that closes the capillary gate, and cellular CO_2 production that opens the capillary gate.[275]

Infarction is rare in youth when capillary beds are intact. The incidence and severity of infarction increases in accord with compromised tissue oxygen delivery. Combinations of accelerated capillary senescence, atherosclerosis, and congestive heart failure invite infarction by compromising tissue perfusion and oxygenation. In marginally oxygenated tissues, infarction can be precipitated by sudden sympathetic nervous activity induced by painful trauma, intense fear such as exposure to earthquakes and gun battles, or acute pain as perceived suddenly during anesthetic emergence in the absence of adequate analgesia. Such sudden, intense nervous activity closes the capillary gate and reduces tissue oxygenation below a critical threshold in tissues with pre-existing hypoxia. Alternatively, atherosclerotic debris can suddenly obstruct downstream oxygen delivery and precipitate infarction.

Infarction occurs most often in tissues with elevated oxygen requirements, such as brain, bowel, heart, and lung tissues. When infarction occurs in the heart, it is called a *heart attack*. When it occurs in the brain, it is called a *stroke*. However, the term stroke is confusing because it encompasses *ischemic stroke* caused by brain hypoxia and *hemorrhagic stroke* caused by brain bleeding. Hemorrhagic strokes are caused by either toxic drugs, such as *Non-Steroidal Anti-Inflammatory drugs* (NSAIDs) or brain aneurysms that cause blood to leak into brain tissue. The distinction is important because powerful anticoagulants that cure ischemic infarction may worsen hemorrhagic infarction. As we will see later, Carbogen is the perfect emergency treatment for both strokes and heart attacks because it directly opens the capillary gate and enhances oxygen release from hemoglobin to relieve cellular hypoxia, but has no effect on blood coagulability and does not exaggerate bleeding.

The heart is the most prominent target organ for infarction, even though it is lacking in tissue factor, and thus generally resistant to systemic stress mechanism hyperactivity. This is because cardiac muscle, like all muscle, requires large quantities of oxygen and nutrients to generate ATP for *both* contraction *and* relaxation,[276] and because specialized nervous conduction pathways coordinate cardiac contractions. Infarction ischemia disrupts the conduction pathways and causes disordered cardiac muscle cell contractions called *dysrhythmias* that paralyze cardiac function and disrupt oxygen transport and delivery.[277] Hypoxic brain damage soon follows unless cardiac function is promptly restored. Those who survive such episodes will likely suffer cell death and permanent damage in the oxygen-starved tissues. Failure to restore perfusion and oxygenation promptly causes cell death and permanent damage in the oxygen-starved tissues.

The History of Anticoagulant Infarction Therapy

William Smith Tillett recognized the anticoagulant effect of streptokinase in 1933, but it was not embraced as a miracle medication for myocardial infarction until the 1980s after extensive clinical trials proved that streptokinase reliably restores perfusion to infarcted tissues and prevents cell death and tissue damage. Urokinase and TPA (Tissue Plasminogen Activator) were likewise proved effective. Regardless of their proven safety and efficacy, these medications remained controversial because their effects are at odds with entrenched medical dogma, which assumes that blood pressure is the driving force of cardiac output, blood flow, and tissue perfusion, and the persuasive notion that atherosclerotic flow obstruction and thrombus formation causes angina and infarction. The shortcomings of these notions have been presented earlier in this book. In reality, anticoagulants must be administered *before* thrombus formation obstructs blood flow to infarcted tissues, and the resulting hypotension reflects successful treatment that reduces flow resistance and restores perfusion to oxygen-starved tissues.

Despite these obstacles, marketing support enabled anticoagulants to became the “gold standard” for treating heart attacks during the 1980s. Sadly, soon after their patents expired, they were supplanted by complex, dangerous, and expensive cardiac bypass surgery. However, bypass surgery was soon succeeded by angioplasty on account of its poor risk/benefit ratio. It became common to transport infarction victims to distant hospitals for emergency angioplasty, but accumulating evidence now indicates that angioplasty is no more beneficial than bypass surgery.[278, 279] It now seems reasonable that anticoagulants, which offer an excellent risk/benefit ratio, should be restored to their status as the preferred treatment for myocardial infarction, even though their patents have expired, now that stress theory explains why their therapeutic benefits cause hypotension. Meanwhile chelation therapy, which offers a safe and effective means to treat atherosclerosis and reduce the risk of infarction, remains ignored.[280]

Stress Theory and Infarction Treatment

From the perspective of stress theory, both coronary artery bypass surgery and angioplasty are founded on false science. They assume that surgical relief of stenotic obstructions can reliably restore perfusion, and they ignore the role of capillary senescence and microvascular ischemia. They may help in some cases by relieving stenotic turbulence, but they cannot reliably restore microvascular oxygenation and halt the infarction process. This is known as the *no reflow phenomenon*.[281, 282, 283, 284] In contrast, Carbogen and anticoagulants reliably open the capillary gate, restore microvascular oxygenation, halt infarction, and prevent cell death and tissue damage.

The best emergency treatment for myocardial infarction is Carbogen, which was an accepted medical treatment 100 years ago but is now abandoned and forgotten.[285] It is an inexpensive mixture of oxygen and carbon dioxide compressed in a tank that can be carried on emergency vehicles and conveniently administered by amateurs outside hospitals. It optimizes all aspects of oxygen transport and delivery, and quickly relieves the tissue hypoxia that causes infarction. It stimulates breathing, releases nitric oxide to open the capillary gate, reduces flow resistance to increase cardiac efficiency, and releases oxygen from the hemoglobin molecule to increase tissue oxygenation. With the infarction thus controlled, Carbogen can be supplanted by anticoagulant treatment in the hospital.

Carbogen also offers a perfect emergency treatment for stroke because it does not affect coagulation. Brain infarction causes most strokes, but about 20% of strokes are caused by brain bleeding (hemorrhagic strokes). Anticoagulants effectively treat ischemic strokes, but they can aggravate hemorrhagic strokes. Thus, anticoagulant stroke treatment necessitates time-consuming brain scans to rule out the possibility of brain bleeding. This delays treatment and invites permanent brain damage. In contrast, Carbogen treatment can be initiated immediately to relieve brain infarction without fear of aggravating brain bleeding. If Carbogen relieves stroke symptoms, then it is safe to conclude that the stroke is caused by infarction, and that anticoagulants can be administered. If Carbogen fails to improve stroke symptoms, the stroke victim can be scheduled for brain scan to confirm brain bleeding.

The second most effective treatment for infarction is powerful anticoagulants that paralyze the capillary gate and restore perfusion and oxygenation to infarcted tissues. Unfortunately, anticoagulants are largely abandoned at present in favor of angioplasty because anticoagulants can cause disconcerting declines in blood pressure, despite their lack of effect on muscle, and regardless of evidence that such hypotension is harmless and beneficial. Stress theory explains why hypotension confirms the therapeutic success of anticoagulant treatments.

Patients suffering acute infarction should be immediately treated and stabilized with Carbogen and anticoagulants. They should never be subjected to alarming ambulance rides to distant locations to obtain unreliable cardiac bypass or angioplasty treatment, for fear of aggravating infarction and inviting death. After they have recovered from the acute infarction, they can be treated with chelation therapy to eradicate atherosclerotic lesions.

In hospital settings, complimentary combinations of Carbogen, anticoagulants, narcotics, and sedatives can optimize infarction treatment. EDTA, trisodium citrate, $MgSO_4$, and sodium nitroprusside may be even more effective than streptokinase, urokinase, and TPA for infarction relief. They do not decrease cardiac output, even though they cause disconcerting decreases in blood pressure that reflects their ability to open the capillary gate, reduce microvascular flow resistance, and restore perfusion and oxygen delivery to infarcted tissues. Thus, properly understood, the hypotension caused by these medications confirms therapeutic success.

Prevention is better than cure. Weight loss slows capillary senescence. Exercise promotes capillary proliferation that counteracts capillary senescence and enhances the ability of blood turbulence to reverse atherosclerosis and restore arterial patency. Chelation therapy likewise exaggerates blood turbulence, disintegrates atherosclerotic lesions, and restores arterial patency. It is safer and arguably as effective as bypass surgery and angioplasty.

MALIGNANCY (CANCER)

Madness is rare in individuals — but in groups, parties, nations, and ages it is the rule.
— Friedrich Nietzsche

“I will abstain from all intentional wrong-doing and harm.”
— Hippocratic Oath

It would be difficult to find a more outrageous example of scientific insanity than the logical inconsistencies and false hopes promoted by the presently prevailing rationale for cancer and its treatments. As wryly observed in *Scientific American*, *"Today's cancer chemotherapy consists of little more than a dismal array of toxic drugs that kill healthy cells along with cancerous ones. Physicians must often play a deadly game of trial and error, hoping to find the right dose of the right medicine before time runs out."*[286] The same may be said of surgery and radiation. All these "treatments" are proven *causes* of cancer, and none of them offers a reliable cure.[287, 288, 289] They can be compared to pouring fuel on a fire: if poured fast enough, it might smother the fire, but it leaves smoldering embers lying in fuel that threatens further fires. Likewise, harmful cancer "treatments" entail pain, disability, debilitation, disfigurement, and distress that temporarily shrinks or excises tumors, only to be followed by recurrence, the inexplicable appearance of seemingly unrelated cancers, and early death from seemingly unrelated causes such as embolus and infarction.[290, 291, 292, 293, 294] On those rare occasions where a patient enjoys spontaneous regression despite these measures, the treatment is celebrated as a "cure."[295, 296, 297, 298, 299, 300, 301, 302, 303, 304]

CANCER HISTORY

Physicians have traditionally regarded cancer as an unexplained systemic illness with localized manifestations that is generally best left untreated; but since the discovery of DNA cancer has come to be regarded as a localized disease with systemic manifestations that can be cured by eliminating the cells rendered malignant by the "defective DNA." Devra Davis, a professional epidemiologist, describes the history that led to today's reversal of traditional medical perspective in her book "The Secret History of the War On Cancer."[305]

During WWII the US Liberty ship SS John Harvey transported an illegal secret cargo of mustard gas to the Italian port of Bari, where it was sunk by German aircraft. The resulting release of mustard gas killed and maimed multitudes of civilians and sailors. Dr. Stewart Francis Alexander investigated the disaster and reported that mustard gas destroys white blood cells. This inspired Louis Goodman and Alfred Gilman, whose wartime research was financed by government subsidies, to investigate the possibility that toxic war gas derivatives might treat incurable leukemia, which is characterized by pathological profusion of white blood cells. By chance they also tested mustard gas derivatives on a single mouse with a solid tumor, causing the tumor to shrink dramatically. Based on this single result, which they were unable to replicate, they treated a dying silversmith. Though the silversmith died days later, they remained convinced that toxic chemicals can cure cancer. Soon afterward, the discovery of DNA inspired the scientifically untenable notion that environmental stress causes cancer by inducing genetic mutations in the DNA of individual cells, causing them to multiply wildly and invade adjacent tissues. Ironically, this was at odds with the famous findings of Max Delbruck and Salvador Luria, the mentors of James Watson, the co-discoverer of DNA. Delbruck and Luria had proved that mutations occur at random, and are not related to environmental stress.[306]

The defective DNA notion was fueled by the advertising skills of Mary and Albert Lasker, who founded the American Cancer Society, created the Lasker Foundation, and lobbied for government cancer research and treatment subsidies, which fueled toxic chemotherapy, mutilating surgery, and destructive radiation therapy that soon became entrenched orthodoxy. Such treatments are typically justified by statistics, which can be manipulated to obscure the harmful side effects of profitable pharmaceuticals. The weary result is documented in countless obituaries. Patients are first terrified by the ominous diagnosis, and then tortured with toxic treatments that confer false hope by temporarily shrinking tumors, but the cancers stubbornly re-appear, along with additional, inexplicably unrelated cancers in distant locations. Such treatments are worse than the disease itself. They often cost up to a million dollars per patient, most of which accrues to pharmaceutical and hospital corporations. Finally, the patient dies prematurely of unknown or unrelated causes.

Cancer seldom kills by invading tissues, disrupting vital organs, or causing catastrophic blood loss. Instead, it exaggerates blood viscosity and coagulability, which causes heart attacks, strokes, pulmonary embolus, and “unexplained” deaths that are seldom attributed to cancer. Spontaneous remission, and/or survival for five years, is arbitrarily attributed to successful treatment.[307, 308, 309] Statistics thus obscure the true lethality of cancer and overstate treatment success. Meanwhile, the incidence of cancer, asthma, autism, diabetes, heart disease, and rheumatoid diseases has risen in accord with increasing industrial and agricultural pollution, automobile exhaust, chlorinated water, and tobacco abuse.[310] The recent small statistical decline in cancer is credited to improved treatments, but it is best explained by the government campaign to discourage cigarette smoking that began during the Clinton administration. It can be anticipated that public health will similarly benefit as electricity replaces petroleum, but much more needs to be done to prevent the pandemic of cancer and chronic illness.

CANCER CELL CLASSIFICATION

Conventional theory attributes cancer to defective DNA, which implies that cancerous tissues and tumors should consist of a single cell type. However, cancerous tumors and tissues invariably consist of various unrelated cell types, all replicating at independent rates. The most common cell type is fibroblast. These are tissue repair cells that are found in all tissues. They differentiate into other cell types and produce collagen to facilitate normal tissue maintenance and tissue repair. Thus, it is impossible to classify cancers with precision, and more than 100 terms are used to describe various cancer tissues. Conventional theory regards each of these terms as if they reflect a unique type of DNA damage. This assumption is at odds with science which seeks the simplest available explanation. In contrast, stress theory explains cancer as tissue repair hyperactivity that has induced a state of self-sustaining tissue repair activity, as follows:

- I hypothesize that malignancy is an abnormal state of self-sustaining mitosis (cell multiplication) induced by excessive thrombin generation.

- I further hypothesize that apoptosis is the antithesis of malignancy: it is harmless cell suicide induced by thrombin starvation that enables embryological development and concludes normal tissue repair.

MALIGNANCY

Recall that tissue damage exposes blood enzymes to tissue factor, which elevates the thrombin generation that energizes tissue repair. Normally the tissue repair process restores the barrier between tissue factor and blood enzymes, and returns thrombin levels to normal levels, whereupon apoptosis concludes the tissue repair process. However, this normal tissue repair process can go haywire in the face of unremitting combinations of environmental stress including emotional distress, toxic chemicals, sepsis, radiation, surgery, trauma, smoking, and obesity that continuously exaggerate thrombin generation. These stresses cause cell proliferation, cell differentiation, inflammation, immune activity, and collagen production that invades adjacent tissues and sheds additional tissue factor into blood.[311, 312, 313, 314, 315, 316, 317, 318, 319, 320, 321, 322, 323, 324, 325, 326] Such abnormal tissue repair activity stimulates nociception that activates factor VIII and sheds tissue factor into systemic circulation that activates factor VII. This in turn accelerates insoluble fibrin generation and harmfully exaggerates blood viscosity, blood coagulability, inflammation, chemotaxis, and mitosis in a self-perpetuating manner.[327, 328] The hyperactivated cells exhibit altered PAR receptor configurations on their surfaces (recall that PAR receptors are thrombin receptors that determine how cells react to thrombin elevations).[329, 330, 331, 332, 333, 334, 335, 336, 337] Such stress mechanism hyperactivity returns to normal if stress subsides, whereupon thrombin starvation causes the hyperactivated cells to undergo spontaneous apoptosis (harmless cell death).[338, 339, 340] However, severe, unremitting stress sometimes causes tissue repair activity to become malignant (i.e. self-sustaining). The malignant hyperactivity produces tumors that consist primarily of fibroblast repair cells and the collagen they produce because fibroblasts are exquisitely sensitive to thrombin, but multiple cell types are incorporated in tumors, all replicating at various accelerated rates, because thrombin energizes mitosis in all types of cells. This observation is at odds with the notion that a single DNA defect causes cancer, for if that were the case, then tumors would consist of a single type of cell.[341, 342]

There is no evidence that malignant cells are different from normal cells. Malignant cells cannot be distinguished from normal tissue repair cells engaged in active tissue repair. In both cases, cells exhibit accelerated mitosis, broken chromosomes, and altered PAR configurations.[343] Cancers are classified by the microscopic appearances of tumor fibroblasts, which are specialized to serve the needs of various tissues and therefore have differences that can be discerned via a microscope. Glandular tissues form tumors called *adenomas*; thus, there are pituitary adenomas, thyroid adenomas, prostate adenomas, bowel adenomas, and so forth. Epithelial (skin) cells produce *carcinomas*. Mesothelial cells produce *sarcomas*. All tumors begin as “benign,” but they become “malignant” when their hyperactivity becomes self-sustaining, depending on the degree and duration of stress mechanism hyperactivity.

Cancer concepts are riddled with untenable assumptions. *There is no evidence that immune activity either causes or prevents cancer, rheumatoid disease, inflammation, or any aspect of tissue repair.* Furthermore, there is no hypothetical mechanism that might explain such a claim. The stress mechanism explains how tissue repair generates thrombin, which energizes and thereby elevates immune activity that prevents infection and remove foreign debris.

Cancer cannot be explained by DNA mutation. Mutations occur at random and are not influenced by environmental stress. This was famously proved by Max Delbruck and Salvador Luria, who were the most famous scientists in the world of medical research before Watson and Crick discovered DNA.[344, 345]

The notion that multiple independent mutations cause distinctly different types of cancer is at odds with science, which seeks the simplest explanation. Stress theory offers the simplest explanation that is consistent with available evidence: unrelenting environmental stress induces malignant stress mechanism hyperactivity.

Metastasis

Malignant stress mechanism hyperactivity exaggerates systemic thrombin generation, which induces systemic inflammation. Tissues rich in tissue factor such as brain, lung, gonads, and cervix are most readily affected by systemic thrombin elevations. This explains why malignancy occurs most often in these tissues. Hyperactivated cells are continually shed from malignant tumors and transported downstream to distant locations by blood and lymph flow. Such *metastatic tumors* thrive best in tissues that are rich in tissue factor. For example, “primary” tumors in the lung, which is rich in tissue factor, preferentially metastasize to the brain, which is directly downstream from the lung and similarly rich in tissue factor. Such metastatic tumors seldom thrive in tissues lacking in tissue factor, such as muscle and connective tissue. Blood and lymph flow also affect metastasis. For example, all bowel venous blood flows through the liver, which filters bowel blood to remove debris and toxins, so that bowel adenomas metastasize to the liver, despite its relative lack of tissue factor. Hodgkin’s lymphoma propagates via the lymph system.

Apoptosis

Apoptosis is the antithesis of malignancy. It is a harmless and normal process of cell “suicide” that enables embryological development and tissue repair. It differs from necrotic or traumatic cell death, which releases harmful cellular substances such as actin.[346, 347, 348, 349, 350, 351, 352, 353, 354, 355, 356] Apoptosis is presently defined as *programmed cell death* but its nature remains unexplained by conventional theory.

Stress theory explains apoptosis as part of the tissue repair mechanism. The stress mechanism elevates thrombin generation to energize fibroblast proliferation and collagen generation, immune cell activity, and angiogenesis (proliferation of vascular endothelium cells to form capillaries) to enable normal tissue repair. Thrombin generation normally declines as tissue repair nears completion and the integrity of the vascular endothelium is restored, causing thrombin starvation, which induces apoptosis that enables wound closure. Excessive and persisting thrombin elevations inhibit apoptosis, prevent resolution of the tissue repair process, and cause malignancy; and extreme thrombin elevation (usually a laboratory phenomenon) kills cells.[357, 358, 359, 360, 361, 362, 363]

A SUMMARY OF CANCER CHARACTERISTICS

Stress mechanism hyperactivity explains the confusing characteristics of cancer:[364]

- Cancer is a systemic illness with localized manifestations. Stress mechanism hyperactivity generates thrombin elevations that energize tumor formation.

- Tumors occur most readily in tissues rich in tissue factor, such as brain, retina, nerves, lung, gonads, cervix, and placenta.[365] Such tissues are "targets" for both primary and "metastatic" tumors.

- Toxic chemotherapy and harmful radiation preferentially kill metabolically active cells. Malignant cells exhibit metabolic hyperactivity due to excessive thrombin elevations, and thus exhibit exaggerated vulnerability to toxic chemotherapy and harmful radiation compared to quiescent cells, but cells that normally exhibit elevated metabolic activity, such as bowel cells, also exhibit exaggerated vulnerability to chemotherapy and radiation.

- Surgery excises tumors, but it counterproductively exaggerates stress mechanism hyperactivity by stimulating nociception and releasing tissue factor into blood circulation, so that it promotes additional cancer.[366]

- Analgesia inhibits cancer caused by surgery.[367, 368]

- The "vicious cycle" of malignancy sustains systemic stress mechanism hyperactivity and inflammation that increases the risk of both metastasis and the appearance of additional, seemingly unrelated types of cancer in locations distant from the original cancer.[369]

- Malignancy disrupts adjacent normal tissues, releases tissue factor into systemic circulation, and increases factor VII activity.

- Malignancy harmfully exaggerates blood viscosity and coagulability by activating nociception that exaggerates sympathetic nervous activity, releases von Willebrand Factor from the vascular endothelium, and increases factor VIII activity.

- Malignancy continuously sheds tissue factor into systemic blood circulation, which activates factor VII, which promotes abnormal tissue repair activity that accelerates cell proliferation.[370]

- Cancer seldom kills by causing fatal bleeding. Instead, the simultaneous activation of factors VII and VIII exaggerates blood viscosity and coagulability, which invites myocardial infarction, strokes, and pulmonary emboli that cause most cancer deaths. Such deaths are seldom attributed to cancer.

- Cancer sometimes culminates in Multi-Organ Failure Syndrome due to positive feedback that grossly exaggerates self-sustaining stress mechanism hyperactivity.[371]

- Malignancy elevates immune activity, but there is no convincing evidence that immune activity affects malignancy.

- Cancer often resolves spontaneously if environmental stresses subside, causing thrombin starvation that disrupts the vicious cycle of malignancy and induces apoptosis, so that the cancerous tissue harmlessly disappears. Thus, apoptosis is the therapeutic objective of cancer treatments.

- Cancer occurs more commonly in humans than less intelligent vertebrates due to their exaggerated sentience, fear, and anxiety.

- Benign tumors differ from malignant tumors only in the degree of stress mechanism hyperactivity. Both are induced by stress mechanism hyperactivity and thrombin elevations, but greater thrombin elevations are necessary to induce self-sustaining malignancy.

- Thrombin energizes cell proliferation in all types of cells thus far tested. Tumors consist of multiple cell types that multiply at individual rates in response to thrombin elevations.

- Cancer cells cannot be distinguished from normal tissue repair cells that are engaged in the most active phase of wound healing. Both cancer cells and normal tissue repair cells exhibit rapid mitosis, broken chromosomes, exaggerated extrusion of tissue factor, and altered PAR configurations.[372]

- Cancer cells are normal cells that are hyperactivated by abnormal thrombin elevations.[373] Cancers are arbitrarily classified in accord with the predominance of specialized cells (mostly fibroblasts, which are specialized tissue repair cells) that predominate in malignant tissues, but this is an arbitrary and misleading distinction. The microscopic appearance of bowel fibroblasts (adenomas) differs from skin fibroblasts (carcinomas) and so forth. Malignant behavior of these normal cells will spontaneously cease when thrombin levels decline to normal, which induces apoptosis, and restores normal cell activity.[374]

- Malignancy continuously stimulates nociception and sheds tissue factor into systemic circulation, which exaggerates systemic thrombin generation. This explains why cancer increases the risk of seemingly unrelated cancers in distant locations.[375]

- Surgery, toxic chemicals, and radiation cannot reliably cure cancer, because they all promote stress mechanism hyperactivity that causes cancer.
- Both primary and metastatic cancers occur most commonly in tissues rich in tissue factor, such as brain, lung, cervix, placenta, and gonads.
- Breast tissue is not rich in tissue factor, but it concentrates blood substances to manufacture milk. As a result, it exhibits exaggerated sensitivity to toxic substances such as chlorine in the water supply.
- Bowel cancer most commonly metastasizes to the liver because blood flow from the bowel drains exclusively to the liver.

CANCER PREVENTION AND TREATMENT

The objective of cancer therapy is apoptosis, which exemplifies cure. Unfortunately, conventional cancer treatment employs toxic chemicals, radiation, and surgery that counterproductively increase stress mechanism hyperactivity, inhibit apoptosis, and promote malignancy. This explains why conventional chemotherapy notoriously accelerates cancer and inhibits life span. Chemotherapy most effectively poisons and kills metabolically hyperactive, rapidly dividing cells. This includes both malignant tumor cells and normal cells that exhibit high levels of mitosis and metabolism, such as bowel and gamete cells. Thus, toxic chemotherapy shrinks tumors and damages normal tissues even as it stimulates the stress mechanism hyperactivity that caused the cancer to begin with. This explains why cancer stubbornly re-appears in the aftermath of chemotherapy treatments, and why both malignancy and chemotherapy exaggerate the risk of additional, seemingly unrelated forms of cancer in distant locations.

Stress theory indicates explains why cancer remission occasionally occurs despite toxic chemotherapy, disfiguring surgery, and harmful radiation.[376, 377] The secret of reliable and effective cure is the opposite of conventional cancer treatments. Cancer remission often occurs spontaneously if environmental stress subsides, whereupon malignant tissue repair hyperactivity normalizes, and thrombin starvation induces apoptosis. Treatments that inhibit stress mechanism hyperactivity can disrupt the vicious cycle of malignancy and cure cancer. *It should be possible to cure cancer quickly, safely, comfortably, and reliably by using combinations of elective endotracheal anesthesia, narcotic analgesia, hypercarbia, and thrombin inhibition with $MgSO_4$, EDTA, or trisodium citrate. However, the available evidence indicates that effective thrombin suppression must be maintained for about fifteen hours to reliably induce apoptosis and cure cancer.* The effective dosage and duration of cancer treatments must be determined by carefully controlled trials, preferably beginning with animal experiments, because stress mechanism activity is essential for life, and excessive stress mechanism suppression is potentially lethal.

In theory, the development of a safe intravenous tissue factor antidote would synergize with sedation and analgesia to control all three pathways of stress mechanism activation, disrupt the vicious cycle of malignancy, and cause malignant tissues to harmlessly shrink and disappear safely and reliably. Such an antidote would be useful to control a wide variety of critical illnesses.

Most cancer can be prevented by eliminating causative environmental stresses including smoking, vaping, chlorinated water, automobile exhaust, pesticides, herbicides, harmful food additives, and toxic industrial waste. Pesticides and herbicides do not harmlessly dissipate in the soil. Instead, they are absorbed into plant tissues, and contaminate the plant food consumed by humans and farm animals. Pollution prevention would minimize chronic illnesses as well as cancer, heart disease, and critical illness. The conversion from petroleum to electricity during the next ten years will predictively benefit public health. Much more can be done by minimizing herbicides and pesticides, eliminating toxic chlorine from drinking water, and minimizing toxic industrial pollution.

TREATMENTS

Stress theory indicates that disease can be cured by controlling harmful stress mechanism hyperactivity. This, however, is easier said than done, because it necessitates the simultaneous control of all three synergistic stress mechanism activation pathways. Furthermore, excessive suppression of stress mechanism activity is potentially lethal. The following discussion of treatments will focus on potentially lethal critical illnesses including the surgical stress syndrome (SSS). These treatments can be employed in synergistic combinations that efficiently halt harmful stress mechanism hyperactivity, inhibit malignancy, promote apoptosis, and restore tissue perfusion, tissue oxygenation, and organ function. However, it should be understood that the synergistic nature of these measures may be so potent as to be lethal. Therefore, they should be carefully evaluated using animal studies before they are cautiously introduced via clinical human trials.

CARBON DIOXIDE: SUPERCHARGING TISSUE PERFUSION, CELLULAR OXYGENATION, AND ORGAN FUNCTION

"Over the oxygen supply of the body carbon dioxide spreads its protecting wings"
— Johannes Friedrich Miescher

The first priority of treatment for critical illness is to optimize oxygen transport and delivery to cells to enhance organ function. This can best be accomplished by maintaining inhaled CO_2 concentrations in the range of 50-100 torr, adjusting inhaled oxygen concentrations to maintain hemoglobin saturation above 90% to assure optimal loading of the hemoglobin molecule with oxygen, and controlling harmful sympathetic nervous activity with generous narcotic analgesia. This will optimize cardiorespiratory function by stimulating respiratory drive, enhancing cardiac output and cardiac efficiency, and elevating tissue oxygen levels. Mechanical hyperventilation that causes hypocarbia is inherently harmful and counterproductive, and should be avoided at all times. A review of carbon dioxide pathophysiology can be found in my published paper called "Four Forgotten Giants of Anesthesia History," that can be downloaded free of charge from my website or the Internet.[378]

ELECTIVE ENDOTRACHEAL INTUBATION

Elective endotracheal intubation (inserting a breathing tube into the windpipe) is fundamental to the management of critical illnesses for the following reasons:

1. It prevents airway obstruction
2. It prevents aspiration of toxic stomach contents and debris into the lungs.
3. It prevents laryngospasm (involuntary closure of vocal cords that obstructs the airway).
4. It enables breathing support that may be needed when patients are treated with powerful sedatives and narcotics, or because critical illness undermines lung function.
5. It enables accurate monitoring, measurement and management of inhaled gas concentrations.
6. It prevents coughing and sneezing that spreads viral and bacterial contagion.

COGNITIVE PATHWAY CONTROL

Hypnotic pharmaceuticals (aka sedatives and tranquilizers) control the cognitive pathway and prevent harmful sympathetic nervous system hyperactivity by inhibiting consciousness and eliminating fear, anxiety, and the perception of nociception as pain. These include anesthetic inhalation agents (Isoforane, Sevoflurane, Desflurane), and intravenously administered benzodiazepines (Valium, Versed, Librium, Xanax), barbiturates (Pentothal, Brevitol, Seconal, Nembutal), Propofol, Etomidate, and beverage alcohol. All hypnotic agents are toxic and addictive, so that their dosage is best minimized. For example, Propofol causes muscle wasting and cardiac toxicity when used for prolonged treatment.[379, 380]

Anesthetic inhalation agents are the most practical choice for controlling the cognitive pathway during critical illness for the following reasons:

1. They are absorbed via the lung and thus do not necessitate intravenous access or unreliable gastrointestinal absorption.
2. They take effect quickly and reliably.
3. They can be maintained at precise concentrations for prolonged periods of time using standardized anesthesia vaporizers, so as to minimize toxicity and maximize therapeutic benefits.

4. There is no need for crude and complex calculations of body surface area.
5. They are minimally affected by organ function.
6. They are less toxic than most alternatives.
7. They can be rapidly rid from the body.

Isoforane is arguably the best available inhalation agent for controlling the cognitive pathway. Desflurane is unacceptable because it stimulates sympathetic nervous activity, which defeats the purpose of stress control, and sometimes causes dangerous bronchoconstriction.[381, 382] Sevoflurane is chemically unstable compared to Isoforane, and far more expensive.[383, 384, 385, 386]

NOCICEPTION PATHWAY CONTROL

There are three means to control the nociception pathway:

1. NSAIDs (Non-Steroidal Anti-Inflammatory drugs) including aspirin, ibuprophin, acetaminophen (Tylenol), Ketorolac, and COX inhibitors (Celebrex, Valdecoxib, Vioxx, naproxen, and so forth) directly inhibit peripheral nociceptors. These medications are inherently toxic. They can disrupt tissue repair, damage organs, and cause bleeding, ulcers, and gastrointestinal perforations. Their studies are tainted by corrupted research, including fabricated information and statistical manipulation. They are best avoided or restricted to short term use.
2. Local analgesics (lidocaine, Marcaine, Ropivacaine, tetracaine, procaine, and so forth) can be used to block peripheral nerves and spinal cord nociception pathways. Such blocks inhibit nociception and improve outcome, but they entail time, specialized skills, expensive equipment, and nursing support, and they invite nerve damage and local anesthetic systemic toxicity (aka *LAST*).
3. Narcotics such as morphine, Demerol, Fentanyl, Sufentanil, Alfentanil and Dilaudid inhibit spinal cord nociception pathways. 100 years ago George Washington Crile used massive intramuscular doses of morphine to successfully treat sepsis and peritonitis, but this powerful medical principle is now abandoned and forgotten.[387] Narcotics obtund respiratory chemoreceptors and can cause lethal respiratory depression when used incorrectly, but they are devoid of tissue toxicity, and if used properly they are the safest class of pharmaceutical medications. Complimentary combinations of inhalation anesthetics, narcotics, hypercarbia, and elective intubation provide a powerful and safe means to control harmful sympathetic nervous activity, mitigate blood viscosity and coagulability, and optimize cardiac output, cardiac efficiency, tissue perfusion, tissue oxygenation, organ function, and antibiotic potency.

TISSUE FACTOR PATHWAY CONTROL

The tissue factor pathway is activated by trauma, surgery, and inflammation, all of which release tissue factor into systemic blood circulation.[388] Unfortunately, there is no treatment presently available for clinical use that can neutralize excess tissue factor in systemic blood circulation despite convincing evidence that such a treatment is possible.[389] Were a safe tissue factor antidote available, it could be combined with anesthesia, analgesia, and hypercarbia to control stress mechanism hyperactivity and revolutionize both surgery and critical care. In the face of such evidence, one must question why pharmaceutical development has failed to produce such a treatment. Fortunately, there are alternatives. It is possible to utilize $MgSO_4$, EDTA, trisodium citrate, and nitroprusside to inhibit thrombin generation and disrupt the positive feedback that produces stress mechanism hyperactivity. All of these chemicals have extensive records of human exposure, so it should be possible to use them with reasonable safety to control the tissue factor pathway.

$MgSO_4$ (magnesium sulfate) is perhaps the most obvious choice for controlling the tissue factor pathway. It appears to be a competitive inhibitor of Ca+. It has been extensively used as a treatment for eclampsia, so that safe protocols for its use already exist. In theory, it could be used to control ARDS, MOFS, SIRS, SARS, MERS, COVID, Ebola, Legionnaire's Disease, cancer, and other forms of critical illness.

EDTA binds directly to Ca+ and therefore may offer a more potent, though possibly more dangerous alternative to $MgSO_4$. It has an extensive record of safe use for "chelation therapy," which has an interesting history. WWII shipyard workers suffered lead poisoning when they inhaled the toxic fumes of lead-based paint. There was no known treatment for lead poisoning, but physicians knew that EDTA binds to Ca+ to prevent coagulation in stored blood. They hypothesized that EDTA might chemically bind to lead and facilitate its excretion from the body. The treatment proved to be effective, and chelation therapy became an accepted treatment for lead poisoning. Controlled studies have subsequently proved that chelation therapy reverses atherosclerosis, enhances antibiotic effectiveness, and mitigates rheumatoid diseases.[390, 391, 392, 393]

Most physicians assume that EDTA exerts its beneficial effects by chelating (binding) to lead and other heavy metals, but this has never been confirmed, and it fails to explain its beneficial effects on atherosclerosis, rheumatoid disease, and antibiotics. Stress theory suggests an alternative explanation: EDTA binds to Ca+ and inhibits thrombin activity. By inhibiting thrombin, EDTA mitigates insoluble fibrin generation, enhances microvascular perfusion, and "flushes" lead from microvascular passageways. EDTA likewise mitigates the pathological tissue repair activity which causes inflammation, fibroblast proliferation, and collagen generation that causes rheumatoid diseases.

Trisodium citrate and sodium nitroprusside are additional alternatives for inhibiting thrombin hyperactivity. Like EDTA, they bind to Ca+ and inhibit thrombin.[394] Trisodium citrate has been used to prevent coagulation in stored blood and during renal dialysis, so it has a record of safe human exposure.[395, 396, 397] Sodium nitroprusside opens the capillary gate, reduces flow resistance, increases cardiac output, and is an established treatment for congestive heart failure.

Optimizing Antibiotic Potency and Penetration

Antibiotics must reach microbes to eradicate them, but this is easier said than done. Physicians choose antibiotics based on their *potency* (i.e., their ability to eradicate microbes) and their *penetration* (i.e., their ability to reach the tissues where the microbes are hiding and thriving).

Gram negative *facultative anaerobes* are bacteria that thrive best in the absence of oxygen but can tolerate low oxygen concentrations. They seldom infest generously perfused and well-oxygenated tissues such as brain, heart, and lung. Instead, they thrive in poorly perfused and oxygenated tissues such as bone, connective tissue, and the prostate, where antibiotic penetration is limited and the partial pressure of oxygen is low. This explains why they are notoriously difficult to treat. Examples include staphylococcus aureus that causes MRSA (methicillin resistant staphylococcus aureus), Borrelia organisms that cause Lyme Disease, and clostridia microbes that cause gas gangrene. Hyperbaric oxygen tanks have been used to treat these infections, but they are not very effective, because hemoglobin, which carries 90% of blood oxygen, is normally 100% saturated in the lung. Thus, hyperbaric oxygen treatment can only saturate small and insignificant additional quantities of oxygen in blood plasma, which only minimally improves tissue oxygenation. Therefore, what is needed is the means to:

1. Optimize tissue perfusion in the infested tissues to maximize antibiotic penetration
2. Optimize the release of oxygen from the hemoglobin molecule to maximize the partial pressure of oxygen in the infested tissues and poison the facultative anaerobes with oxygen, thus maximizing antibiotic potency.
3. Antibiotic treatment that targets the causative microbes.

Carbogen enhances both tissue perfusion and tissue oxygenation because carbon dioxide simultaneously opens the capillary gate and releases oxygen from the hemoglobin molecule. It therefore simultaneously optimizes antibiotic potency and penetration, and when combined with appropriate antibiotics it provides a powerful, simple, inexpensive means to optimize antibiotic treatment.

This author has used transcutaneous O_2/CO_2 monitoring and microvascular Doppler monitoring to confirm that general anesthesia supplemented by hypercarbia and narcotics can enhance tissue perfusion and elevate the partial pressure of oxygen in peripheral tissues from a normal level of 70 torr to above 200 torr by releasing oxygen from hemoglobin. The enhanced tissue perfusion should optimize antibiotic penetration, and the elevated tissue oxygen levels should optimize antibiotic potency.

Alternatively, treatment with $MgSO_4$, trisodium citrate, or EDTA can paralyze the capillary gate, maximize microvascular perfusion, and optimize antibiotic penetration.[398, 399, 400] This author has successfully treated his own prostate infection by supplementing intravenous Ancef with intravenous EDTA (chelation therapy) to eradicate a stubborn *P. Mirabilis* infestation that had persisted despite multiple treatments with Cipro and Ancef.[401, 402]

Life-threatening sepsis and organ failure can be treated with combinations of general endotracheal anesthesia, hypercarbia, narcotic supplementation, and anticoagulation with EDTA, $MgSO_4$, or trisodium citrate to maximize tissue perfusion, tissue oxygenation, and antibiotic potency and penetration.

Mycobacteria, such as the tuberculosis bacilli, are difficult to treat for the opposite reason. They are *obligate anaerobes* that thrive in the lung, where oxygen is abundant, and they induce stress mechanism activity that causes caseation that surrounds them with insoluble fibrin that protects them from antibiotic exposure. Stress theory indicates that chelation therapy combined with appropriate antibiotics can efficiently cure mycobacterial infestations by disintegrating the caseation barrier and directly exposing the mycobacteria to antibiotics. In theory, oxygen starvation should further enhance the susceptibility of mycobacteria to antibiotics. Such oxygen starvation could be induced via general anesthesia with compressed atmospheric air (to minimize oxygen concentrations in the lung) and mechanical hyperventilation to deplete CO_2 tissue reserves and inhibit the release of oxygen from the hemoglobin molecule into tissues. However, such a treatment would be potentially harmful to the patient as well as the microbes.

Cancer Treatment

Stress theory explains malignancy as self-perpetuating tissue repair hyperactivity. This implies that thrombin suppression can disrupt the "vicious cycle" of nociception and tissue factor release that perpetuates the malignant process. Stress theory suggests that cancer patients could be cured safely and efficiently using a combination of elective endotracheal intubation, generous narcotic supplementation, hypercarbia in the range of 100 torr, and intravenous treatment with $MgSO_4$, EDTA, or trisodium citrate. The available evidence suggests that if this treatment regimen were maintained for about 15 hours it would induce apoptosis and cure the cancer.[403] A safe antidote for tissue factor in blood circulation would theoretically be safer and easier to used than EDTA, $MgSO_4$, and trisodium citrate.

CONCLUSION

The discovery of Selye's mammalian stress mechanism fulfills the long-anticipated promise of stress theory. It enables Selye's posited *Unified Theory of Medicine* that explains the nature of disease and confers universal treatments that safely and efficiently prevent, control, and cure all types of disease. It provides a framework for productive medical research and profitable pharmaceutical development that promises to introduce an unprecedented era of health, longevity, and prosperity, free from the eternal curse of disease and premature death. However, much work remains to be done. The newly discovered stress mechanism must be acknowledged, tested, verified, and clarified. Effective pharmaceuticals must be developed, and physicians must learn to use them.

NOTES

1 S. E. Luria, M. Delbruck, Mutations of Bacteria from Virus Sensitivity to Virus Resistance. *Genetics* 28, 491-511 (1943).

2 J. A. Erukhimov *et al.*, Actin-containing sera from patients with adult respiratory distress syndrome are toxic to sheep pulmonary endothelial cells. *American journal of respiratory and critical care medicine* 162, 288-294 (2000).

3 P. A. Janmey, J. A. Lamb, R. M. Ezzell, S. Hvidt, S. E. Lind, Effects of actin filaments on fibrin clot structure and lysis. *Blood* 80, 928-936 (1992).

4 J. R. Aluoch, Fibrinolytic activity in adult Kenyan patients with sickle cell disease. *East Afr Med J* 75, 351-352 (1998).

5 G. J. Kato, M. H. Steinberg, M. T. Gladwin, Intravascular hemolysis and the pathophysiology of sickle cell disease. *The Journal of clinical investigation* 127, 750-760 (2017).

6 S. K. Ballas, Lactate dehydrogenase and hemolysis in sickle cell disease. *Blood* 121, 243-244 (2013).

7 A. Bugeja, P. Blanco, E. G. Clark, M. M. Sood, Sickle cell disease: a case report of renal amyloidosis. *BMC Nephrol* 19, 256 (2018).

8 D. V. Devine, T. R. Kinney, P. F. Thomas, W. F. Rosse, C. S. Greenberg, Fragment D-dimer levels: an objective marker of vaso-occlusive crisis and other complications of sickle cell disease. *Blood* 68, 317-319 (1986).

9 M. A. Elsharawy, K. M. Moghazy, M. A. Shawarby, Atherosclerosis in sickle cell disease - a review. *Int J Angiol* 18, 62-66 (2009).

10 A. A. Famodu, Plasma fibrinolytic activity in sickle cell disease. *Trop Geogr Med* 40, 331-333 (1988).

11 R. B. Francis, Jr., Elevated fibrin D-dimer fragment in sickle cell anemia: evidence for activation of coagulation during the steady state as well as in painful crisis. *Haemostasis* 19, 105-111 (1989).

12 M. F. Whelihan *et al.*, Thrombin generation and cell-dependent hypercoagulability in sickle cell disease. *J Thromb Haemost* 14, 1941-1952 (2016).

13 J. E. Salvaggio, C. A. Arnold, C. H. Banov, Long-term anti-coagulation in sickle-cell disease. A clinical study. *N Engl J Med* 269, 182-186 (1963).

14 L. R. Solomon, Treatment and prevention of pain due to vaso-occlusive crises in adults with sickle cell disease: an educational void. *Blood* 111, 997-1003 (2008).

15 M. Grunewald *et al.*, Absence of compensatory platelet activation in patients with severe haemophilia, but evidence for a platelet collagen-activation defect. *Platelets* 13, 451-458 (2002).

16 P. Markakis, N. Papajoannou, G. Kambouruglou, [Fractures in hemophilia patients]. *Unfallchirurgie* 10, 173-176 (1984).

17 B. Koch, F. M. Galioto, Jr., J. Kelleher, D. Goldstein, Physical fitness in children with hemophilia. *Arch Phys Med Rehabil* 65, 324-326 (1984).

18 F. Bilora *et al.*, Do hemophilia A and von Willebrand disease protect against carotid atherosclerosis? A comparative study between coagulopathics and normal subjects by means of carotid echo-color Doppler scan. *Clin Appl Thromb Hemost* 5, 232-235 (1999).

19 F. Bilora, V. Boccioletti, E. Zanon, F. Petrobelli, A. Girolami, Hemophilia A, von Willebrand disease, and atherosclerosis of abdominal aorta and leg arteries: factor VIII and von Willebrand Factor defects appear to protect abdominal aorta and leg arteries from atherosclerosis. *Clin Appl Thromb Hemost* 7, 311-313 (2001).

20 L. Dintenfass, P. Castaldi, A Study In Viscosity And Clotting Of Blood In Haemophiliacs. *Thrombosis et diathesis haemorrhagica* 11, 267-274 (1964).

21 L. Dintenfass, C. D. Forbes, Viscosity of blood in patients with myocardial infarction, haemophilia and thyroid diseases. Effect of fibrinogen, albumin and globulin. *Biorheology* 10, 457-462 (1973).

22 L. Dintenfass, C. D. Forbes, Effect of fibrinogen on aggregation of red cells and on apparent viscosity of artificial thrombi in haemophilia, myocardial infarction, thyroid disease, cancer and control systems: effect of ABO blood groups. *Microvasc Res* 9, 107-118 (1975).

23 F. R. Rosendaal *et al.*, Haemophilia protects against ischaemic heart disease: a study of risk factors. *British journal of haematology* 75, 525-530 (1990).

24 S. Butenas, K. E. Brummel, S. G. Paradis, K. G. Mann, Influence of factor VIIa and phospholipids on coagulation in "acquired" hemophilia. *Arterioscler Thromb Vasc Biol* 23, 123-129 (2003).

25 S. Butenas, K. E. Brummel, R. F. Branda, S. G. Paradis, K. G. Mann, Mechanism of factor VIIa-dependent coagulation in hemophilia blood. *Blood* 99, 923-930 (2002).

26 I. N. Basso, D. Keeling, Myocardial infarction following recombinant activated factor VII in a patient with type 2A von Willebrand disease. *Blood Coagul Fibrinolysis* 15, 503-504 (2004).

27 M. Franchini, Haemophilia and cancer: a personal perspective. *Blood Transfus* 11, 26-31 (2013).

28 J. A. Hovinga *et al.*, ADAMTS-13, von Willebrand factor and related parameters in severe sepsis and septic shock. *J Thromb Haemost*, (2007).

29 A. M. Muller, C. Skrzynski, G. Skipka, K. M. Muller, Expression of von Willebrand factor by human pulmonary endothelial cells in vivo. *Respiration* 69, 526-533 (2002).

30 F. Rodeghiero, von Willebrand disease: still an intriguing disorder in the era of molecular medicine. *Haemophilia* 8, 292-300 (2002).

31 F. Bilora *et al.*, Do hemophilia A and von Willebrand disease protect against carotid atherosclerosis? A comparative study between coagulopathics and normal subjects by means of carotid echo-color Doppler scan. *Clin Appl Thromb Hemost* 5, 232-235 (1999).

32 F. Bilora, V. Boccioletti, E. Zanon, F. Petrobelli, A. Girolami, Hemophilia A, von Willebrand disease, and atherosclerosis of abdominal aorta and leg arteries: factor VIII and von Willebrand factor defects appear to protect abdominal aorta and leg arteries from atherosclerosis. *Clin Appl Thromb Hemost* 7, 311-313 (2001).

33 A. B. Federici, The factor VIII/von Willebrand factor complex: basic and clinical issues. *Haematologica* 88, EREP02 (2003).

34 B. R. Gudmundsdottir, V. J. Marder, P. T. Onundarson, Risk of excessive bleeding associated with marginally low von Willebrand factor and mild platelet dysfunction. *J Thromb Haemost* 5, 274-281 (2007).

35 J. K. Koscielny *et al.*, Capillary microscopic and rheological dimensions for the diagnosis of von Willebrand disease in comparison to other haemorrhagic diatheses. *Thromb Haemost* 84, 981-988 (2000).

36 A. A. Kurth, G. Ludwig, I. Scharrer, [Prevalence, pathophysiology, diagnosis and treatment of von Willebrand syndrome in orthopedic trauma patients]. *Orthopade* 28, 366-374 (1999).

37 T. K. Rosborough, W. R. Swaim, Acquired von Willebrand's disease, platelet-release defect and angiodysplasia. *Am J Med* 65, 96-100 (1978).

38 A. Veyradier *et al.*, Abnormal von Willebrand factor in bleeding angiodysplasias of the digestive tract. *Gastroenterology* 120, 346-353 (2001).

39 T. E. Warkentin, J. C. Moore, S. S. Anand, E. M. Lonn, D. G. Morgan, Gastrointestinal bleeding, angiodysplasia, cardiovascular disease, and acquired von Willebrand syndrome. *Transfusion medicine reviews* 17, 272-286 (2003).

40 A. Casonato *et al.*, Plasma and platelet von Willebrand factor abnormalities in patients with uremia: lack of correlation with uremic bleeding. *Clin Appl Thromb Hemost* 7, 81-86 (2001).

41 A. Casonato *et al.*, Plasma and platelet von Willebrand factor abnormalities in patients with uremia: lack of correlation with uremic bleeding. *Clin Appl Thromb Hemost* 7, 81-86 (2001).

42 L. S. Coleman, A capillary hemostasis mechanism regulated by sympathetic tone and activity via factor VIII or von Willebrand's factor may function as a "capillary gate" and may explain angiodysplasia, angioneurotic edema, and variations in systemic vascular resistance. *Med Hypotheses* 66, 773-775 (2006).

43 L. S. Coleman, A capillary hemostasis mechanism regulated by sympathetic tone and activity via factor VIII or von Willebrand's factor may function as a "capillary gate" and may explain angiodysplasia, angioneurotic edema, and variations in systemic vascular resistance. *Med Hypotheses* 66, 773-775 (2006).

44 A. Casonato *et al.*, Fainting induces an acute increase in the concentration of plasma factor VIII and von Willebrand factor. *Haematologica* 88, 688-693 (2003).

45 L. S. Coleman, A capillary hemostasis mechanism regulated by sympathetic tone and activity via factor VIII or von Willebrand's factor may function as a "capillary gate" and may explain angiodysplasia, angioneurotic edema, and variations in systemic vascular resistance. *Med Hypotheses*, (2005).

46 L. S. Coleman, To the Editor: Is von Willebrand Factor a Hormone that Regulates a Coagulation Mechanism? *World J Surg*, (2006).

47 A. Kuitunen *et al.*, Anaesthesia affects plasma concentrations of vasopressin, von Willebrand factor and coagulation factor VIII in cardiac surgical patients. *Br J Anaesth* 70, 173-180 (1993).

48 D. B. Rubin *et al.*, Elevated von Willebrand factor antigen is an early plasma predictor of acute lung injury in nonpulmonary sepsis syndrome. *The Journal of clinical investigation* 86, 474-480 (1990).

49 D. B. Rubin *et al.*, Elevated von Willebrand factor antigen is an early plasma predictor of acute lung injury in nonpulmonary sepsis syndrome. *The Journal of clinical investigation* 86, 474-480 (1990).

50 L. W. Crile GW, *Anoci-association.* (Saunders, Philadelphia, 1914), pp. 223-225.

51 L. Bergamaschini *et al.*, Posttransfusion anaphylactic reactions in a patient with severe von Willebrand disease: role of complement and alloantibodies to von Willebrand factor. *J Lab Clin Med* 125, 348-355 (1995).

52 F. R. Rickles, L. W. Hoyer, M. E. Rick, D. J. Ahr, The effects of epinephrine infusion in patients with von Willebrand's disease. *The Journal of clinical investigation* 57, 1618-1625 (1976).

53 U. M. Vischer, C. B. Wollheim, Epinephrine induces von Willebrand factor release from cultured endothelial cells: involvement of cyclic AMP-dependent signalling in exocytosis. *Thromb Haemost* 77, 1182-1188 (1997).

54 A. B. Federici, S. D. Berkowitz, T. S. Zimmerman, P. M. Mannucci, Proteolysis of von Willebrand factor after thrombolytic therapy in patients with acute myocardial infarction. *Blood* 79, 38-44 (1992).

55 E. Hernandez *et al.*, Elevation of von Willebrand factor levels in patients with IgA nephropathy: effect of ACE inhibition. *American journal of kidney diseases : the official journal of the National Kidney Foundation* 30, 397-403 (1997).

56 Y. Hojo, U. Ikeda, K. Shimada, Effect of angiotensin converting enzyme inhibitor on plasma von Willebrand factor level in patients with coronary artery disease. *Cardiovasc Drugs Ther* 15, 91-92 (2001).

57 J. C. Ngo, M. Huang, D. A. Roth, B. C. Furie, B. Furie, Crystal structure of human factor VIII: implications for the formation of the factor IXa-factor VIIIa complex. *Structure* 16, 597-606 (2008).

58 D. Y. Shi, S. J. Wang, Advances of Coagulation Factor XIII. *Chin Med J (Engl)* 130, 219-223 (2017).

59 P. Wiland, R. Wojtala, J. Goodacre, J. Szechinski, The prevalence of subclinical amyloidosis in Polish patients with rheumatoid arthritis. *Clin Rheumatol* 23, 193-198 (2004).

60 G. M. Hirschfield, P. N. Hawkins, Amyloidosis: new strategies for treatment. *Int J Biochem Cell Biol* 35, 1608-1613 (2003).

61 B. V. Shekhonin, V. S. Rukosuev, [On the participation of fibrinogen (fibrin) in the formation of amyloid in experimental amyloidosis]. *Biull Eksp Biol Med* 65, 119-122 (1968).

62 A. Bohler, B. Lammle, [Decreased Quick percentage, acquired factor X deficiency, hemarthrosis and ecchymosis: amyloidosis]. *Ther Umsch* 56, 523-525 (1999).

63 F. V. Beardell, M. Varma, J. Martinez, Normalization of plasma factor X levels in amyloidosis after plasma exchange. *Am J Hematol* 54, 68-71 (1997).

64 A. Chandra, Role of amyloid from a multiple sclerosis perspective: a literature review. *Neuroimmunomodulation* 22, 343-346 (2015).

65 E. B. Choufani *et al.*, Acquired factor X deficiency in patients with amyloid light-chain amyloidosis: incidence, bleeding manifestations, and response to high-dose chemotherapy. *Blood* 97, 1885-1887 (2001).
66 D. Foguel, J. L. Silva, New insights into the mechanisms of protein misfolding and aggregation in amyloidogenic diseases derived from pressure studies. *Biochemistry* 43, 11361-11370 (2004).
67 B. Furie, E. Greene, B. C. Furie, Syndrome of acquired factor X deficiency and systemic amyloidosis in vivo studies of the metabolic fate of factor X. *N Engl J Med* 297, 81-85 (1977).
68 G. Gamba *et al.*, Clotting alterations in primary systemic amyloidosis. *Haematologica* 85, 289-292 (2000).
69 G. Gamba *et al.*, Abnormalities in thrombin-antithrombin pathway in AL amyloidosis. *Amyloid* 6, 273-277 (1999).
70 D. A. Gastineau, M. A. Gertz, T. M. Daniels, R. A. Kyle, E. J. Bowie, Inhibitor of the thrombin time in systemic amyloidosis: a common coagulation abnormality. *Blood* 77, 2637-2640 (1991).
71 P. R. Greipp, R. A. Kyle, E. J. Bowie, Factor-X deficiency in amyloidosis: a critical review. *Am J Hematol* 11, 443-450 (1981).
72 M. Marcatti *et al.*, Unusual bleeding manifestations in a case of primary amyloidosis with factor X deficiency but elevations of in vivo markers of thrombin formation and activity. *Thromb Res* 80, 333-337 (1995).
73 S. Idell, Endothelium and disordered fibrin turnover in the injured lung: newly recognized pathways. *Critical care medicine* 30, S274-280 (2002).
74 R. Holemans, Enhancement of Fibrinolysis in the Dog by Injection of Vasoactive Drugs. *The American journal of physiology* 208, 511-520 (1965).
75 G. M. Hirschfield, P. N. Hawkins, Amyloidosis: new strategies for treatment. *Int J Biochem Cell Biol* 35, 1608-1613 (2003).
76 G. M. Hirschfield, P. N. Hawkins, Amyloidosis: new strategies for treatment. *Int J Biochem Cell Biol* 35, 1608-1613 (2003).
77 B. V. Shekhonin, V. S. Rukosuev, [On the participation of fibrinogen (fibrin) in the formation of amyloid in experimental amyloidosis]. *Biull Eksp Biol Med* 65, 119-122 (1968).
78 A. Bohler, B. Lammle, [Decreased Quick percentage, acquired factor X deficiency, hemarthrosis and ecchymosis: amyloidosis]. *Ther Umsch* 56, 523-525 (1999).
79 F. V. Beardell, M. Varma, J. Martinez, Normalization of plasma factor X levels in amyloidosis after plasma exchange. *Am J Hematol* 54, 68-71 (1997).
80 A. Chandra, Role of amyloid from a multiple sclerosis perspective: a literature review. *Neuroimmunomodulation* 22, 343-346 (2015).
81 E. B. Choufani *et al.*, Acquired factor X deficiency in patients with amyloid light-chain amyloidosis: incidence, bleeding manifestations, and response to high-dose chemotherapy. *Blood* 97, 1885-1887 (2001).
82 D. Foguel, J. L. Silva, New insights into the mechanisms of protein misfolding and aggregation in amyloidogenic diseases derived from pressure studies. *Biochemistry* 43, 11361-11370 (2004).

83 B. Furie, E. Greene, B. C. Furie, Syndrome of acquired factor X deficiency and systemic amyloidosis in vivo studies of the metabolic fate of factor X. *N Engl J Med* 297, 81-85 (1977).
84 G. Gamba *et al.*, Clotting alterations in primary systemic amyloidosis. *Haematologica* 85, 289-292 (2000).
85 G. Gamba *et al.*, Abnormalities in thrombin-antithrombin pathway in AL amyloidosis. *Amyloid* 6, 273-277 (1999).
86 D. A. Gastineau, M. A. Gertz, T. M. Daniels, R. A. Kyle, E. J. Bowie, Inhibitor of the thrombin time in systemic amyloidosis: a common coagulation abnormality. *Blood* 77, 2637-2640 (1991).
87 P. R. Greipp, R. A. Kyle, E. J. Bowie, Factor-X deficiency in amyloidosis: a critical review. *Am J Hematol* 11, 443-450 (1981).
88 M. Marcatti *et al.*, Unusual bleeding manifestations in a case of primary amyloidosis with factor X deficiency but elevations of in vivo markers of thrombin formation and activity. *Thromb Res* 80, 333-337 (1995).
89 A. Bugeja, P. Blanco, E. G. Clark, M. M. Sood, Sickle cell disease: a case report of renal amyloidosis. *BMC Nephrol* 19, 256 (2018).
90 Y. A. Goffin, E. Gruys, G. D. Sorenson, F. Wellens, Amyloid deposits in bioprosthetic cardiac valves after long-term implantation in man. A new localization of amyloidosis. *The American journal of pathology* 114, 431-442 (1984).
91 S. Iqbal, S. Reehana, D. Lawrence, Unique type of isolated cardiac valvular amyloidosis. *J Cardiothorac Surg* 1, 38 (2006).
92 A. Jabor *et al.*, [Serum amyloid A as an effective marker for the assessment of surgical trauma and risk of post-operative complications]. *Ceska Gynekol* 71, 131-136 (2006).
93 C. Jacquot, C. d'Auzac, P. Loirat, J. Bariety, [Dramatic aggravation of renal amyloidosis after surgery. Three cases (author's transl)]. *Nouv Presse Med* 10, 3389-3390, 3395 (1981).
94 L. Ho *et al.*, Diet-induced insulin resistance promotes amyloidosis in a transgenic mouse model of Alzheimer's disease. *FASEB journal : official publication of the Federation of American Societies for Experimental Biology* 18, 902-904 (2004).
95 S. J. Kang *et al.*, Secondary amyloidosis associated with multiple sclerosis. *J Clin Neurol* 5, 146-148 (2009).
96 U. Mousa, S. Aktas, H. Uner, Amyloid accumulation in the toxic nodule of the thyroid gland in a patient with end stage renal failure. *Case Rep Endocrinol* 2012, 741754 (2012).
97 Y. Wang, X. Feng, B. Shen, J. Ma, W. Zhao, Is Vascular Amyloidosis Intertwined with Arterial Aging, Hypertension and Atherosclerosis? *Front Genet* 8, 126 (2017).
98 P. Wiland, R. Wojtala, J. Goodacre, J. Szechinski, The prevalence of subclinical amyloidosis in Polish patients with rheumatoid arthritis. *Clin Rheumatol* 23, 193-198 (2004).
99 G. M. Hirschfield, P. N. Hawkins, Amyloidosis: new strategies for treatment. *Int J Biochem Cell Biol* 35, 1608-1613 (2003).

100 D. Hall, H. Edskes, Silent prions lying in wait: a two-hit model of prion/ amyloid formation and infection. *J Mol Biol* 336, 775-786 (2004).

101 P. Schwartz, [Amyloidosis, expression and cause of presenile and senile mental and physical regression. A revision of the amyloid problem (author's transl)]. *Zentralbl Allg Pathol* 119, 533-548 (1975).

102 P. Schwartz, [Amyloidosis as a manifestation and origin of presenile and senile degeneration]. *Fortschr Med* 94, 890-896 (1976).

103 I. Schwarzinger *et al.*, Recurrent, isolated factor X deficiency in myeloma: repeated normalization of factor X levels after cytostatic chemotherapy followed by late treatment failure associated with the development of systemic amyloidosis. *Thromb Haemost* 68, 648-651 (1992).

104 P. Milani *et al.*, The lung in amyloidosis. *Eur Respir Rev* 26, (2017).

105 M. P. Vitek *et al.*, Advanced glycation end products contribute to amyloidosis in Alzheimer disease. *Proceedings of the National Academy of Sciences of the United States of America* 91, 4766-4770 (1994).

106 R. A. Armstrong, Classic beta-amyloid deposits cluster around large diameter blood vessels rather than capillaries in sporadic Alzheimer's disease. *Curr Neurovasc Res* 3, 289-294 (2006).

107 K. Dahlback, H. Lofberg, B. Dahlback, Immunohistochemical demonstration of vitronectin in association with elastin and amyloid deposits in human kidney. *Histochemistry* 87, 511-515 (1987).

108 K. Dahlback, H. Lofberg, B. Dahlback, Immunohistochemical studies on vitronectin in elastic tissue disorders, cutaneous amyloidosis, lichen ruber planus and porphyria. *Acta Derm Venereol* 68, 107-115 (1988).

109 S. Sahoo, W. Reeves, R. M. DeMay, Amyloid tumor: a clinical and cytomorphologic study. *Diagn Cytopathol* 28, 325-328 (2003).

110 A. Alwitry, E. T. Brackenbury, F. D. Beggs, I. Soomro, Vascular amyloidosis causing spontaneous mediastinal haemorrhage with haemothorax. *Eur J Cardiothorac Surg* 20, 871-873 (2001).

111 A. R. Mizutani, C. F. Ward, Amyloidosis associated bleeding diatheses in the surgical patient. *Can J Anaesth* 37, 910-912 (1990).

112 A. D. Mumford *et al.*, Bleeding symptoms and coagulation abnormalities in 337 patients with AL-amyloidosis. *British journal of haematology* 110, 454-460 (2000).

113 A. Genovese *et al.*, Inhibition of hypoxia-induced cardiac hypertrophy by verapamil in rats. *Basic Res Cardiol* 75, 757-763 (1980).

114 A. Genovese, M. Chiariello, S. Latte, W. DeAlfieri, M. Condorelli, Bilateral ventricular hypertrophy in rats exposed to acute or chronic hypobaric hypoxia. *Respiration* 44, 289-293 (1983).

115 A. Genovese, W. De Alfieri, S. Latte, M. Chiariello, M. Condorelli, Regression of myocardial hypertrophy in the rat following removal of acute or chronic hypobaric hypoxia. *Eur Heart J* 3 Suppl A, 161-164 (1982).

116 A. Genovese, S. Latte, M. Bozzaotre, M. Chiariello, Response of the left ventricular connective tissue to hypoxia. *Res Exp Med (Berl)* 183, 111-115 (1983).

117 L. S. Coleman, An Improved Explanation of Atherosclerosis. *PLoS Medicine* 2(4):e98, (2005).

118 L. S. Coleman, Atherosclerosis may be caused by inadequate levels of turbulence and mixing. *World J Surg* 30, 638-639 (2006).
119 N. Anim-Nyame *et al.*, Evidence of impaired microvascular function in pre-eclampsia: a non-invasive study. *Clin Sci (Lond)* 104, 405-412 (2003).
120 O. Erez *et al.*, Tissue factor activity in women with preeclampsia or SGA: a potential explanation for the excessive thrombin generation in these syndromes. *J Matern Fetal Neonatal Med* 31, 1568-1577 (2018).
121 T. C. Inglis, J. Stuart, A. J. George, A. J. Davies, Haemostatic and rheological changes in normal pregnancy and pre-eclampsia. *British journal of haematology* 50, 461-465 (1982).
122 M. G. Newman, M. K. Lindsay, W. Graves, Cigarette smoking and pre-eclampsia: their association and effects on clinical outcomes. *J Matern Fetal Med* 10, 166-170 (2001).
123 T. Ogawa, Y. Suzuki, S. Sayama, H. Soma, Possible relationship between the ratio of Factor VIII complex and placental insufficiency in preeclampsia. *Biol Res Pregnancy Perinatol* 4, 155-157 (1983).
124 J. M. Roberts *et al.*, Preeclampsia: an endothelial cell disorder. *Am J Obstet Gynecol* 161, 1200-1204 (1989).
125 K. K. Roy, N. Malhotra, K. Banerjee, Recurrent eclampsia in a woman with chronic pyelonephritis. *Eur J Obstet Gynecol Reprod Biol* 94, 307-308 (2001).
126 G. T. Sukhikh *et al.*, Differences of glycocalyx composition in the structural elements of placenta in preeclampsia. *Placenta* 43, 69-76 (2016).
127 Y. Wang, X. Feng, B. Shen, J. Ma, W. Zhao, Is Vascular Amyloidosis Intertwined with Arterial Aging, Hypertension and Atherosclerosis? *Front Genet* 8, 126 (2017).
128 H. A. Liebman *et al.*, Excessive fibrinolysis in amyloidosis associated with elevated plasma single-chain urokinase. *Am J Clin Pathol* 98, 534-541 (1992).
129 J. M. Frank, N. J. Palomino, Primary amyloidosis with diffuse splenic infiltration presenting as fulminant pneumococcal sepsis. *Am J Clin Pathol* 87, 405-407 (1987).
130 J. Tian, J. Shi, K. Bailey, D. M. Mann, Relationships between arteriosclerosis, cerebral amyloid angiopathy and myelin loss from cerebral cortical white matter in Alzheimer's disease. *Neuropathol Appl Neurobiol* 30, 46-56 (2004).
131 W. EshKalback *et al.*, Atherosclerosis, vascular amyloidosis and brain hypoperfusion in the pathogenesis of sporadic Alzheimer's disease. *Neurol Res* 26, 525-539 (2004).
132 I. Schwarzinger *et al.*, Recurrent, isolated factor X deficiency in myeloma: repeated normalization of factor X levels after cytostatic chemotherapy followed by late treatment failure associated with the development of systemic amyloidosis. *Thromb Haemost* 68, 648-651 (1992).
133 N. J. Bahlis, H. M. Lazarus, Multiple myeloma-associated AL amyloidosis: is a distinctive therapeutic approach warranted? *Bone Marrow Transplant* 38, 7-15 (2006).
134 M. J. Sherman, J. B. Morales, E. D. Bayrd, W. D. Schierman, Amyloid nephrosis secondary to Hodgkin's disease; report of a case. *AMA Arch Intern Med* 95, 618-621 (1955).
135 P. Milani *et al.*, The lung in amyloidosis. *Eur Respir Rev* 26, (2017).

136 S. Beji *et al.*, [Renal amyloidosis complicating the outcome of chronic inflammatory colitis]. *Presse Med* 33, 862-865 (2004).

137 A. Clark, M. R. Nilsson, Islet amyloid: a complication of islet dysfunction or an aetiological factor in Type 2 diabetes? *Diabetologia* 47, 157-169 (2004).

138 A. Genovese *et al.*, Inhibition of hypoxia-induced cardiac hypertrophy by verapamil in rats. *Basic Res Cardiol* 75, 757-763 (1980).

139 A. Genovese, M. Chiariello, S. Latte, W. DeAlfieri, M. Condorelli, Bilateral ventricular hypertrophy in rats exposed to acute or chronic hypobaric hypoxia. *Respiration* 44, 289-293 (1983).

140 A. Genovese, W. De Alfieri, S. Latte, M. Chiariello, M. Condorelli, Regression of myocardial hypertrophy in the rat following removal of acute or chronic hypobaric hypoxia. *Eur Heart J* 3 Suppl A, 161-164 (1982).

141 A. Genovese, S. Latte, M. Bozzaotre, M. Chiariello, Response of the left ventricular connective tissue to hypoxia. *Res Exp Med (Berl)* 183, 111-115 (1983).

142 A. Genovese *et al.*, Inhibition of hypoxia-induced cardiac hypertrophy by verapamil in rats. *Basic Res Cardiol* 75, 757-763 (1980).

143 A. Genovese, M. Chiariello, S. Latte, W. DeAlfieri, M. Condorelli, Bilateral ventricular hypertrophy in rats exposed to acute or chronic hypobaric hypoxia. *Respiration* 44, 289-293 (1983).

144 A. Genovese, W. De Alfieri, S. Latte, M. Chiariello, M. Condorelli, Regression of myocardial hypertrophy in the rat following removal of acute or chronic hypobaric hypoxia. *Eur Heart J* 3 Suppl A, 161-164 (1982).

145 A. Genovese, S. Latte, M. Bozzaotre, M. Chiariello, Response of the left ventricular connective tissue to hypoxia. *Res Exp Med (Berl)* 183, 111-115 (1983).

146 W. EshKalback *et al.*, Atherosclerosis, vascular amyloidosis and brain hypoperfusion in the pathogenesis of sporadic Alzheimer's disease. *Neurol Res* 26, 525-539 (2004).

147 G. Leblebisatan, I. Sasmaz, B. Antmen, M. Ergin, Y. Kilinc, Concomitance of idiopathic myelofibrosis and amyloidosis. *Hematol Oncol Stem Cell Ther* 3, 206-207 (2010).

148 G. Herbai, Scleroderma (progressive systemic sclerosis, PSS); pathophysiological, clinical and pharmacological aspects of the syndrome. *Acta Med Acad Sci Hung* 35, 201-211 (1978).

149 C. C. Mok, T. H. Kwan, L. Chow, Scleroderma renal crisis sine scleroderma during pregnancy. *Scand J Rheumatol* 32, 55-57 (2003).

150 A. Chandra, Role of amyloid from a multiple sclerosis perspective: a literature review. *Neuroimmunomodulation* 22, 343-346 (2015).

151 S. J. Kang *et al.*, Secondary amyloidosis associated with multiple sclerosis. *J Clin Neurol* 5, 146-148 (2009).

152 A. Chandra, Role of amyloid from a multiple sclerosis perspective: a literature review. *Neuroimmunomodulation* 22, 343-346 (2015).

153 W. EshKalback *et al.*, Atherosclerosis, vascular amyloidosis and brain hypoperfusion in the pathogenesis of sporadic Alzheimer's disease. *Neurol Res* 26, 525-539 (2004).

154 R. Deane, Z. Wu, B. V. Zlokovic, RAGE (yin) versus LRP (yang) balance regulates alzheimer amyloid beta-peptide clearance through transport across the blood-brain barrier. *Stroke* 35, 2628-2631 (2004).

155 R. Nortley *et al.*, Amyloid beta oligomers constrict human capillaries in Alzheimer's disease via signaling to pericytes. *Science* 365, (2019).
156 G. R. Smith, K. E. Tymms, M. Falk, Etanercept treatment of renal amyloidosis complicating rheumatoid arthritis. *Intern Med J* 34, 570-572 (2004).
157 J. L. Benumof, Obesity, sleep apnea, the airway and anesthesia. *Current opinion in anaesthesiology* 17, 21-30 (2004).
158 U. Berglund, L. Wallentin, H. von Schenck, Platelet function and plasma fibrinogen and their relations to gender, smoking habits, obesity and beta-blocker treatment in young survivors of myocardial infarction. *Thromb Haemost* 60, 21-24 (1988).
159 S. V. Brodsky *et al.*, Prevention and reversal of premature endothelial cell senescence and vasculopathy in obesity-induced diabetes by ebselen. *Circ Res* 94, 377-384 (2004).
160 K. A. Cooney, S. B. Gruber, Hyperglycemia, obesity, and cancer risks on the horizon. *Jama* 293, 235-236 (2005).
161 B. Kabon *et al.*, Obesity decreases perioperative tissue oxygenation. *Anesthesiology* 100, 274-280 (2004).
162 R. Kawamoto, N. Ohtsuka, D. Ninomiya, S. Nakamura, Association of obesity and visceral fat distribution with intima-media thickness of carotid arteries in middle-aged and older persons. *Intern Med* 47, 143-149 (2008).
163 B. Mokhlesi, A. Tulaimat, Recent advances in obesity hypoventilation syndrome. *Chest* 132, 1322-1336 (2007).
164 N. Blank *et al.*, Obesity is a significant susceptibility factor for idiopathic AA amyloidosis. *Amyloid* 25, 37-45 (2018).
165 L. J. Page, M. E. Huff, J. W. Kelly, W. E. Balch, Ca2+ binding protects against gelsolin amyloidosis. *Biochemical and biophysical research communications* 322, 1105-1110 (2004).
166 L. R. Otterbein, C. Cosio, P. Graceffa, R. Dominguez, Crystal structures of the vitamin D-binding protein and its complex with actin: structural basis of the actin-scavenger system. *Proceedings of the National Academy of Sciences of the United States of America* 99, 8003-8008 (2002).
167 S. E. Lind, D. B. Smith, P. A. Janmey, T. P. Stossel, Role of plasma gelsolin and the vitamin D-binding protein in clearing actin from the circulation. *The Journal of clinical investigation* 78, 736-742 (1986).
168 S. E. Lind, D. B. Smith, P. A. Janmey, T. P. Stossel, Depression of gelsolin levels and detection of gelsolin-actin complexes in plasma of patients with acute lung injury. *Am Rev Respir Dis* 138, 429-434 (1988).
169 D. B. Smith, P. A. Janmey, J. A. Sherwood, R. J. Howard, S. E. Lind, Decreased plasma gelsolin levels in patients with Plasmodium falciparum malaria: a consequence of hemolysis? *Blood* 72, 214-218 (1988).
170 J. A. Erukhimov *et al.*, Actin-containing sera from patients with adult respiratory distress syndrome are toxic to sheep pulmonary endothelial cells. *American journal of respiratory and critical care medicine* 162, 288-294 (2000).
171 P. A. Janmey, J. A. Lamb, R. M. Ezzell, S. Hvidt, S. E. Lind, Effects of actin filaments on fibrin clot structure and lysis. *Blood* 80, 928-936 (1992).
172 S. E. Lind, C. J. Smith, Actin is a noncompetitive plasmin inhibitor. *J Biol Chem* 266, 5273-5278 (1991).

173 S. E. Lind, C. J. Smith, Actin accelerates plasmin generation by tissue plasminogen activator. *J Biol Chem* 266, 17673-17678 (1991).
174 L. R. Otterbein, C. Cosio, P. Graceffa, R. Dominguez, Crystal structures of the vitamin D-binding protein and its complex with actin: structural basis of the actin-scavenger system. *Proceedings of the National Academy of Sciences of the United States of America* 99, 8003-8008 (2002).
175 S. E. Lind, D. B. Smith, P. A. Janmey, T. P. Stossel, Role of plasma gelsolin and the vitamin D-binding protein in clearing actin from the circulation. *The Journal of clinical investigation* 78, 736-742 (1986).
176 S. E. Lind, D. B. Smith, P. A. Janmey, T. P. Stossel, Depression of gelsolin levels and detection of gelsolin-actin complexes in plasma of patients with acute lung injury. *Am Rev Respir Dis* 138, 429-434 (1988).
177 M. T. Mizwicki *et al.*, Genomic and nongenomic signaling induced by 1alpha,25(OH)2-vitamin D3 promotes the recovery of amyloid-beta phagocytosis by Alzheimer's disease macrophages. *J Alzheimers Dis* 29, 51-62 (2012).
178 C. J. Chalmers, K. Balmanno, K. Hadfield, R. Ley, S. J. Cook, Thrombin inhibits Bim (Bcl-2-interacting mediator of cell death) expression and prevents serum-withdrawal-induced apoptosis via protease-activated receptor 1. *Biochem J* 375, 99-109 (2003).
179 S. E. Lind, P. A. Janmey, C. Chaponnier, T. J. Herbert, T. P. Stossel, Reversible binding of actin to gelsolin and profilin in human platelet extracts. *J Cell Biol* 105, 833-842 (1987).
180 H. Orbach *et al.*, Novel biomarkers in autoimmune diseases: prolactin, ferritin, vitamin D, and TPA levels in autoimmune diseases. *Annals of the New York Academy of Sciences* 1109, 385-400 (2007).
181 A. Banerjee *et al.*, Vitamin D and Alzheimer's Disease: Neurocognition to Therapeutics. *Int J Alzheimers Dis* 2015, 192747 (2015).
182 S. E. Lind, P. A. Janmey, C. Chaponnier, T. J. Herbert, T. P. Stossel, Reversible binding of actin to gelsolin and profilin in human platelet extracts. *J Cell Biol* 105, 833-842 (1987).
183 P. A. Janmey *et al.*, Interactions of gelsolin and gelsolin-actin complexes with actin. Effects of calcium on actin nucleation, filament severing, and end blocking. *Biochemistry* 24, 3714-3723 (1985).
184 S. E. Lind, P. A. Janmey, Human plasma gelsolin binds to fibronectin. *J Biol Chem* 259, 13262-13266 (1984).
185 D. B. Smith, P. A. Janmey, T. J. Herbert, S. E. Lind, Quantitative measurement of plasma gelsolin and its incorporation into fibrin clots. *J Lab Clin Med* 110, 189-195 (1987).
186 D. B. Smith, P. A. Janmey, S. E. Lind, Circulating actin-gelsolin complexes following oleic acid-induced lung injury. *The American journal of pathology* 130, 261-267 (1988).
187 G. T. Sukhikh *et al.*, Differences of glycocalyx composition in the structural elements of placenta in preeclampsia. *Placenta* 43, 69-76 (2016).
188 S. Reitsma, D. W. Slaaf, H. Vink, M. A. van Zandvoort, M. G. oude Egbrink, The endothelial glycocalyx: composition, functions, and visualization. *Pflugers Arch* 454, 345-359 (2007).

189 G. T. Sukhikh *et al.*, Differences of glycocalyx composition in the structural elements of placenta in preeclampsia. *Placenta* 43, 69-76 (2016).
190 S. Reitsma, D. W. Slaaf, H. Vink, M. A. van Zandvoort, M. G. oude Egbrink, The endothelial glycocalyx: composition, functions, and visualization. *Pflugers Arch* 454, 345-359 (2007).
191 J. J. M. Zeron Ghazarian, Raminderjit Sekhon, Andrew Staron, Neil Parikh, Michael Hanna, M.A. Khan, in *American Thoracic Society 2015 International Conference*. (ATS Journals, 2015).
192 H. Ahmadzai *et al.*, Sarcoidosis: a state of the art review from the Thoracic Society of Australia and New Zealand. *Med J Aust* 208, 499-504 (2018).
193 L. R. Otterbein, C. Cosio, P. Graceffa, R. Dominguez, Crystal structures of the vitamin D-binding protein and its complex with actin: structural basis of the actin-scavenger system. *Proceedings of the National Academy of Sciences of the United States of America* 99, 8003-8008 (2002).
194 S. E. Lind, D. B. Smith, P. A. Janmey, T. P. Stossel, Role of plasma gelsolin and the vitamin D-binding protein in clearing actin from the circulation. *The Journal of clinical investigation* 78, 736-742 (1986).
195 M. T. Mizwicki *et al.*, Genomic and nongenomic signaling induced by 1alpha,25(OH)2-vitamin D3 promotes the recovery of amyloid-beta phagocytosis by Alzheimer's disease macrophages. *J Alzheimers Dis* 29, 51-62 (2012).
196 H. Orbach *et al.*, Novel biomarkers in autoimmune diseases: prolactin, ferritin, vitamin D, and TPA levels in autoimmune diseases. *Annals of the New York Academy of Sciences* 1109, 385-400 (2007).
197 G. Devereux *et al.*, Maternal vitamin D intake during pregnancy and early childhood wheezing. *The American journal of clinical nutrition* 85, 853-859 (2007).
198 W. B. Grant, Epidemiology of disease risks in relation to vitamin D insufficiency. *Prog Biophys Mol Biol* 92, 65-79 (2006).
199 L. Yan *et al.*, The Impact of Vitamin D on Cognitive Dysfunction in Mice with Systemic Lupus Erythematosus. *Med Sci Monit* 25, 4716-4722 (2019).
200 C. F. Garland *et al.*, The role of vitamin D in cancer prevention. *Am J Public Health* 96, 252-261 (2006).
201 L. J. Leipzig, A. J. Boyle, D. S. McCann, Case histories of rheumatoid arthritis treated with sodium or magnesium EDTA. *J Chronic Dis* 22, 553-563 (1970).
202 S. Finnegan, S. L. Percival, EDTA: An Antimicrobial and Antibiofilm Agent for Use in Wound Care. *Adv Wound Care (New Rochelle)* 4, 415-421 (2015).
203 R. E. Wooley, M. S. Jones, Action of EDTA-Tris and antimicrobial agent combinations on selected pathogenic bacteria. *Veterinary microbiology* 8, 271-280 (1983).
204 S. T. Ali, N. I. Rakkah, Neurophysiological role of sildenafil citrate (Viagra) on seminal parameters in diabetic males with and without neuropathy. *Pak J Pharm Sci* 20, 36-42 (2007).
205 K. Przyklenk, R. A. Kloner, Sildenafil citrate (Viagra) does not exacerbate myocardial ischemia in canine models of coronary artery stenosis. *J Am Coll Cardiol* 37, 286-292 (2001).

206 I. Eardley, J. Brooks, P. K. Yates, P. Ellis, M. Boolell, Sildenafil citrate (VIAGRA): an oral treatment for erectile function with activity for up to four hours' duration. *International journal of clinical practice. Supplement* 102, 32-34 (1999).

207 J. E. Ayala *et al.*, Chronic treatment with sildenafil improves energy balance and insulin action in high fat-fed conscious mice. *Diabetes* 56, 1025-1033 (2007).

208 D. Behr-Roussel *et al.*, Daily Treatment with Sildenafil Reverses Endothelial Dysfunction and Oxidative Stress in an Animal Model of Insulin Resistance. *Eur Urol*, (2007).

209 S. Korom *et al.*, Sildenafil extends survival and graft function in a large animal lung transplantation model. *Eur J Cardiothorac Surg* 29, 288-293 (2006).

210 G. D. Lewis *et al.*, Sildenafil Improves Exercise Capacity and Quality of Life in Patients With Systolic Heart Failure and Secondary Pulmonary Hypertension. *Circulation*, (2007).

211 J. D. Parker *et al.*, Safety of intravenous nitroglycerin after administration of sildenafil citrate to men with coronary artery disease: A double-blind, placebo-controlled, randomized, crossover trial. *Critical care medicine* 35, 1863-1868 (2007).

212 A. Schafer *et al.*, Improvement of vascular function by acute and chronic treatment with the PDE-5 inhibitor sildenafil in experimental diabetes mellitus. *Br J Pharmacol* 153, 886-893 (2008).

213 A. L. Trachte *et al.*, Oral sildenafil reduces pulmonary hypertension after cardiac surgery. *Ann Thorac Surg* 79, 194-197; discussion 194-197 (2005).

214 D. J. Webb *et al.*, Sildenafil citrate potentiates the hypotensive effects of nitric oxide donor drugs in male patients with stable angina. *J Am Coll Cardiol* 36, 25-31 (2000).

215 X. Zhang, J. Tack, J. Janssens, D. A. Sifrim, Effect of sildenafil, a phosphodiesterase-5 inhibitor, on oesophageal peristalsis and lower oesophageal sphincter function in cats. *Neurogastroenterol Motil* 13, 325-331 (2001).

216 O. Erez *et al.*, Tissue factor activity in women with preeclampsia or SGA: a potential explanation for the excessive thrombin generation in these syndromes. *J Matern Fetal Neonatal Med* 31, 1568-1577 (2018).

217 R. A. Fleck, L. V. Rao, S. I. Rapaport, N. Varki, Localization of human tissue factor antigen by immunostaining with monospecific, polyclonal anti-human tissue factor antibody. *Thromb Res* 59, 421-437 (1990).

218 R. S. Kasthuri, M. B. Taubman, N. Mackman, Role of tissue factor in cancer. *J Clin Oncol* 27, 4834-4838 (2009).

219 M. Levi, T. van der Poll, H. Ten Cate, Tissue factor in infection and severe inflammation. *Semin Thromb Hemost* 32, 33-39 (2006).

220 J. H. McVey, Tissue factor pathway. *Baillieres Best Pract Res Clin Haematol* 12, 361-372 (1999).

221 M. Tsai, M. K. Byun, J. Shin, L. E. C. Alexander, Effects of E-cigarettes and vaping devices on cardiac and pulmonary physiology. *The Journal of physiology*, (2020).

222 E. Kipnis *et al.*, Massive alveolar thrombin activation in Pseudomonas aeruginosa-induced acute lung injury. *Shock* 21, 444-451 (2004).
223 K. Inoue, H. Takano, R. Yanagisawa, M. Sakurai, T. Yoshikawa, Surgical stress in ARDS open-lung biopsy. *Chest* 126, 1383-1384; author reply 1384 (2004).
224 D. C. Howell *et al.*, Direct thrombin inhibition reduces lung collagen, accumulation, and connective tissue growth factor mRNA levels in bleomycin-induced pulmonary fibrosis. *The American journal of pathology* 159, 1383-1395 (2001).
225 C. Adembri *et al.*, Pulmonary injury follows systemic inflammatory reaction in infrarenal aortic surgery. *Critical care medicine* 32, 1170-1177 (2004).
226 P. A. Modesti *et al.*, Response of Serum Proteome in Patients Undergoing Infrarenal Aortic Aneurysm Repair. *Anesthesiology*, (2009).
227 T. G. Monk, V. Saini, B. C. Weldon, J. C. Sigl, Anesthetic management and one-year mortality after noncardiac surgery. *Anesth Analg* 100, 4-10 (2005).
228 T. G. Monk, B. C. Weldon, Anesthetic depth is a predictor of mortality: it's time to take the next step. *Anesthesiology* 112, 1070-1072 (2010).
229 L. W. Crile GW, *Anoci-association*. (Saunders, Philadelphia, 1914), pp. 223-225.
230 E. Lowenstein *et al.*, Cardiovascular response to large doses of intravenous morphine in man. *N Engl J Med* 281, 1389-1393 (1969).
231 S. N. Raja, E. Lowenstein, The birth of opioid anesthesia. *Anesthesiology* 100, 1013-1015 (2004).
232 M. G. Davies, P. O. Hagen, Systemic inflammatory response syndrome. *Br J Surg* 84, 920-935 (1997).
233 Y. G. Liu, C. S. Li, [Coagulation dysfunctions in patients with systemic inflammatory response syndrome]. *Zhonghua Nei Ke Za Zhi* 43, 94-97 (2004).
234 K. Takenaka, E. Ogawa, H. Wada, T. Hirata, Systemic inflammatory response syndrome and surgical stress in thoracic surgery. *J Crit Care* 21, 48-53; discussion 53-45 (2006).
235 J. M. Barry, *The great influenza : the epic story of the deadliest plague in history*. (Penguin Books, New York, 2005), pp. 546 p.
236 L. J. Saif, Animal coronaviruses: what can they teach us about the severe acute respiratory syndrome? *Rev Sci Tech* 23, 643-660 (2004).
237 L. J. Saif, K. Jung, Comparative Pathogenesis of Bovine and Porcine Respiratory Coronaviruses in the Animal Host Species and SARS-CoV-2 in Humans. *J Clin Microbiol* 58, (2020).
238 J. Interlandi, A man-made contagion. *Sci Am* 306, 14 (2012).
239 L. S. Coleman, Four Forgotten Giants of Anesthesia History. *Journal of Anesthesia and Surgery* 3, 1-17 (2015).
240 W. Rath, A. Faridi, J. W. Dudenhausen, HELLP syndrome. *J Perinat Med* 28, 249-260 (2000).
241 W. Rath, A. Faridi, J. W. Dudenhausen, HELLP syndrome. *J Perinat Med* 28, 249-260 (2000).
242 R. S. Kasthuri, M. B. Taubman, N. Mackman, Role of tissue factor in cancer. *J Clin Oncol* 27, 4834-4838 (2009).
243 A. A. Khorana, Cancer and coagulation. *Am J Hematol* 87 Suppl 1, S82-87 (2012).

244 G. F. von Tempelhoff, F. Nieman, L. Heilmann, G. Hommel, Association between blood rheology, thrombosis and cancer survival in patients with gynecologic malignancy. *Clin Hemorheol Microcirc* 22, 107-130 (2000).

245 N. Sikri, A. Bardia, A history of streptokinase use in acute myocardial infarction. *Tex Heart Inst J* 34, 318-327 (2007).

246 N. Sikri, A. Bardia, A history of streptokinase use in acute myocardial infarction. *Tex Heart Inst J* 34, 318-327 (2007).

247 H. D. Neuhof H, Glaser E, Wolf H, Lasch H.G, Streptokinase Therapy in Patients with Acute Myocardial Infarction. *Intensive Care Medicine* Volume 1, 27-30 (1975).

248 B. J. Smith, Thrombolysis in acute myocardial infarction: analysis of studies comparing accelerated t-PA and streptokinase. *J Accid Emerg Med* 16, 407-411 (1999).

249 G. Tatu-Chitoiu *et al.*, Streptokinase-induced hypotension has no detrimental effect on patients with thrombolytic treatment for acute myocardial infarction. A substudy of the Romanian Study for Accelerated Streptokinase in Acute Myocardial Infarction (ASK-ROMANIA). *Rom J Intern Med* 42, 557-573 (2004).

250 G. Landesberg *et al.*, Myocardial infarction after vascular surgery: the role of prolonged stress-induced, ST depression-type ischemia. *J Am Coll Cardiol* 37, 1839-1845 (2001).

251 T. Matsuo, S. Suzuki, K. Kario, H. Kobayashi, [Acute myocardial infarction in the 1995 Hanshin-Awaji Earthquake]. *Rinsho Byori* Suppl 104, 133-141 (1997).

252 S. Tarhan, E. A. Moffitt, W. F. Taylor, E. R. Giuliani, Myocardial infarction after general anesthesia. *Anesth Analg* 56, 455-461 (1977).

253 W. S. Beattie, N. H. Badner, P. Choi, Epidural analgesia reduces postoperative myocardial infarction: a meta-analysis. *Anesth Analg* 93, 853-858 (2001).

254 L. Dintenfass, C. D. Forbes, Viscosity of blood in patients with myocardial infarction, haemophilia and thyroid diseases. Effect of fibrinogen, albumin and globulin. *Biorheology* 10, 457-462 (1973).

255 L. Dintenfass, C. D. Forbes, Effect of fibrinogen on aggregation of red cells and on apparent viscosity of artificial thrombi in haemophilia, myocardial infarction, thyroid disease, cancer and control systems: effect of ABO blood groups. *Microvasc Res* 9, 107-118 (1975).

256 I. N. Basso, D. Keeling, Myocardial infarction following recombinant activated factor VII in a patient with type 2A von Willebrand disease. *Blood Coagul Fibrinolysis* 15, 503-504 (2004).

257 H. Ito, K. Iwakura, Assessing the relation between coronary reflow and myocardial reflow. *Am J Cardiol* 81, 8G-12G (1998).

258 R. A. Kloner, C. E. Ganote, R. B. Jennings, The “no-reflow” phenomenon after temporary coronary occlusion in the dog. *The Journal of clinical investigation* 54, 1496-1508 (1974).

259 I. Morishima *et al.*, Angiographic no-reflow phenomenon as a predictor of adverse long-term outcome in patients treated with percutaneous transluminal coronary angioplasty for first acute myocardial infarction. *J Am Coll Cardiol* 36, 1202-1209 (2000).

260 S. H. Rezkalla, R. A. Kloner, No-reflow phenomenon. *Circulation* 105, 656-662 (2002).
261 H. R. Hellstrom, Coronary artery stasis after induced myocardial infarction in the dog. *Cardiovasc Res* 5, 371-375 (1971).
262 S. V. Brodsky *et al.*, Prevention and reversal of premature endothelial cell senescence and vasculopathy in obesity-induced diabetes by ebselen. *Circ Res* 94, 377-384 (2004).
263 J. Chen, M. S. Goligorsky, Premature senescence of endothelial cells: Methusaleh's dilemma. *Am J Physiol Heart Circ Physiol* 290, H1729-1739 (2006).
264 L. S. Coleman, The Mammalian Stress Mechanism (MSM) Explains Radial Artery Spasm (RAS). J *Anes Surg* 5, 81-94 (2018).
265 A. Genovese et al., Inhibition of hypoxia-induced cardiac hypertrophy by verapamil in rats. *Basic Res Cardiol* 75, 757-763 (1980).
266 A. Genovese, M. Chiariello, S. Latte, W. DeAlfieri, M. Condorelli, Bilateral ventricular hypertrophy in rats exposed to acute or chronic hypobaric hypoxia. *Respiration* 44, 289-293 (1983).
267 A. Genovese, W. De Alfieri, S. Latte, M. Chiariello, M. Condorelli, Regression of myocardial hypertrophy in the rat following removal of acute or chronic hypobaric hypoxia. *Eur Heart J* 3 Suppl A, 161-164 (1982).
268 A. Genovese, S. Latte, M. Bozzaotre, M. Chiariello, Response of the left ventricular connective tissue to hypoxia. *Res Exp Med (Berl)* 183, 111-115 (1983).
269 L. S. Coleman, Atherosclerosis may be caused by inadequate levels of turbulence and mixing. *World J Surg* 30, 638-639 (2006).
270 M. H. Kwon, M. Cevasco, J. D. Schmitto, F. Y. Chen, Ventricular restraint therapy for heart failure: A review, summary of state of the art, and future directions. *J Thorac Cardiovasc Surg* 144, 771-777 e771 (2012).
271 L. S. Lee *et al.*, Ventricular restraint therapy for heart failure: the right ventricle is different from the left ventricle. *J Thorac Cardiovasc Surg* 139, 1012-1018 (2010).
272 L. S. Lee *et al.*, Optimized ventricular restraint therapy: adjustable restraint is superior to standard restraint in an ovine model of ischemic cardiomyopathy. *J Thorac Cardiovasc Surg* 145, 824-831 (2013).
273 Y. Henderson, Resuscitation with Carbon Dioxide. *Science* 83, 399-402 (1936).
274 K. J. Boniface, J. M. Brown, Effect of carbon dioxide excess on contractile force of heart, in situ. *The American journal of physiology* 172, 752-756 (1953).
275 L. S. Coleman, Bupivacaine and ventricular fibrillation. *Anesth Analg* 99, 1269; author reply 1269 (2004).
276 L. S. Coleman, The Mammalian Stress Mechanism (MSM) Explains Radial Artery Spasm (RAS). *J Anes Surg* 5, 81-94 (2018).
277 K. M. Kent, S. E. Epstein, Neural basis for the genesis and control of arrhythmias associated with myocardial infarction. *Cardiology* 61, 61-74 (1976).
278 I. Morishima *et al.*, Angiographic no-reflow phenomenon as a predictor of adverse long-term outcome in patients treated with percutaneous transluminal coronary angioplasty for first acute myocardial infarction. *J Am Coll Cardiol* 36, 1202-1209 (2000).

279 H. Ito *et al.*, Myocardial perfusion patterns related to thrombolysis in myocardial infarction perfusion grades after coronary angioplasty in patients with acute anterior wall myocardial infarction. *Circulation* 93, 1993-1999 (1996).
280 E. Escolar *et al.*, The Effect of an EDTA-based Chelation Regimen on Patients With Diabetes Mellitus and Prior Myocardial Infarction in the Trial to Assess Chelation Therapy (TACT). *Circulation. Cardiovascular quality and outcomes*, (2013).
281 H. Ito, K. Iwakura, Assessing the relation between coronary reflow and myocardial reflow. *Am J Cardiol* 81, 8G-12G (1998).
282 R. A. Kloner, C. E. Ganote, R. B. Jennings, The "no-reflow" phenomenon after temporary coronary occlusion in the dog. *The Journal of clinical investigation* 54, 1496-1508 (1974).
283 I. Morishima *et al.*, Angiographic no-reflow phenomenon as a predictor of adverse long-term outcome in patients treated with percutaneous transluminal coronary angioplasty for first acute myocardial infarction. *J Am Coll Cardiol* 36, 1202-1209 (2000).
284 S. H. Rezkalla, R. A. Kloner, No-reflow phenomenon. *Circulation* 105, 656-662 (2002).
285 L. S. Coleman, Four Forgotten Giants of Anesthesia History. *Journal of Anesthesia and Surgery* 3, 1-17 (2015).
286 J. Interlandi, in *Scientific American*. (2007), chap. 24, pp. 1.
287 M. B. Sporn, Carcinogenesis and cancer: different perspectives on the same disease. *Cancer Res* 51, 6215-6218 (1991).
288 J. F. a. M. P. O. H. Leon Bradlow, *Cancer: Genetics and the Environment*. J. F. a. M. P. O. H. Leon Bradlow, Ed., Annals of the New York Academy of Sciences (New York Academy of Sciences, 1997), vol. 833, pp. 209.
289 C. Maltoni, M. A. Mehlman, *Carcinogenesis bioassays and protecting public health : commemorating the lifework of Cesare Maltoni and colleagues*. Annals of the New York Academy of Sciences, (New York Academy of Sciences, New York, 2002), pp. xii, 231 p.
290 A. A. Khorana, Cancer and coagulation. *Am J Hematol* 87 Suppl 1, S82-87 (2012).
291 G. F. von Tempelhoff, F. Nieman, L. Heilmann, G. Hommel, Association between blood rheology, thrombosis and cancer survival in patients with gynecologic malignancy. *Clin Hemorheol Microcirc* 22, 107-130 (2000).
292 A. Angelov, S. Vicheva, [Thrombohemorrhagic syndrome as the 1st manifestation of stomach cancer with multiple tumor embolisms in the microcirculation]. *Vutr Boles* 20, 8-14 (1981).
293 G. F. von Tempelhoff, N. Schonmann, L. Heilmann, Thrombosis — a clue of poor prognosis in primary non-metastatic breast cancer? *Breast Cancer Res Treat* 73, 275-277 (2002).
294 X. Wang, E. Wang, J. J. Kavanagh, R. S. Freedman, Ovarian cancer, the coagulation pathway, and inflammation. *J Transl Med* 3, 25 (2005).
295 R. J. Collins *et al.*, Spontaneous programmed death (apoptosis) of B-chronic lymphocytic leukaemia cells following their culture in vitro. *British journal of haematology* 71, 343-350 (1989).

296 A. D. Janiszewska, S. Poletajew, A. Wasiutynski, Spontaneous regression of renal cell carcinoma. *Contemporary oncology* 17, 123-127 (2013).

297 T. Jessy, Immunity over inability: The spontaneous regression of cancer. *J Nat Sci Biol Med* 2, 43-49 (2011).

298 R. M. Kluck *et al.*, Spontaneous apoptosis in NS-1 myeloma cultures: effects of cell density, conditioned medium and acid pH. *Immunobiology* 188, 124-133 (1993).

299 D. H. MacAdam, *Spontaneous Regression: Cancer And The Immune System.* (Xlibris, 2003), pp. 160.

300 J. Searle, D. J. Collins, B. Harmon, J. F. Kerr, The spontaneous occurrence of apoptosis in squamous carcinomas of the uterine cervix. *Pathology* 5, 163-169 (1973).

301 M. Baum, R. Demicheli, W. Hrushesky, M. Retsky, Does surgery unfavourably perturb the "natural history" of early breast cancer by accelerating the appearance of distant metastases? *Eur J Cancer* 41, 508-515 (2005).

302 S. Bhatia *et al.*, Breast cancer and other second neoplasms after childhood Hodgkin's disease. *N Engl J Med* 334, 745-751 (1996).

303 P. H. Zahl, P. C. Gotzsche, J. Maehlen, Natural history of breast cancers detected in the Swedish mammography screening programme: a cohort study. *Lancet Oncol* 12, 1118-1124 (2011).

304 P. H. Zahl, J. Maehlen, H. G. Welch, The natural history of invasive breast cancers detected by screening mammography. *Arch Intern Med* 168, 2311-2316 (2008).

305 D. Davis, *The Secret History of the War on Cancer.* (Perseus Books Group, New York, N.Y., 2007).

306 S. E. Luria, M. Delbruck, Mutations of Bacteria from Virus Sensitivity to Virus Resistance. *Genetics* 28, 491-511 (1943).

307 D. H. MacAdam, *Spontaneous Regression: Cancer And The Immune System.* (Xlibris, 2003), pp. 160.

308 P. H. Zahl, P. C. Gotzsche, J. Maehlen, Natural history of breast cancers detected in the Swedish mammography screening programme: a cohort study. *Lancet Oncol* 12, 1118-1124 (2011).

309 P. H. Zahl, J. Maehlen, H. G. Welch, The natural history of invasive breast cancers detected by screening mammography. *Arch Intern Med* 168, 2311-2316 (2008).

310 S. H. Rezkalla, R. A. Kloner, No-reflow phenomenon. *Circulation* 105, 656-662 (2002).

311 R. A. Fleck, L. V. Rao, S. I. Rapaport, N. Varki, Localization of human tissue factor antigen by immunostaining with monospecific, polyclonal anti-human tissue factor antibody. *Thromb Res* 59, 421-437 (1990).

312 R. S. Kasthuri, M. B. Taubman, N. Mackman, Role of tissue factor in cancer. J *Clin Oncol* 27, 4834-4838 (2009).

313 J. H. McVey, Tissue factor pathway. *Baillieres Best Pract Res Clin Haematol* 12, 361-372 (1999).

314 P. Cirillo *et al.*, Tissue factor binding of activated factor VII triggers smooth muscle cell proliferation via extracellular signal-regulated kinase activation. *Circulation* 109, 2911-2916 (2004).

315 S. Gando, Tissue factor in trauma and organ dysfunction. *Semin Thromb Hemost* 32, 48-53 (2006).
316 S. Gando, S. Nanzaki, S. Sasaki, O. Kemmotsu, Significant correlations between tissue factor and thrombin markers in trauma and septic patients with Disseminated Intravascular Coagulation. *Thromb Haemost* 79, 1111-1115 (1998).
317 G. M. Hjortoe et al., Tissue factor-factor VIIa-specific up-regulation of IL-8 expression in MDA-MB-231 cells is mediated by PAR-2 and results in increased cell migration. *Blood* 103, 3029-3037 (2004).
318 T. Imamura, [Tissue factor expression at the site of inflammation: a cross-talk between inflammation and the blood coagulation system]. *Rinsho Byori* 52, 342-349 (2004).
319 A. A. Khorana et al., Tissue factor expression, angiogenesis, and thrombosis in pancreatic cancer. *Clin Cancer Res* 13, 2870-2875 (2007).
320 F. R. Rickles, S. Patierno, P. M. Fernandez, Tissue factor, thrombin, and cancer. *Chest* 124, 58S-68S (2003).
321 S. Weijer et al., Inhibition of the tissue factor/factor VIIa pathway does not influence the inflammatory or antibacterial response to abdominal sepsis induced by Escherichia coli in mice. J *Infect Dis* 189, 2308-2317 (2004).
322 J. L. Yao et al., Tissue factor and VEGF expression in prostate carcinoma: a tissue microarray study. *Cancer Invest* 27, 430-434 (2009).
323 A. Siegbahn, Cellular consequences upon factor VIIa binding to tissue factor. *Haemostasis* 30 Suppl 2, 41-47 (2000).
324 A. Siegbahn et al., Binding of factor VIIa to tissue factor on human fibroblasts leads to activation of phospholipase C and enhanced PDGF-BB-stimulated chemotaxis. *Blood* 96, 3452-3458 (2000).
325 T. Ueno, M. Toi, M. Koike, S. Nakamura, T. Tominaga, Tissue factor expression in breast cancer tissues: its correlation with prognosis and plasma concentration. *Br J Cancer* 83, 164-170 (2000).
326 H. J. Weiss, V. T. Turitto, H. R. Baumgartner, Y. Nemerson, T. Hoffmann, Evidence for the presence of tissue factor activity on subendothelium. *Blood* 73, 968-975 (1989).
327 A. Angelov, S. Vicheva, [Thrombohemorrhagic syndrome as the 1st manifestation of stomach cancer with multiple tumor embolisms in the microcirculation]. *Vutr Boles* 20, 8-14 (1981).
328 L. M. Coussens, Z. Werb, Inflammation and cancer. *Nature* 420, 860-867 (2002).
329 E. Archiniegas, C. Y. Neves, D. Candelle, J. E. Cardier, Thrombin and its protease-activated receptor-1 (PAR1) participate in the endothelial-mesenchymal transdifferentiation process. *DNA Cell Biol* 23, 815-825 (2004).
330 E. Bretschneider et al., Evidence for proteinase-activated receptor-2 (PAR-2)-mediated mitogenesis in coronary artery smooth muscle cells. *Br J Pharmacol* 126, 1735-1740 (1999).
331 S. Fiorucci et al., PAR1 antagonism protects against experimental liver fibrosis. Role of proteinase receptors in stellate cell activation. *Hepatology* 39, 365-375 (2004).

332 E. Jin et al., Protease-activated receptor (PAR)-1 and PAR-2 participate in the cell growth of alveolar capillary endothelium in primary lung adenocarcinomas. *Cancer* 97, 703-713 (2003).
333 W. Luo, Y. Wang, G. Reiser, Two types of protease-activated receptors (PAR-1 and PAR-2) mediate calcium signaling in rat retinal ganglion cells RGC-5. *Brain Res* 1047, 159-167 (2005).
334 A. Naldini, C. Bernini, A. Pucci, F. Carraro, Thrombin-mediated IL-10 up-regulation involves protease-activated receptor (PAR)-1 expression in human mononuclear leukocytes. J Leukoc Biol 78, 736-744 (2005).
335 K. Tantivejkul et al., PAR1-mediated NFkappaB activation promotes survival of prostate cancer cells through a Bcl-xL-dependent mechanism. *J Cell Biochem* 96, 641-652 (2005).
336 T. Tran, A. G. Stewart, Protease-activated receptor (PAR)-independent growth and pro-inflammatory actions of thrombin on human cultured airway smooth muscle. *Br J Pharmacol* 138, 865-875 (2003).
337 P. Zania, S. Kritikou, C. S. Flordellis, M. E. Maragoudakis, N. E. Tsopanoglou, Blockade of angiogenesis by small molecule antagonists to protease-activated receptor-1 (PAR-1): Association with endothelial cell growth suppression and induction of apoptosis. J Pharmacol Exp Ther, (2006).
338 J. F. Kerr, C. M. Winterford, B. V. Harmon, Apoptosis. Its significance in cancer and cancer therapy. *Cancer* 73, 2013-2026 (1994).
339 J. F. Kerr, A. H. Wyllie, A. R. Currie, Apoptosis: a basic biological phenomenon with wide-ranging implications in tissue kinetics. *British Journal of Cancer* 26, 239-257 (1972).
340 C. N. Pagel et al., Inhibition of osteoblast apoptosis by thrombin. *Bone* 33, 733-743 (2003).
341 G. Poste, R. Greig, On the genesis and regulation of cellular heterogeneity in malignant tumors. *Invasion Metastasis* 2, 137-176 (1982).
342 H. Schiller et al., Thrombin as a survival factor for cancer cells: thrombin activation in malignant effusions in vivo and inhibition of idarubicin-induced cell death in vitro. Int J *Clin Pharmacol Ther* 40, 329-335 (2002).
343 P. Duesberg, Chromosomal chaos and cancer. *Sci Am* 296, 52-59 (2007).
344 S. E. Luria, M. Delbruck, Mutations of Bacteria from Virus Sensitivity to Virus Resistance. *Genetics* 28, 491-511 (1943).
345 W. W. Gibbs, Untangling the roots of cancer. *Sci Am* 289, 56-65 (2003).
346 J. A. Erukhimov *et al.*, Actin-containing sera from patients with adult respiratory distress syndrome are toxic to sheep pulmonary endothelial cells. *American journal of respiratory and critical care medicine* 162, 288-294 (2000).
347 P. A. Janmey, J. A. Lamb, R. M. Ezzell, S. Hvidt, S. E. Lind, Effects of actin filaments on fibrin clot structure and lysis. *Blood* 80, 928-936 (1992).
348 L. R. Otterbein, C. Cosio, P. Graceffa, R. Dominguez, Crystal structures of the vitamin D-binding protein and its complex with actin: structural basis of the actin-scavenger system. *Proceedings of the National Academy of Sciences of the United States of America* 99, 8003-8008 (2002).
349 S. E. Lind, D. B. Smith, P. A. Janmey, T. P. Stossel, Role of plasma gelsolin and the vitamin D-binding protein in clearing actin from the circulation. *The Journal of clinical investigation* 78, 736-742 (1986).

350 S. E. Lind, C. J. Smith, Actin is a noncompetitive plasmin inhibitor. *J Biol Chem* 266, 5273-5278 (1991).

351 S. E. Lind, C. J. Smith, Actin accelerates plasmin generation by tissue plasminogen activator. *J Biol Chem* 266, 17673-17678 (1991).

352 S. E. Lind, P. A. Janmey, C. Chaponnier, T. J. Herbert, T. P. Stossel, Reversible binding of actin to gelsolin and profilin in human platelet extracts. *J Cell Biol* 105, 833-842 (1987).

353 P. A. Janmey *et al.*, Interactions of gelsolin and gelsolin-actin complexes with actin. Effects of calcium on actin nucleation, filament severing, and end blocking. *Biochemistry* 24, 3714-3723 (1985).

354 D. B. Smith, P. A. Janmey, S. E. Lind, Circulating actin-gelsolin complexes following oleic acid-induced lung injury. *The American journal of pathology* 130, 261-267 (1988).

355 P. A. Janmey, S. E. Lind, Capacity of human serum to depolymerize actin filaments. *Blood* 70, 524-530 (1987).

356 P. A. Janmey, S. E. Lind, H. L. Yin, T. P. Stossel, Effects of semi-dilute actin solutions on the mobility of fibrin protofibrils during clot formation. *Biochim Biophys Acta* 841, 151-158 (1985).

357 A. D. Janiszewska, S. Poletajew, A. Wasiutynski, Spontaneous regression of renal cell carcinoma. *Contemporary oncology* 17, 123-127 (2013).

358 T. Jessy, Immunity over inability: The spontaneous regression of cancer. *J Nat Sci Biol Med* 2, 43-49 (2011).

359 R. M. Kluck *et al.*, Spontaneous apoptosis in NS-1 myeloma cultures: effects of cell density, conditioned medium and acid pH. *Immunobiology* 188, 124-133 (1993).

360 D. H. MacAdam, *Spontaneous Regression: Cancer And The Immune System*. (Xlibris, 2003), pp. 160.

361 J. Searle, D. J. Collins, B. Harmon, J. F. Kerr, The spontaneous occurrence of apoptosis in squamous carcinomas of the uterine cervix. *Pathology* 5, 163-169 (1973).

362 M. Baum, R. Demicheli, W. Hrushesky, M. Retsky, Does surgery unfavourably perturb the "natural history" of early breast cancer by accelerating the appearance of distant metastases? *Eur J Cancer* 41, 508-515 (2005).

363 S. Bhatia *et al.*, Breast cancer and other second neoplasms after childhood Hodgkin's disease. *N Engl J Med* 334, 745-751 (1996).

364 P. H. Zahl, P. C. Gotzsche, J. Maehlen, Natural history of breast cancers detected in the Swedish mammography screening programme: a cohort study. *Lancet Oncol* 12, 1118-1124 (2011).

365 R. A. Fleck, L. V. Rao, S. I. Rapaport, N. Varki, Localization of human tissue factor antigen by immunostaining with monospecific, polyclonal anti-human tissue factor antibody. *Thromb Res* 59, 421-437 (1990).

366 M. Baum, R. Demicheli, W. Hrushesky, M. Retsky, Does surgery unfavourably perturb the "natural history" of early breast cancer by accelerating the appearance of distant metastases? *Eur J Cancer* 41, 508-515 (2005).

367 B. Biki *et al.*, Anesthetic technique for radical prostatectomy surgery affects cancer recurrence: a retrospective analysis. *Anesthesiology* 109, 180-187 (2008).

368 A. K. Exadaktylos, D. J. Buggy, D. C. Moriarty, E. Mascha, D. I. Sessler, Can anesthetic technique for primary breast cancer surgery affect recurrence or metastasis? *Anesthesiology* 105, 660-664 (2006).

369 S. Bhatia *et al.*, Breast cancer and other second neoplasms after childhood Hodgkin's disease. *N Engl J Med* 334, 745-751 (1996).

370 R. S. Kasthuri, M. B. Taubman, N. Mackman, Role of tissue factor in cancer. *J Clin Oncol* 27, 4834-4838 (2009).

371 M. E. McFadden, S. E. Sartorius, Multiple systems organ failure in the patient with cancer. Part I: pathophysiologic perspectives. *Oncol Nurs Forum* 19, 719-724 (1992).

372 P. Duesberg, Chromosomal chaos and cancer. *Sci Am* 296, 52-59 (2007).

373 H. Schiller *et al.*, Thrombin as a survival factor for cancer cells: thrombin activation in malignant effusions in vivo and inhibition of idarubicin-induced cell death in vitro. *Int J Clin Pharmacol Ther* 40, 329-335 (2002).

374 F. R. Rickles, S. Patierno, P. M. Fernandez, Tissue factor, thrombin, and cancer. *Chest* 124, 58S-68S (2003).

375 F. R. Rickles, S. Patierno, P. M. Fernandez, Tissue factor, thrombin, and cancer. *Chest* 124, 58S-68S (2003).

376 P. H. Zahl, P. C. Gotzsche, J. Maehlen, Natural history of breast cancers detected in the Swedish mammography screening programme: a cohort study. *Lancet Oncol* 12, 1118-1124 (2011).

377 P. H. Zahl, J. Maehlen, H. G. Welch, The natural history of invasive breast cancers detected by screening mammography. *Arch Intern Med* 168, 2311-2316 (2008).

378 https://www.ommegaonline.org/article-details/Four-Forgotten-Giants-of-Anesthesia-History/468

379 E. A. Bermudez, M. H. Chen, Cardiac arrest associated with intravenous propofol during transesophageal echocardiography before DC cardioversion. *Heart Dis* 4, 355-357 (2002).

380 B. Vasile, F. Rasulo, A. Candiani, N. Latronico, The pathophysiology of propofol infusion syndrome: a simple name for a complex syndrome. *Intensive Care Med* 29, 1417-1425 (2003).

381 G. G. Pacentine, M. Muzi, T. J. Ebert, Effects of fentanyl on sympathetic activation associated with the administration of desflurane. *Anesthesiology* 82, 823-831 (1995).

382 M. J. Goff, S. R. Arain, D. J. Ficke, T. D. Uhrich, T. J. Ebert, Absence of bronchodilation during desflurane anesthesia: a comparison to sevoflurane and thiopental. *Anesthesiology* 93, 404-408 (2000).

383 J. P. Cravero, M. Beach, B. Thyr, K. Whalen, The effect of small dose fentanyl on the emergence characteristics of pediatric patients after sevoflurane anesthesia without surgery. *Anesth Analg* 97, 364-367, table of contents (2003).

384 C. Keijzer, R. S. Perez, J. J. De Lange, Carbon monoxide production from five volatile anesthetics in dry sodalime in a patient model: halothane and sevoflurane do produce carbon monoxide; temperature is a poor predictor of carbon monoxide production. *BMC Anesthesiol* 5, 6 (2005).

385 M. J. Laster, E. I. Eger, 2nd, Temperatures in soda lime during degradation of desflurane, isoflurane, and sevoflurane by desiccated soda lime. *Anesth Analg* 101, 753-757, table of contents (2005).

386 G. Moriwaki, H. Bito, K. Ikeda, Partly exhausted soda lime or soda lime with water added, inhibits the increase in compound A concentration in the circle system during low-flow sevoflurane anaesthesia. *Br J Anaesth* 79, 782-786 (1997).

387 L. W. Crile GW, *Anoci-association.* (Saunders, Philadelphia, 1914), pp. 223-225.

388 L. S. Coleman, 30 Years Lost in Anesthesia Theory. *Cardiovasc Hematol Agents Med Chem* 10, 31-49 (2012).

389 K. E. Welty-Wolf *et al.*, Coagulation blockade prevents sepsis-induced respiratory and renal failure in baboons. *American journal of respiratory and critical care medicine* 164, 1988-1996 (2001).

390 L. J. Leipzig, A. J. Boyle, D. S. McCann, Case histories of rheumatoid arthritis treated with sodium or magnesium EDTA. *J Chronic Dis* 22, 553-563 (1970).

391 S. Finnegan, S. L. Percival, EDTA: An Antimicrobial and Antibiofilm Agent for Use in Wound Care. *Adv Wound Care (New Rochelle)* 4, 415-421 (2015).

392 R. E. Wooley, M. S. Jones, Action of EDTA-Tris and antimicrobial agent combinations on selected pathogenic bacteria. *Veterinary microbiology* 8, 271-280 (1983).

393 E. Escolar *et al.*, The Effect of an EDTA-based Chelation Regimen on Patients With Diabetes Mellitus and Prior Myocardial Infarction in the Trial to Assess Chelation Therapy (TACT). *Circulation. Cardiovascular quality and outcomes*, (2013).

394 A. Zarbock *et al.*, Effect of Regional Citrate Anticoagulation vs Systemic Heparin Anticoagulation During Continuous Kidney Replacement Therapy on Dialysis Filter Life Span and Mortality Among Critically Ill Patients With Acute Kidney Injury: A Randomized Clinical Trial. *JAMA* 324, 1629-1639 (2020).

395 F. Mariano *et al.*, Citrate anticoagulation for continuous renal replacement therapy in critically ill patients: success and limits. *International journal of nephrology* 2011, 748320 (2011).

396 S. Morgera *et al.*, A safe citrate anticoagulation protocol with variable treatment efficacy and excellent control of the acid-base status. *Critical care medicine* 37, 2018-2024 (2009).

397 A. J. Tolwani, M. B. Prendergast, R. R. Speer, B. S. Stofan, K. M. Wille, A practical citrate anticoagulation continuous venovenous hemodiafiltration protocol for metabolic control and high solute clearance. *Clinical journal of the American Society of Nephrology : CJASN* 1, 79-87 (2006).

398 J. C. Bos *et al.*, Low polymorphonuclear cell degranulation during citrate anticoagulation: a comparison between citrate and heparin dialysis. *Nephrology, dialysis, transplantation : official publication of the European Dialysis and Transplant Association - European Renal Association* 12, 1387-1393 (1997).

399 M. Gritters *et al.*, Citrate anticoagulation abolishes degranulation of polymorphonuclear cells and platelets and reduces oxidative stress during haemodialysis. *Nephrology, dialysis, transplantation : official publication of the*

European Dialysis and Transplant Association - European Renal Association 21, 153-159 (2006).
400 K. Kottke-Marchant, J. M. Anderson, A. Rabinovitch, R. A. Huskey, R. Herzig, The effect of heparin vs. citrate on the interaction of platelets with vascular graft materials. *Thromb Haemost* 54, 842-848 (1985).
401 S. Finnegan, S. L. Percival, EDTA: An Antimicrobial and Antibiofilm Agent for Use in Wound Care. *Adv Wound Care (New Rochelle)* 4, 415-421 (2015).
402 R. E. Wooley, M. S. Jones, Action of EDTA-Tris and antimicrobial agent combinations on selected pathogenic bacteria. *Veterinary microbiology* 8, 271-280 (1983).
403 H. Schiller *et al.*, Thrombin as a survival factor for cancer cells: thrombin activation in malignant effusions in vivo and inhibition of idarubicin-induced cell death in vitro. *Int J Clin Pharmacol Ther* 40, 329-335 (2002).

Chapter 17

Unified Theory of Biology

"It is the function of science to discover the existence of a general reign of order in nature and to find the causes governing this order. And this refers in equal measure to the relations of man — social and political — and to the entire universe as a whole."
— Dmitri Mendeleev

Selye's theory exceeds the bounds of medicine. It confers a *Unified Theory of Biology* that explains the origin and nature of life, including botany, zoology, microbiology, parasitology, embryology, evolution, taxonomy, paleontology, anatomy, physiology, ethology, intelligence, the Cambrian Explosion, and dinosaurs.

Each class of complex multicellular animals and plants is characterized by its own unique extra-cellular stress mechanism that is adapted to a specific set of environmental circumstances. Each stress mechanism requires compatible combinations of anatomy, physiology, intelligence, and behavior that determine its food requirements, temperature tolerance, exercise capacity, toxin tolerance, and other attributes common to the class.

Each stress mechanism is accompanied by its own unique genome that is retained within every cell. The sex mechanism continually collects fresh mutations, culls harmful mutations, retains potentially useful mutations, and distributes fresh characteristics throughout the breeding population, so that genomes gradually grow more capable and complex with time, as observed by Lamarck. Speciation selects complimentary combinations of available genomic attributes to enable the stress mechanism to adapt to changing environmental circumstances, as observed by Darwin. The extinction of a species is of little greater consequence than the death of an individual, because stress mechanisms and their closely associated genomes persist in the surviving species of the same class.

Selye's theory further implies that unique intracellular stress mechanisms characterize classes of free-living cells. These maintain individual cell structure, regulate cell chemistry, and determine cell anatomy, motility, nutrition requirements, toxin resistance, temperature tolerance, and so forth to enable adaptation to specific sets of environmental circumstances.

Microbial intra-cellular stress mechanisms flourish in great diversity, but stress theory suggests that all animals consist of identical or closely similar eukaryotic cells as it seems unlikely that intracellular and extracellular stress mechanisms could evolve in concert. Nevertheless, the genome enables extensive cell specialization that facilitates specialized organs and tissues. The same principle applies to plants.

Structurally complex multicellular plants may have preceded animals on land, but extracellular animal stress mechanisms may have evolved in rift zones deep in the oceans long before the appearance of multicellular plants.

STRESS THEORY AND BIOLOGY

This section presents a series of short essays that illustrate how stress theory clarifies biology. It is intended as a heuristic inspiration for further research rather than an attempt to explain all aspects of biology. At present, the mammalian stress mechanism is the only stress mechanism that can be described with any assurance, and its description remains crude and untested. Hopefully these speculations will inspire its clarification and a search for additional stress mechanisms.

Obviously, much in this section is speculative. We do not know with any certainty, for example, whether or how life began on Earth, and there are serious suggestions that life did not begin on Earth. Barring any second data point from elsewhere in the cosmos (not even from Mars, as of this writing), we need to work with the data we have. What *is* certain is that life is abundant on Earth, that it thrives in a wide variety of environments, and that it has evolved here on Earth over several billions of years. Therefore, the simplest assumption is that Earth's life originated on Earth, has evolved in the absence of alien influence, and that evolution proceeds with the appearance, exploitation, and extinction of stress mechanisms in accord with environmental change and natural selection.

THE ORIGIN OF LIFE

Human egocentricity has focused conventional theories of life's origin on the Earth's primordial surface,[1] but this fails to explain food sources that might sustain life or the source of protective oceans and breathable atmospheric gases, without which the Earth's surface would be exposed to the inimical vacuum of space, drastic temperature fluctuations, harmful radiation, and bombardment by comets and meteors. Meanwhile, the teeming mass of microbes that thrives in the seemingly inhospitable subterranean environment remains unappreciated, even though the Kola Superdeep Borehole revealed an unexpected abundance of microbial life, water, hydrogen, carbon dioxide, and other atmospheric gases lurking deep beneath the Earth's surface. Furthermore, strange new microbial and multicellular life forms that are strikingly similar in appearance to familiar sea creatures have been discovered thriving in abundance in the boiling hot, hypoxic, and seemingly toxic volcanic rift zones that ring the Earth between tectonic plates like the seams of a baseball. Also, a previously unsuspected third domain of microbes called *Archaea* has been discovered thriving in volcanic pools where life was previously assumed to be impossible (the other domains being Bacteria and Eukarya). From the perspective of stress theory, these observations provide powerful clues to the origin and nature of life.

I hypothesize that life is a spontaneous phenomenon that appears when environmental circumstances are favorable, and that the Earth's life originated deep beneath the Earth's surface, where the nuclear core provides a stable, hot environment that provides energy-rich chemicals that function as food and heat promotes chemical reactions. These conditions are vastly more favorable for the prospective origin of life than the barren, hostile, and chaotic surface environment. I thus propose that life originated as self-sustaining chemical reactions that eventually evolved enclosing cell walls that revolutionized their survival, along with chemical-based information storage systems (DNA/RNA) to extract nutrients from their environment and sustain their survival. There is growing suspicion that free-living cells possess some form of intelligence that enables them to evolve complex molecular machinery that enhances their survival.[2] The resulting microbes subsequently evolved over eons into a vast and diverse subterranean biomass with numerous intra-cellular stress mechanisms that maintain, repair, and reproduce their structures; multiple metabolic pathways that extract energy from the rich stew of energetic chemicals in their environment; and complex genomes that continually capture and accumulate fresh capabilities and characteristics.

The incessant activities of the diverse biomass of subterranean unicellular life continuously generates the water that fills Earth's oceans, gases that enable the Earth's atmosphere, and oil that oozes to Earth's surface — where surface life forms devour it, thereby providing the foundation of the food chain that sustains marine and terrestrial life on the Earth's surface.[3]

Subterranean microbial activity thus explains why water appears in the middle of deserts; why inexhaustible reserves of oil appear everywhere beneath Earth's crust; why oil emerges from the ground saturated with CO_2; why CO_2 accumulates in caves; and why volcanic activity belches forth vast quantities of carbon dioxide. Unceasing subterranean microbial metabolic activity produces gases and water that replenish and maintain Earth's atmosphere and oceans at equilibrium with entropy and gravity. Earth's own nuclear core generates most of Earth's heat. Oceanic and atmospheric currents retain and redistribute this heat energy.

In this way, Earth stabilized its surface environment and enabled the evolution of multicellular terrestrial life forms. The Earth's multicellular surface life thus reflects the following:

1. Radiant energy from the sun
2. Core generation of energy-rich chemicals
3. The vast subterranean biomass that converts the energy rich chemicals into water, oil, and atmospheric gases
4. Gravity that prevents atmospheric gases from dissipating into space via entropy
5. Core generation of the "magnetic bottle" that protects atmospheric gases from the solar wind

The presence of Earth's multicellular surface life reflects a combination of radiant energy received from the sun, core generation of heat and energy-rich chemicals, conversion of energy-rich chemicals into water, oil and atmospheric gases by the vast subterranean biomass, and gravity that offsets entropy and retains surface substances and atmospheric gases.

Intracellular stress mechanisms evolved within the earliest organisms as a survival response to environmental pressures. Those organisms best able to incorporate nutrients, regulate their internal conditions, and repair themselves survived.

Those that could not, perished.

This hypothesis implies that Earth's life is literally digesting its own habitat, and will render the Earth a dying planet like Mars, long before the decline of the sun.

MARS AS THE ABODE OF LIFE

By all accounts, Mars once was a blue-water planet like Earth.[4] Its core no longer sustains a magnetosphere like the Earth's that protects against the constant predation of the solar wind. Its fading nuclear fires no longer replenish oceans, atmosphere, and surface life, and these have been swept away in eons past. The thin CO_2 atmosphere of Mars, the water ice accumulations discovered at its poles, and methane plumes detected by spectroscopic scanning are most simply explained by surviving subterranean microbial activity.

NASA states that methane plumes have been detected in Mars' northern hemisphere. Methane in the current Martian atmosphere has a chemical lifetime of no more than 600 years, which indicates that the methane detected there by spectrogram could not have been present since Mars' formation more than four billion years ago. Some ongoing process is creating fresh methane. Both biological and non-biological processes can create methane. Biological methane and non-biological methane differ only in the ratio of isotopes of carbon and hydrogen (life favors the lighter isotopes), and these would show up in any vented gases. Unfortunately, the spectrum of the elements is the same regardless of isotope. Until we have human or robotic boots on the ground to sample and analyze the gas, we can't know whether it is biologically produced.[5]

A single meteorite chipped from Mars by an ancient collision, and which was retrieved by scientists in Antarctica, seems to include an indication of a Martian microbe. Small amounts of water still on the surface of Mars and a thin Martian atmosphere consisting of carbon dioxide and methane suggest lingering subterranean microbial activity.[6, 7, 8]

The confirmation of a separately evolved subterranean microbial life form from a planet as close and accessible to us as Mars would imply that life is common throughout the universe and that it probably occurs wherever circumstances are favorable. However, surface life is probably less common than subterranean life.

Should subterranean life eventually be discovered on Mars, the odds are that it will display stress mechanisms similar to those of life forms that have evolved on Earth. The simplest explanation is that like environments should produce like responses.

Time (and exploration) will tell. Extraterrestrial life may be the great discovery of the 21st century, much as nuclear energy was of the 20th.

THE RISE OF CELLULAR LIFE ON EARTH

We now know of three domains of cells that thrive on Earth: *Bacteria*, *Archaea* (both of which are prokaryotes), and *Eukarya*. These three cell types encompass all presently known living creatures on Earth.

Prokaryotes employ their cellular membranes for respiration. This limits them to single-cell existence, small size, simple internal structures, and shapes that maximize surface area relative to internal volume to optimize respiration.

Eukaryotes are more complex than prokaryotes. They comprise all multi-celled organisms. Eukaryotic cells are believed to have originated when parent cells engulfed previously free-living prokaryotes and incorporated them as symbiotic cytoplasmic organelles. Organelle types include mitochondria and chloroplasts that retain their own independently replicating DNA. The parent cell maintains its DNA in the form of chromosomes that are isolated from organelle DNA by a thick nuclear membrane. Organelle DNA floats freely within the organelle and employs the *transcription/translation* mechanism, but chromosomal DNA employs an undiscovered mechanism that transmits its information outside the nucleus.

Mitochondria are universally present in eukaryotic cells, but chloroplasts are unique to eukaryotic plant cells. Both provide internal sources of respiration energy that enable eukaryotic cells to grow larger and develop complex internal structures, alternate modes of locomotion, specialized anatomy, and the ability to form the complex multicellular structures of animals and plants.

The mechanism that converts the chromosomal DNA information in the cellular nucleus into specialized and multi-cellular structures (i.e. embryological development) remains unknown. Present DNA understanding derives primarily from prokaryote studies that are mostly irrelevant to eukaryotic cells. Chromosomal DNA does not transmit its genetic information via simple RNA transcription/translation as in the case of prokaryotic DNA. Furthermore, there is no evidence that either chromosomal DNA or cytoplasmic DNA (from organelles) specifically regulates tissue repair or hemodynamic physiology.

These tasks are managed by the organism's extracellular stress mechanism.

MAINTENANCE OF THE INTERNAL MILIEU

Stress mechanisms govern hemodynamic physiology to maintain an internal milieu in the multi-cellular body that optimizes cell survival and function. Optimal eukaryotic animal cell mitosis and metabolism, and therefore vertebrate embryological growth, development, maturation, brain function, and exercise capacity, all necessitate temperature, oxygenation, nutrition, and waste removal that are compatible with cellular survival and function.

Eukaryotic cell metabolism declines progressively at temperatures below 37 degrees Celsius. Mammalian brains become dysfunctional when heated above 42 degrees Celsius or chilled below 32 degrees Celsius. Temperature around 37.7 degrees Celsius optimizes mitosis, exercise capacity, growth, maturation, organ function, and evolution in both reptiles and mammals.[9] Food availability, cardiorespiratory function, and body temperature are all essential to cell function, and therefore affect exercise capacity, growth, size, maturation, and evolution in all vertebrates.

Effective cardiopulmonary function is essential to transport oxygen and nutrients to cells to enable mitochondria to generate ATP that energizes cell and enzyme activities. Mild oxygen starvation inhibits eukaryotic cell metabolism and mitosis. Moderate oxygen starvation causes eukaryotic cells to revert to inefficient anaerobic respiration that can only briefly sustain intense cellular activity, and causes painful lactic acid accumulation in tissues. Severe oxygen starvation causes eukaryotic cell death.

LIPOPROTEIN PATHOPHYSIOLOGY

Proteins and carbohydrates are easily digested, dissolved in blood, and transported to cells, but they deliver only four calories/gram.[10] Fat delivers nine calories/gram and is an essential element in vertebrate diets. However, fat floats on water, is insoluble in blood, cannot be transported to cells in its natural state, and is a difficult to digest. The combined effects of pancreatic lipases, intestinal enzymes, and hepatic bile salts are necessary to digest fat to small molecules. These must be attached to proteins to form lipoproteins before they can be suspended in blood and transported to cells. Mammalian intestines produce large lipoproteins called chylomicrons that are transported via portal circulation to the liver, where they are reduced to smaller size and released into systemic circulation. Congestive heart failure impairs cardiac efficiency, undermines lipoprotein transportation to cells, and exaggerates hypertriglyceridemia after meals.[11]

Lipoproteins are temperature-sensitive. At or above mammalian body temperature they liquefy, blend seamlessly with blood serum, and have negligible effect on blood viscosity.[12] Below mammalian body temperature (37.7 degrees Celsius) they solidify, exaggerate blood viscosity, decrease cardiac efficiency, deposit on the inner walls of arteries, and accelerate atherosclerosis.[13] Lipoprotein temperature sensitivity is readily demonstrated when human blood samples are obtained immediately after a meal. Lipoproteins solidify, rise to the top of the blood sample, and form a white layer as the blood samples cool to room temperature. During emergency surgery, portal veins sometimes appear milky white due to lipoprotein solidification when intestines are cooled by exposure to room air. This suggests the utility of capillary-to-capillary portal vein systems that drain blood from the intestine to the liver; these prevent solidified chylomicrons from harmfully entering systemic circulation when the intestines are cooled below the threshold of lipoprotein liquefaction.

A FRESH THEORY OF EMBRYOLOGY

The nature of embryological development has eluded science. During embryological development cellular mitosis, metabolism, differentiation, and apoptosis proceeds considerably faster than at any other time of life, and the appearance and subsequent disappearance of ancient structures provides compelling evidence of evolution. The cells of complex animals possess the ability to spontaneously organize themselves into functional organs and tissues without extra-cellular instruction or induction, which implies that each cell is governed from within by its own chromosomal copy of the genome, but the mechanism of this spontaneous organization remains unknown.[14, 15]

The previous generation of stress researchers hypothesized that Selye's stress mechanism functions as a companion mechanism that works closely with DNA to convert genetic information into embryological development; that it subsequently remains active throughout life to maintain and repair mature structures; and that DNA resumes quiescence once embryological development is complete. These ideas still represent the simplest description of embryological development, and they were never refuted, but they remain unconfirmed. As a result, the idea that DNA directly determines physiology, pathology, and stress in the manner of prokaryotic DNA transcription/translation mechanisms became the presently prevailing paradigm of medical and biological research. This is known as the "Central Dogma of Molecular Biology."

The DNA structure described by Watson and Crick explains the ability of chromosomes to store vast amounts of information, but the DNA mechanism by itself cannot explain how genetic information governs embryological development. The protein-encoding portions of DNA in the eukaryotic cells of adult animals consists of short segments that are interspersed with large sections of seemingly inert DNA. Researchers initially referred to this inactive DNA as "junk" because they assumed it only codes for characteristics that were retained in the genome from the distant past but are no longer needed. There is little evidence that this "junk" DNA produces messenger RNA (transcription) or proteins (translation) in the manner of prokaryotic DNA, either during embryological development or during the subsequent life of the mature individual, so there is no convincing evidence that DNA directly governs tissue maintenance, tissue repair, or hemodynamic physiology (which we have just shown is done by the stress mechanism). Furthermore, recent research reveals that at least some portions of seemingly inert "junk" DNA determines embryological development. Because of this, some members of the current generation of researchers suspect the presence of an undiscovered cytoplasmic mechanism that explains how nuclear DNA controls embryological development.[16, 17, 18, 19]

As predicted by previous stress researchers, I hypothesize that an undiscovered mechanism works closely with the mammalian stress mechanism and chromosomal DNA to convert the genetic blueprint into embryological development. The undiscovered mechanism does this by determining the expression of PAR (thrombin) receptors and tissue factor on the cell surface in accord with DNA instructions. The tissue factor interacts with factors VII and X to generate thrombin that energizes cell division and differentiation, and the PAR numbers and proportions determine how the cells react to thrombin elevations.

I hypothesize that the mammalian stress mechanism converts chromosomal information into embryological development as predicted by previous stress researchers. It does this by controlling the expression of PAR (thrombin) receptors and tissue factor on the cell surface. All eukaryotic cells possess the ability to differentiate into specialized cell forms. They can also de-differentiate and re-differentiate into alternate specialized cell forms.[20, 21] Such cell specialization is an essential element of embryological development. Eukaryotic cells also possess the ability to organize themselves into organ tissues during embryological development, and to re-organize when separated from one another. All eukaryotic cells possess a precise timing mechanism and the ability to secrete tissue factor on their surface. By secreting tissue factor, cells can induce localized thrombin generation. Thrombin receptors (Protease Activated Receptors, aka PAR) are present on the outer surface of all animal cells thus far tested, and they determine how cells respond to thrombin.[22, 23, 24] Stable PAR configurations characterize mature cells, but cells can alter their thrombin receptors during tissue repair and malignancy, and thereby alter their responses to thrombin. The simplest explanation of embryological development is that the nuclear DNA of eukaryotic cells alters PAR receptors and tissue factor expression.

Therefore, the simplest explanation of embryological development is that the nuclear DNA of eukaryotic cells alters PAR and tissue factor expression at precise time intervals and locations. This alteration controls cell mitosis, differentiation, and the release of chemokines, cytokines, prostaglandins, and other cellular hormones. Such changes and releases enable cell-to-cell communications and allow cells to develop three-dimensional structures.

Tissue factor expression activates Factors VII and X to generate thrombin, and the PAR configurations determine how the cells react to thrombin elevations. This hypothetical mechanism would appear to represent the simplest explanation of embryological development. It explains why nuclear DNA does not produce telltale messenger RNA and proteins that determine embryological development, tissue repair, or hemodynamic physiology. It explains why most nuclear DNA remains quiescent once embryological development is complete, as predicted by the previous generation of stress researchers. It explains why defects that disrupt the enzymatic interaction of tissue factor with Factors VII and X lethally disrupt embryological development, while defects in Factors VIII, IX, and XIII that affect only hemostasis are survivable. It explains why toxins such as Coumadin and thalidomide that disrupt the enzymatic interaction of tissue factor with Factors VII and X undermine tissue repair and cause catastrophic disruption of embryological development.[25, 26, 27] It also explains why coagulopathies that affect only the function of coagulation Factors VIII, IX, and XIII have minimal effect on embryological development, tissue maintenance, and tissue repair, and why anticoagulant medications that affect only these factors, such as Plavix and heparin, have negligible effect on embryological development and tissue repair.

I further hypothesize that genomic information that defines anatomy, intelligence, and other attributes is encoded in the form of chaotic equations called fractals that involve more than one gene. This would explain how DNA efficiently stores complex anatomical shapes and other characteristics that can be efficiently altered in accord with changing environmental circumstances. Deciphering such information will require a new generation of computer technology.

STRESS MECHANISMS AND EVOLUTION

Multicellular organisms are incompatible with the subterranean environment. Extracellular stress mechanisms appeared after eukaryotic cells adapted to the open ocean environment in the vicinity of hydrothermal vents, where multicellular structures become possible.

Much like the seams of a baseball, hydrothermal vents ring the Earth at the bottoms of oceans in expansion zones between tectonic plates, where volcanic material wells up from the hot interior. These hydrothermal vents continuously spew forth hot, sulfurous, energy-rich chemicals that sustain both prokaryotes and eukaryotes. These organisms thrive in such abundance that they visibly coat the ocean floor of this singular environment.[28, 29]

I hypothesize that the microbes we see surrounding the hydrothermal vents originated from subterranean microbial ancestors. Their intra-cellular stress mechanisms evolved from that of their underground forebears to allow them to survive in the hydrothermal vent environment. These organisms presumably forage on accumulating volcanic chemicals. They have evolved stress mechanisms that can survive in distant cold-ocean environments, and ultimately on exposed land surfaces.

The first complex multicellular animals originated and evolved in the hydrothermal vent environment after eukaryotic cells developed the first extra-cellular stress mechanism. For billions of years in the hydrothermal vent environments, these organisms slowly differentiated and accumulated complex genomes consisting of alternative anatomical shapes, behaviors, and physiological mechanisms. Today, hundreds of distinct forms of complex multi-cellular animal life thrive in this hot, seemingly toxic, marine environment, including fish, crabs, eels, mussels, tubeworms, and shrimp. More are discovered every year. They are probably the precursors of more familiar sea creatures, but their physiology remains dependent on the hot, Sulphur-rich hydrothermal vent environment, where flourishing microbes and nutritious chemicals provide abundant food. These multicellular animals presumably originated from a single cellular ancestor, in accord with Darwin's theory.

The first primitive multi-cellular creatures that survived distant from the hydrothermal vents probably were sponges and hydras. These consisted of eukaryotic cells that had evolved the ability to aggregate themselves into simple structures. They foraged on the microbes that had previously adapted to the colder ocean environment. Marine worms followed and left a legacy of tunnels on the ocean floor.[30] The genome of these wormlike creatures presumably possessed numerous anatomical structures, behaviors, and metabolic pathways that had evolved and accumulated in the hydrothermal vent environment. In the presence of abundant food and in the absence of competition, predation, and parasitism, these creatures rapidly evolved alternative stress mechanisms that are reflected by the sudden appearance of fresh marine animal classes. This dramatic biological expansion into the as-yet untapped resources of the open ocean explains the Cambrian Explosion, when a wide variety of complex oceanic life suddenly appeared in the geological record throughout the world.

It is generally surmised that animals came before plants, but the available evidence suggests otherwise. Sponges, for example, lack organs, muscles, and nerves, but are nevertheless multicellular animals that reproduce sexually. Geological traces of sponges predate those of plants by 200 million years.[31]

Plants and animals undoubtedly share a common eukaryotic ancestor, but only plants possess chloroplasts. While it might seem intuitively obvious that plants appeared first and provided a food source for a subsequent animal evolution, I hypothesize that the original extra-cellular plant stress mechanism arose later than the original animal extra-cellular stress mechanism — after free-living eukaryotic cells adapted to the warm ocean surface environment, incorporated symbiotic chloroplasts, and finally evolved extracellular stress mechanisms that enable complex multicellular plants such as Sargasso and kelp that thrive in the oceans and in terrestrial plants.

Cyanobacteria came first and transformed Earth's atmosphere.

Plant activity explains the present oxygen-rich atmosphere, which in turn enabled and influenced terrestrial animal evolution. Cyanobacteria are abundant photosynthetic, nitrogen-fixing prokaryotes that exist in wide variety on the Earth's surface. They have transformed Earth's atmosphere by generating oxygen as a by-product of their photosynthesis. Eukaryotic cells presumably incorporated symbiotic cyanobacteria and became algae that evolved into complex multicellular plants.

THE EVOLUTION OF STRESS MECHANISMS

The characteristics that define animal classes and species tend to endure for as long as environmental conditions remain stable, but environmental change precipitates sudden evolutionary changes at irregular distant intervals (saltation).[32, 33] It seems counter-intuitive that one animal can suddenly replace another in the geologic record, and conventional evolutionary theory has never explained saltation.

Stress theory suggests that evolution proceeds via three independent pathways:

1. **Intra-cellular stress mechanism alteration.** This mechanism enabled the subterranean *extremophile* microbes to adapt to a multitude of subterranean environments, and then adapt to the hydrothermal vent environment, and finally to the cold ocean and terrestrial environments distant from the hydrothermal vents. Intra-cellular evolution (the accumulation of random internal cellular mutations that survival sorts out over time) seems unlikely in complex animals, because it would likely render cells incompatible with their own evolved extra-cellular stress mechanisms, but it may explain the divergence of plants and animals in the distant past. It may also be related to the ability of eukaryotic animal and plant cells to differentiate into specialized cell types, and to de-differentiate and re-differentiate.
2. **Extra-cellular stress mechanism alteration.** This explains saltation as the evolution and extinction of complex animal and plant classes. The appearance of new extra-cellular stress mechanisms is rare, and is usually precipitated by prolonged, severe environmental stress. Once they appear, stress mechanisms tend to persist, for they are widely distributed among

numerous species included in the class. Examples include the evolution of angiosperms from gymnosperms; amphibians from fish; reptiles from amphibians; mammals and archosaurs from reptiles; dinosaurs from archosaurs; and birds from flying dinosaurs. The extinction of stress mechanisms, such as the extinction of dinosaurs, is even less common.

3. **Speciation,** which consists of alterations in the expression, suppression, and scope of individual genomic characteristics to produce combinations of intelligence, behavior, and anatomy that are compatible with the extracellular stress mechanism and that refine environmental adaptation. This occurs via small changes that occur at distant intervals and persist because they facilitate survival ("survival of the fittest"). This is the most common mechanism of evolution, and it explains why Darwin believed that evolution proceeds slowly via small changes.

Changing environmental circumstances render most species extinct over time, but genomes and stress mechanisms endure in surviving members of their class. Animal husbandry illustrates the innate flexibility of genomes. Selective breeding has evolved numerous dog species from the wolf and has evolved a wide variety of domestic cattle from the aurochs. Such synthetic speciation endures for as long as humans maintain artificial environments to support them, but domestic dogs retain the canine genome and would presumably revert to a wolf phenotype if left to survive in the wild, as cattle would revert to the aurochs phenotype.

Successful adaptations persist while environmental circumstances remain favorable, but speciation is inherently plastic and can readily be altered by environmental change. For example, polar bears are grizzly bears adapted to arctic climate, including clear (not white) fur that allows sunlight to reach the bear's skin and produce vitamin D. Polar bears are disappearing as global warming destroys their arctic environment, but their characteristics will survive as suppressed traits of the grizzly bear genome, and they will reappear with the onset of the next ice age. Woolly mammoths were elephants that evolved fur, gained size, and flourished during the most recent ice age, but they disappeared when global warming eliminated the environment that was compatible with their characteristics. Smaller hairless elephants endure in Africa, India, and Southeast Asia. Similarly, over a period of nearly 500,000 years, humans evolved Neanderthal characteristics that enhanced their survival during several successive glaciations, including mesomorph body shapes to retain body heat, large noses, exaggerated muscle mass, larger brains, and increased intelligence to capture and subdue their prey — which were formidable mammals that grew larger and more dangerous in the cold environment. Neanderthals probably developed hypertrichosis in response to the cold environment. As the ice age retreated and global warming ensued, humans gradually lost their body hair and became taller and thinner so as to optimize heat dissipation rather than heat retention. I hypothesize that the exaggerated human intelligence induced by ice age stress explains the appearance of civilization and farming in its aftermath.

Extra-cellular stress mechanisms rarely suffer extinction, but the dinosaur stress mechanism is an exception. It represented an adaptation to the unceasingly hot terrestrial climate of Pangaea, but could not survive cold climates or water immersion that depletes body heat. However, most elements of the dinosaur stress mechanism presumably survive in the genomes of their enduring archosaur ancestors (alligators, crocodiles, and monitor lizards) and their avian descendants, so that the dinosaur stress mechanism would presumably reappear if continuously hot environments were to recur for a sufficient length of time.

A WORD ABOUT SEX

DNA retains vast reserves of information, and each eukaryotic cell processes a complete copy of the unique genome that characterizes its class and complements its particular stress mechanism. In multi-cellular animals and plants, *genome* includes a wide variety of alternative structural, metabolic, and behavioral characteristics. I hypothesize that the DNA molecule efficiently stores a huge library of past characteristics in the form of fractals (chaotic equations that use recursion to produce structures that are similar across different scales). Multiple genes may be required to describe an individual fractal shape or characteristic, and this may explain how complex life forms are able to alter their shapes, sizes, and behaviors in accord with changing environmental circumstances.[34]

Max Delbruck and Salvador Luria proved that mutations occur at random, and that environmental stress has no effect on the rate of mutation.[35] This is a basic tenet of such studies as the Eve hypothesis, which attempts to trace the evolution of the human mitochondrial genome backwards over time. An uneven rate of mutation over time would make such a study impossible.

The mechanism of sex facilitates environmental adaptability by continuously capturing fresh mutations, preserving them in the haploid state, distributing them throughout the breeding population, and culling counterproductive characteristics when they occur in the diploid state. This explains why genomes become more complex over time, as observed by Lamarck. However, *genetic drifting* occurs when populations become separated, evolve in different environments, develop incompatible characteristics, and cease to interbreed. This explains why most genomes share numerous common genes, and why the genomes of closely related species are nearly identical. For example, the genomes of humans and chimpanzees differ by only 5%. Over time, such genomic differences inexorably increase.

In many ways, it is true that "ontogeny recapitulates phylogeny" — our embryological development shows the progression of previous stages in our evolution. The genome retains but suppresses this information. Notochords and primitive renal systems appear temporarily during human embryological development as vestiges of evolution. Most genomic characteristics normally remain suppressed by mechanisms that have yet to be identified. Their latent presence is revealed by anomalies caused by incomplete suppression. For example, useless legs sometimes appear on whales. Abnormal fingers, toes, horns, breasts, and tails occasionally appear in humans. Human skin is covered with invisible lines and patches called Blaschko's Lines, that may be the genetic remnant of a camouflaging fur pattern like a zebra's (the skin of a zebra also is patterned, not just its fur).[36] Hypertrichosis reflects the latent human ability to cover the body with hair. The author personally has witnessed a human who possessed a patch of true fur (not hair).

Anatomical and behavioral characteristics must be compatible with an organism's stress mechanism, or at least must be benign in its presence. For example, euthermia would be lethal to a reptile, because it would harmfully exaggerate food requirements in the absence of adequate intelligence and exercise tolerance. (Reptiles are ambush hunters that attack what happens by, with little ability to chase their prey long distances in the manner of mammals.)

Stress theory explains why similar genotypes — like that of chimp and man — may be expressed or suppressed differently by different animals in their phenotypes. Classical Mendelian genetics fails to explain why genotype differs from phenotype. Under a stress theory interpretation, individual common genes are selectively suppressed to enable speciation without affecting the genome. The mechanism that enables such suppression remains to be elucidated.

BIOCHEMISTRY, BOTANY, AND ZOOLOGY

Biochemical similarities reflect the common origin of plants and animals. The eukaryotic cells of both animals and plants possess the Krebs cycle, anaerobic respiration pathways, actin muscle fibers, and intra-cellular ATPase enzyme that energizes muscle fiber contraction.[37]

The plant enzyme auxin may be analogous to the animal enzyme thrombin. Both appear to be *Universal Enzymes of Energy Transduction* that convert ATP energy into cell and enzyme activities. Like thrombin, auxin enables cell proliferation, tissue repair, and embryological development.[38] Also like thrombin, auxin activity is dependent on Ca+ and ATP.[39] Auxin activates intra-cellular ATPase enzyme to energize plant cell metabolism and mitosis,[40] and thrombin most likely energizes animal cell activity via the same mechanism.[41]

Both plants and animals retain the von Willebrand characteristic. Plants produce calcium-dependent "copine" proteins that bind to membranes to enable structural development and disease resistance. Copines possess the von Willebrand Factor A domain.[42] The CTCK domain of von Willebrand Factor appears in diverse protein families involved in barrier function, eye and inner ear development, the process of insect blood coagulation and immunity, axon guidance, and signaling in extra-cellular matrices.[43]

Like animals, complex plants possess circulatory systems that transport nutrients, enzymes, and waste products to facilitate cellular survival and function. A protective layer of cellulose surrounds plant cells as opposed to the vascular endothelia that isolate animal circulation from extravascular tissues. I hypothesize that damage to this cellulose barrier exposes plant cell membranes to circulating enzymes that generate auxin to energize tissue repair. This appears to offer the simplest explanation of tissue repair in plants.

Both the hemoglobin molecule that transports oxygen in animals and the chlorophyll molecule that converts sunlight to sugar in plants are remarkably similar porphyrin rings.[44]

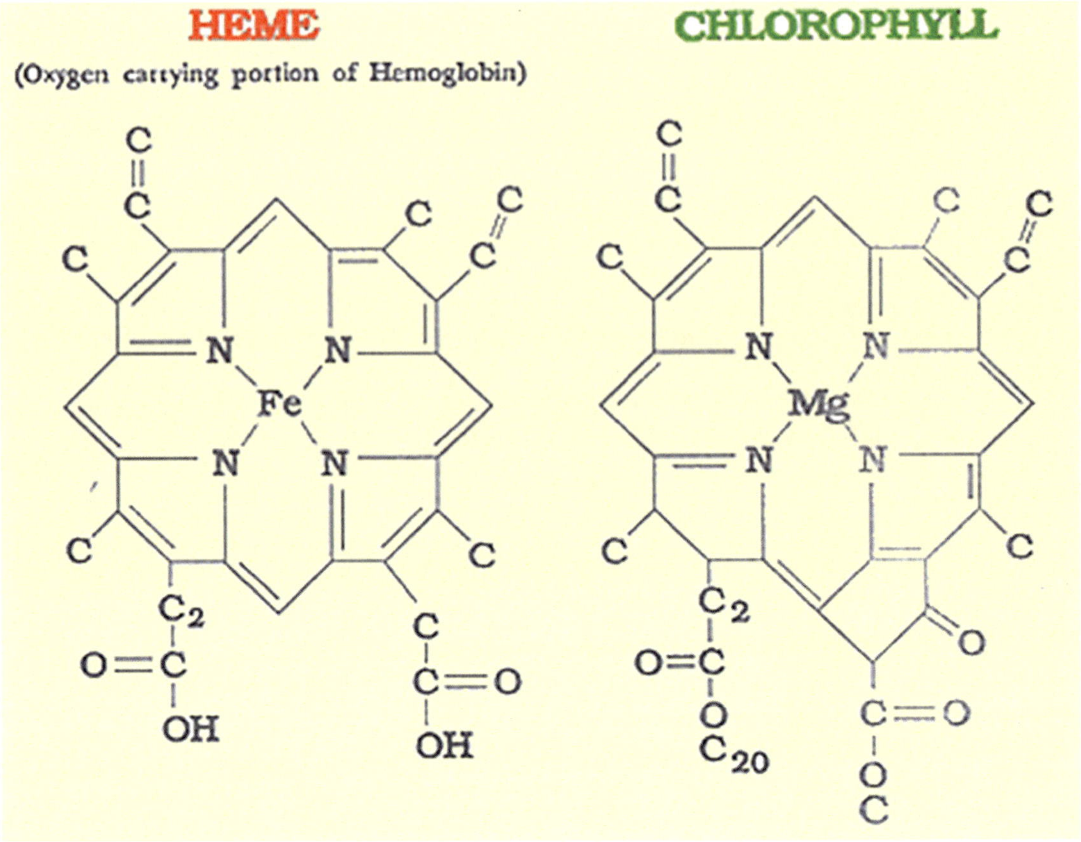

Figure 17.1: Heme and chlorophyll.
Note the chemical and bonding similarities between heme — the oxygen-carrying portion of hemoglobin with iron at its core — and chlorophyll with magnesium at its core.

ACCUMULATION OF PHYSICAL CHARACTERISTICS — LAMARCK VERSUS DARWIN

Stress theory resolves the superficial disparity between Darwin and Lamarck.[45] Genomes become more complex as they accumulate fresh characteristics via random mutation. This explains Lamarck's observation that animal classes become increasingly complex over time. Latent characteristics retained within genomes enable environmental adaptation via speciation.

Stress theory explains convergent evolution; wherein similar features appear in seemingly unrelated lineages at distant time intervals. The vast information storage capacity of DNA enables the indefinite retention of potentially useful characteristics within genomes, and these can appear in similar combinations that enhance survival in similar environmental circumstances. For example, the overt anatomy of mosasaurs closely resembled that of mammalian dolphins. Like dolphins, mosasaurs were air-breathing vertebrates that thrived in salt water. However, dolphins are euthermic mammals with superior exercise- and cold-tolerance that thrive in both cold oceans and freshwater environments, while poikilothermic mosasaurs could not survive in the open ocean due to their limited exercise– and cold-tolerance. Mosasaurs were archosaurs that thrived in warm shallow inland seas during the era of Pangaea and were closely related to alligators, crocodiles, and monitor lizards.

EVOLUTION — LAMARCK AND DARWIN

Environmentalists and geneticists bewail extinction as eradicating irreplaceable genetic information. They believe that evolution can proceed only in a forward direction via DNA mutations. The evidence suggests otherwise.

If evolution depended solely on the slow tick of mutations, then evolutionary changes would occur gradually at regular intervals.[46] Instead, animal species and classes remain stable for prolonged periods of time, and evolutionary changes occur suddenly at *irregular* distant intervals in accord with environmental stress.[47]

Stephen Jay Gould called this idea *punctuated equilibrium.*[48] The fossil record repeatedly reflects the appearance, disappearance, and reappearance of distinctive characteristics, and this is not readily explained by mutation. The simplest explanation is that the vertebrate genome retains an enormous library of structural blueprints and characteristics, most of which are suppressed at any given time.[49, 50] The appearance and subsequent disappearance of ancient structures such as notochords, primitive urinary systems, and pharyngeal clefts during human embryological development reflects the latent presence of these ancient blueprints in the vertebrate genome. Sometimes the blueprints are incompletely suppressed, so that useless hind legs appear on whales, extra fingers and toes appear on humans, and so forth. Environmental stress that kills most "normal" members of a species simultaneously increases the odds that latent characteristics that enhance survival in the altered environment will emerge in homozygous form and spread rapidly among surviving members of the species.

Red cell morphology, homeothermy, and alternate breathing mechanisms are examples of potentially useful physiological characteristics that are "hidden" in the vertebrate genome and can be activated or suppressed either independently or in concert. When activated, they alter physiology in ways that necessitate compatible alterations in anatomy and behavior. Thus, stress mechanisms are the occult foundation of taxonomy because they indirectly determine compatible combinations of physiology, anatomy, intelligence, and ethology that characterize classes. Cats and dogs may be enemies that compete for the same food sources, but they are closely related cousins that share the mammalian stress mechanism. They have evolved alternate combinations of anatomy and ethology that enable different predation strategies.

THE BALDWIN EFFECT

Stress theory also explains the Baldwin Effect.

Lamarck proposed that living creatures acquire characteristics and pass them to their offspring. While no mechanism for this process has been demonstrated, the Baldwin Effect comes close. It posits that an organism's ability to respond to its environment and learn new behaviors — for example, in overcoming a stressor such as an impending ice age — directly affects the organism's ability to preserve its genome. Thus, while the genome is not altered by the learned behavior *per se*, the ability to learn enhances genomic survival.

Animal genomes incorporate a wide variety of inherited behaviors as well as anatomical shapes that enhance survival in the presence of changing environmental circumstances. Natural selection favors functional combinations of anatomy, intelligence, and behavior that are compatible with stress mechanism limitations and requirements. The emotional mechanism alters behavior in changing environmental circumstances by continually reassessing emotional memory via the dreaming mechanism to identify dangerous circumstances, avoid them, and survive to evolve more sophisticated forms of intelligence and consciousness.

STRESS THEORY CLARIFIES TAXONOMY

Taxonomy is the conception, naming, and classification of organisms. Modern taxonomy classifies organisms in the following hierarchy: Life, Domain, Kingdom, Phylum, Class, Order, Family, Genus, and Species.[51]

Taxonomists have long suspected the presence of occult physiological mechanisms that determine visible characteristics, but since no such mechanisms have been previously identified, they have been obliged to base their classifications on superficial characteristics. As a result, there is considerable debate regarding nomenclature and definition of terms, including the definition of taxonomy. The discovery of additional stress mechanisms will clarify taxonomy. The discovery of mechanisms that facilitate speciation will further refine taxonomy.

THE CAMBRIAN EXPLOSION

Primitive unicellular life arrived on the Earth's surface more than 3850 million years ago. Sponges and other simple animal forms were present on the ocean floor at least 750 million years ago. The cells of these primitive animals already had the ability spontaneously to re-arrange themselves into complex structures, even when the animal was crushed into a pulp and passed through a fine filter.[52] Similar cell behavior has been confirmed in higher animals.[53] Complex animals did not appear in the fossil record until 540 million years ago, when precursors of all modern animal phyla suddenly appeared on the ocean floor. This is called the Cambrian Explosion. It is also called "Darwin's Dilemma" because Darwin's theory postulates that all animal life can be traced to a single animal ancestor. Until recently, there was no clue of pre-existing animal ancestors that could explain the Cambrian Explosion.

The subterranean environment is incompatible with multicellular existence. Subterranean eukaryotic extremophile cells initially adapted to the transitional environment of hydrothermal vents, where the hot subterranean environment interfaces with the cold ocean environment.[54, 55] Chemosynthetic bacteria abound in these unique transitional environments, including both prokaryotes and eukaryotes.[56]

Chemosynthetic bacteria thrive on the nutritious volcanic chemicals produced by hydrothermal vents. The development of the first multicellular stress mechanism enabled eukaryotic cells to build, repair, and maintain complex structures, and these became the ancestors of animals. A variety of complex animals subsequently evolved in the hydrothermal vent environment. Chaos Theory may explain the ability of their DNA to efficiently retain complex structural blueprints that facilitate adaptation and evolution; chaotic equations that describe anatomical shapes and forms are encoded into gene combinations that describe *strange attractors*, a term from Chaos Theory.[57] Chemosynthetic bacteria and nutritious volcanic chemicals presently sustain hundreds of species of complex animals, including worms, crabs, shrimp, mussels, snails, fish, and octopi that have presumably evolved for millions of years in the nutrient-rich environment of hydrothermal vents but remained unable to survive in the, cold, nutrient poor open ocean.[58]

The viability of the original complex animals was probably limited for billions of years to the hot, sulfurous environment near hydrothermal vents, where extremophile bacteria flourished and provided plentiful food. Eventually these animals evolved new extra-cellular stress mechanisms that enabled survival in the cold, oxygen rich waters of the open ocean. The first such creatures were probably worms and sponges that thrived on microbes that subsisted on nutritious volcanic chemical deposits on the ocean floor.[59] In the presence of abundant, unexploited food sources and the absence of predation, parasitism, and competition, the hydrothermal vent species utilized their collections of genomic structures and behaviors and rapidly evolved alternate extra-cellular stress mechanisms, and created the various phyla that became the precursors of presently prevailing animal life. This explains the Cambrian Explosion, corroborates Darwin's theory, and extends its implications.[60] Eventually these new complex animals evolved additional stress mechanisms that enabled amphibians, predation, parasitism, and terrestrial existence.

MAMMALS AND REPTILES

Mammals and reptiles differ strikingly in behavior and appearance even though mammals evolved directly from reptiles, and both are composed of identical eukaryotic cells. Their visible differences are explained by the differences between their stress mechanisms. The differences are surprisingly simple. They involve alterations in metabolism and red cell morphology that adapt to the problem of lipoprotein solidification in different climate zones. These seemingly simple alterations entail complex "tradeoffs" in atherosclerosis resistance, cardiac efficiency, exercise tolerance, cold tolerance, heat tolerance, and food requirements. The resulting alterations in physiology necessitate compatible combinations of behavior, intelligence, and anatomy that explain the visible differences between mammals and reptiles and characterize their classes.

The mammalian stress mechanism optimizes exercise capacity, cold tolerance, and atherosclerosis resistance at the expense of increased food requirements and decreased heat tolerance. Mammals maintain their body temperature slightly above the threshold of lipoprotein liquefaction (37.7 degrees Celsius). This neutralizes the exaggerated blood viscosity and atherosclerosis acceleration caused by lipoprotein solidification,[61, 62] and enables mammals to employ biconcave red cells that simultaneously optimize cardiac efficiency and atherosclerosis resistance. The resulting mammalian cardiac efficiency enables human athletes and racehorses to dramatically increase their exercise capacity via exercise training that proliferates capillaries in affected muscles. The capillary proliferation reduces blood viscosity (flow resistance) and enhances cardiac output, cardiac efficiency, tissue perfusion, and oxygen delivery with negligible increase in heart size and strength.

Mammals employ an elevated metabolic rate to maintain their body temperature slightly above the threshold of lipoprotein liquefaction. As a result, they require roughly ten times as many calories as reptiles of equivalent size. The exaggerated food requirements are offset by increased exercise capacity, intelligence, and cold tolerance that enhance food acquisition. The increased exercise capacity of mammalian herbivores enables them to engage in long-distance seasonal migrations to obtain food, and it enables mammalian predators to chase and overpower their prey. Mammals search for food at night when cool temperatures favor heat dissipation, and when reptiles are quiescent. They thrive in cold climates where reptiles cannot survive. Their exaggerated food requirements are reflected in predator-to-prey ratios as low as 0.1%. By contrast, the predator-to-prey ratio of cold-blooded reptiles and spiders is 10–20%.[63]

The increased food requirement also explains why mammals generally have larger brains than reptiles: they require enhanced intelligence to satisfy their food requirements. Also, the exaggerated metabolism of brain tissue necessitates continuous perfusion and oxygenation, stable warm temperature, and increased calories that mammalian physiology provides, but reptile physiology cannot accommodate and do not require. This explains why reptile brains remain small and primitive.

Eukaryotic cell mitosis and metabolism are temperature–, oxygen–, and energy-dependent. The elevated body temperature, enhanced food intake, and superior tissue oxygenation of mammals explains why they attain larger size and grow, mature, diversify, and evolve faster than reptiles.[64] Because of their rapid growth and development, mammals develop Haversian Canals in their long bones, and they exhibit high rates of evolution. However, the increased food requirement of mammals renders them more vulnerable to starvation and extinction.[65]

Note: Haversian Canals are distinctive microscopic structures that surround neurovascular bundles in bone. They are closely associated with rapid bone growth and development. They commonly appear in the long bones of fast-growing warm-blooded protomammals, mammals, and birds, but are rare in slow-growing reptiles. They also appear in the bones of dinosaur fossils, and some experts regard this as convincing evidence that dinosaurs were euthermic, like mammals and birds. This idea is at odds with the gigantic size of dinosaurs, which would cause heat dissipation problems in the hot climate of the dinosaur era if dinosaurs were warm blooded. It is also at odds with the small mouths of dinosaur herbivores, which would not accommodate the large quantities of food needed to maintain euthermic physiology. Haversian Canals reflect rapid growth and development rather than euthermia per se. Dinosaurs developed a unique poikilothermic physiology that enabled rapid growth and development in their hot climate without the need for euthermia, and this explains the presence of Haversian Canals in dinosaur fossils.

The importance of mammalian red cells is reflected by the latent capacity for extra-medullary hematopoiesis during stress, and the fact that red cells constitute 25% of all human cells.[66] Mammalian red cells dispose of their nucleus, Golgi bodies, mitochondria, and ribosomes shortly before they mature and are released into systemic blood circulation. This enables their distinctive biconcave shape and small size that optimizes cardiac efficiency, but without these organelles, mammalian red cells are bereft of aerobic metabolism, and their life span is attenuated. Furthermore, they are unable to de-differentiate and re-differentiate to participate in tissue repair, so that mammalian bone and tissue repair is inferior to that of reptiles and amphibians.[67]

Mammals are covered with insulating hair that retains body heat and reduces food requirements. They are often nocturnal because cool night temperatures enhance mammalian body heat dissipation, and reptile predators are inactive at night. They thrive in cold environments where food sources are rich and reptile competition is rare, but their exaggerated caloric requirements and heat intolerance are disadvantageous in hot climates.

In the present era of temperate climate, mammalian herbivores are the largest terrestrial animals. Their superior exercise tolerance and exaggerated intelligence enables them to seek and find large quantities of low-quality plant food. Even so, a combination of exaggerated metabolic rate, heat intolerance, and limited food availability constrains their size. Despite these constraints, the mammalian blue whale is the largest of all known animals, including dinosaurs. It attains its gigantic size despite its elevated mammalian metabolism because it lives in cold water that dissipates its body heat, and because it has access to abundant sources of rich animal food and a gigantic mouth that enables it to ingest vast quantities of that food.

In contrast to mammals, reptile physiology minimizes food requirements and optimizes atherosclerosis resistance and heat tolerance at the expense of exercise capacity and cold tolerance. Reptiles conserve food energy by allowing their body temperature to fluctuate in accord with the external environment. Unlike mammals, they do not expend food energy to maintain their body temperature above environmental temperatures, so their food requirements are minimal. Their metabolic efficiency reduces their vulnerability to starvation and extinction. However, it allows lipoprotein solidification when cool environmental temperatures cause reptile body temperature to fall below 37.7 degrees Celsius (mammalian body temperature). Solidified lipoproteins increase blood viscosity, inhibit cardiac efficiency, undermine oxygen delivery, reduce exercise capacity, paralyze brain function, and accelerate atherosclerosis, which limits longevity. Cool temperatures also reduce reptile cellular metabolism, which further inhibits brain function, exercise tolerance, growth, development, size, and evolution. Because of their slow growth rates, reptiles seldom develop Haversian Canals in their long bones.

Large, nucleated, oblong reptile red cells exaggerate systolic blood turbulence to offset atherosclerosis acceleration caused by lipoprotein solidification at cool temperatures. This enhances reptile longevity, but it exaggerates systolic turbulent blood flow resistance (viscosity), which constrains cardiac efficiency, cardiac output, tissue oxygenation, and exercise capacity. Reptiles function best when their body temperature is elevated above the threshold of lipoprotein liquefaction (mammalian body temperature), but even at optimal temperatures the systolic turbulent flow resistance imparted by reptilian red cells inhibits cardiac output to the point that tissue perfusion and oxygenation is marginal at rest. Thus, vigorous or prolonged exercise rapidly causes tissue oxygen shortages by increasing oxygen consumption in the marginally perfused reptilian muscles. This limits exercise tolerance and explains the quiescent behavior of reptiles. It also explains why most reptiles are predators that prefer rich animal food, why true reptile herbivores are rare, and why reptiles do not engage in long-distance seasonal migrations in search of food. Plant food is less nutritious than animal food, and reptiles lack the exercise capacity needed to seek and find the relatively large quantities of plant food necessary to sustain herbivore existence. The familiar iguana, which is found in tropical environments rich in plant food, is not a true reptile. It is a monitor lizard that is closely related to alligators, crocodiles, and Komodo dragons, with enhanced exercise capacity that enables it to obtain adequate quantities of plant food.[68]

The exaggerated oxygen and energy dependence, fragility, and temperature sensitivity of brain tissue explains why reptile brains remain primitive. Marginal reptilian exercise tolerance and limited intelligence explains the simple, efficient ambush strategy employed by most reptiles. They lack the exercise capacity and intelligence to seek food, outsmart it, chase after it, or engage in prolonged struggles to subdue it. They make extensive use of camouflage and ambush. Their low food requirements and minimal intelligence enable them to wait patiently for months, until a rich source of animal food blunders within striking range. They utilize anaerobic respiration to energize an efficient burst of activity that overcomes their prey, and then resume immobility while they slowly digest their victim. For example, python snakes strike their victim in the face, quickly wrap themselves around its body, and efficiently squeeze it to death, using the victim's own act of exhalation as a weapon against it. Rattlesnakes inject venom into their victim, and then wait patiently until the venom kills the prey and pre-digests its tissues. The limited exercise capacity of reptiles explains why they have never attained the ability to fly or become truly bipedal.

Reptiles suffer sharp reductions in cardiac output, exercise capacity, and brain function at night when temperatures fall below the threshold of lipoprotein solidification and reduce cellular perfusion and metabolism. Therefore, most reptiles seek safe hiding places at night. They typically emerge in the morning and bask in the sun until their internal temperature climbs above the threshold of lipoprotein liquefaction, which restores optimal oxygen delivery, cell function, and exercise capacity. Most reptiles are covered with dark-colored plates or scales that optimize heat absorption to facilitate restoration and maintenance of their internal body temperature above the threshold of lipoprotein solidification. Some reptiles living in cool climates, such as the "finback" reptiles of the Permian era, evolved large "sails" that enhanced their absorption of sunlight energy.

Reptiles cannot function at cold temperatures, and thus cannot thrive in cold climates, but their metabolic efficiency and heat tolerance confers a competitive advantage in warm climates. Reptiles proliferate and extend their territory during eras of "global warming," but they are constrained to equatorial existence during "ice ages."

Inadequate nutrition and low metabolic rate normally constrain the growth, development, intelligence, and size of reptiles. Most wild reptiles exist in a state of chronic starvation. Surveys of wild animals demonstrate that most mammal stomachs contain food, while reptile stomachs are usually empty. This explains why reptilian growth, maturation, and mass are constrained compared to mammals. However, reptiles can grow nearly as fast as mammals if they are continuously maintained in warm environments with unlimited food.[69] The modest metabolic requirements of poikilothermic animals explain their latent potential to attain exaggerated size when environmental conditions are favorable. Gigantic python snakes and alligators have appeared in the Florida Everglades and Far East during the present era of global warming.

PANGAEA AND EVOLUTION

By combining Chaos Theory[70, 71, 72] with the stress theory of Hans Selye (1907-1982), the tectonic plate theory of Alfred Wegener (1880-1930), and paleontology,[73] it is possible to explain how geological events induced the sequential evolution of amphibians, reptiles, mammals, archosaurs, dinosaurs, and birds.[74, 75, 76]

Like Selye, Kepler, and most other scientific visionaries, Alfred Wegener did not live to witness the acceptance of his revolutionary tectonic plate theory. Wegener perceived that the Earth's continents continuously drift on *tectonic plates*. Their movements and collisions cause earthquakes and tidal waves, trigger volcanoes, and heave up mountain ranges. During glaciations, the polar ice caps periodically extend towards the equator and then retreat, causing sea levels to fluctuate. Continents drift into different climatic zones, causing drastic environmental changes. Over billions of years Earth's continents have more than once merged into a single gigantic supercontinent that remains intact for millions of years and then rifts apart into individual continents. Wegener named the most recent supercontinent *Pangaea*.

Evolution proceeds in response to unceasing environmental stress. Supercontinent formation causes environmental perturbations of Biblical proportions. Pangaea extended from the North Pole to the South Pole. Today, ocean currents flow comparatively unimpeded around the world, redistributing the planet's heat, but with a single continent blocking ocean currents from pole to pole, such distributing currents could not flow. The collision of all continents into one threw up Himalaya-sized mountain ranges that altered wind patterns and rainfall. The continental interior was so vast that atmospheric moisture rarely reached it.

Pangaea's formation gradually disrupted global weather zones and ocean currents, exaggerated ocean temperatures, melted the polar ice caps, elevated sea levels, and flooded vast interior land surfaces with warm, shallow inland seas. Terrestrial temperatures in the continental interior rose, and remained abnormally hot for 100 million years. It was a "hot age" in contrast to an "ice age."

The eventual rifting of Pangaea into individual continents caused equally drastic environmental upheaval as it gradually restored normal global weather patterns, ocean currents, and temperatures. These drastic climate changes induced the appearance of new plant and animal stress mechanisms. The drastic environmental disturbances caused by the formation and disintegration of Pangaea occurred in the midst of the saga of multicellular life on Earth, and thus induced the evolution of mammals, archosaurs, dinosaurs, birds, and angiosperms as well as the mass extinctions of animal and plant life that are reflected in the fossil record. The resulting paleontological evidence is customarily divided into four Megadynasties that will be discussed in detail:[77]

- **Megadynasty I:** Primitive amphibians and reptiles that flourished during the Coal Age and Early Permian.
- **Megadynasty II:** So-called *Protomammals* that are known as the Order Therapsida dominated.
- **Megadynasty III:** The Age of Dinosaurs.
- **Megadynasty IV:** The present Age of Mammals.

MEGADYNASTY I

Pangaea formation began about 300 million years ago, at the beginning of the Permian era, when the terrestrial animal life of Megadynasty I consisted of primitive amphibians and reptiles that thrived in warm rain forests. Most plants were gymnosperms (conifers). Angiosperms (flowering plants) had not yet evolved. Mammals, archosaurs, dinosaurs, and birds had not yet appeared, and animal diversity was limited. Diadectes, a reptile-like amphibian, was the most common herbivore. Dimetrodon, a reptile, was the most common predator. Consistent with its poikilothermic physiology, Dimetrodon was an inefficient predator, and its predator-to-prey ratios were high. Haversian Canals, which indicate fast growth in animals, were rare in the bones of these slow-growing, sedentary, poikilothermic animals.

MEGADYNASTY II

The formation of Pangaea was a slow process that lasted around 40 million years. Temperature and rainfall declined in landmasses that slowly drifted away from the equator and moved toward their ultimate destinations in the supercontinent. The cooling environment on these landmasses caused the *Age of Finbacks* as reptiles and amphibians evolved spectacular sails that captured sunlight energy to elevate and maintain their body temperature. The fossil remains of these creatures have been found mostly in what is now North America and Germany.

The increasing cold-stress eventually caused close relatives of Dimetrodon to evolve euthermia to prevent lipoprotein solidification at cool temperatures, and enable the animal to remain active despite the increasingly cold climate.[78] The evolution of biconcave red cells presumably occurred later, and produced the first true mammals. Alternatively, it is conceivable that red cell morphology and metabolic rate are somehow linked in the vertebrate genome as complementary characteristics that can be simultaneously turned *on* and *off*.[79] The fossil record indicates that protomammals were quite energetic, which suggests that they had evolved biconcave red cells as well as euthermia, and were true mammals.[80]

Regardless of these details, the superior cold tolerance and exercise capacity of protomammals caused the *Kazanian Revolution* that brought a sudden end to the golden era of finbacks and Megadynasty I.

Megadynasty II began as the protomammals displaced reptiles and amphibians, and became the dominant predators and herbivores. The physiological similarity of protomammals to modern mammals is reflected by their large size, low predator-to-prey ratio, energetic behavior, and the abundance of Haversian Canals in their bones.[81] Whether or when the protomammals developed fur remains unclear but the simplest explanation is that the protomammals were actually true mammals that have persisted continuously to the present.

Other landmasses experienced warming climates with abundant rainfall as they moved toward their final destinations in Pangaea. The resulting tropical environment fostered the appearance of pareiasaurs, which were herbivorous reptiles with small heads that grew to the size of a hippopotamus in the presence of abundant food and the absence of competition or predation. Today the fossil remains of these creatures are found in China, Russia, and South Africa.

INTERLUDE: THE PERMIAN-TRIASSIC EXTINCTION

Protomammals and pareiasaurs flourished only briefly as Pangaea formed. They declined as the supercontinent formation process neared completion, which disrupted global ocean currents and weather patterns, warmed the oceans, and melted the polar ice caps. Shallow saltwater seas flooded the interior of the supercontinent. Terrestrial temperatures rose above the threshold of lipoprotein liquefaction. This drastic environmental disruption explains the *Permian-Triassic extinction* that occurred about 252 million years ago and killed up to 96% of all marine species and 70% of terrestrial vertebrates.[82] Even insects were devastated. For the next 100 million years the ocean temperatures remained elevated, and exposed lands in the interior of Pangaea remained abnormally hot and dry, until Pangaea rifted apart.[83]

This hot climate caused the large protomammals to become extinct because of their heat dissipation problems. The surviving mammals shrank in size and retreated to cooler climates more compatible with their physiology. Similarly, the herbivorous pareiasaurs perished as their plant food sources disappeared.

MEGADYNASTY III

Megadynasty III began in the aftermath of the Permian-Triassic Extinction. Exotic new plants and animals evolved and thrived in the unceasingly hot terrestrial climate and the vast, warm, stagnant, shallow inland seas of Pangaea. Palm trees flourished on lands that are now near the North Pole.[84]

Ammonites were giant mollusks that produced segmented spiral shells; the animal lived in the largest segment and inflated the smaller segments with air so that it floated on the surface of the inland seas. They fed on plankton. Belemnites were squid relatives that developed a bony sheath that protected them from predators in the shallow seas. Elevated temperatures killed coral animals in coastal waters, because they thrive only in cold water. Rudists replaced them, because bivalves are better adapted to the warm water. Angiosperms (flowering plants) evolved via a new plant stress mechanism that utilized conduction tubes in their roots, stems, and leaves. This enabled them to displace previously prevailing gymnosperms.

THE RISE OF THE ARCHOSAURS

The hot climate of Pangaea eliminated the problem of lipoprotein solidification, so that reptile metabolism and exercise tolerance were continuously optimized. Plankton and plant life flourished in the warm waters of the inland seas, and provided rich food sources. The combination of abundant food and optimal temperature caused a Golden Era of Reptiles during the early years of Pangaea, as the surviving reptiles grew rapidly, matured sooner, became larger, and evolved faster.[85, 86]

Archosaurs (sometimes called "Crimson crocodiles") evolved from reptiles by developing a new stress mechanism with a four-chambered heart, air sacs in the abdomen, and hollow bone shafts that enabled their lungs to be continuously suffused with fresh air (Figure 17.2 on page 263 and Figure 17.3 on page 264).

This archosaur stress mechanism enhanced respiratory efficiency, exercise tolerance, and buoyancy, and conferred the ability to lurk underwater. It retained the modest poikilothermic reptilian food requirements and ability to survive cool temperatures.[87, 88, 89] These characteristics enabled archosaurs to become the dominant predators in the warm, shallow saltwater seas of Pangaea. They evolved into gigantic alligators, crocodiles, mosasaurs, plesiosaurs, ichthyosaurs, and monitor lizards. Some were herbivorous, which reflects their enhanced exercise tolerance. The mosasaurs, plesiosaurs, and ichthyosaurs became extinct when the warm, shallow inland seas of Pangaea subsided. The alligators, crocodiles and monitor lizards, including the Komodo dragon, persist to the present, but they are mistakenly classified as reptiles.[90, 91]

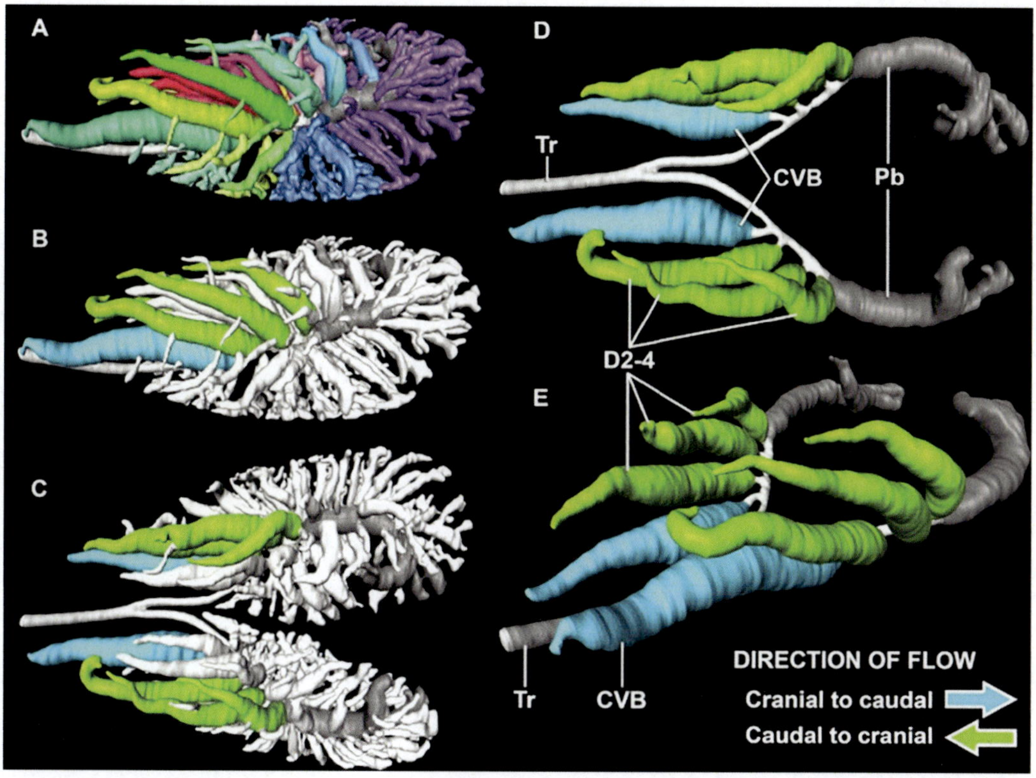

Figure 17.2: Breathing system of Nile crocodile.

These drawings depict the breathing system of the Nile crocodile. Air sacs in the chest and abdomen enhance the buoyancy of the animal and enable it to lurk just beneath the surface. The breathing system also enhances breathing efficiency and exercise tolerance in these formidable predators. Recent evidence reveals that similar pulmonary physiology was present in archosaurs and their descendants, including dinosaurs, plesiosaurs, mosasaurs, ichthyosaurs, monitor lizards, alligators, crocodiles, and birds. (Drawings by Emma Schachter reproduced courtesy of PeerJ under the prevailing CC-BY license).

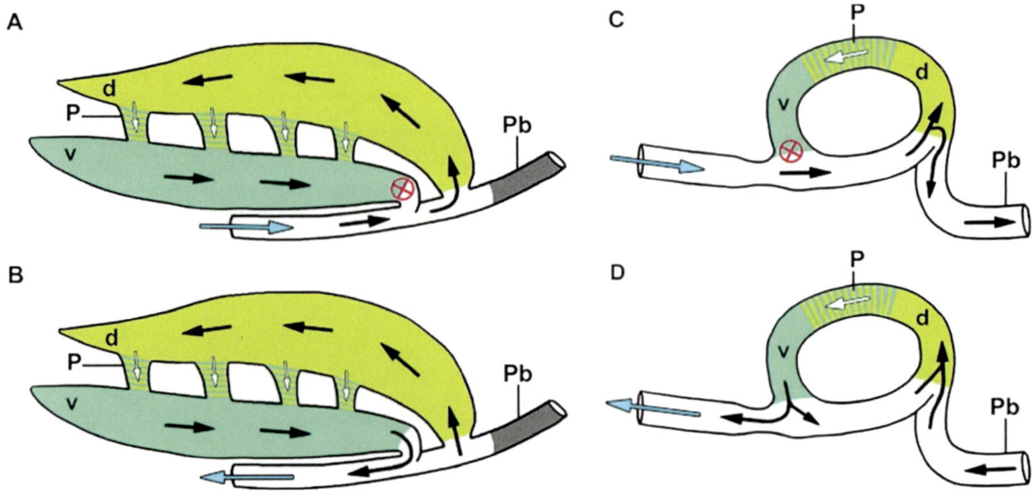

Figure 17.3: Pulmonary structure of Nile crocodile.

Drawings that illustrate the pulmonary structure and airflow pattern of the Nile crocodile. Recent evidence reveals that this pulmonary mechanism, which originated in archosaurs, is common to archosaurs, dinosaurs, alligators, crocodiles, monitor lizards (including the Komodo dragon) and birds.[92, 93, 94] (Drawings by Emma Schachter reproduced courtesy of PeerJ under the prevailing CC-BY license.[95, 96]

THE RISE — AND FALL — OF THE DINOSAURS

Dinosaurs are the most dramatic example of the accelerated evolution induced by the hot climate of Pangaea. Dinosaurs evolved from archosaurs, which had evolved from reptiles, but dinosaurs were unique in several respects. Reptile herbivores are rare, but dinosaur herbivores were common, and they became the largest terrestrial vertebrates on record. Like plesiosaurs, dinosaur herbivores had small heads and mouths relative to their gigantic size. Bipedal locomotion is rare and restricted in reptiles and unknown in archosaurs, but bipedal dinosaur predators weighing two tons could attain speeds of 40 mph (according to calculations based on preserved mud footprints). Dinosaurs grew much faster than ordinary reptiles and archosaurs, and developed Haversian Canals in their long bones — like mammals and birds. These observations have inspired considerable speculation that dinosaurs were "warm-blooded" like mammals.[97] However, their gigantic size would have exaggerated their food requirements and entailed heat dissipation problems if they were euthermic. Also, dinosaur herbivores had small mouths that limited food intake, and dinosaurs failed to thrive in water. The circumstances of dinosaur extinctions remain unexplained. This evidence suggests that a simple explanation has been overlooked.

The simplest explanation is that the continuously hot climate of Pangaea eliminated the problem of lipoprotein solidification, and therefore eliminated the necessity of a reptilian red cell morphology that exaggerates systolic blood turbulence. Thus, by altering red cell shape to improve flow efficiency, dinosaurs enjoyed exercise tolerance superior to that of archosaurs. However, without systolic blood turbulence they simultaneously sacrificed the ability to survive temperatures below the threshold of lipoprotein solidification. This explains why they did not thrive in water, which rapidly extracts heat, reduces body temperature, and precipitates lipoprotein solidification. It also explains why dinosaurs developed feathers that retained their body heat at night, when temperatures decline. The insulating dinosaur feathers, decreased dinosaur bone density (due to air passages in their bones), and the exaggerated dinosaur exercise capacity paved the path to dinosaur flight.

Schweitzer *et al* recently discovered intact blood vessels containing apparent red cell remnants in dinosaur bone marrow.[98] Some suspect that dinosaurs evolved biconcave, enucleated red cells similar to those of mammals.[99] This would conveniently explain their enhanced exercise tolerance as well as their inability to tolerate cool temperatures. Unfortunately, it may never be possible to confirm the morphology of dinosaur red cells. Moreover, since birds evolved directly from dinosaurs, have nucleated red cells, are euthermic, and enjoy enhanced exercise tolerance, the simplest explanation is that dinosaur red cells were likewise nucleated. The shape of bird red cells differs markedly from the shape of crocodile (Archosaur) red cells but they may be identical to those of flying dinosaurs that were their immediate ancestors. The heart size of birds and bats is greater than the heart size of comparably sized ground-dwelling mammals,[100] which presumably reflects the demands of flight. Lastly, birds exhibit enhanced longevity compared to mammals. Based on these limited facts, the simplest explanation would appear to be that bird red cell shape confers an alternate diastolic turbulence pattern that confers superior atherosclerosis resistance, which would explain enhanced bird longevity, but this seems at odds with their exercise capacity. Further studies will be needed to explain how birds enjoy both enhanced longevity and exaggerated exercise capacity.

The increased dinosaur exercise capacity enhanced their food acquisition. Unlike mammals, dinosaurs did not expend food energy to maintain their body temperature above the threshold of lipoprotein solidification, so their food requirements remained modest. Like poikilothermic reptiles, there was no physiological limit on their body size other than inadequate food. Therefore, the increased food intake exaggerated their body size, and they became the biggest terrestrial vertebrates on record. Their low metabolic rate negated heat dissipation difficulties in the hot climate despite their exaggerated size. Since the hot climate abolished lipoprotein solidification, they were not threatened by atherosclerosis, and had no need for a red cell shape that generated turbulence. The hot climate optimized eukaryotic cell mitosis and metabolism without the need for euthermia. This combination of exaggerated exercise capacity, enhanced food acquisition, and superior metabolic efficiency explains the accelerated dinosaur growth, maturation, diversification, and exaggerated size. It explains why dinosaurs attained greater size than terrestrial reptiles, archosaurs, and mammals.[101] Their rapid growth rate explains the presence of Haversian Canals in their long bones, which are common in fast-growing, warm-blooded mammals and birds but rare in slow-growing reptiles, amphibians, and turtles.

Dinosaur herbivores became the biggest terrestrial vertebrates on record. To achieve comparable size, mammalian herbivores would theoretically require vastly greater food intake, which in turn would necessitate gigantic mammalian heads and mouths. Heat dissipation problems also inhibit the size of Euthermic terrestrial mammals. Mammals survived the era of Pangaea, but their size became small because of the hot climate. This explains why mammalian fossils from the dinosaur era are rare. They still enjoyed some advantage at night, when cool temperatures inhibited dinosaur and reptile activity.

Limited exercise capacity explains why bipedal locomotion is rare and restricted in reptiles and archosaurs. Enhanced exercise capacity explains why bipedal locomotion is common in mammals and was likewise common in dinosaurs.

The combination of low food requirements and enhanced exercise tolerance explains why dinosaurs had small brains. They probably retained the simple ambush predation strategy they inherited from reptiles. Their low food requirements and limited intelligence explains why dinosaur predator-to-prey ratios remained higher than mammals, while their enhanced exercise capacity explains why their predator-to-prey ratios were lower than reptiles.

Dinosaurs failed to thrive in water, which rapidly extracts heat from vertebrate bodies, because their physiology could not tolerate lipoprotein solidification precipitated by water immersion. The water-dwelling snakes, alligators, crocodiles, plesiosaurs, mosasaurs, ichthyosaurs, and monitor lizards thrived in the shallow saltwater seas of Pangaea, where they were safe from dinosaur predators.

The presence of dinosaur fossils from polar regions (those lands that once were at the geographic pole, but have since moved) is explained by summer migration when the days were long, food was abundant, and temperatures were hospitable. Such migration can be accomplished with limited intelligence, as demonstrated by modern bird migration. The absence of turtle and reptile fossils in these regions is explained by their inferior exercise tolerance that prevented migration.

The simplest explanation of the spinal plates of stegosaurs, which were otherwise useless, is that these plates enhanced solar energy absorption to maintain body temperature, like the "sails" that reptiles had developed during the Permian era. In addition to being an example of how the genome retains successful adaptations in its library, this suggests that stegosaurs thrived in relatively cool regions of Pangaea, and their body carriage (pelvic structure) evolved to allow them to pivot to an upright position to reach elevated vegetation.[102]

BIRDS

Flying dinosaurs (pterosaurs) were the direct ancestors of birds. Their unique cardiopulmonary anatomy can be traced via fossil evidence from archosaurs to dinosaurs to birds. This physiology enhanced oxygen delivery to peripheral tissues and reduced their bone density, and explains why archosaurs were more energetic than reptiles. Archosaurs, however, failed to fly. Dinosaurs evolved from archosaurs, and their feathers, reduced bone density, and exaggerated exercise capacity paved the path for flight.

The pterosaurs flew low and slow in the hot, humid air above the vast inland seas of Pangaea and the later archaic continents to utilize the ground effect phenomenon that exaggerates wing lift. They feasted on fish, and attained the size of light aircraft. Their fish diet and leisurely flight explains their curved, elongated necks, and their teeth. Such enormous size is impossible in euthermic birds, whose metabolic rate, caloric requirements, and heat dissipation limitations are even greater than mammals. Modern ostriches can reach 300 pounds and stand six feet high; the so-called terror birds of the Cenozoic Era grew 10 feet in height courtesy of the cold climate. Neither ostriches nor terror birds could fly, which illustrates the flight limitations encountered by large, heavy birds. Though they lost the ability to fly, these large birds thrived in the cold climate due to their superior exercise capacity, cold tolerance, and gizzards that enabled them to digest a wide variety of foods. The modern condor may have a wingspan of 11 feet, but that is only a third the spread of a large pterosaur. The temperature-dependent physiology of pterosaurs prevented them from attaining high altitudes where the air is cold. The simplest explanation is that pterosaurs eventually evolved euthermia, and transformed into birds that fly at cold high altitudes and penguins that thrive in frigid waters.

Birds retain the latent ability to produce teeth, but no longer need them because they have evolved the ability to use their claws to capture fish and other game. This enables shorter necks that enhance aerodynamic efficiency; shorter necks minimize air resistance and increase flight speed.

Birds have increased heart size compared to mammals of similar size. This presumably reflects their adaptation to flight, but it is also possible that this was a trait shared with their dinosaur ancestors. Birds are notoriously sensitive to smoke and Teflon fumes because of their efficient respiratory physiology. This is why canaries were carried into coalmines to detect toxic gas before it became lethal to humans, and why Teflon fumes are toxic to birds but tolerated by humans.

Birds maintain their body temperature around 38.2–40 degrees Celsius, which is slightly higher than mammalian body temperature. This may represent an adaptation to flight at cold high altitudes. The exaggerated euthermic cellular metabolism causes birds to grow, mature, and evolve even faster than protomammals, mammals, and dinosaurs, and therefore develop Haversian Canals in their long bones.

The ability to fly enhances food acquisition, expansion into unexploited territories, and the ability to escape danger. It negates the need for a fragile, heavy, flight-restricting, energy-consuming large brain. This explains the limited intelligence of birds.

Some birds have retained their migratory patterns since they evolved shortly after Pangaea broke apart into Gondwanaland and Laurasia, even though their destinations have drifted to distant locations. This explains why some birds migrate as far as 26,000 miles.

Several unique bird characteristics presumably enhance their flight capabilities. Their erythrocytes have a unique orange color and morphology. Their heart size is generally bigger than mammals of similar size. Mammalian heart size is generally proportional to body size, but small birds have proportionately bigger hearts than large birds. The red cells of both reptiles and birds are nucleated, but reptilian red cells are larger, flatter, and thinner than avian erythrocytes. Birds presumably inherited most or all their red-cell characteristics from their dinosaur predecessors along with advanced cardiopulmonary physiology. Better understanding of these characteristics and their effects on blood turbulence and viscosity might indirectly clarify the nature of dinosaur physiology.

THE CRETACEOUS-TERTIARY (KT) EXTINCTION EVENT

Ocean currents, global weather patterns, ocean temperatures, and terrestrial climates gradually changed towards the global conditions we experience now as Pangaea slowly rifted apart beginning 200 million years ago. Ocean currents and global weather patterns changed; ocean temperatures declined; polar ice re-appeared; sea levels fell; the vast shallow inland seas receded; and terrestrial temperature and rainfall fell. Many areas of the disintegrating supercontinent retained their hot, terrestrial climates, but new water barriers disrupted dinosaur migration and created cold waves that precipitated a series of partial dinosaur extinctions. The surviving dinosaurs temporarily recovered and even diversified after these setbacks, but all dinosaurs except the flying dinosaurs (which evolved into birds) eventually became extinct as terrestrial temperatures inexorably declined in all portions of the rearranging world.

Large collections of fossilized dinosaur skeletons have been found, and their commonly contorted postures suggest that their deaths were both sudden and agonizing. The doomed dinosaurs may have huddled together for warmth as lipoprotein solidification undermined their cardiac efficiency, disrupted tissue perfusion and oxygenation, inhibited aerobic metabolism, induced anaerobic cellular respiration, and caused painful lactic acid accumulation in their muscles, so that muscular cramping caused their bodies to assume twisted and contorted postures in death. Increased volcanic activity that accompanied Pangaea's disintegration was coincidental with the dinosaur extinction but was not the cause.

Hot temperatures and abundant food persisted for millions of years in some areas as continents drifted around the surface of the globe. Gigantic reptiles and archosaurs endured long after the dinosaurs disappeared, because their physiology enabled them to survive intermittent episodes of cold. Plesiosaurs, mosasaurs, and ichthyosaurs perished as the polar ice returned, sea levels lowered, and their warm, shallow inland sea environments disappeared, but monitor lizards, alligators, and crocodiles adapted to the altered environment and survived. Likewise, the specialized ammonites, belemnites, and plants that thrived in the inland seas became extinct as their environment disappeared. Rudists retreated and coral animals recovered as marine temperatures declined. Reptiles, birds, archosaurs, and mammals survived the mass extinction because their physiology was compatible with cooler temperatures.

The KT extinction wherein a gigantic meteor struck the Earth is the presently prevailing explanation for dinosaur extinction.[103] This explanation is weak because it fails to explain the four partial dinosaur extinctions that preceded the final dinosaur extinction, and it fails to explain why only dinosaurs, plesiosaurs, ichthyosaurs, and mosasaurs were rendered extinct, while reptiles, amphibians, mammals, birds, alligators, crocodiles and monitor lizards survived. The shallow saltwater inland seas are known to have persisted as late as 65 million years ago in what are now Kansas and Missouri. This implies that the Pangaea phenomenon continued to disrupt ocean currents and global airflow patterns and sustain the hot environments that enabled the survival of dinosaurs, plesiosaurs, ichthyosaurs, and mosasaurs for millions of years after Pangaea began to disintegrate into distant continents, and ocean and atmospheric currents changed marine and terrestrial temperatures to more closely resemble those we enjoy today. The KT event coincides with the final extinction of the dinosaurs and may have contributed to or caused their final demise, but the dinosaurs, ichthyosaurs, mosasaurs, and plesiosaurs were doomed by the destruction of the warm environments that sustained them, regardless of the KT event.

MEGADYNASTY IV (THE PRESENT)

Megadynasty IV persists to the present. It began about 175 million years ago as Pangaea rifted into individual pieces that became recognizable modern continents, and environmental conditions gradually changed to our current conditions.

The cooling temperatures and increased rainfall restored mammalian terrestrial dominance. Birds flourished, humans evolved, and reptiles endured. Alligators, crocodiles, and monitor lizards also persisted, though they are mistakenly classified as reptiles.

Environmental change continues to exaggerate evolution via cyclical ice ages that are interspersed by periods of global warming, which is presumably caused by the movements of heavenly bodies, variable output qualities of the Sun, the wobbling tilt of Earth's axis, and possibly other factors. Euthermic mammals and birds flourish, gain size, and extend their territories during the ice ages, while reptiles and archosaurs flourish, gain size, and extend their territories during periods of global warming.

During the most recent ice age that began about 70,000 years ago, woolly mammoths, mastodons, giant ground sloths, saber-toothed cats, flightless terror birds and other unusually large euthermic species evolved and flourished.

Humans evolved Neanderthal characteristics including mesomorph shape that enhanced heat retention, exaggerated muscle strength, enlarged mandibles, heavier bones, and bigger brains that facilitated survival in the cold climate. Most of these characteristics subsided as global warming undermined their utility, but the sudden appearance of civilization — with written language, animal husbandry, technology, and agriculture — may be due to residual intelligence induced by the most recent ice age. Ostriches, emus, and polar bears are also vestiges of the most recent ice age. During the present period of global warming, reptiles and archosaurs are increasing their size and range, while mammals are decreasing in size and numbers, coral reefs are in retreat, and polar bears are disappearing. With the onset of the next ice age, euthermic mammals and birds will once again flourish at the expense of poikilothermic reptiles and archosaurs. Meanwhile, the surviving genomes of plants, animals, and microbes continue to slowly accumulate fresh characteristics that enhance their complexity and adaptability to ever-challenging environmental turmoil.

SUMMARY

Stress theory resolves the disparities and deficiencies of Darwin and Lamarck. It explains dinosaurs and the Cambrian Explosion. The survival and evolution of complex animals necessitates extra-cellular stress mechanisms that convert chromosomal information into embryological development, regulate hemodynamic physiology to optimize cell survival and function, and maintain and repair multicellular structures. Each stress mechanism provides physiological characteristics that promote survival in a specific set of environmental circumstances, and each is associated with a specific genomic collection of alternative characteristics that enables speciation, which refines adaptation to environmental circumstances by selecting compatible combinations of anatomy, intelligence, and ethology. The sex mechanism enables genomes to gradually embellish their collections of useful characteristics. Environmental change induces evolutionary adaptation that gradually increases biological complexity as genomes accumulate additional characteristics and new stress mechanisms evolve. Each stress mechanism has advantages and disadvantages, and none is innately superior. Environmental circumstances determine the success and longevity of species, classes, and stress mechanisms. The fossil record provides mute evidence that the extinction of species is as essential to evolution as the death of individuals. The dinosaur stress mechanism, or something similar, would reappear if the environmental circumstances that produced it were to return, because its fundamental elements remain latent in surviving archosaur and bird genomes.

Stress theory clarifies taxonomy and explains its nature. For example, camels, llamas and alpacas may not be true mammals, because they lack biconcave red cells. This explains their inferior exercise tolerance compared to true mammals. Similarly, alligators, crocodiles and monitor lizards are presently classified as reptiles, but they might better be classified as archosaurs, because their countercurrent cardiopulmonary physiology distinguishes them from reptiles and explains why they thrive in water, while reptiles are restricted to land.

CONCLUSION

The ability to explain disparate phenomena is the hallmark of effective theory. Selye's long-anticipated mechanism fulfills all the predictions and expectations of his medical theory. In addition, it confers a Unified Theory of Biology that clarifies the origin and nature of life. Selye's visionary theory is now poised to fulfill its promise to complement and extend the meaning of DNA and provide a new paradigm for medical and biological research. Much work remains to be done, but the stress mechanism indicates that the data acquired via the recent Genome Project will one day yield to inexorably advancing computer technology that will clarify chromosomal genetic information storage and enable humans to alter evolution in ways that presently reside in the realm of science fiction.

NOTES

1 S. L. Miller, A production of amino acids under possible primitive earth conditions. *Science* 117, 528-529 (1953).

2 M. J. Behe, *Darwin devolves : the new science about DNA that challenges evolution*. (HarperOne, an imprint of HarperCollins Publishers, New York, NY, ed. First edition., 2019), pp. 342 pages.

3 J. M. Blair, *The control of oil*. (Pantheon Books, New York, ed. 1st, 1976), pp. xxii, 441 p.

4 http://www.bbc.com/news/science-environment-33067582

5 http://exploration.esa.int/mars/46038-methane-on-mars/

6 H. G. Hill, J. A. Nuth, The catalytic potential of cosmic dust: implications for prebiotic chemistry in the solar nebula and other protoplanetary systems. *Astrobiology* 3, 291-304 (2003).

7 A. Koestler, *The sleepwalkers; a history of man's changing vision of the universe*. (Hutchinson, London, 1959), pp. 624 p.

8 S. Goldman, A unified theory of biology and physics. *Journal of Social and Biological Structures* 3, 331-360 (1980).

9 R. T. Bakker, *The Dinosaur Heresies: New Theories Unlocking the Mystery of the Dinosaurs and Their Extinction*. (Citadel, 2001), pp. 480.

10 K. A. Ross, *Science Knowledge and the Environment*. R. K. A. a. L. E. Littledyke M., Ed., (David Fulton Publishers, London, 2000).

11 S. J. Rim *et al.*, Decrease in coronary blood flow reserve during hyperlipidemia is secondary to an increase in blood viscosity. *Circulation* 104, 2704-2709 (2001).

12 L. Dintenfass, Some Observations on the Viscosity of Pathological Human Blood Plasma. *Thrombosis et diathesis haemorrhagica* 13, 492-499 (1965).

13 S. J. Rim *et al.*, Decrease in coronary blood flow reserve during hyperlipidemia is secondary to an increase in blood viscosity. *Circulation* 104, 2704-2709 (2001).

14 P. Weiss, A. C. Taylor, Reconstitution of Complete Organs from Single-Cell Suspensions of Chick Embryos in Advanced Stages of Differentiation. *Proceedings of the National Academy of Sciences of the United States of America* 46, 1177-1185 (1960).

15 H. V. Wilson, On the behavior of the dissociated cells in hydroids, alcyonaria, and Asterias. *Journal of Experimental Zoology* 11, 281-338 (1911).

16 V. B. Mahajan, K. S. Pai, A. Lau, D. D. Cunningham, Creatine kinase, an ATP-generating enzyme, is required for thrombin receptor signaling to the cytoskeleton. *Proceedings of the National Academy of Sciences of the United States of America* 97, 12062-12067 (2000).

17 J. S. Mattick, Challenging the dogma: the hidden layer of non-protein-coding RNAs in complex organisms. *Bioessays* 25, 930-939 (2003).

18 J. S. Mattick, The hidden genetic program of complex organisms. *Sci Am* 291, 60-67 (2004).

19 J. S. Mattick, The regulatory architecture of the human genome. *Asia Pac J Clin Nutr* 13, S14 (2004).

20 R. O. Becker, G. Selden, *The body electric : electromagnetism and the foundation of life*. (Quill, New York, ed. 1st Quill, 1985), pp. 364 p.

21 C. R. Woese, G. E. Fox, Phylogenetic structure of the prokaryotic domain: the primary kingdoms. *Proceedings of the National Academy of Sciences of the United States of America* 74, 5088-5090 (1977).

22 M. W. Kirschner, J. C. Gerhart, *The Plausibility of Life: Resolving Darwin's Dilemma*. (Vail-Ballou Press, 2005).

23 K. Tantivejkul *et al.*, PAR1-mediated NFkappaB activation promotes survival of prostate cancer cells through a Bcl-xL-dependent mechanism. *J Cell Biochem* 96, 641-652 (2005).

24 W. Luo, Y. Wang, G. Reiser, Two types of protease-activated receptors (PAR-1 and PAR-2) mediate calcium signaling in rat retinal ganglion cells RGC-5. *Brain Res* 1047, 159-167 (2005).

25 J. B. Marriott *et al.*, A novel subclass of thalidomide analogue with anti-solid tumor activity in which caspase-dependent apoptosis is associated with altered expression of bcl-2 family proteins. *Cancer Res* 63, 593-599 (2003).

26 C. J. Sutherland, J. R. Schurman, Complications associated with warfarin prophylaxis in total knee arthroplasty. *Clin Orthop*, 158-162 (1987).

27 S. J. Gould, The evolution of life on the earth. *Sci Am* 271, 84-91 (1994).

28 M. T. Duret *et al.*, Size-fractionated diversity of eukaryotic microbial communities in the Eastern Tropical North Pacific oxygen minimum zone. *FEMS Microbiol Ecol* 91, (2015).

29 V. P. Edgcomb, D. T. Kysela, A. Teske, A. de Vera Gomez, M. L. Sogin, Benthic eukaryotic diversity in the Guaymas Basin hydrothermal vent environment. *Proceedings of the National Academy of Sciences of the United States of America* 99, 7658-7662 (2002).

30 M. W. Kirschner, J. C. Gerhart, *The Plausibility of Life: Resolving Darwin's Dilemma*. (Vail-Ballou Press, 2005).

31 https://www.quora.com/Which-evolved-first-plants-or-animals

32 R. T. Bakker, *The Dinosaur Heresies: New Theories Unlocking the Mystery of the Dinosaurs and Their Extinction*. (Citadel, 2001), pp. 480.

33 S. J. Gould, & Eldredge, Niles "Punctuated equilibria: the tempo and mode of evolution reconsidered." *Paleobiology* 3, 115-151 (1977).

34 M. F. Barnsley, P. Massopust, H. Strickland, A. D. Sloan, Fractal modeling of biological structures. *Annals of the New York Academy of Sciences* 504, 179-194 (1987).

35 S. E. Luria, M. Delbruck, Mutations of Bacteria from Virus Sensitivity to Virus Resistance. *Genetics* 28, 491-511 (1943).

36 http://mentalfloss.com/article/65092/our-skin-covered-invisible-stripes

37 P. Nick, Probing the actin-auxin oscillator. *Plant Signal Behav* 5, 94-98 (2010).

38 B. Moller, D. Weijers, Auxin control of embryo patterning. *Cold Spring Harb Perspect Biol* 1, a001545 (2009).

39 S. Vanneste, J. Friml, Calcium: The Missing Link in Auxin Action. *Plants (Basel)* 2, 650-675 (2013).

40 K. Takahashi, K. Hayashi, T. Kinoshita, Auxin activates the plasma membrane H+-ATPase by phosphorylation during hypocotyl elongation in Arabidopsis. *Plant Physiol* 159, 632-641 (2012).

41 J. G. Meszaros, N. J. Karin, Inhibitors of ER Ca(2+)-ATPase activity deplete the ATP- and thrombin-sensitive Ca2+ pool in UMR 106-01 osteosarcoma cells. *J Bone Miner Res* 10, 704-710 (1995).

42 B. Zou *et al.*, Identification and analysis of copine/BONZAI proteins among evolutionarily diverse plant species. *Genome* 59, 565-573 (2016).

43 Y. F. Zhou, T. A. Springer, Highly reinforced structure of a C-terminal dimerization domain in von Willebrand Factor. *Blood* 123, 1785-1793 (2014).

44 https://www.reddit.com/r/askscience/comments/26louo/why_do_chlorophyll_and_haemoglobin_molecules_look/

45 C. R. Woese, G. E. Fox, Phylogenetic structure of the prokaryotic domain: the primary kingdoms. *Proceedings of the National Academy of Sciences of the United States of America* 74, 5088-5090 (1977).

46 S. E. Luria, M. Delbruck, Mutations of Bacteria from Virus Sensitivity to Virus Resistance. *Genetics* 28, 491-511 (1943).

47 S. J. Gould, & Eldredge, Niles "Punctuated equilibria: the tempo and mode of evolution reconsidered." *Paleobiology* 3, 115-151 (1977).

48 https://en.wikipedia.org/wiki/Punctuated_equilibrium

49 R. T. Bakker, *The Dinosaur Heresies: New Theories Unlocking the Mystery of the Dinosaurs and Their Extinction.* (Citadel, 2001), pp. 480.

50 M. W. Kirschner, J. C. Gerhart, *The Plausibility of Life: Resolving Darwin's Dilemma.* (Vail-Ballou Press, 2005).

51 V. P. Edgcomb, D. T. Kysela, A. Teske, A. de Vera Gomez, M. L. Sogin, Benthic eukaryotic diversity in the Guaymas Basin hydrothermal vent environment. *Proceedings of the National Academy of Sciences of the United States of America* 99, 7658-7662 (2002).

52 H. V. Wilson, On the behavior of the dissociated cells in hydroids, alcyonaria, and Asterias. *Journal of Experimental Zoology* 11, 281-338 (1911).

53 P. Weiss, A. C. Taylor, Reconstitution of Complete Organs from Single-Cell Suspensions of Chick Embryos in Advanced Stages of Differentiation. *Proceedings of the National Academy of Sciences of the United States of America* 46, 1177-1185 (1960).

54 G. Wachtershauser, Before enzymes and templates: theory of surface metabolism. *Microbiological reviews* 52, 452-484 (1988).

55 G. Wachtershauser, Evolution of the first metabolic cycles. *Proceedings of the National Academy of Sciences of the United States of America* 87, 200-204 (1990).

56 F. Noguchi *et al.*, A novel alveolate in bivalves with chemosynthetic bacteria inhabiting deep-sea methane seeps. *The Journal of eukaryotic microbiology* 60, 158-165 (2013).

57 S. Havlin *et al.*, Fractals in biology and medicine. *Chaos, solitons, and fractals* 6, 171-201 (1995).

58 G. Roeselers, I. L. Newton, On the evolutionary ecology of symbioses between chemosynthetic bacteria and bivalves. *Applied microbiology and biotechnology* 94, 1-10 (2012).

59 M. W. Kirschner, J. C. Gerhart, *The Plausibility of Life: Resolving Darwin's Dilemma.* (Vail-Ballou Press, 2005).

60 M. W. Kirschner, J. C. Gerhart, *The Plausibility of Life: Resolving Darwin's Dilemma*. (Vail-Ballou Press, 2005).
61 S. J. Rim *et al.*, Decrease in coronary blood flow reserve during hyperlipidemia is secondary to an increase in blood viscosity. *Circulation* 104, 2704-2709 (2001).
62 L. Dintenfass, Some Observations on the Viscosity of Pathological Human Blood Plasma. *Thrombosis et diathesis haemorrhagica* 13, 492-499 (1965).
63 R. T. Bakker, *The Dinosaur Heresies: New Theories Unlocking the Mystery of the Dinosaurs and Their Extinction*. (Citadel, 2001), pp. 480.
64 R. T. Bakker, *The Dinosaur Heresies: New Theories Unlocking the Mystery of the Dinosaurs and Their Extinction*. (Citadel, 2001), pp. 480.
65 R. T. Bakker, *The Dinosaur Heresies: New Theories Unlocking the Mystery of the Dinosaurs and Their Extinction*. (Citadel, 2001), pp. 480.
66 M. A. Al-Drees *et al.*, Making Blood: The Haematopoietic Niche throughout Ontogeny. *Stem Cells Int* 2015, 571893 (2015).
67 R. O. Becker, G. Selden, *The body electric : electromagnetism and the foundation of life*. (Quill, New York, ed. 1st Quill, 1985), pp. 364 p.
68 R. L. Cieri, B. A. Craven, E. R. Schachner, C. G. Farmer, New insight into the evolution of the vertebrate respiratory system and the discovery of unidirectional airflow in iguana lungs. *Proceedings of the National Academy of Sciences of the United States of America* 111, 17218-17223 (2014).
69 R. T. Bakker, *The Dinosaur Heresies: New Theories Unlocking the Mystery of the Dinosaurs and Their Extinction*. (Citadel, 2001), pp. 480.
70 M. F. Barnsley, P. Massopust, H. Strickland, A. D. Sloan, Fractal modeling of biological structures. *Annals of the New York Academy of Sciences* 504, 179-194 (1987).
71 S. Havlin *et al.*, Fractals in biology and medicine. *Chaos, solitons, and fractals* 6, 171-201 (1995).
72 J. Gleick, *Chaos: Making a New Science*. (1988).
73 R. T. Bakker, *The Dinosaur Heresies: New Theories Unlocking the Mystery of the Dinosaurs and Their Extinction*. (Citadel, 2001), pp. 480.
74 J. S. Mattick, Challenging the dogma: the hidden layer of non-protein-coding RNAs in complex organisms. *Bioessays* 25, 930-939 (2003).
75 J. S. Mattick, The hidden genetic program of complex organisms. *Sci Am* 291, 60-67 (2004).
76 J. S. Mattick, The regulatory architecture of the human genome. *Asia Pac J Clin Nutr* 13, S14 (2004).
77 R. T. Bakker, *The Dinosaur Heresies: New Theories Unlocking the Mystery of the Dinosaurs and Their Extinction*. (Citadel, 2001), pp. 480.
78 S. J. Rim *et al.*, Decrease in coronary blood flow reserve during hyperlipidemia is secondary to an increase in blood viscosity. *Circulation* 104, 2704-2709 (2001).
79 M. W. Kirschner, J. C. Gerhart, *The Plausibility of Life: Resolving Darwin's Dilemma*. (Vail-Ballou Press, 2005).
80 R. T. Bakker, *The Dinosaur Heresies: New Theories Unlocking the Mystery of the Dinosaurs and Their Extinction*. (Citadel, 2001), pp. 480.

81 R. T. Bakker, *The Dinosaur Heresies: New Theories Unlocking the Mystery of the Dinosaurs and Their Extinction.* (Citadel, 2001), pp. 480.

82 S. Sahney, M. J. Benton, Recovery from the most profound mass extinction of all time. *Proc Biol Sci* 275, 759-765 (2008).

83 S. M. Stanley, *Earth system history.* (W.H. Freeman, New York, ed. 3rd, 2009), pp. xxi, 551 p.

84 P. Nick, Probing the actin-auxin oscillator. *Plant Signal Behav* 5, 94-98 (2010).

85 B. Moller, D. Weijers, Auxin control of embryo patterning. *Cold Spring Harb Perspect Biol* 1, a001545 (2009).

86 J. J. Head *et al.*, Giant boid snake from the Palaeocene neotropics reveals hotter past equatorial temperatures. *Nature* 457, 715-717 (2009).

87 R. T. Bakker, *The Dinosaur Heresies: New Theories Unlocking the Mystery of the Dinosaurs and Their Extinction.* (Citadel, 2001), pp. 480.

88 E. R. Schachner, R. L. Cieri, J. P. Butler, C. G. Farmer, Unidirectional pulmonary airflow patterns in the savannah monitor lizard. *Nature* 506, 367-370 (2014).

89 E. R. Schachner, J. R. Hutchinson, C. Farmer, Pulmonary anatomy in the Nile crocodile and the evolution of unidirectional airflow in Archosauria. *PeerJ* 1, e60 (2013).

90 S. Havlin *et al.*, Fractals in biology and medicine. *Chaos, solitons, and fractals* 6, 171-201 (1995).

91 E. R. Schachner, J. R. Hutchinson, C. Farmer, Pulmonary anatomy in the Nile crocodile and the evolution of unidirectional airflow in Archosauria. *PeerJ* 1, e60 (2013).

92 R. L. Cieri, B. A. Craven, E. R. Schachner, C. G. Farmer, New insight into the evolution of the vertebrate respiratory system and the discovery of unidirectional airflow in iguana lungs. *Proceedings of the National Academy of Sciences of the United States of America* 111, 17218-17223 (2014).

93 E. R. Schachner, R. L. Cieri, J. P. Butler, C. G. Farmer, Unidirectional pulmonary airflow patterns in the savannah monitor lizard. *Nature* 506, 367-370 (2014).

94 E. R. Schachner, J. R. Hutchinson, C. Farmer, Pulmonary anatomy in the Nile crocodile and the evolution of unidirectional airflow in Archosauria. *PeerJ* 1, e60 (2013).

95 E. R. Schachner, R. L. Cieri, J. P. Butler, C. G. Farmer, Unidirectional pulmonary airflow patterns in the savannah monitor lizard. *Nature* 506, 367-370 (2014).

96 E. R. Schachner, J. R. Hutchinson, C. Farmer, Pulmonary anatomy in the Nile crocodile and the evolution of unidirectional airflow in Archosauria. *PeerJ* 1, e60 (2013).

97 A. J. Desmond, *The hot-blooded dinosaurs : a revolution in palaeontology.* (Dial Press, New York, 1976), pp. 238 p.

98 M. H. Schweitzer, J. L. Wittmeyer, J. R. Horner, J. K. Toporski, Soft-tissue vessels and cellular preservation in Tyrannosaurus rex. *Science* 307, 1952-1955 (2005).

99 http://www.bbc.com/news/science-environment-33067582

100 C. A. Mauricio Canals, Bruno Grossi, and José Iriarte-Díaz, Relative Sizes of hearts and lungs of small bats. *Acta Chiropterologica* 7(1):65-72. 2005, (2005).

101 C. A. Mauricio Canals, Bruno Grossi, and José Iriarte-Díaz, Relative Sizes of hearts and lungs of small bats. *Acta Chiropterologica* 7(1):65-72. 2005, (2005).

102 R. T. Bakker, *The Dinosaur Heresies: New Theories Unlocking the Mystery of the Dinosaurs and Their Extinction.* (Citadel, 2001), pp. 480.

103 https://en.wikipedia.org/wiki/Cretaceous–Paleogene_extinction_event

Chapter 18

The Great Medical Hoax of the 20th Century

"Real courage is when you know you're licked before you begin, but you begin anyway and see it through no matter what."
—Harper Lee, To Kill a Mockingbird

The medical profession in the United States ceased, very largely, to be a profession of the fatherly confessors and unprofessing humanitarians and became one of the largest groups of hardheaded petty-bourgeois hustlers in the United States, and their professional association became the most ruthlessly materialistic lobbying association of any professional group.
—Carrol Quigley

Modern anesthesia practice is founded on a cunning hoax that has killed countless patients.[1, 2, 3, 4] The purpose of this appendix is to destroy the hoax and restore medical progress by explaining how it was created and why it is promoted and protected by corporations, publications, and professions. It's important to understand the history of the hoax, where it came from, and why it originally was created, because it affects us still. That information forms Part I of this appendix. Part II details its multifarious malignant manifestations.

The hoax — a deliberate misunderstanding of the nature of CO_2 — worked. It devastated the reputation of the nurse-anesthetists who dominated anesthesia after WWI, replaced them with MD anesthesiologists, and founded the MD anesthesiology profession. However, it exacted a terrible price. It has rendered at least three generations of anesthesiologists ignorant of the scientific foundation of their profession. It has escaped the bounds of anesthesia and wreaked havoc throughout medicine. It derailed the efforts of unsuspecting stress researchers who were so close to revolutionizing medicine that they could almost taste it.

But there is worse, yet.

Almost everyone knows that cancer, heart disease, chronic illness, and medical costs keep climbing despite persistent promises that DNA discoveries will soon produce miraculous cures, and nobody wishes to think that the medical services they seek might be their worst enemy. Everyone faces the prospect of surgery with advancing age, but few appreciate that anesthesia prevents agonizing death *after* surgery, or suspect that anesthetic mismanagement exaggerates surgical morbidity and mortality. Almost nobody understands the covert ways that powerful corporations promote the hoax for the sake of profit and at the price of public health. There is no hope for a cure for cancer or any other medical progress until the hoax is eradicated.

Think of this, the next time you go under the knife.

I have no leaked interoffice memos to wave in angry evidence of this. I cannot point to a specific individual or company and allege that they support the hoax *because* it results in additional morbidity and mortality — and income — but readers can judge for themselves the evidence that is presented in this appendix.

I apologize in advance for the length of this appendix, but I believe its detail is essential to explain the chaos and confusion caused by the hoax, expose its consequences, and explain why it stubbornly resists revelation and reform.

Note: In this appendix, the author's explanation of and interpolation within lengthy quotations are set off by square brackets, bold type, and Italics: [ ]

PART I: MY STRUGGLE TO UNDERSTAND ANESTHESIA

"The key to wisdom is this — constant and frequent questioning, for by doubting we are led to question and by questioning we arrive at the truth"

—Peter Abelard

My personal experience may illustrate the power and persistence of the anesthesia hoax. As mentioned in the preface, CO_2 confusion dogged my education and career. I knew in high school that the Krebs cycle continuously consumes oxygen and produces carbon dioxide in all animal cells. In college I was instructed to anesthetize cockroaches with carbon dioxide for my honors research project. This impressed me as illogical. How could CO_2 possess anesthetic properties when it is produced by every cell in the body? If it did, we would all be drunk! I suspected that I was suffocating the cockroaches with CO_2 and feared that harmful hypoxia might disrupt my experiment. However, one hesitates to challenge one's professors, so I kept my questions to myself.

Figure 18.1: "Anesthetizing" roaches

This could have been me during my undergraduate research project at the Ohio State University. This graduate student is venting carbon dioxide into a beaker to anesthetize insects. In reality he is suffocating the insects. This is standard practice in insect and small animal research. The practice originated with the publications of Dr. Ralph Waters, who alleged that CO_2 has anesthetic properties that are satisfactory for animals but not humans.

Photo courtesy of Dr. David Denlinger, Department of Entomology, the Ohio State University https://entomology.osu.edu/our-people/david-denlinger

Over the years I have wondered how it happened that carbon dioxide is routinely used by researchers throughout the world for both euthanasia and anesthesia without a thought to the logical inconsistency, cruelty, or perturbation of research results.[5, 6, 7, 8, 9] How could trained physicians and scientists believe that CO_2 has both toxic and anesthetic properties when it is continuously produced by every cell in the body and saturates body fluids and tissues?

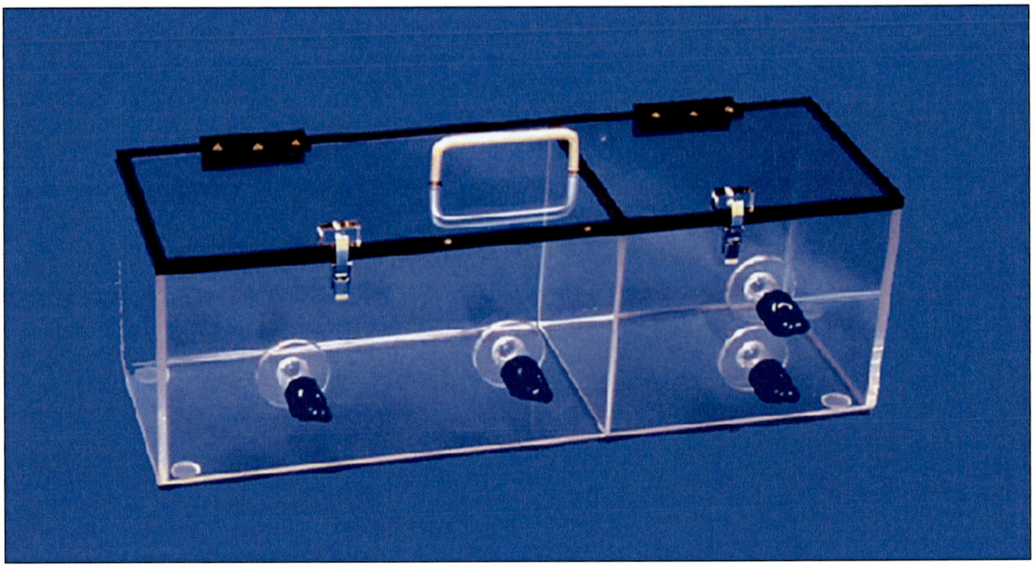

Figure 18.2: Anesthesia chamber
A typical CO_2 chamber used for either anesthesia or euthanasia of research animals.

Madness is rare in individuals — but in groups, parties, nations, and ages it is the rule.
— Friedrich Nietzsche

In medical school, the lectures of Dr. Marc Peisch, a CO_2 researcher, explained how the unique chemistry of carbon dioxide enables oxygen transport and delivery, so that CO_2 is as essential for life as oxygen. He also explained how CO_2 asphyxiation can mimic both general anesthesia and toxicity (see Chapter 14: Oxygen Transport and Delivery, on page 137). This confirmed my suspicions that I had asphyxiated my cockroaches. However, Dr. Peisch failed to mention the practical therapeutic properties of carbon dioxide or the important history of its use as a treatment.

I confronted the specter of CO_2 confusion yet again when I began my anesthesiology training, but I didn't realize until much later that I had blundered into its source. I found myself surrounded by residents and professors who assumed that carbon dioxide is "toxic waste, like urine" that must be "rid from the body" using mechanical hyperventilation lest it cause dangerous toxic and narcotic (anesthetic) effects. This was at odds with everything I knew about carbon dioxide and medical science. In those days there was no means to monitor and measure gas concentrations, so we were instructed to use a crude hyperventilation algorithm that assured effective riddance of carbon dioxide. All the anesthesia machines were prominently equipped with canisters of soda lime that absorbed carbon dioxide from the breathing circuit, and we were instructed to replace the soda lime whenever it became saturated with carbon dioxide, as if this represented a serious safety hazard.[10, 11] In contrast, the residents and staff freely discussed *apneic oxygenation* studies and accident reports documenting harmless exposure to CO_2 levels at least as high as 500 mm Hg (normal being 40 mm Hg), as if the disparity between dogma and data didn't exist.[12, 13] Furthermore, Dr. J. W. Bellville, a prominent CO_2

researcher on the UCLA anesthesiology staff, had published studies demonstrating that hypercarbia beneficially counteracts narcotic respiratory depression, and the relationships of sleep, sedation, and narcotics.[14] Nobody seemed to notice the logical inconsistencies between research, practical experience, and anesthesia practice.

For me, the residency was a Kafkaesque experience, as if I had been plunged into a lunatic asylum where my perception of reality was at odds with everyone else. I couldn't imagine that I was the only person in possession of his wits, while everyone else was deluded. I thought that I must have somehow overlooked some important scientific information that everyone else took for granted. In retrospect, this was my first confrontation with the gullibility of human beings that enables politics, propaganda,[15] Ponzi schemes, hoaxes,[16] and salesmanship.

The individual has always had to struggle to keep from being overwhelmed by the tribe. If you try it, you will be lonely often, and sometimes frightened. But no price is too high to pay for the privilege of owning yourself.
— Friedrich Nietzsche

The confusion was not confined to carbon dioxide. The terms *anesthesia* (ablation of consciousness) and *analgesia* (reduction of pain) implied distinctly different phenomena, but the terms were used interchangeably. For example, analgesic agents such as lidocaine, Novocain, tetracaine, and prilocaine that were used to block nerves were called *local anesthetics* even though they didn't abolish consciousness in the manner of general anesthesia. Everyone assumed that anesthetic inhalation agents possessed analgesic properties that lowered blood pressure by beneficially inhibiting harmful nervous activity, despite evidence that their toxicity poisons cardiac contractility. Drugs as different as Demerol (a narcotic) and pentothal (a sedative) were used interchangeably to "deepen" anesthesia, as if their effects were equivalent. Everyone embraced Cannon's concept that blood pressure is the driving force of blood flow and tissue perfusion, and that the heart is the "Charles Atlas" of the body with cardiac reserve that enables ten-fold increases in cardiac output, but my medical school stress theory lectures had demolished those conventional notions (see Chapter 8 on page 59 and Chapter 13 on page 119), and current research had demonstrated that cardiac *output* varies with flow resistance, while cardiac *contractility* remains constant. Regardless, blood pressure was the mainstay of anesthesia monitoring and management. It was slavishly recorded every five minutes, and if it fell below "normal," we were expected to treat patients with vasopressors that counterproductively exaggerated flow resistance and decreased cardiac output.[17] If blood pressure rose above normal, we increased the "overpressure" of toxic inhalation agents. Everyone assumed that both respiratory and metabolic acidosis were equally harmful, and that alkalosis is innately beneficial (see Chapter 14: Oxygen Transport and Delivery, on page 137). Harmless acidosis was routinely treated with sodium bicarbonate, while harmful alkalosis confirmed competent anesthetic management. Intravenous fluids were administered with abandon, based on the presumption that fluids "expand intravascular volume" and improve cardiac output, and regardless of evidence that they undermine coagulation, exaggerate blood loss, and escape from circulation to the bowel, where they cause the bowel to swell like a sausage, compromise breathing, and disrupt surgical repairs.

The UCLA training prioritized (1.) Abolition of consciousness (based on the assumption that this controls all forms of surgical stress). (2.) Secure airway (elective endotracheal intubation). (3.) Surgical convenience (paralysis that prevents untoward movements and muscle tension). Patient safety was an afterthought. The training emphasized what I now call the *Waters technique*.[18] This began with rapid intravenous induction with sodium pentothal (a short-acting hypnotic agent that abolishes consciousness) immediately accompanied by paralyzing agents that enable elective endotracheal intubation to “secure the airway.” Patients were then maintained with inhalation agents. Mechanical hyperventilation was used to “rid the body” of carbon dioxide, and toxic inhalation agent overpressure was used to counteract the hypertension caused by surgical stress, based on the belief that inhalation agents possess analgesic properties that beneficially prevent surgical stress. Paralyzing agents prevented untoward muscle tension that inconvenienced surgeons.

I suspected from the start that there must be a better means to manage anesthesia. The inhalation agents didn’t appear to possess analgesic properties, because patients suffered intense pain, tachycardia, hypertension, dysrhythmias, dementia, delirium, laryngospasm, nausea, vomiting, and (thankfully rare) strokes and heart attacks as they emerged from anesthesia.[19] Residual postoperative paralysis was a persistent problem that pervaded anesthesia publications, and there was no reliable way to assess and correct it.[20, 21, 22, 23, 24] Narcotic supplementation seemed to offer an obvious solution, but narcotics were forbidden for fear of respiratory depression.

Thankfully, the older attendings remembered the superior safety of mask induction and regarded mask management as fundamental to safe practice. They taught simple techniques of airway management that are now largely forgotten. I am forever grateful that Dr. Herman Schwartz, an older attending, taught me the fundamentals of Crile’s forgotten “Anoci-Association” technique wherein respiratory rate was maintained between 8–12 breaths per minute using narcotic supplementation[25] — though he did this without mentioning Crile. This method produced excellent outcomes, but its utility was limited because most surgeries necessitated paralysis and mechanical hyperventilation.

Group pressure prevails in human relationships, and I was no exception. I was so intimidated by the disparity between my understanding of science and the beliefs and practices embraced by everyone else that I thought I had failed to grasp some obvious scientific secret. I was afraid to reveal my embarrassing ignorance to the attendings. In desperation, I cautiously asked a fellow resident if he knew anything about the mysterious evidence that carbon dioxide has toxic and narcotic properties. He recommended the dog study of Edmund Eger, which I sought and read.[26] I found this study illogical, unconvincing, and difficult to read, as if the authors were deliberately muddling their message. They initially anesthetized their dogs with Halothane instead of carbon dioxide. If carbon dioxide possessed anesthetic properties, then why was it necessary to anesthetize the dogs with Halothane? They explained how the Halothane was gradually supplanted by increasing concentrations of carbon dioxide, and how the anesthetic effect of carbon dioxide began suddenly when CO_2 concentrations approached 30% (with 70% oxygen). This *anesthetic effect* was closely followed by the sudden appearance of acidosis in blood and cerebrospinal fluids, and only slightly higher CO_2 concentrations caused convulsions and deaths in the dogs. Such sudden onset of acidosis, convulsions, and death is not a characteristic of anesthesia. These observations were most simply explained by CO_2 asphyxiation, which undermines brain function and causes eukaryotic cells to revert to *anaerobic metabolism* pathways that produce lactic acid when oxygen is lacking. Inexplicably, the Eger study never mentioned this possibility. Furthermore, even though Eger *et al* sampled arterial blood, venous blood, and cerebrospinal fluid for blood gas analysis, they inexplicably omitted blood oxygen data from their report. Instead, they concluded that acidosis, as opposed to brain hypoxia, had caused the *anesthetic* effects observed in the dogs — never mind that there is no evidence that acidosis causes anesthesia, or that anesthesia causes convulsions and deaths. I was puzzled that these inconsistencies had failed to attract scientific scrutiny, but I feared to pursue the matter. I just added it to my long list of doubts.

One day, while exploring the anesthesia equipment rooms at UCLA, I stumbled upon a closet filled with old anesthesia machines. An elegant, curved lever was mounted on the right-hand side of one of these old machines. I asked one of the younger attendings, Dr. George Herr, if he knew the purpose of this lever. He did. He explained that it operated a valve that injected carbon dioxide into the breathing circuit, but that such machines were dangerous and were therefore no longer used. This was a powerful clue to the history of anesthesia, which starkly contrasted with the modern UCLA dogma that CO_2 is "toxic waste, like urine" that must be "rid from the body" using mechanical hyperventilation. As a result, I concluded that CO_2 must possess some sort of therapeutic property that had been appreciated in the forgotten past to the point that it had been a standard feature of anesthesia machines. This further piqued my curiosity, but I had no idea of how to pursue the matter, so I put it aside — but I didn't forget it. In those days, literature research remained an onerous task for lack of computers and Internet. Many years later I attempted to locate evidence of such machines, and even wrote to the Wood Library and the Smithsonian Institution, but such machines seem to have disappeared without a trace. The machine I saw was discarded when the UCLA anesthesia department moved to better quarters after I left. In retrospect, I suspect that this was no accident.

Stress research seems to have lingered longer in anesthesiology than in other medical specialties, and during my residency there were ongoing clinical studies that evaluated the ability of narcotic supplementation to improve cardiac surgery outcome. These studies utilized large doses of the new generation of short-acting synthetic narcotics (fentanyl, Sufentanil, Alfentanil, etc.) that were not yet accepted for use in non-cardiac surgery.[27] Much later I grasped the reason: patients were routinely maintained on mechanical ventilation *for hours* after cardiac surgery, which prevented respiratory depression caused by the narcotics. However, as an ordinary resident I was allowed to use only morphine and Demerol, which also depressed respiration.

Having escaped the residency, passed my board examinations, and entered private practice, I was finally free to pursue knowledge for my own purposes. I began by exploring the new generation of anesthesia textbooks that became available around that time, in hope of resolving the conceptual conflicts that had confused my education and training, and thereby improve my clinical results. I learned a lot about managing arcane medical conditions, but found no explanation of CO_2 consequences, or of the relationship between CO_2 and narcotics, or any of the other issues previously noted. Meanwhile, I didn't dare depart from the universal habit of hyperventilation because private practice demanded conformity even more than the residency.

In retrospect, I entered private practice just as anesthesia entered a period of rapid change. Automatic blood pressure machines and pulse oximeters were introduced. Invasive monitoring techniques — central venous pressure lines, arterial cannulas, and Swan-Ganz heart catheters — escaped from cardiac anesthesia into daily practice, and I was obliged to master these tricky and hazardous techniques with essentially no help or guidance. New paralyzing agents, induction agents, inhalation agents, sedatives, and resuscitation drugs appeared. These were not always improvements. For example, Propofol (the Michael Jackson drug) was far more toxic than the sodium pentothal it replaced.

SYNTHETIC NARCOTICS

In the early 1980s, soon after I began private practice, the Janssen Corporation began to promote its short-acting synthetic narcotics for non-cardiac anesthesia. I tested fentanyl with the recommended dosage with two young, healthy patients a few days apart, and was impressed by the dramatic reduction in postoperative pain, delirium, tachycardia, hypertension, and so forth. However, both patients unexpectedly stopped breathing about 30 minutes after I left them in the recovery room. Thankfully, pulse oximetry had become standard by that time, and that plus the alert nursing staff averted disaster. However, these episodes were so unnerving that I abandoned narcotic supplementation and resumed the Waters technique. In retrospect, the CO_2 hoax caused these problems. Capnography had not yet arrived, and everyone, including me, hyperventilated their patients, as this was the *de-facto* standard of practice. Others encountered similar problems and were similarly baffled. Soon articles appeared in the literature attributing these episodes to *re-narcotization syndrome* or *opioid hypersensitivity*. Nobody recognized that the underlying cause of this problem was mechanical hyperventilation that depleted CO_2 tissue reserves, obtunded respiratory chemoreceptors, and exaggerated opioid respiratory depression.[28, 29, 30] Nevertheless, anesthesiologists were impressed by the benefits of the new narcotics, and adapted to them by imitating the practice of the cardiac anesthesiologists: they began leaving their patients intubated and mechanically ventilated for hours after surgery. Soon armies of respiratory therapists materialized to manage the difficult "weaning" from mechanical ventilation so that the anesthesiologists could return to the operating room. This escaped my notice, because I was working alone at a small hospital when this change in practice took place.

PERMISSIVE HYPERCARBIA

In the late 1980s critical care specialists inadvertently discovered that hypercarbia reduced the morbidity and mortality of ARDS (Acute Respiratory Distress Syndrome), prevents lung damage, and promotes respiratory drive, cardiac output, and tissue oxygenation.[31, 32, 33, 34, 35, 36, 37, 38, 39] This caused considerable excitement, because CO_2 was supposed to be toxic and dangerous. In deference to the CO_2 hoax, they called this *permissive hypercarbia* (sometimes also called *permissive hypercapnia*). Nobody understood that they had re-discovered the classic principles of Yandell Henderson, nor did they understand that hypercarbia enhances oxygen transport and delivery, which — of course — was absent from medical textbooks and literature on account of the hoax. I read their enthusiastic reports with interest, but since capnography (monitors that measure CO_2 concentration in exhaled air) was not yet available, I focused my efforts on spinal and epidural analgesia techniques, with or without general anesthesia, to improve the outcome of my patients. I remained reluctant to take chances with narcotics except in rare cases when patients could be allowed to breathe spontaneously.

CAPNOGRAPHY

During my residency training, anesthesiology was regarded as the most dangerous of all medical specialties. The greatest source of anesthesia risk was esophageal intubation, where the anesthesiologist accidentally inserted the endotracheal tube in the esophagus instead of the trachea. In retrospect, this was yet another problem related to the hoax, because patients were routinely paralyzed, intubated, and placed on mechanical ventilation. Listening for breath sounds with a stethoscope was treacherous, because the sound of air moving back and forth in the esophagus was closely similar to the sound of air moving into and out of the lungs. Thus, brain damage or death due to these accidents was fairly common. Pulse oximetry was introduced early in the 1980s and this reduced the risk of such problems but did not eliminate them. Then, in 1985, private companies founded the *Anesthesia Patient Safety Foundation* (APSF) and placed a prominent anesthesiology professor, Dr. Robert Stoelting, in charge. Their motive was to promote universal acceptance of newly-developed capnograph monitors, which could monitor and measure exhaled carbon dioxide. With the help of APSF promotion, these new machines soon became ubiquitous, and they practically eliminated the problem of esophageal intubation overnight. This, in turn, drastically reduced anesthesia risk and the terrible cost of anesthesia malpractice insurance.

I first encountered the new capnography technology when I accepted a job in Bakersfield, CA in the early 1990s. Capnography confirmed that I was hyperventilating my patients, and I immediately corrected my ventilator settings to maintain CO_2 in the normal range of 40 mm Hg. Then, having been reassured of the safety and benefits of hypercarbia by the critical care evidence, I reasoned that I could employ mild hypercarbia to counteract the respiratory depression caused by the new generation of short-acting narcotics, and thereby improve my anesthetic outcomes. I cautiously combined *permissive hypercarbia* with the short-acting synthetic narcotics, using capnography to monitor CO_2 levels and pulse oximetry to assure effective red cell oxygen saturation. I reasoned that as long as pulse oximetry confirmed that I was saturating the patients' red cells with oxygen, there should be no problems with tissue oxygenation, and I knew from medical school that hypercarbia should exaggerate the release of oxygen from the red cells into tissues. I was immediately gratified by the improvement in outcome and the absence of unexpected postoperative respiratory depression. It seemed that narcotics and hypercarbia went together like love and marriage; the greater the hypercarbia and narcotic dosages, the better the surgical outcome. This strategy drastically reduced paralysis requirements, lingering postoperative paralysis, toxic paralysis reversal requirements, and postoperative pain, fever, tachycardia, hypertension, dysrhythmias, laryngospasm, delirium, and dementia. Nausea and vomiting were no worse than before. Best of all, there were no episodes of unexpected respiratory arrest. I had discovered that hyperventilation was the cause of that problem. I had no idea that I had re-discovered the original scientific anesthesia technique pioneered by Crile, Henderson, and the nurse-anesthetists.

THE TRIUMPH OF THE HOAX

The other anesthesiologists in Bakersfield ignored the critical care advances, and likewise ignored the ability of their new capnography monitors to measure exhaled CO_2 concentrations. They continued to hyperventilate their patients as before, even after anesthesiology research confirmed the benefits of permissive hypercarbia during anesthesia.[40, 41, 42, 43, 44, 45] In fact, they threatened me with potential repercussions. None of them ever asked me to explain my reasons. On one occasion, the Chief of Anesthesia, who was a cardiac anesthesiologist trained at the Cleveland Clinic and therefore familiar with large doses of narcotics, insisted on assuming care for my C-section patient so that I could return to the operating room for other duties, whereupon he "corrected" my hypercarbia ventilator settings and hyperventilated the patient, and then couldn't figure out why she wouldn't breathe when the case was finished. Then he proceeded to blame me for his ignorance behind my back. After that, I dared not allow the other anesthesiologists to assume care of my cases, as was customary in those days, for fear that they would hyperventilate them and cause a death. I later learned that another of the anesthesiologists tried to imitate my technique, but he couldn't understand it either, even though I charted my vital signs, narcotic dosages, and hypercarbia for all to see. The hoax reigned supreme.

When the human race has once acquired a superstition, nothing short of death is ever likely to remove it.
—Mark Twain

THE POWER OF MONEY

Anesthesia practice dangles in the wind of reimbursements. One can only wonder that if capnography had been introduced earlier in the 1980s, during the heyday of synthetic opioid promotion, then the complimentary relationship between opioids and hypercarbia might have been re-discovered. This would have revolutionized anesthesia practice.[46, 47, 48] Instead, the fickle finger of fate blindsided anesthesia progress from two directions at once. First, capnography and pulse oximetry largely abolished airway disasters, causing malpractice insurance rates to plummet. This fostered the fiction that further improvements in anesthesia outcome were impossible.[49] Second, insurance companies suddenly altered reimbursements, supposedly to restrict the expensive postoperative weaning necessitated by incompatible combinations of opioids and hyperventilation.[50] As a consequence, outpatient surgery centers proliferated; Crile's principles of early admission, premedication, delayed discharge, and narcotic supplementation were all abandoned; and anesthesiologists ignored permissive hypercarbia, re-abandoned opioids, and re-embraced the Waters technique.[51] Even the cardiac anesthesia specialists, who receive lavish fees for their expertise, promptly abandoned their narcotic-based cardiac anesthesia technique that had been famously pioneered by Dr. Edward Lowenstein, and resumed the Waters technique.[52, 53] By 1997 the results of inadequate narcotic supplementation during bypass surgery manifested as the previously unknown Systemic Inflammatory Response Syndrome (SIRS), but the cause was ignored.[54, 55, 56, 57] In reality, this was the same stress reaction that had been mitigated by Lowenstein's discovery of morphine supplementation. That, of course, was long forgotten. The tragedy became complete in 2005, when Dr. Terri Monk published her landmark study that unexpectedly revealed a 30% increase in morbidity and mortality in the distant aftermath of surgery that is influenced by anesthetic management.[58] This should have inspired fresh interest in balanced anesthesia (narcotic supplementation). Instead, Dr. Monk was pressured into silence; the morbidity and mortality due to inadequate narcotic supplementation was ignored; anesthesia literature suddenly emphasized the dangers of narcotics; prominent professors and specious studies promoted the preposterous premise that narcotics cause cancer (narcotics prevent postoperative cancer and heart disease);[59, 60] and new regulations restricted clinical research to the purview of powerful pharmaceutical companies.

By 2010, the hoax appeared in yet another form. By this time, anesthesiologists had largely abandoned narcotic supplementation but continued to hyperventilate patients. This causes a slightly different version of the original anesthetic problem of spontaneous hyperventilation that had plagued ether anesthesia in 1910, and which depleted CO_2 tissue reserves and caused patients to unexpectedly stop breathing and die after surgery. The only difference was that this time the patients were paralyzed and mechanically hyperventilated. It became evident that seemingly healthy geriatric patients were dying mysteriously of unexplained respiratory arrest hours after uneventful surgery, usually soon after treatment with small doses of opioids and sedatives in their private rooms.[61, 62, 63] The mysterious death of Andy Warhol hours after his gall bladder surgery is a classic example. This should have focused attention on the harmful effects of hyperventilation and the need for guidelines and standards for CO_2 management, but instead the problem was dismissed as “narcotic hypersensitivity,” and the issue vanished from view.[64]

While my colleagues abandoned narcotics to maintain their reimbursements, I continued to employ combinations of hypercarbia and narcotic supplementation to optimize my outcomes. With this technique, I was able to discharge my patients the same day with minimal complications. I thought that I had discovered an important new anesthetic principle. Finally, I decided to pursue clinical research to prove its benefits. To that end, I arranged a personal meeting with Dr. Ted Stanley at the University of Utah, who was the most prominent academic advocate of narcotic supplementation during anesthesia.

On that same occasion, I attended Ted’s annual “skiing conference” in Park City, Utah. Here the fickle finger of fate intervened again. The very first lecture by Dr. Charise T. Petrovitch described fresh information that coagulation factor VIII is a *chimera* consisting of enzymatic factor VIIIC produced continuously by the liver and von Willebrand Factor (VWF) produced by the vascular endothelium and released into blood under nervous control. These two seemingly unrelated proteins bind together in circulation to form a gigantic molecular complex called *factor VIII* so that they circulate together and exert their effects in concert. VWF stabilizes VIIIC to enable its enzymatic effects, so that nervous activity determines the enzymatic activity of factor VIII. In addition to its stabilizing effects on VIIIC, VWF maintains capillary structure and prevents flow-related damage called angiodysplasia.

This information changed my life. I knew that biological chimeras are exceedingly rare, and I was immediately impressed that factor VIII must represent an important but unrecognized physiological mechanism that links blood enzymes to nervous activity. I became so curious that I began spending all my spare time for the next six years exploring published research to learn more about this fascinating phenomenon. By this time, the Internet and personal computers — which had been only a distant dream during my education and training — had become ubiquitous, so I was able to utilize advanced computer search techniques to efficiently review thousands of research publications. Such an effort would have been impossible during the days of stress research.

The peculiar characteristics of factor VIII functioned as a Rosetta Stone that gradually enabled me to decipher the intricacies of coagulation enzyme interactions with nervous activity and discover the long-sought *mammalian stress mechanism* postulated by Hans Selye. I struggled to publish my findings and investigate their extended implications. During this process Ted Stanley proved to be a wonderful friend and source of support until his death from cancer in 2017. He endorsed my anesthesia technique, critiqued my fledgling publications, arranged for me to present my early findings before the research committee at the University of Utah, and helped in numerous other ways, but he warned me that my dream of returning to the academic environment to pursue clinical research was unrealistic on account of my age (I was only 53 at that point) and lack of academic qualifications (never mind that I was a board-certified anesthesiologist with years of clinical experience). His last gift shortly before he died was his warning that my carbon dioxide viewpoint was likely to engender hostility. He wasn’t joking. I am thankful that before Ted died, I completed the first version of my book describing the stress mechanism, and I sent him a copy of the final chapter that presented my *Unified Theory of Biology* based on stress theory. Ted called me one afternoon to tell me that my manuscript was the most interesting paper he had ever read. (You can read it as Chapter 17: Unified Theory of Biology, on page 239.) I will always cherish that memory. It was then that I realized that one day, long after I am gone, stress theory will be embraced as the most important advance in the history of medicine and biology.

Finally, in the process of preparing this book, I sought the source of the confusing carbon dioxide concepts that had bedeviled my career. I did this by following the references in modern anesthesia publications backward in time. The effort confirmed my suspicions that my own profession of anesthesiology was the source of the CO_2 hoax, and it revealed a treasure trove of scientific literature left behind by Crile, Henderson, Lundy, Jackson, and others that documents the discovery of modern anesthesia science that is now forgotten due to the anesthesia hoax. I have published a preliminary version the results of this investigation in a review paper entitled "Four Forgotten Giants of Anesthesia History" that is available free of charge via its publisher at:

https://www.ommegaonline.org/article-details/Four-Forgotten-Giants-of-Anesthesia-History/468#METHODS

I recommend this as a source of additional detail.

What follows is my review of the discovery of modern anesthesia science and my explanation of the salient circumstances that produced the anesthesia hoax.[65] I have included supporting pictorial evidence that was not available until now.

THE CONFUSING CHEMISTRY OF CARBON DIOXIDE

Carbon dioxide chemistry is different from that of other atmospheric gases. It causes a frightful sense of suffocation that warns of its presence when it is present in concentrations as low as 1% (the normal concentration of CO_2 is only 0.03% in atmospheric air). More importantly, carbon dioxide inhibits the binding of oxygen to the hemoglobin molecule that carries oxygen from the lungs to cells deep within the body. This property plays a focal role in the mechanism of oxygen transport and delivery that is essential for vertebrate life, which is described in detail in Chapter 14: Oxygen Transport and Delivery, on page 137. As a result of its odd chemistry, CO_2 can cause asphyxiation (inadequate oxygen delivery to cells and tissues) at much lower concentrations than can nitrogen, argon, hydrogen, methane, and the other gases found in the atmosphere. Nitrogen, for example, constitutes 70% of earth's atmosphere, but it doesn't affect hemoglobin, so it doesn't cause asphyxiation unless its concentration rises above 90%, at which point it dilutes oxygen below 10%, which disrupts oxygen transport and delivery, and causes asphyxiation. By contrast, carbon dioxide concentrations as low as 10% mixed with atmospheric air can cause potentially lethal asphyxiation. Because of these peculiar characteristics, CO_2 can *mimic* both anesthesia (narcosis) and toxicity but is neither anesthetic nor toxic.

THE MYTH OF CO_2 NARCOSIS

The myth of carbon dioxide "narcosis" can be traced to the crude CO_2 experiments of Henry Hill Hickman, an English physician, during the earliest days of gas research in the 1800s, when the ability of carbon dioxide to interfere with hemoglobin was not yet understood. He placed animals in boxes and flooded the boxes with pure carbon dioxide. He also placed animals in closed containers until they consumed all available oxygen and collapsed. Then he performed gruesome surgical procedures on the animals while they remained obtunded. He was surprised to note that minimal blood loss accompanied these procedures. In retrospect it is obvious that the combined stress of surgery and hypoxia exaggerated blood coagulability in the animals, as explained in this book. Most of the animals recovered when they were restored to room air, but several of them died, whereupon Hickman discarded them from his report. He mistakenly concluded that he had discovered an important means to prevent surgical pain. His published reports were harshly ridiculed by his English and French contemporaries, who recognized the nature of gas suffocation. Dr. Hickman died at the age of 30, supposedly of syphilis, but the circumstances suggest suicide because there is no evidence that he suffered any symptoms of syphilis. Regardless, he is now remembered as a "father of anesthesia." He might better be remembered as the first person to demonstrate the ability of CO_2 asphyxiation to mimic anesthesia.

THE MYTH OF CO_2 TOXICITY

The legend of carbon dioxide toxicity probably originated during the earliest days of gas discovery. Carbon dioxide was the first gas isolated from the atmosphere and identified as an independent gas in the early 1600s. It was discovered by a Flemish physician named Jan Baptist van Helmont, who recognized its similarity to "fermenting must," which was known to render the air of caves unbreathable. Atmospheric CO_2 originates primarily from microbe metabolism deep beneath the earth's surface, and its peculiar chemistry causes it to accumulate in caves and mines, so that it was probably blamed for the toxic effects of other noxious gases in English coal mines. Its reputation for toxicity may also have originated from the experiments of Humphry Davy, who was unable to inhale CO_2 concentrations above 30% because of the intense sense of suffocation and laryngospasm it causes.[66] This is easily misinterpreted as toxicity, but it doesn't harm tissues (and so is not *toxic*). In addition, Anton Lavoisier equated the asphyxiating carbon dioxide produced by wood-burning to that produced by animal metabolism. These observations conveyed the impression that CO_2 is toxic, but today we know that it is chemically inert, incapable of damaging tissues, and devoid of toxicity. Its ability to mimic anesthesia and toxicity derives from its unique ability to cause asphyxiation at lower concentrations than other atmospheric gases by inhibiting the binding of oxygen to blood hemoglobin, as explained in Chapter 14.

THE ORIGIN OF MODERN ANESTHESIA

At the beginning of the 20th century, American medical schools were relatively independent of one another and pursued various schools of thought (homeopathy, osteopathy, etc.). American medical students often traveled to Europe to learn the latest medical ideas and techniques. War being the mainspring of medical advance — if only because disease is deadlier than bullets and bombs — the United States prepared for the impending apocalypse of WWI by founding Johns Hopkins Medical School, promoting medical research, and standardizing medical education via the Flexner report. In this fertile environment, modern anesthesia emerged as a distinctly American phenomenon.

THE FIRST ANESTHETIC SAFETY PROBLEM: SPONTANEOUS HYPERVENTILATION

The nervous pathways that stimulate breathing are confusing, poorly understood, and surprisingly fragile and vulnerable to dangerous disruption. There are three sources of respiratory drive that stimulate breathing:

1. **Primary respiratory drive of consciousness.** Specialized acid sensors detect acid levels in the cerebrospinal fluid, which rapidly equilibrates with CO_2 levels in blood. This form of respiratory drive resists hyperventilation and CO_2 depletion as long as consciousness remains intact.
2. **Secondary respiratory drive of chemoreceptors**. Specialized chemoreceptors detect the partial pressure of carbon dioxide in blood and tissues. These chemoreceptors sustain breathing during normal sleep or anesthesia, both of which extinguishes primary respiratory drive. This explains why normal sleep is accompanied by mild hypoxia and hypercarbia. However, the hypercarbia improves cardiac efficiency, tissue perfusion, and oxygen delivery, so that oxygen delivery to cells is safely sustained. Respiratory chemoreceptor activity is paralyzed by hypocarbia, so that hyperventilation during surgery depletes CO_2 tissue reserves and paralyzes respiratory chemoreceptors. Hyperventilated patients appear to breathe normally when their consciousness is restored, even though their chemoreceptors are paralyzed by CO_2 depletion. In this condition, if the patient loses consciousness for any reason, he or she may unexpectedly stop breathing and die.
3. **Nociception**. Specialized tissue disruption sensors in the skin and internal organs generate nervous activity called *nociception*. Nociception is perceived by consciousness as pain. Nociception stimulates breathing even in the absence of consciousness. Thus, surgical nociception can stimulate hyperventilation that depletes CO_2 tissue reserves and paralyzes respiratory chemoreceptors. This causes the same CO_2 depletion problem noted in item 2, wherein patients emerging from anesthesia may suddenly stop breathing if they lose consciousness.

Shallow water blackout syndrome illustrates the subtle relationships of primary and secondary respiratory drive and their vulnerabilities. Every year a few healthy young swimmers accidentally kill themselves by voluntarily hyperventilating themselves in preparation for swimming under water. Hyperventilation cannot increase oxygen reserves in the body, but instead it quickly depletes CO_2 tissue reserves and undermines both primary and secondary respiratory drive, which enables the swimmer to remain under water longer before the urgency to breathe causes him to return above the water surface. However, severe hyperventilation can obtund respiratory drive longer than oxygen reserves can sustain consciousness. The resulting brain hypoxia undermines judgment, so that the unsuspecting swimmer falls asleep and drowns at the bottom of the pool. Similar problems can occur during surgery, when surgical stimulation causes hyperventilation that depletes CO_2 tissue reserves and paralyzes respiratory chemoreceptors.

Ether Hyperventilation Syndrome. Despite its obnoxious smell, ether initially became the most popular anesthetic agent because its pulmonary toxicity stimulates breathing despite overdose and it didn't cause liver failure like chloroform, so that it was perceived as safer than contemporary alternatives. However, its toxic form of safety came at a price. Today it is forgotten that the combination of surgical stimulation plus the toxic stimulation of lung tissue by ether caused hyperventilation during surgery that depleted CO_2 tissue reserves, which obtunded respiratory chemoreceptors and deceptively undermined respiratory drive. As a result, patients recovering from otherwise uneventful ether anesthesia sometimes suffered unexpected respiratory arrest if they fell asleep and extinguished the primary respiratory drive of consciousness.[67] This baffling phenomenon of unexpected postoperative respiratory arrest became the first serious problem to be attributed to anesthesia. In retrospect, the hyperventilation also caused deaths *during* anesthesia (as well as after). Victims were most likely older patients with compromised coronary circulation.

THE SCIENTISTS

During my exploration of old literature, I was impressed that the early world of anesthesia was very different from today. Advances in knowledge and practice proceeded with remarkable speed in the absence of government regulation. Most of the important advances were introduced by a handful of idealistic, scientifically sophisticated physicians and researchers with impressive intelligence and integrity who communicated clearly and left a priceless legacy. Four of them stood out as major contributors to anesthesia advance: George Washington Crile, Yandell Henderson, John Silas Lundy, and Dennis Jackson. Mention of their contributions is assiduously avoided in today's anesthesia literature. They deserve to be remembered.

The nurse-anesthetists embraced the emerging anesthesia science, but they lacked modern machines, monitors, and medications. The problem was, they had no means to measure and monitor CO_2 levels, so they were obliged to rely on crude clinical observations, so that they had no effective warning when CO_2 levels were approaching dangerous elevations.

GEORGE WASHINGTON CRILE

George Washington Crile was famous in his own time and should be remembered as the Father of Anesthesia and one of the most important physicians who ever lived; instead, he is forgotten and his important contributions are abandoned.

Figure 18.3: George Washington Crile (1864-1943)

Crile reviled war, but he served as a physician in both the Spanish American War and WWI. He became a trauma surgeon, built his own animal research laboratory, and performed extensive animal and human clinical experiments. He performed the first successful blood transfusion, invented cardiopulmonary resuscitation, and proved that preventing perioperative fear and supplementing anesthesia with analgesia optimizes surgical outcome. He introduced preoperative visitation to prevent harmful fear, and postoperative visitation to evaluate problems. These proven policies that enhanced anesthesia safety were abolished around 1990 by arbitrary insurance company edict.

Crile recognized the role of nervous activity in disease, and I was astonished to learn that long before the discovery of antibiotics, he cured sepsis and peritonitis using intramuscular morphine that rendered patients unconscious for a week. He understood essentially everything we know today about surgical stress and how to prevent it, with the exception of capillary gate operation. Here is the passage from his classic book called "Anoci-Association" that describes his principle:

ANOCI-ASSOCIATION Principle. General Technique; Morphine and Scopolamine; Nitrous-Oxid Oxygen; Novocain; Quinin and Urea Hydrochloride; Gentle Manipulations; Sharp Dissection.

PRINCIPLE On the Kinetic Theory already enunciated in the foregoing chapters of this book a new principle of operative surgery has been founded, the "paramount object of this new technique (being) to reduce the toxic action of the general anesthetic and the traumatic factor of the operative manipulations to a minimum" (Bloodgood). ***[In other words, Crile understood that the combination of ether toxicity and surgery increased surgical risk.]*** As already stated, every adequate stimulus with or without inhalation anesthesia, whether from trauma or emotion, activates the kinetic system, causing the brain-cells to discharge some of their stored energy — that is to say, the sight of the operating room, the spoken word implying danger, the taking of the anesthetic, the instrumental injury of tissues in the course of the operation, and the traction of the stitches after the operation, all are adequate stimuli. Obviously the only practical method of preventing the consumption of the energy stored in the brain-cells is the development of an operative technique which will exclude from the brain the stimuli of the special senses and the stimuli of common sensation. The principle of anoci-association may be illustrated by the wrecked Titanic. The story of the stress and the psychic strain of the survivors is known, that of the lost may be easily imagined; the future haunting memory of this experience in the minds of the survivors may be safely predicted. Now, if a survivor of this ship had been so skillfully anesthetized in his bed just before the accident that he knew nothing of the impending disaster, and if he then had been gently carried up on deck, lowered into a lifeboat, and taken aboard the rescue ship without being allowed to awaken from his anesthesia until he was safely in bed again in a comfortable state-room, -if then he had been told that he had been transferred from the sinking ship, but that now he was safe and would soon see his home , he would have passed through the accident in a state of anoci-association. Is there a single anesthetic that will exclude all nocuous or harmful physical and psychic stimuli from the brain? By blocking nerve conduction local anesthetics protect the brain from the effects of local operative injury, but they do not protect the brain against destructive psychic strain; inhalation anesthetics exclude the psychic stimulation of the brain-cells, but do not exclude the operative stimulation; and general anesthetics introduced hypodermically, being uncontrollable, are excluded on principle. Each anesthetic covers a part of the field, but there is no single agent that alone can produce anoci-association, which is the goal of operative surgery. We, therefore, do not advocate ether alone, nor chloroform alone, nor nitrous-oxide-oxygen alone; we do not advocate local anesthesia alone, nor morphine and scopolamine alone, nor spinal anesthesia alone, but through selection and combination of anesthetics we aim to attain the anesthesia that in the case in hand will exclude all stimuli from the brain, and thereby attain anoci-association.

Crile introduced the principle of narcotic supplementation of general anesthesia to optimize surgical outcome, as described in this passage from "Anoci-Association":

> GENERAL TECHNIQUE MORPHIN AND SCOPOLAMIN To mitigate the preoperative dread and to facilitate the induction of anesthesia, a solacing dose of morphine and scopolamine (usually morphine, a gr., Scopolamine, 150 gr.) is given an hour before the operation to all patients excepting the aged, the very young, and those whose feeble condition contraindicates the use of these narcotics. The use of morphine serves the double purpose of diminishing the pre-operative psychic strain and of actually preventing, to some extent, the damage to the organs of the kinetic system by the trauma of the operation. Laboratory experiments have shown that in morphinized animals subjected to trauma, the changes in the cells of the brain, the suprarenals, and the liver are less than in traumatized animals without this protection. That deep morphinization will almost completely prevent shock has been abundantly proved in both the laboratory and the clinic. ***[In other words, Crile understood the dose-dependent ability of narcotic supplementation to improve surgical outcome 100 years ago.]*** A striking confirmation of this fact is found in the following case: The patient was a woman 46 years of age with obstruction of the bowels from a large solid tumor which occupied the entire abdomen and caused distention greater even than pregnancy at term. This tumor had been explored six months previously by a skillful colleague who had wisely judged the case to be inoperable, the urgent condition of obstruction not being present at that time. When the patient came to us, she was weak, emaciated and was running a slight daily fever. Under anoci-association the abdomen was opened and a careful search was made for the obstruction. The widespread adhesions and the actual inclusions of the viscera in the growth soon showed us that we could not relieve the obstruction without attempting the apparently impossible removal of the tumor. When this decision was reached, the pulse had mounted to 142 and the respiration to 38. The completion of the operation would mean resection of the bladder and of part of the small intestine. Nerve blocking in these deep fused structures was impossible. Transfusion would be insufficient to relieve the situation. Death was certain if we retreated and apparently equally certain if we proceeded. In no case could a new resource be more urgently needed. It was at once decided to apply our laboratory findings regarding the protective effects of morphine. The operation was interrupted — the patient being kept under the lightest possible nitrous-oxide-oxygen anesthesia — while morphine was given in quarter-grain doses at ten-minute intervals until the respiration had fallen from 32 to 12 per minute. We then resumed the tedious and difficult task of separating the huge, solid, adherent, semi-malignant tumor by means of sharp dissection. To accomplish this separation, it was necessary to resect a portion of the small intestine, the entire fundus of the bladder and several inches of the right ureter, the proximal end of which was implanted into the back by Bottomley's method. The tumor had been unable to gain a sufficient blood supply through its original pedicle and had, therefore, made

> many vascular connections with the neighboring organs and consequently much blood was unavoidably lost during the operation. To counteract the resultant anemia a transfusion of blood from the patient's sister was given. At the close of the operation and the transfusion, the respiration ranged from 6 to 10 per minute and the pulse rate was 134. The patient was placed in bed in a modified Fowler position and sodium bicarbonate and glucose were given immediately by the drop method per rectum, while sufficient morphine was given in repeated doses during the first 24 hours to hold the respiration to 12 per minute. The patient made a splendid recovery from the operation. The protective effect of morphine is remarkably exhibited also in those cases of exophthalmic goiter in which some exceptional local condition causes a break in the complete anoci-association of the patient, as a consequence of which the pulse and respiration increase markedly during or after the operation. In these cases, if morphine be given in repeated doses until the respiration and pulse are held stationary or fall, the dangerous exhaustion of the patient will be avoided. The morphine may be given at any time during or after the operation when it is seen that the patient's energy is being expended at too rapid a rate. ***[In other words, Crile recognized 100 years ago that additional narcotic dosage must be administered as needed during surgery to maintain respiratory rate between 8–12 breaths per minute to control surgical nociception and optimize surgical outcome.]*** Morphine is especially useful also in those cases of acute infection in which emergency operations must be performed. In such cases morphine affords a double protection — it protects the brain against both the infection and the operative trauma, the effects of which are increased, because during the activations of a toxin, the brain thresholds are greatly lowered. Here also morphine should not be given in one dose, but in repeated doses until the physiological effect is produced. This point will be indicated by the reduction of the respiration to the normal rate or less. In brief, by proper use, morphine to a large extent controls the metabolic processes. It should be added that it is not our intention to suggest an increase in the use of morphine in average cases, but to emphasize its usefulness when employed in physiologic dosage in certain exceptional cases. ***[This patient would have suffered excruciating pain, delirium, fever, and death within 24 hours after this extensive procedure were it not for the protective effects of morphine.]***

Crile hated the toxic odor of ether and used nitrous oxide supplemented with scopolamine, morphine, and prilocaine injections to prevent harmful surgical nociception, hyperventilation, and muscle tension. His findings are summarized in his book called "Anoci-Association" that should be "must reading" for all medical professionals.[68] It is available free of charge on the Internet.

Despite his distaste for war, Crile and his nurse-anesthetist, Agatha Hudgens, served in the American medical corps in WWI, and shared their knowledge with European physicians and nurses. American physicians being in short supply after WWI, Crile helped Agatha Hudgens establish the first school of nurse anesthesia. The nurses embraced the scientific principles of Crile, Henderson, and Lundy, and became famous for their excellent results. They dominated anesthesia practice in the United States until the establishment of the MD anesthesiology profession, which reduced them to subservient status.

Crile and his adoring wife are buried with unobtrusive markers not far from John D. Rockefeller's gigantic phallic monument in Cleveland's famous Lakeside Cemetery. Agatha Hudgens never married and is buried in Chatham, MA. Money talks, while merit walks.

Figure 18.4: Crile gravestones.

The gravestones of George Washington Crile, and his adoring wife in Lakeside Cemetery in Cleveland, Ohio. Crile's nurse-anesthetist, Agatha Hudgens (1877-1945), founded the first school of nurse anesthesia.

YANDELL HENDERSON

Dr. Yandell Henderson was the PhD Director of the Yale Laboratory of Applied Physiology. He led a team of international experts who studied respiratory physiology at the top of Pike's Peak in 1910 and he was considered the world's foremost expert on gas toxicity and pathophysiology. His sojourn at Yale probably accounts for the enduring eminence of its medical school.[69]

Figure 18.5: Yandell Henderson, PhD (1873-1944)

Yandell Henderson, PhD was considered the world's foremost expert in gas pathophysiology at the turn of the previous century. He was an idealist who believed that science should serve the public interest, and he opposed American involvement in WWI. He testified before congress to protest the addition of toxic tetraethyl lead to gasoline.[70] He was ignored at the time, but lead was quietly removed from gasoline many years later. This picture shows him in 1932 testifying before a U.S. Senate subcommittee that beers containing 3-4% alcohol by volume were not intoxicating. Credit: Wellcome Collection. Attribution 4.0 International (CC BY 4.0) [https://wellcomecollection.org/works/p4jat3sy and https://creativecommons.org/licenses/by/4.0/]

Dr. Henderson investigated the mysterious postoperative respiratory arrests and recommended that patients breathe a mixture of 5% CO_2 and 95% oxygen (Carbogen) during surgery to maintain the normal concentration of 5% CO_2 in the lung, and thereby prevent depletion of CO_2 tissue reserves. This eliminated the problem of postoperative respiratory arrest. In addition, it sharply reduced the risk of death during surgery, which had been largely overlooked.[71, 72] With the help of today's understanding of stress theory, the simplest explanation of these intraoperative deaths is that spontaneous hyperventilation during surgery depleted CO_2 tissue reserves, which closed the capillary gate, increased cardiac work, and decreased oxygen delivery to the heart and other organs. Most of the intra-operative deaths were undoubtedly caused by heart attacks in patients with pre-existing compromise in their coronary circulation. Similarly, oxygen starvation in bowel tissue is the reason patients suffer nausea, vomiting, and bowel ileus after surgery.

Henderson subsequently demonstrated that CO_2 supplementation prevents postoperative nausea, vomiting, bowel ileus, asthma, atelectasis, pneumonia, and laryngospasm, and that it provides a powerful treatment for asthma, atelectasis, pneumonia, drowning, smoke inhalation, heart attacks, strokes, alcohol inebriation, carbon monoxide poisoning, and breathing problems in newborn babies.

Henderson was known for his practicality. For example, in one of his most important experiments he devised a cardiometer by enclosing a dog's heart within an airtight toy rubber ball that he connected to a tobacco can lid using a thick rubber tube.[73] With this makeshift cardiometer he was able to demonstrate that CO_2 supplementation enhances venous return and cardiac output. However, due to the relatively primitive understanding of basic medical science during that era, he could not have understood that CO_2 increases venous return and cardiac output by opening the capillary gate and decreasing microvascular flow resistance. Instead, he speculated that "muscle tonus" somehow accounted for the increased venous blood return to the heart. In addition, he could not have appreciated that his rubber ball substituted for the tough pericardial sac that prevents the heart from becoming "overdistended" and damaged by excessive venous return. Even today this critical function of the pericardium is often overlooked. Like all muscle tissues, heart muscle can easily be "overstretched," causing permanent damage. This understanding of muscle pathophysiology was unclear until the era of stress research many years later; cardiac surgeons are only now realizing that they can palliate congestive heart failure by enclosing the failing heart in a cloth sack that prevents overdistension of the heart muscle. As we shall see momentarily, this lack of appreciation for the function of the pericardium contributed to CO_2 confusion in the study of Boniface and Brown some thirty years later.

Because of the stubborn, widespread conviction of CO_2 toxicity, Henderson made little headway in his efforts to persuade physicians to accept CO_2 as a treatment until John Lundy, a prominent physician at the Mayo Clinic, demonstrated that CO_2 supplementation accelerates both ether induction and emergence.[74] The nurse-anesthetists who dominated anesthesia practice combined Lundy's report with the research of Crile and Henderson using Dennis Jackson's revolutionary anesthesia machines and became famous for their excellent anesthetic outcomes. The nurses thus employed a synergistic, complimentary combination of general anesthesia, hypercarbia, and narcotic supplementation that minimized surgical stress and optimized surgical outcome. Their success belatedly gained the attention of physicians, whereupon Carbogen (pressurized mixtures of oxygen and carbon dioxide contained in tanks) gained acceptance as a treatment and became commonplace on fire trucks and in hospitals and medical offices. It saved countless lives (note the pictures and descriptions of Carbogen equipment located at the conclusion of this appendix).[75, 76, 77, 78]

JOHN SILAS LUNDY

Dr. John Silas Lundy was chairman of the anesthesia department at the Mayo Clinic. He developed an anesthesia training program for physicians and produced numerous papers and a textbook of anesthesia. He introduced intravenous barbiturate induction, mechanical ventilators, ventilator vests, oxygen tents, nasal oxygen, and the concept of *balanced anesthesia* (morphine supplementation of ether anesthesia). As mentioned before, he demonstrated that CO_2 supplementation accelerates ether induction and emergence. He developed anesthesia machines, and needles and syringes for spinal anesthesia. He established the first blood bank, the first anatomy lab, and the first recovery room. He produced numerous papers and a textbook of anesthesia. He established an anesthesia training program that was attended by Dr. Ralph Waters, who later established the anesthesiology profession.

Figure 18.6: John Silas Lundy (1894-1973)

DENNIS JACKSON

Dr. Dennis Jackson was a medical physician with a PhD in pharmacology who sympathized with patients who could not afford the high price of medical gases, which were wasted by "non-rebreathing" equipment (such devices were still available for residents to use when I trained at UCLA; and Jackson's machines were clearly superior).[79] Jackson devised the first *circle system* anesthesia machines that employed an efficient "re-breathing" circuit to prevent waste.[80, 81] His machines enabled positive-pressure respiratory support and minimized the risk of ether explosions in the operating room. His "closed circuit" machines quickly gained universal acceptance and became the precursors of today's "open circuit" anesthesia machines.

Figure 18.7: Dennis Jackson (1878-1980)
Courtesy Wood Library Museum of Anesthesiology History

Like nearly everyone, Jackson assumed that carbon dioxide was dangerous, so his early machines incorporated milk bottles filled with sodium hydroxide solutions that turned white as they absorbed carbon dioxide from the breathing circuit. They necessitated electric motors that circulated gases through the sodium hydroxide liquid. The motors generated sparks that invited devastating ether explosions, as Jackson acknowledged in his memoirs:

> I never had any explosions or fires in any of my work. No doubt God stood by sometimes and put out my fires or stopped my explosions before they occurred. But I was always very cautious and careful. There was always some element of danger in pumps, electric motors and oil. But wherever there is ether or explosive mixtures of gases there will always be some danger.[82]

The milk bottles were soon supplanted by soda lime (small rocks coated with sodium hydroxide), which eliminated the electric motor.

Figure 18.8: Closed-circuit machine

This image shows an early version of Jackson's closed circuit anesthesia machine that utilized bottles filled with sodium hydroxide to absorb carbon dioxide. This necessitated dangerous electric motors to pump gases through the liquid soda lime.

THE CRILE/HENDERSON ANESTHESIA TECHNIQUE

The nurse-anesthetists faced forgotten problems. They lacked modern monitoring, machines, and medications. Ether was smelly, toxic, slow, and explosive. Morphine was plagued by itching, nausea, and prolonged respiratory depression. Capnography and pulse oximetry didn't arrive until decades later. Primitive needle technology prevented intravenous access. Endotracheal intubation was difficult, so they managed most cases by mask with spontaneous breathing, but mask management was impossible in prone position, and unsatisfactory for dental and ENT procedures. Vomiting, aspiration, airway obstruction, and laryngospasm were constant threats. However, they possessed superior scientific knowledge. They combined the science of Crile, Henderson, and Lundy with Jackson's anesthesia machines to create what I call the *Crile/Henderson* anesthesia technique. They utilized pre-medication with morphine and scopolamine to prevent preoperative pain and fear and obtund the smell of ether. Mask induction without paralysis avoided unexpected anatomical airway problems.[83] They used CO_2 supplementation to accelerate ether induction and emergence, which was extremely slow compared to modern inhalation agents. During surgery, they used *balanced anesthesia* (intramuscular morphine supplementation) to prevent the surgical nociception that caused morbidity, mortality, untoward muscle tension, spontaneous hyperventilation, heart attacks, bowel ileus, and other manifestations of surgical stress. They added *bypass valves*[84] to Jackson's anesthesia machines to isolate soda lime from the breathing circuit and facilitate the CO_2 supplementation to accelerate ether induction and emergence; prevent ether explosions and fires; counteract morphine respiratory depression; optimize cardiac output, tissue oxygenation, and organ protection; prevent infarction; and minimize postoperative pain, nausea, vomiting, bowel ileus, asthma, atelectasis, pneumonia, laryngospasm, delirium, dementia, dysrhythmias, fever, tachycardia, and hypertension.

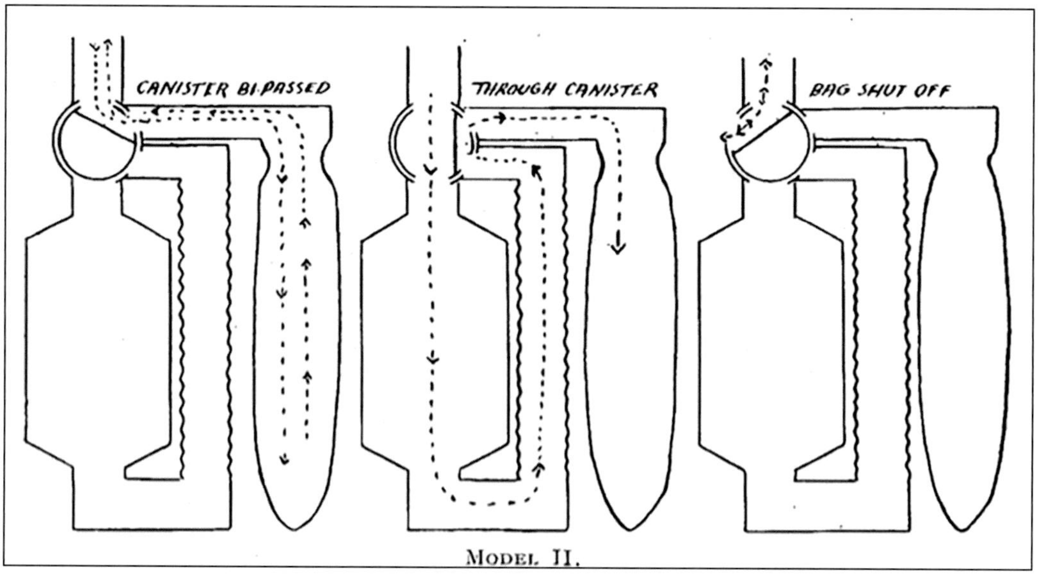

Figure 18.9: Bypass valve
Diagram of a CO_2 Bypass Valve. These valves were installed on Jackson's anesthesia machines to temporarily bypass the soda lime canister to enable CO_2 accumulation in the inhaled gas mixture. Auxiliary CO_2 supplementation systems were also added to Jackson's machines to elevate CO_2 concentrations in the inhaled gas mixture. Courtesy Wood Library Museum of Anesthesiology History

SHORTCOMINGS OF THE CRILE/HENDERSON TECHNIQUE

The balanced anesthesia technique utilized by the nurses produced excellent outcomes but suffered significant limitations. Mask management could not "secure the airway" to prevent airway obstruction, laryngospasm, and aspiration, or enable mechanical ventilation, prone position, or dental and ENT procedures.

By far the most alarming problem was unexpected convulsions during surgery. Though these events were uncommon, their mortality approached 20%. Worse yet, most of the victims were children.[85] At first these problems were mysterious, but the role of overenthusiastic CO_2 supplementation soon became obvious.[86, 87] At first, the convulsions were thought to be caused by respiratory acidosis induced by carbon dioxide.[88] Today we understand that acidosis *per se* is harmless, but CO_2 causes asphyxiation at much lower concentrations than other atmospheric gases, and children are more susceptible to oxygen starvation than adults on account of their elevated metabolism (see Chapter 14: Oxygen Transport and Delivery, on page 137). However, these nuances were not obvious in that earlier time. The asphyxiation events had an unnerving "Dr. Jekyll/Mr. Hyde" characteristic. CO_2 supplementation during anesthesia would improve tissue oxygenation and cardiac output up to a critical threshold, whereupon it suddenly disrupted oxygen transport and caused cyanosis, fever, and hypoxic brain convulsions. Without capnography there was no means of monitoring and measuring CO_2 concentrations in the breathing circuit, and therefore no warning. As we shall see, this was the "Achilles Heel" of the nurses that led to the undoing of their profession.[89]

THE POLITICIANS

Power, politics, and privilege prevail over all forms of human activity, and medicine is no exception. While Crile, Henderson, Jackson, and Lundy toiled to clarify the nature of anesthesia, analgesia, gas pathophysiology, and practical anesthesia machinery, others focused their efforts on practical politics.

FRANCIS HOEFFER MCMECHAN

Dr. Francis Hoeffer McMechan founded and edited the "Yellow Journal" (*Anesthesia and Analgesia*) and became one of the most influential figures in the creation of the anesthesiology profession. He excelled as a student, and his physician father allowed him to anesthetize patients during emergencies. He obtained his master's degree at age 18 but because he was too young to enter medical school, he became a reporter for the Cincinnati Press, where he learned to write terse, effective prose, which prepared him for his subsequent career as an editor. At age 21 he entered the Medical College of Cincinnati and graduated in three years rather than the usual four and passed the newly introduced Medical Board Examination. He then pursued a career in anesthesia and was soon considered an expert. Soon after he became self-supporting, he met and married Laurette van Varseveld. Within 18 months of his marriage, he developed devastating rheumatoid arthritis that left him wheelchair-bound, with his joints fused into a sitting position. Laurette became his enabler and provided for his every need. Thus, unable to pursue his career in anesthesia practice, he devoted the remainder of his life to establishing professional anesthesiology.[90]

Figure 18.10: Dr. Francis Hoeffer McMechan (1879-1939) and Mrs. McMechan

McMechan was aghast that nurses dominated anesthesia service: He was adamant that only physicians should administer anesthesia. He was staunchly opposed to the administration of anesthesia by nonphysicians. Addressing a joint session of the Council on Medical Education and Hospitals of the AMA and the Federation of State Medical Licensing Boards in 1935 about the necessity of licensing anesthesiologists to practice medicine, he wrote:

> Anesthesia was the gift of pioneer doctors and dentists to suffering humanity, and every significant advance in its science and practice has been contributed by doctors, dentists, and research workers of similar standing. In contrast, technicians have added nothing of any consequence. Anesthetics are among the most potent and dangerous drugs used in the practice of medicine; they penetrate to every cell and organ of the body and may cause almost instant or delayed death by their toxic effects. Dosage of general inhalation anesthetics cannot be prescribed in advance but must be

> determined from moment to moment during administration. The dosage of local and other anesthetics must be determined by the risk of the patient, the nature and duration of the operation to be done — certainly a challenge to the knowledge and experience of the keenest doctor. No patient should ever be given an anesthetic whose condition and risk has not been diagnosed in advance of the operation, so that every resource of medical science can be used to lessen the risk and make the recovery more reassuring. Certainly, in this preoperative evaluation and the more selection of the safest anesthetic and best method of administration, the medical anesthetist is more in a position to act as a consultant than a technician. The safety of the patient demands that the anesthetist be able to treat every complication that may arise from the anesthetic itself by the use of methods of treatment that may be indicated. The medical anesthetist can do this, the technician cannot. More recent developments have extended the field of medical anesthesia to include resuscitation, oxygen therapy and therapeutic nerve block for injectable pain and treatment of various conditions of disease and the rehabilitation of the disabled — all fields of practice quite beyond the capacity of the technician.

During the 1920s Dr. McMechan engaged in several actions promoting anesthesia administration by physicians only. For example, in 1925 he supported efforts by John Henry Evans to remove nurse anesthetists from the Buffalo, NY teaching hospitals. He also tried unsuccessfully to have the AMA require physician anesthesia as a criterion for rating hospitals.

Dr. McMechan's advocacy of physician-only anesthesia encountered powerful opposition. Hospitals found that hiring nurses was financially advantageous, for they could charge an anesthetic fee to the patient for the nurses' services. Revenue from these fees was far in excess of the costs of the nurse. Some surgeons engaged in this same practice, hiring nurses and charging anesthetic fees well in excess of the nurses' salary. Even general practitioners opposed his efforts, because in return for referring cases to a surgeon, they were allowed to administer anesthesia and then bill for their services.[91]

Dr. McMechan worked with Dr. Ralph Waters, Dr. John Lundy, and others to pave the path for the creation of the anesthesiology specialty but died shortly before his dream was realized. Some believed that the AMA hesitated to sanction the certification of anesthesiology as an independent profession while he was alive for fear that he would control it.[92]

RALPH WATERS AND CHAUNCEY LEAKE

The facts I have discovered have driven me to the conclusion that Dr. Chauncey Leake colluded with Dr. Ralph Waters to fabricate deceptive animal experiments that deliberately confused CO_2 asphyxiation with inhalation anesthesia to discredit the nurse anesthesia profession and pave the path for founding the anesthesiology profession. This is a sobering allegation, but considering the mayhem caused by their mischief, readers are encouraged to examine the following evidence with care.

Figure 18.11: Dr. Chauncey Leake (L) (1896-1978).
Dr. Leake is shown with Dr. Morris Fishbein, the controversial editor of the Journal of the American Medical Association. Source: National Institutes of Health, U.S. National Library of Medicine, public domain

"Every violation of truth is not only a sort of suicide in the liar, but is a stab at the health of human society."
—Ralph Waldo Emerson

Dr. Chauncey Leake was awarded bachelor's degrees in biology, chemistry, and philosophy in questionable circumstances by Princeton University.[93] He obtained a PhD at the University of Wisconsin, became its Chairman of Pharmacology, and led a team of researchers that studied toxic war gases during WWI. Immediately after completing his experiments with Dr. Waters, Dr. Leake departed from Wisconsin to accept a series of prestigious academic positions at UC San Francisco, the University of Texas at Galveston, and the Ohio State University before returning to UC San Francisco to supervise research. Though his career consisted of prestigious appointments, awards, and accolades, he seems to have left no record of research other than his animal experiments with Dr. Waters.

Soon after Dr. Leake's return to UC San Francisco, a team of researchers including Drs. Edmund Eger, Larry Seidman, and John Severinghaus (the son of Elmer Severinghaus, a biochemist who had been a member of Leake's research team at Wisconsin) began a series of animal experiments that promoted the CO_2 hoax. Their experiments were suspiciously similar to the EARLIER reports of Drs. Leake and Waters, because they also confused CO_2 asphyxiation with anesthesia.

While Dr. Waters preached professionalism, integrity, and camaraderie, Dr. Leake was gracious, generous, and never uttered an unkind word. In 1940 and again in 1954 he served as vice-president, member of the Board of Directors, and then was elected president of the American Association for the Advancement of Science (AAAS) in 1959. On that occasion he was described as "...a pharmacologist, chemist, physiologist, administrator, writer, editor, historian, philosopher, and orator. Above all, he is a humanist and catalyst with a fine intuition with a fine intuition for what is truly important and tolerant disregard for dull trivia. For him, science is a product of the human mind, inseparable from considerations of human ethics, human politics, and human aspirations, and dependent upon untrammeled communication of ideas." He signed the Humanist Manifesto in 1973 and was awarded the UCSF medal in 1975. What did he do to merit such praise?

The Devil speaks with a silver tongue; birds of a feather flock together; and actions speak louder than words.

Figure 18.12: Dr. Ralph Waters (1883-1979)

There are horrible people who, instead of solving a problem, tangle it up and make it harder to solve for anyone who wants to deal with it. Whoever does not know how to hit the nail on the head should be asked not to hit it at all.
— Friedrich Nietzsche

Dr. Ralph Waters was the revered founder of the Anesthesiology profession, but his fame is founded on malignant mythology rather than beneficial science. After medical school at Western Reserve University in Cleveland he specialized in obstetric anesthesia. He left private practice for three years to study regional anesthesia (spinal or epidural anesthesia) with Dr. John Lundy at the Mayo Clinic. Afterward he demonstrated the utility of Arthur Guedel's endotracheal tubes at medical conventions by plunging his pet dog "Airway" into a tank of water.[94] He also introduced cyclopropane and soda lime, and patented his profitable *Waters Canister* that absorbed CO_2 from anesthesia circuits (see Figure 18.20 on page 352).

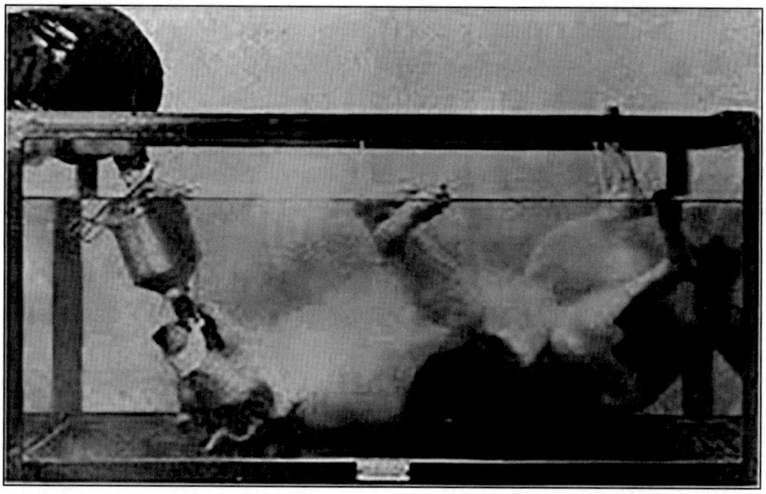

Figure 18.13: "Airway"
A photograph of "Airway," the dog that Dr. Waters famously submerged in water to demonstrate Guedel's endotracheal tubes.[95]

Dr. Waters revealed his determination to establish anesthesiology as an independent medical specialty in 1919, when he published an essay entitled "Why the Professional Anesthesiologist?"[96] In 1927 Dr. Charles Bardeen, the Dean of the Medical School at the University of Wisconsin, selected Dr. Waters to become the first chairman of a university anesthesia department in the world. This conferred considerable prestige, which Waters exploited to establish the anesthesiology profession.

Dr. Bardeen himself was the son of a wealthy New York Publisher. He was the first graduate from Johns Hopkins Medical School in 1897, and he was selected to become dean of the Wisconsin medical school in 1907. He later said, "Ralph Waters was the first person the university hired to put people to sleep, but, instead, he awakened a worldwide interest in anesthesia."

Dr. Waters always insisted that he was attracted to Wisconsin by its pharmacology department. Even before his arrival at Wisconsin, he exchanged curiously effusive personal letters with Dr. Chauncey Leake, the Chairman of the Pharmacology Department at Wisconsin, that suggested an intimate camaraderie between the two men, as if they had been childhood companions. Immediately after Dr. Waters arrived at Wisconsin, they performed animal experiments that deliberately confused CO_2 asphyxiation with general anesthesia to fabricate false evidence that carbon dioxide has anesthetic properties.[97, 98]

Meanwhile, Dr. Waters expressed alarm about the CO_2 asphyxiation disasters. He also had a personal stake in promoting his patented Waters canister that absorbed CO_2 from anesthesia breathing circuits (see pictures at the end of this appendix) The sales of his cannister were implicitly threatened by the increasing acceptance of CO_2 supplementation at that time. Whatever his motives, he began publishing papers excoriating CO_2 dangers immediately after he arrived at Wisconsin. These fell on fertile ground, and they tarnished the reputation of the nurse-anesthetists, because many physicians and patients still believed that carbon dioxide was dangerous despite Henderson's research that had made Carbogen and CO_2 treatments commonplace (see the photographs of Carbogen equipment following).[99, 100, 101, 102, 103, 104, 105, 106] Waters' papers were blatantly biased. He persistently characterized CO_2 as "toxic waste, like urine" that must be "rid from the body," and ignored its proven therapeutic properties. He fabricated imaginary events to further his objectives, as well as deliberately misrepresenting the results of his animal experiments with Dr. Leake.

The First Paper: "CARBON DIOXIDE: ITS PLACE IN ANESTHESIA" (1927)[107]

This paper was published immediately after Dr. Waters became Chairman at Wisconsin. He began by citing Hickman's "careful animal experiments" as "proof" that CO_2 has anesthetic effects.[108, 109, 110] He ignored the harsh criticisms of Hickman's British and French contemporaries. Hickman's use of carbon dioxide might have been reasonable within the context of the limited knowledge in his time (the 1820s), but the same cannot be said of Dr. Waters a century later, who undoubtedly understood the difference between anesthesia and asphyxiation.[111] Dr. Waters incorrectly attributed tachycardia, hypertension, and blood hypercoagulability to CO_2 toxicity, but these effects are caused by the alarming sensation of suffocation caused by CO_2, which activates sympathetic nervous activity. Dr. Waters cited public demonstrations of "CO_2 anesthesia" as evidence of CO_2 narcotic effects, but as a trained physician, he surely knew that CO_2 asphyxiation caused these effects. He incorrectly attributed corn silo poisoning to CO_2 toxicity, but corn silo poisoning (asphyxiation caused by the accumulation of concentrated methane in silos as a by-product of microbial activity) is caused by methane rather than CO_2. Finally, he attacked Yandell Henderson's CO_2 research:

> Fortunately, and unfortunately, the researches in carbon dioxide of the New Haven school of physiologists (i.e., Yandell Henderson) have entered the literature of anesthesia in the past decade: fortunately because these researches are brilliant and have taught us many facts of the greatest importance; unfortunately, because the facts have been made use of by practical anesthetists in some quarters to make easy their work without sufficient investigation as to what by-effects may accompany the altered physiology which their methods induce. ... Let us not allow the spectacular results of "stunt flying" in anesthesia to make us forget ... sound principles.

> The mere fact that anesthesia can be more quickly and conveniently induced and terminated by the use of carbon dioxide is not sufficient ground for administering that gas. ***[Elisions by the author. This vague and contradictory condemnation of carbon dioxide supplementation as "stunt flying" ignored its important benefits, including its ability to counteract narcotic respiratory depression, prevent depletion of CO_2 tissue reserves, optimize cardiac output and tissue oxygenation, prevent infarction, nausea, vomiting, asthma, atelectasis, pneumonia, and so forth.]***

At the end of his paper, Dr. Waters offered the following conclusions:

> 1. Changing the amount of carbon dioxide inspired by sick patients is interfering with fundamental physiology and should not be undertaken lightly. ***[He didn't detail the "fundamental physiology."]*** 2. The use of carbon dioxide by the anesthetist probably can be put on a rational basis. ***[He failed to explain "rational basis."]*** 3. Until more knowledge is available, conservatism in the use of carbon dioxide by the anesthetist should be practiced." ***[Caution may be wise in the absence of knowledge, but there was abundant proof that CO_2 benefits justified its use.]***

This passage is deceitful, because Dr. Waters knew full well of the nature of asphyxiation, as reflected by the cunning design of his animal experiments.

The Second Paper: "THE ANESTHETIC PROPERTIES OF CARBON DIOXIDE" (1929)[112]

Soon after he arrived at Wisconsin, Dr. Waters performed animal experiments with the help of Dr. Chauncey Leake, the Chairman of the Pharmacology Department. These experiments confused CO_2 asphyxiation with general anesthesia.[113] The experiments were similar to Hickman's experiments of 100 years earlier, except that they asphyxiated rabbits and dogs with various mixtures of Carbogen (mixtures of CO_2 and oxygen in pressurized containers) instead of asphyxiating the animals with pure CO_2 or suffocating them in closed containers. The published description of their experiment was vague and sorely lacking in detail, but Dr. Leake confirmed their use of Carbogen on at least two later occasions (see below). They deceitfully made no mention of the possibility of CO_2 asphyxiation, and ignored the fact that animal pigment and fur obscures hypoxic cyanosis — you cannot observe the skin. They began their report as follows:

> Aside from Hickman's effort a century ago, and the neglected records of B.W. Richardson, ***[Waters failed to provide a proper reference for these records. When I searched for Richardson's reference, the only mention that I could find was in a paper published in 1925 by — guess who? — Dr. Chauncey Leake, in which Leake confirmed their obscurity!![114]]*** little significance has been reported on the anesthetic powers of carbon dioxide. ***[On the contrary, the ability of CO_2 asphyxiation to mimic general anesthesia was well known to science, and commonly used for fake anesthesia machine demonstrations***

> ***as well as used in circus side shows that faked general anesthesia with CO_2 using audience "shills."]*** We have found that 30 to 50 percent carbon dioxide with oxygen will anesthetize rabbits in about a minute's time, without apparently causing much excitement, and usually without struggling. ***[This is true ONLY if the CO_2 is supplemented with pure oxygen, as in the case of Carbogen. CO_2 concentrations of 10% mixed with ordinary air causes reversible hypoxia that mimics anesthesia, and CO_2 levels only slightly higher than 10% in atmospheric air cause lethal asphyxiation.]*** Placed in chambers to which such a mixture is supplied, these animals (rabbits) will move about sniffing and breathing more deeply than usual. Muscular activity is gradually depressed and, when anesthesia comes on, muscular relaxation is marked. At this time the respiration is much deeper, but the rate is about the same as normally. Upon removal from the chamber into a normal atmosphere, recovery is prompt and apparently without after-effect. ***[Mild hypoxic brain damage is easily overlooked in both rabbits and humans.]***

Next, the report switched from rabbits to dogs, which is confusing because of species differences. The dogs displayed greater distress than the rabbits when they were exposed to CO_2. The remainder of the report was founded on dog observations. Their report is bereft of detail, but they apparently exposed the dogs to various concentrations of Carbogen (pressurized mixtures of oxygen and carbon dioxide in tanks), which was readily available in that era. The oxygen in Carbogen counteracts CO_2 asphyxiation, so that animals and humans can tolerate higher concentrations of CO_2.

The sense of suffocation caused by 10–20 percent Carbogen concentrations explains the agitation and hypertension observed in the dogs. Thirty percent Carbogen caused asphyxiation that lowered brain hypoxia below a critical threshold, obtunded brain function, and masqueraded as general anesthesia. Unsurprisingly, the asphyxiation was often accompanied by hypoxic brain convulsions and deaths. This hypoxic condition enabled surgical exposure of the intestines. They further reported that 40 percent Carbogen severely depressed cardiorespiratory function, and that Carbogen concentrations above 80 percent caused lethal hypoxia that manifested as a sudden rise of blood pressure, slowing of the heart, and stimulation of breathing that was soon followed by a drastic drop in blood pressure, cardiorespiratory collapse, and death. In contrast, genuine general anesthesia is pleasant, euphoric, and addictive. It causes a gradual, dose-related loss of consciousness that is accompanied by a dose-related decline in blood pressure and pulse rate but does not cause asphyxiation or hypoxia. Dr. Waters had extensive previous experience anesthetizing his pet dog, Airway, with ether to demonstrate Guedel's endotracheal tubes, so he surely knew the difference between anesthesia and CO_2 asphyxiation.

Leake and Waters failed to acknowledge that asphyxiation is counteracted by elevated oxygen concentrations. Without such oxygen supplementation, CO_2 mixed with room air in concentrations as low as 10% causes asphyxiation that mimics general anesthesia.[115, 116] Many years later Dr. Leakey tacitly confessed this in an interview at UC San Francisco:

> Well, in 1828, Henry Hill Hickman, in England, had reported on the anesthetic properties of carbon dioxide. Nobody had ever looked at it since that time, so Ralph Waters and I thought, well, we ought to take a look at it, and see if he was right. ***[Presumably Dr. Leake didn't bother to investigate Hickman's original reports or the criticisms of his contemporaries.]*** So, a hundred years later, 1928, we did put out a report on carbon dioxide as an anesthetic agent. It is 30% carbon dioxide and seventy percent oxygen, ***[Here Leake confirmed that they used Carbogen (CO_2 mixed with oxygen) but he cunningly overlooked the role of oxygen in counteracting CO_2 asphyxiation.]*** so there is no possibility of asphyxiation; that is anesthetic. ***[This is "smoking gun" evidence that Leake and Waters knew full well that the oxygen in Carbogen counteracted CO_2 asphyxiation. Note how he emphasized the notion that "there is no possibility of asphyxiation." If CO_2 had anesthetic properties, it would not have needed oxygen enrichment; without oxygen, 30% CO_2 would have caused lethal asphyxiation. Leake was a PhD in pharmacology, and Waters was a trained MD physician. Both surely knew that CO_2 inhibits the binding of hemoglobin to oxygen and causes asphyxiation at lower concentrations than do other atmospheric gases! This was fundamental scientific knowledge that was well understood in their time. This is damning evidence that their experiment was a cunning fraud, and by implication both were nefarious prevaricators.]*** One can maintain, as we did, animals, dogs, rabbits, and so on, eight and ten hours in carbon dioxide anesthesia without any difficulty at all. There is always a little trouble when they go over, a little neck jerking and so on. ***[No wonder: they were suffering hypoxic brain convulsions.]*** We tried it in humans but decided not to use it because it upset the surgeons, and all. ***[There is no known evidence that either of them performed such human trials, and they offered none.]*** The animals' neck muscles would twitch; but that would be over pretty quickly. We never had guts enough to hold it in humans. This would be fully physiological: I haven't any idea how on earth it works. ***[Leake was a PhD expert in gas chemistry, and yet he claims that he doesn't understand CO_2 asphyxiation!!]*** Nobody knows how anesthetic agents work, as yet; there are all kinds of theories."[117] ***[Asphyxiation is NOT "fully physiological." We do not know how anesthesia works, but we DO know how CO_2 asphyxiation works, and so did Dr. Leake! This is reflected by his cunning experimental circumstances. He understood the nature of CO_2 asphyxiation well enough to make sure he mixed CO_2 with oxygen instead of atmospheric air. There is only one explanation for this specious statement: Dr. Leake was lying!]***

In retrospect, it seems significant that neither Dr. Leake nor Dr. Waters left any record of scientific research other than this rigged dog experiment. It likewise seems significant that on at least two occasions Dr. Leake denied that there was any possibility that the results of his animal experiments with Dr. Waters could have been caused by CO_2 asphyxiation.[118, 119] *He thus implicitly admitted that he understood the nature of asphyxiation.*

In the opinion of this author, it is impossible to believe that a PhD scientist and a medical physician, both of whom were scientifically sophisticated, were unable to comprehend that asphyxiation was the cause of these observations. In my opinion, this study was an inexcusably cruel, abusive, specious, bold, and cunning fraud that deliberately confused CO_2 asphyxiation with anesthesia for the sake of politics at the price of science. What is perhaps even more shocking is that this outrageous fraud has never before been called to account.

In 1928, immediately after completing the dog experiments with Dr. Waters, Dr. Leake departed Wisconsin to establish a department of pharmacology at the UC San Francisco medical school. While there he continued to exchange effusive letters with Dr. Waters, expressing his regret for leaving Wisconsin, and Waters responded in kind. He left San Francisco in 1942 to become executive vice-president of the University of Texas Medical School at Galveston. He left Galveston in 1956 to become professor of pharmacology and assistant dean at the Ohio State University. In 1962 he returned to UC San Francisco to coordinate student research. While occupying these prestigious positions he seems to have devoted most of his time to promoting library construction and investigating arcane medical history.

In 1978 Dr. Leake died of a ruptured aneurysm. As Shakespeare wrote: "Nothing in his life became him like the leaving of it."

The Third Paper: "TOXIC EFFECTS OF CARBON DIOXIDE" (1933)[120]

Based on animal experiments and peer-reviewed reports that have never been refuted, Dr. Yandell Henderson recommended 5% CO_2 supplementation during anesthesia to prevent postoperative respiratory depression, atelectasis, pneumonia, and nausea, but practicing anesthetists quickly embraced mixtures of 25% CO_2 and 75% O_2 to accelerate ether induction and emergence.[121, 122, 123] Reports of overenthusiastic CO_2 supplementation that resulted in convulsions and deaths soon appeared,[124] but these "ether fits" were initially attributed to fever and occult medical problems.[125, 126] Dr. Waters alleged that CO_2 toxicity caused these disasters. As noted previously, a trained physician should have recognized CO_2 asphyxiation as the cause, and we know from the admission of Chauncey Leake cited above that both Leake and Waters were confirmed liars. *Therefore, all their statements, particularly those pertaining to carbon dioxide, must be regarded as lies until proven otherwise.* Instead, he cited several examples as evidence of CO_2 toxicity. All his examples reflected the ability of carbon dioxide asphyxiation to mimic toxicity and/or anesthesia under specific circumstances.

Dr. Waters' first cited the volcanic *Grotto del Cane* (aka "Dog's Cave") near Naples, Italy that has been known since antiquity.[127] The *Grotto del Cane* is not a cave, but rather a small passage where a volcanic fumarole emits a continuous discharge of carbon dioxide that forms an invisible lake of asphyxiating CO_2 in the depressed floor of the passage. People who walked through the passage are not affected because their heads remain above the level of CO_2, but dogs were drowned in the CO_2 and quickly became unconscious. To impress tourists, the same dogs were repeatedly exposed to the CO_2 without harm, which reflects the absence of CO_2 toxicity. A truly toxic agent, such as mustard gas or sulfur dioxide, would produce permanent tissue damage after repeated exposure.[128] Dr. Waters visited England only once, and he never visited the European continent in his life, so he never saw the *Grotto del Cane*.

Next, Dr. Waters presented this melodramatic description of the CO_2 "toxicity" syndrome provided by an *"anonymous practitioner"*:

The patient was a well-developed girl of fifteen ***[Most victims were young children.]*** who entered the hospital with a chief complaint of tenderness localized at McBurney's point ***[McBurney's point refers to the point on the lower right quadrant of the abdomen where tenderness is maximal in acute appendicitis.]***. Her temperature was 101.8 F.; her pulse was about 85, and respirations 20. Chest examination was negative. An appendectomy was advised. To produce anesthesia, I employed nitrous oxide induction followed by ether, with the absorption technique. ***[The "absorption technique" refers to the use of soda lime to absorb carbon dioxide.]*** The machine did not absorb carbon dioxide very well at any time. ***[This suggests that the soda lime was fully saturated with carbon dioxide. Waters seems to imply that poor CO_2 absorption contributed to the problem.]*** The anesthesia was well established with ether, and the operation begun, the patient having excellent color, respirations about 26, pulse 90, pupils small and extrinsic muscles of the eyeball paralyzed, when she began to have small constant twitchings of the face. ***[There was no mention of CO_2 supplementation up to this point, as if spontaneous CO_2 accumulation caused the supposed CO_2 toxicity. Dangerous CO_2 accumulation is unlikely without CO_2 supplementation. Also note that CO_2 asphyxiation causes a blue-gray skin color due to cyanosis. The anesthetist should have immediately flushed his breathing circuit with pure oxygen and shut off all sources of carbon dioxide at the first sign of convulsions. Failure to do so reflects incompetence.]*** A short time later there appeared a small clonic body convulsion. I thought of anoxemia as a cause even though respirations were not depressed, pupils small, pulse good and color good. Oxygen was administered, ether was deepened, then abandoned, then reapplied and various other procedures tried in an attempt to relieve the situation. These efforts included giving a lot of carbon dioxide because of possible alkalosis. ***[The "anonymous practitioner" would have been an idiot to supplement with CO_2 under such circumstances; the first measure taken would be to flush all gases other than oxygen from the machine.]*** The convulsions continued, getting steadily more violent, her pulse rose to 120, and her temperature obviously increased. A half hour after the onset of the convulsions, she died. This case has been widely discussed among our hospital staff members, and no adequate explanation offered. The surgeon says it was a case of too much ether, ***[Ether does not cause convulsions, and any surgeon would have known this.]*** our anesthetic technician believes it to have been anoxemia, ***[That is basically correct; CO_2 asphyxiation causes lethal brain hypoxia.]*** and the pathology department, after a complete postmortem examination including the brain, concluded that it must have been asphyxia, largely because of the essentially normal findings. ***[This makes no sense. If CO_2 asphyxiation caused this problem, then the brain would exhibit hypoxic tissue damage that is easy to detect with the naked eye at autopsy.]***

This account was obviously concocted. It is dramatized by someone who knew enough to fabricate the account, but wasn't smart enough to avoid revealing blunders — or perhaps was so arrogant as to assume that his statements would be accepted as divine Gospel by scientific illiterates. Here is what leads me to believe that the account was a complete fabrication:

1. No physician would refer to a female patient as "well-developed" because that would imply prurient interest or manipulation.
2. CO_2 asphyxiation causes lethal oxygen starvation, but brain hypoxia severe enough to cause death causes characteristic brain damage that is visible to the naked eye at autopsy. This is inconsistent with Waters' claim that the brain was normal at autopsy. Either Dr. Waters was ignorant of this fact or assumed that nobody would notice.
3. The practitioner reported "excellent color" in the presence of convulsions. CO_2 asphyxiation severe enough to cause convulsions would have been preceded by cyanosis (bluish skin color) rather than "excellent color."
4. The supposedly "anonymous" practitioner continued to supplement with CO_2 in the face of these alarming symptoms. The first measure taken under such circumstances would be to shut off all gases but oxygen. This practitioner (assuming he existed) would have been damning himself by providing written proof of his incompetence, which is unlikely.
5. This incident cannot be confirmed. There are no letters from the "anonymous" practitioner or mention of this "anonymous" source anywhere in Dr. Waters' carefully conserved correspondence. There is no evidence that this anonymous source ever existed.
6. The account describes the "Jekyll/Hyde" nature of CO_2 asphyxiation. During anesthesia, beneficial CO_2 supplementation improves cardiac output, tissue perfusion, and tissue oxygenation, so that the limbs would be warm, pulse strong, blood pressure stable, and skin color pink. This reassuring situation would change suddenly when CO_2 concentration rose above a critical threshold (around 30% in the presence of oxygen) and disrupted oxygen transport and delivery, causing the sudden, unexpected convulsions, fever, and cyanosis. The cyanosis would reliably appear *before* brain oxygen starvation became severe enough to cause convulsions.
7. The account does not prove CO_2 "toxicity" as alleged by Dr. Waters. We know that there is no such thing as CO_2 toxicity, as explained in Chapter 14, on page 137, that reviews oxygen transport and delivery. Modern hyperbaric chamber studies have proved that the brain can tolerate fantastic CO_2 concentrations without damage as long as cellular oxygenation is sustained. The nature of CO_2 asphyxiation was understood in Waters' time even without modern evidence.
8. Interestingly, the account suggests that practitioners of that early era correctly regarded respiratory alkalosis as harmful and dangerous (it is caused by hyperventilation that depletes CO_2 tissue reserves and undermines cardiorespiratory function) and that it can be corrected using CO_2 supplementation.

Dr. Waters next cited the *same* "anonymous source" in this report of "CO_2 toxicity" in an infant. It is important to note that in Waters' day, gases would be delivered by enclosing an infant's bassinet in a tent (today we would use a cannula).

However, Dr. Waters does not describe how the gas was administered or even if the child was in a bassinet; we are left to assume. What he does do, is attempt to sell the prospect that *any* amount of carbon dioxide would be toxic.

> Early this morning I was called to the ward to see a premature infant that was blue and hardly breathing in spite of a steady stream of carbon dioxide 5% and oxygen 95% or Carbogen. It gave occasional twitching gasps. ***[This claim is illogical and unbelievable. Carbogen is a stable gas mixture that is not affected by gravity and would not "pool" at the bottom of a bassinet or tent, as would pure CO_2 fresh, cold, and unmixed from a tank. The 5% CO_2 would have been harmless and beneficial, especially when combined with 95% oxygen. See Chapter 14.]*** This time I recognized carbon dioxide toxemia, so gave lots of pure oxygen at once, and the infant rapidly improved. ***[This event might have happened — if — an inexperienced nurse mistakenly replaced a tank of Carbogen with a tank of pure carbon dioxide. Pure CO_2 released from such a tank readily forms a cool, odorless, invisible cloud that is temporarily affected by gravity and fills dependent spaces. This would explain how CO_2 filled the baby's bassinet and suffocated the baby. Carbogen, however, consists mostly of oxygen, and is not affected by gravity. On the contrary it would have improved oxygen transport and delivery, even in a premature infant.]*** After being given some saline subcutaneously and one-half hour of pure oxygen, it was crying lustily and in good condition. ***[There is no evidence that crystalloids — the saline — confer benefit. Crile had long-since proved that crystalloids confer no benefit and are potentially harmful. Plus, this treatment is irrelevant to the account.]*** It seemed spectacular to the nurses how oxygen, not Carbogen, was effective. ... In this hospital, I have been able to count fifty Carbogen cylinders and only two oxygen cylinders available for therapeutic purposes. ***[This affirms that tanks of Carbogen were commonplace in hospitals of that era; of course, Waters' point was that "toxic" and "harmful" Carbogen was available in abundance. He has an axe to grind, and he grinds it.]***

Dr. Waters' concocted commentary was as follows:

> If this man's experience was an isolated instance, it would be of little significance. On the contrary, his letter is one of many received in my office describing fatal and non-fatal cases of carbon dioxide poisoning. ***[There is no evidence of any such correspondence in any of the three libraries that preserve Dr. Waters' papers, and this is the SECOND case provided by the same "anonymous" author.]*** It is my belief that atmospheres containing excess carbon dioxide are frequently toxic and often fatal to certain individuals. If illness has interfered with normal metabolism, or if the transport of this waste product ***[Waste product??? CO_2 is essential for life, as explained in Chapter 14, and this was well-understood in Waters' time.]*** from the cells of the body to the environmental atmosphere is handicapped by biochemical abnormality, by respiratory or cardiac deficiency, or by drug effect, any increase in the tension of carbon dioxide in the atmosphere

> inspired may prove detrimental. ***[This statement can only be explained by incredible ignorance or prevarication. The essential role of carbon dioxide in oxygen transport and delivery was clarified before Waters was born.]***

This account is illogical, unbelievable, and unverifiable. The assertion that 5% Carbogen toxicity caused these symptoms is impossible. Carbogen at 5% is a stable mixture of 95% oxygen and 5% CO_2 prepared and stored at pressures below 750 torr, so that both gases remain in a uniform mixture, are not affected by gravity, and there is no experimental evidence that 5% CO_2 causes any harmful effects. On the contrary, such a mixture of Carbogen is a proven treatment for asthma, atelectasis, pneumonia, carbon monoxide poisoning, drowning, smoke inhalation, myocardial infarction, and stroke. The described symptoms could only be explained if an inexperienced nurse mistakenly replaced a tank of 5% Carbogen with a tank of pure carbon dioxide. Medical grade CO_2 is typically delivered in tanks pressurized above 760 torr so as to maintain the CO_2 in liquid state and optimize tank storage capacity. Pure CO_2 has refrigerant properties so that when it is released from such a tank it forms a cool, odorless, invisible cloud that is temporarily affected by gravity (because of its temporary and temperature-dependent greater density than air) and fills dependent spaces. This would hypothetically explain how CO_2 might fill a baby's bassinet and cause asphyxiation.[129] *This would not happen with Carbogen.*

Given the historical significance of these supposedly anonymous reports, this author dispatched inquiries to the Wood Library, UC San Francisco, and the University of Wisconsin, all of which maintain archives of Waters' correspondence. None of them possessed copies of these anonymous letters. This suggests that Dr. Waters fabricated these incidents.

After citing his earlier dog study as "proof" of the narcotic properties of CO_2,[130] Dr. Waters casually described his subsequent endeavor to anesthetize humans using CO_2:

> We then attempted to administer 30 percent in oxygen as an anesthetic agent to three apparently normal women, with the following result. The first was satisfactorily anesthetized for thirteen minutes with no untoward result during inhalation. The second had a convulsion on the operating table, during which the ward nurse appeared in the doorway to ask, "Doctor, what shall I do? The first patient is in a convulsion." The third patient tolerated ten minutes of carbon dioxide anesthesia, which was then changed to nitrous oxide. That evening, none of the three appeared to be worse for the experience.

This concocted account is hilarious. First, "satisfactory" anesthesia to Woman 1 would not result in convulsions after she was off the table and in recovery — she would go into convulsions as soon as her brain became hypoxic, and not a moment later. It is not a condition that reveals itself slowly. Second, the convulsions of Woman 2 on the table would indicate hypoxia right then and there, and cyanosis would confirm it. Both such patients would end up being "worse" for the experience.

I daresay that any "normal" woman would have objected vociferously to being subjected to the alarming sense of suffocation induced by even 1% concentrations of carbon dioxide, let alone 30%. Even a man would protest such abuse.

There is no existing evidence that these events ever occurred. Dr. Waters offers no evidence of charts, consents, dates, or anesthesia records. The patients would have complained on account of the sense of suffocation caused by carbon dioxide. The nurses would have documented the convulsions and the complaints. In the unlikely event that Dr. Waters casually conducted this outrageous clinical trial despite his documented assertions that CO_2 is toxic, then one should question both his integrity and his sanity. Such a wanton experiment would invite hypoxic brain damage, but Dr. Waters never mentioned that possibility. Instead, he attributed the convulsions to toxicity. These events are blatant concocted fiction. Dr. Waters was a confirmed liar, as previously demonstrated, so all his accounts must be dismissed as fiction until proven otherwise.

The Fourth Paper: "CARBON DIOXIDE" (1938)[131]

In this paper, Dr. Waters directly attacked the research of Yandell Henderson.[132]

> Clinical misapprehensions regarding the physiology of carbon dioxide seem to have had their inception in the work of Yandell Henderson in the early years of the present century. ***[Seem to have had?? Why didn't he cite specific examples? This is the language of politics and underhanded persuasion, not science!]*** Henderson deduced from his experiments that surgical shock results from low carbon dioxide in the blood and tissues; that low carbon dioxide in the blood and tissues is a usual accompaniment of anaesthesia; and that carbon dioxide and oxygen mixtures should prevent post-operative pulmonary atelectasis. All three of these deductions are fallacious. ***[This is a bold bluff and a mangled misrepresentation of Dr. Henderson's recommendations. Waters goes on to say:]*** I make this statement based upon my own personal experience in clinic and laboratory, verified amply by the experience and experiments of others. ***[There is no evidence that Dr. Waters ever performed any controlled scientific experiments in his life other than the specious animal asphyxiation experiment described earlier.]*** The belief that shock is entirely dependent on low carbon dioxide is disproved by the fact that Seevers has hyperventilated dogs for many hours, reducing the blood carbon dioxide to points below 15 volumes per cent, without extreme drops in blood pressure and with no evidences of shock. ***[Henderson never claimed that hyperventilation was the sole source of shock. Henderson's report remains consistent with known science and has never been disproved. Hypocarbia closes the capillary gate, increases microvascular flow resistance, and maintains blood pressure even as it undermines cardiorespiratory function.]*** Carbon dioxide added to the inspired air of both men and animals whose blood pressures were low as a result of profound spinal anaesthesia has, in certain cases, resulted in a further drop of blood pressure, and has never resulted in a more sustained rise in blood pressure than could be produced by ventilating the lungs with oxygen alone. ***[Here Waters assumes that blood pressure is the driving force of blood flow, but***

> ***we now know that the main variable that determines cardiac output is flow resistance. Carbon dioxide supplementation beneficially opens the capillary gate, decreases microvascular flow resistance, increases cardiac output, improves cardiorespiratory function, enhances tissue oxygenation, and lowers blood pressure.]*** Eastman and Kreiselman have recently made a cinematographic record showing that, in reestablishing respiration following oxygen crisis or a complete apnoea due to oxygen want, carbon dioxide added to the oxygen used in artificial respiration has, on certain occasions, caused a further drop in blood pressure and always a less satisfactory circulatory recovery than was secured by artificial respiration with oxygen alone. ***[This plays on the widespread belief that blood pressure is the driving force of blood flow; this commonplace assumption has been questioned by researchers since the beginning of modern research.]*** Henderson's assumption that anaesthesia is accompanied by low carbon dioxide in alveoli, blood and tissues may have had a basis during the last century. Such is far from the case in anaesthesia of the present day. ***[It is Waters' assumptions, not Henderson's, that were founded in the previous century.]***

Dr. Waters also stated:

> First of all, we must not lose sight of the fact that it (CO_2) is a waste product of body metabolism, just as are the constituents of urine. Reduction in the physiological efficiency of respiration tends to dam back carbon dioxide in the tissues just as physiological disturbances of the kidney tend to dam back the excretory products usually eliminated through that organ.[133] ***[This deranged drivel defies courteous comment.]***

Waters' assault on Dr. Henderson's research is unsubstantiated, unethical, unprofessional, and slanderous. Dr. Henderson's international fame was founded on an extensive body of careful, published, and peer-reviewed experiments that have withstood the test of time. Most of them merit "five-star" ratings on PubMed despite their antiquity. Henderson published several authoritative books, many of which remain in print today. In contrast, Dr. Waters published only worthless screeds, and his only "research experience" consisted of asphyxiating dogs with various concentrations of Carbogen, and deliberately misrepresenting the results.[134] Fortunately for him, dogs don't complain or exhibit telltale cyanosis (they're covered in fur).

Waters' claim of "personal experience in clinic and laboratory, verified amply by the experience and experiments of others" is pompous myth. He misinterprets the research of Seevers, Eastman, and Kreiselman, because CO_2 depletion closes the capillary gate and maintains blood pressure, even as it harmfully inhibits tissue perfusion and oxygenation. In contrast, hypercarbia opens the capillary gate and lowers blood pressure, which explains why combinations of spinal anesthesia and hypercarbia harmlessly lower blood pressure and increase tissue perfusion and oxygenation.[135, 136, 137] Moreover, blood pressure is a counterintuitive reflection of cardiovascular performance, as elaborated in Chapter 13: The Turbulence Mechanism, on page 119. Reductions in blood pressure normally are accompanied by beneficial improvements in tissue perfusion and oxygenation, as in trained athletes, while tachycardia and hypertension invariably reflect harmful reductions in cardiac efficiency, tissue perfusion, and tissue oxygenation. Lastly, Waters' histrionic characterization of carbon dioxide as a "toxic waste, like urine" are the words of a salesman, as opposed to a scientist. His malignant manipulations have no place in scientific discourse.

Two years later, in 1940, Dr. Henderson published a scholarly summary of the studies that founded his viewpoint. This was undoubtedly intended to serve as a rebuttal of the Waters paper bearing the same title.[138, 139] Although he did not dignify Dr. Water's attacks with a direct response, he undoubtedly had them in mind when he composed this prelude:

> The human mind is inherently inclined to take moralistic view of nature. Prior to the modern scientific era, which only goes back a generation or two, if indeed it can be said as yet even to have begun in popular thought, nearly every problem was viewed as an alternative between good and evil, righteousness and sin, God and the Devil. This superstitious slant still distorts the conceptions of health and disease; indeed, it is mainly derived from the experience of physical suffering. Lavoisier contributed unintentionally to this conception when he defined the life supporting character of oxygen and the suffocating power of carbon dioxide. Accordingly, for more than a century after his death, and even now in the field of respiration and related functions, oxygen typifies the Good and carbon dioxide is still regarded as a spirit of Evil. There could scarcely be a greater misconception of the true biological relations of these gases.

History has declared the winner of this contest between a medical physician and a PhD researcher. Henderson was a gentleman and a scientist, and no match for Dr. Waters, who was a master of politics and persuasion. Dr. Waters indoctrinated his residents with his flawed ideas, guided them to prestigious board certification and influential academic careers, and thereby paved the path for anesthesiologists to displace nurse-anesthetists. Waters states:

> My ambition is for the men who spend some time with me here to get eventually in teaching positions in other universities because I think that is the only way we can hope to improve the specialty in the future. It has therefore been a disappointment to me each time that one of my boys has gone to private practice.[140]

Of the original 60 anesthesia residents at Wisconsin, 40 occupied teaching positions in academic centers for a major portion of their careers, and half of those became chairpersons or directors of academic programs in medical schools of the United States and elsewhere. The *Aqualumni* (the "aqua" referring to "Waters") were a cohesive group, remarkably loyal to the Chief. They faithfully attended yearly meetings in Madison, where they refreshed their relationships with one another and Dr. Waters. Even today thousands of practicing anesthesiologists can claim linkage to the Ralph Waters professional lineage through their own teachers and teachers' teachers.[141]

DR. RALPH WATERS AND THE HOAX

Seldom has a single civilian inflicted so much harm upon so many as Dr. Ralph Waters. He is presently venerated as the political genius who founded the anesthesiology profession, but history will not likely be so kind. He was a confirmed liar who fabricated and manipulated evidence for the sake of his political agenda — the replacement of the nurse anesthetists. His attacks on Dr. Henderson were scurrilous. He founded the anesthesiology profession on false science — an intentional misrepresentation of carbon dioxide and a misrepresentation that has killed countless patients. His false science, persuasive publications, and political stratagems created a devastating hoax that has abolished the therapeutic benefits of carbon dioxide, entrenched the harmful habit of hyperventilation, discouraged opioid treatment, exaggerated surgical morbidity and mortality, derailed medical research, reversed professional progress, and established inhumane CO_2 asphyxiation as a *de facto* standard of animal research.[142, 143] His hoax now threatens the very survival of the profession he created, because it has progressively perverted anesthesia beliefs and practices and prevented professional progress, so that anesthesiologists cannot claim to provide superior service.

Unfortunately, neither Dr. Waters nor Dr. Leake are available to explain their motives or the shortcomings of their infamous animal experiment.[144] These cannot be dismissed as ignorance, because CO_2 asphyxiation was well understood in their time.[145] Furthermore, they surely knew that CO_2 supplementation reduced the risk of disastrous ether and cyclopropane explosions that were a major hazard for patients and practitioners alike in their time. All available evidence indicates that they deliberately devised their experiments to foster the fiction that CO_2 has narcotic properties for the purpose of establishing the anesthesiology profession, undermining the reputation of the nurse-anesthetists, and replacing them with MD anesthesiologists.

In theory, Dr. Waters might have believed that CO_2 is harmful. If so, the kindest thing that can be said is that his salesmanship superseded his science. My hunch is that he assumed that his promotion of mechanical hyperventilation would be harmless because practically all practitioners then-utilized Jackson's closed circuit machines. With those machines, hyperventilation didn't deplete carbon dioxide from the patients' bodies, and narcotic treatments were not affected. He may thus have regarded his prank as a harmless means to a worthy end. If so, he didn't anticipate that the anesthesia machine manufacturers would introduce a new generation of open circuit machines with continuous fresh gas inflows to eliminate any chance of CO_2 accumulation within the breathing circuit. They surely succeeded, because there have been no reports of "CO_2 toxicity" since open circuit machines replaced Jackson's closed-circuit machines. However, the new generation of open circuit machines introduced a deadly new problem: occult CO_2 depletion that paralyzes respiratory chemoreceptors and causes unexpected postoperative respiratory arrest.[146, 147] This problem has killed countless patients.[148, 149] The realization of this new hazard may explain why Dr. Waters promptly retired at the age of 65 and thereafter avoided all contact with the profession he created until he died in 1979 at the age of 96. The available evidence indicates that he was a moral and mental midget whose mischief serves as a timeless warning of the dangers of elevating politics above science. Regardless, it is time to correct his disastrous mistakes for the sake of everyone everywhere.

The Waters Technique as Derived from His Papers

Advancing needle technology in the early 1930s facilitated intravenous access, and this enabled Dr. Waters to introduce his practical new technique of general anesthesia that eliminated the shortcomings of the Crile/Henderson mask induction technique. It utilized intravenous barbiturate induction that was faster and more pleasant than ether induction by mask. The pentothal was followed by intravenous curare paralysis, which enabled elective endotracheal intubation that prevented laryngospasm, airway obstruction, and aspiration of vomit and other foreign materials into the lungs. The curare paralysis also optimized surgical convenience. Intubation enabled mechanical ventilation (essential in paralyzed patients), prone and lateral patient positioning, ENT surgery, and dental procedures, all of which were impractical with mask maintenance. CO_2 supplementation was no longer needed to accelerate induction and emergence, and it was dismissed as dangerous. This eliminated the problem of CO_2 asphyxiation.

Shortcomings of the Waters Technique

Recall how Chapter 10: The Autonomic Nervous System, on page 85 of this book explained how ether anesthesia mimics analgesia, and why general anesthesia must be supplemented by analgesia (narcotics or analgesic nerve blocks) to optimize surgical outcome. Dr. Waters seized upon the false notion that ether possesses analgesic properties, so that morphine was not needed to control harmful pain. This seemed reasonable because increasing ether concentrations lowered blood pressure the same as narcotics. However, ether lowers blood pressure via its toxic depression of cardiac contractility, and has no effect on spinal cord nociception pathways, while morphine is a genuine analgesic that inhibits spinal cord nociception. Thus, eliminating morphine supplementation exaggerated morbidity and mortality. The contributions of Crile were thus sacrificed on the altar of convenience.

Dr. Waters also advocated mechanical hyperventilation with enriched oxygen mixtures to optimize tissue oxygenation, but this too is a false notion. Hyperventilation cannot enhance tissue oxygenation, which is limited by hemoglobin oxygen saturation. On the contrary, it depletes CO_2 tissue reserves, closes the capillary gate, increases microvascular flow resistance, decreases cardiac output, and undermines oxygen delivery to tissues. Either Dr. Waters didn't understand physiology, or he valued salesmanship over science. The contributions of Henderson were thus sacrificed on the altar of convenience along with the contributions of Crile.

DR. WATERS' RETIREMENT AND THE HOAX

Dr. Waters retired to Florida in 1949 at the age of 65 and thereafter refused any contact with the profession he created until his death in 1979 at the age of 96. This highly unusual behavior has been the subject of considerable curiosity. In my opinion, the most likely explanation is that he feared that the impending Beecher study would focus attention on his fictions and fabrications, and he wished to escape questioning. He may also have comprehended the consequences of the new generation of open circuit anesthesia machines that were being introduced at that time. Unfortunately, he is no longer available to clarify these questions.

THE BEECHER STUDY AND THE HOAX

The Waters technique didn't escape controversy. Surgeons complained that it increased morbidity and mortality.[150, 151, 152] The uproar inspired the famous "Beecher Study," which was a review of some 600,000 surgical deaths in ten university hospitals during a period of ten years. It issued its findings in 1954.[153] The study had shortcomings, as might be expected in such a complex and extensive endeavor. It did not assess the role of morphine supplementation or CO_2 supplementation. It did, however, demonstrate that anesthesiologist deaths were twice as common as nurse-anesthetist deaths, and that curare paralysis was associated with disproportionate deaths. Since the nurses didn't use curare, this was powerful evidence that the Waters technique exaggerated mortality. The cause should have been obvious: the anesthesiologists were using ether anesthesia instead of morphine analgesia to control surgical nociception. Unfortunately, that escaped notice. Medical memory is short, and Crile's contributions were forgotten. The controversy dissipated with the passage of time, and the Waters technique became the *de facto* standard of general anesthesia throughout the world. But the problems didn't end there.

PART II: THE CRUX OF THE HOAX

Dr. Waters had deliberately founded his profession on false science, and, whether he knew it or not, he had created a subtle, deceptive, and deadly hoax that consisted of a Gordian knot of perverted beliefs, principles, practices, and confusion.

His new anesthesia technique emphasized mechanical hyperventilation to "rid the body" of carbon dioxide to eliminate the problem of CO_2 "toxicity."[154] However, Jackson's closed circuit anesthesia machines, which dominated anesthesia practice at that time, were expressly designed to conserve then-expensive medical gases. With these old machines, mechanical hyperventilation did not "rid the body" of anything. On the contrary, CO_2 produced by normal metabolism beneficially accumulated within the breathing circuit, with or without mechanical hyperventilation. Respiratory chemoreceptor function remained intact, so there were no problems with narcotic hypersensitivity or respiratory depression when patients were treated with narcotics to control postoperative pain.[155, 156] Everybody was happy. Dr. Waters had heroically eliminated the problem of "CO_2 toxicity," and his anesthesiologists began to replace nurse-anesthetists throughout the world. However, nobody can predict the future, including Dr. Waters.

ANESTHESIA MACHINES AND THE HOAX

By the time of Waters' retirement, the cost of medical gases had declined substantially, and the hoax (that CO_2 was toxic and narcotic) inspired anesthesia machine manufacturers to introduce a new generation of open circuit anesthesia machines that utilized continuous fresh gas inflows *to eliminate any possibility* that CO_2 accumulation within the breathing circuit might cause "CO_2 toxicity."[157, 158] These new machines quickly replaced the old Jackson machines. However, the new anesthesia machines introduced an even worse CO_2 problem. Because of the hoax, mechanical hyperventilation had become a *de-facto* standard of anesthesia practice throughout the world. Unlike the old Jackson machines, hyperventilation with the new machines *depleted CO_2 tissue reserves* in the bodies of patients. Without CO_2 reserves, oxygen cannot be released from hemoglobin to tissues. This dangerously undermined cardiorespiratory function and disrupted oxygen transport and delivery during surgery, but this result was completely overlooked, with disastrous consequences, as the perverted principles, practices, and beliefs of the hoax had propagated throughout all medical specialties throughout the world like a malignant disease. The following examples will illustrate how this happened.

HALOTHANE AND THE HOAX

Halothane was the first modern halogenated inhalation agent. It was introduced soon after Dr. Waters retired, and it quickly replaced ether because it was non-explosive, sweet-smelling, and far less soluble in blood than ether. It provided a final nail in the coffin of nurse-anesthesia because it eliminated the need for CO_2 supplementation to accelerate induction, speed emergence, and prevent explosions.

Halothane had a unique characteristic that exaggerated the hoax: it causes *nodal dysrhythmias* that are suppressed by hyperventilation. Why Halothane causes these dysrhythmias, and how hyperventilation prevents them remains unclear. Regardless, use of Halothane promoted the notion that hyperventilation and respiratory alkalosis are beneficial, while acidosis is innately dangerous. This further entrenched the harmful habit of hyperventilation — and the hoax.

Halothane dominated anesthesia during my residency, but it caused liver toxicity, and Isoforane replaced it. Many years later, after I had discovered the benefits of hypercarbia and narcotic supplementation using Isoforane, I attempted to switch from Isoforane to Halothane in the presence of hypercarbia. I had forgotten that Halothane causes frightening nodal dysrhythmias in the presence of hypercarbia. That was the last time that I ever used Halothane.

RESEARCH AND THE HOAX

Anesthesia research has inadvertently or deliberately reinforced the Waters hoax. Here are three examples:

Boniface and Brown[159]

Recall that Yandell Henderson used a child's rubber ball to enclose a dog's heart while he discovered that hypercarbia exaggerates venous blood return to the heart and increases cardiac output. At the time, Henderson didn't realize that the rubber ball prevented heart damage due to excessive venous return to the heart.

In the early 1950s Boniface and Brown performed a similar dog experiment in which they opened the dog's pericardium and exposed its heart to the atmosphere.[160] Unlike Henderson, they did not enclose the dog's heart in a protective rubber ball. Instead, they sewed a strain gauge to the exposed heart to assess changes in cardiac contractility caused by carbon dioxide. They expected that increased cardiac contractility caused increased cardiac output (which reflected prevailing beliefs at that time). When the dog inhaled carbon dioxide, the unprotected heart became so dramatically engorged by excessive venous return that it protruded from the dog's chest and disrupted their strain gauge. This caused excessive cardiac muscle stretching that damaged the heart muscle, so that it no longer contracted properly. As a result, Boniface and Brown mistakenly concluded that carbon dioxide causes cardiac toxicity, rather than open the capillary gate to increase blood flow. Like Henderson, they did not understand the function of the capillary gate or of the pericardial sac that prevents excessive stretching of cardiac muscle when carbon dioxide increases venous return to the heart. As a result, their flawed conclusion corroborated Waters' fiction that carbon dioxide is toxic, and further entrenched the hoax.

Eger *et al* #1[161]

After Dr. Leake returned to UCSF in 1962, a team of researchers including John Severinghaus (the son of Elmer Severinghaus, Leake's biochemist at Wisconsin), Edmund Eger, and Larry Seidman conducted sham dog studies at UC San Francisco. These experiments were suspiciously similar to Leake's specious dog studies with Dr. Waters in that they deliberately confused CO_2 asphyxiation with general anesthesia. The 1967 report of Eisle, Eger, and Muallem entitled "Narcotic Properties of Carbon Dioxide in the Dog" began with this statement: *"In 1820 the narcotic properties of CO_2 were demonstrated by Hickman, and a century later anesthesia was produced in man with 30 per cent CO_2 and O_2."* Obviously, these researchers assumed that the fallacious studies of Hickman, Waters, and Leake had proved the anesthetic properties of CO_2 beyond question.[162] Their supposed objective was to clarify the relationship between CO_2 anesthesia (aka narcosis) and acidosis. It was well understood that CO_2 asphyxiation causes cellular hypoxia, which causes cells to revert to "anaerobic pathways" that produce harmless lactic acid, but the lactic acid lacks anesthetic effects.

They anesthetized dogs with Halothane and installed cannulas to sample arterial blood, venous blood, and cerebrospinal fluid (CSF). Then they exposed the dogs to increasing concentrations of carbon dioxide while simultaneously decreasing the Halothane concentration. If CO_2 had anesthetic properties, then why did they need Halothane? They were able to eliminate the Halothane altogether as $FICO_2$ (fractional-inspired CO_2) approached 30%, at which point they observed sudden reductions in nervous reflex activity equivalent to 1 MAC Halothane (i.e., the dogs did not move when subjected to painful surgical incision or tail clamping). Unsurprisingly, these effects, which were unquestionably due to CO_2 asphyxiation, were accompanied by seizures, deaths, and (lactic) acidosis due to anaerobic cellular respiration. The acidosis appeared first in brain tissue, which has the highest oxygen requirements of all tissues, and thus appeared in CSF shortly before it appeared in arterial and venous blood.

Inexcusably, these researchers reported only $PaCO_2$ (partial pressure CO_2) and pH results, *and deliberately omitted PaO_2 and O_2 saturation data.* Blood gas analysis directly measures arterial oxygen tension (PaO_2 or partial pressure O_2), carbon dioxide tension ($PaCO_2$), and acidity (pH), from which oxygen saturation (SaO_2 or saturation of O_2) is derived; *therefore, they obviously had the PaO_2 measurements in their possession.*[163] They ignored the obvious explanation that CO_2 asphyxiation had caused hypoxia, anaerobic respiration, and lactic acidosis. The missing PaO_2 and/or SaO_2 data would have confirmed this, which explains why they omitted this information from their report. Furthermore, Eger et al proved that they possessed sophisticated understanding of CO_2 pathophysiology and CO_2 asphyxiation in their subsequent 1971 study of the effects of extreme hypocapnia in dogs.[164] I asked Dr. Eger via email if CO_2 asphyxiation might explain these observations. He replied "I THINK NOT!!!" When I asked him why PaO_2 data was excluded from the report, he stated that he "could not recall," but he insisted that his study reflected CO_2 narcosis rather than asphyxiation. In my opinion, Dr. Eger is a nefarious prevaricator, and these researchers deliberately rigged their experiment to reinforce the Waters hoax. Readers can judge for themselves.

Despite these defects, the study provided valuable information: it confirmed that potentially lethal CO_2 asphyxiation occurs suddenly in dogs when CO_2 concentrations rise above 30% in oxygen.[165] Presumably humans are closely similar to dogs in this respect; but of course, similar human experiments would be unethical.

Eger *et al* #2[166]

In 2007, Dr. Eger *et al* produced yet another specious study entitled "Anesthetic Properties of Carbon Dioxide in the Rat." As before, this report opened by stating: "The anesthetic effects of CO_2 have long been recognized" and cited Hickman's report as "proof."[167] As in the previous study, asphyxiation was never mentioned. The study neglected oxygen saturation data that could easily have been obtained using pulse oximetry or blood gas sampling, but this data would have proved the presence of asphyxiation and refuted the notion of CO_2 "narcosis" that was claimed by these researchers.

These devious studies of Eger *et al* were published in prominent anesthesiology journals, but the editors and reviewers failed to object to their obvious oversights.

CARDIAC ANESTHESIA, NARCOTICS, AND THE HOAX

Heart-lung bypass technology was developed after WWII to enable the correction of congenital cardiac defects. The machinery substitutes for the heart and lungs while surgeons open the heart and repair its defects. The technology remained impractical until the1960s, when Lowenstein discovered that morphine supplementation drastically reduced the morbidity and mortality caused by the heart-lung machinery.[168] Doctors were shocked by this discovery because they had come to regard large doses of morphine as lethal on account of the hoax.

In essence, Lowenstein had re-discovered Crile's principle that supplementation of general anesthesia with analgesia lowers morbidity and mortality.[169] Unfortunately, Lowenstein did not re-discover Henderson's principle that hypercarbia counteracts narcotic respiratory suppression and optimizes cardiorespiratory function.[170, 171, 172, 173, 174, 175, 176] As a result, cardiac anesthesiologists and bypass pump technicians continued to maintain hypocarbia in accord with the hoax, so that the large doses of morphine necessitated prolonged postoperative "weaning" from mechanical ventilation. The weaning allowed the body to replace its CO_2, which had been "blown off" during surgical anesthesia. Prolonged postoperative mechanical ventilation thus became a *de-facto* "standard" of cardiac anesthesia practice that distinguished cardiac anesthesia from non-cardiac surgery. Nevertheless, the morphine supplementation reduced morbidity and mortality to the point that bypass surgery became practical.

The Janssen Corporation developed a new generation of powerful, short-acting synthetic narcotics and developed a substantial body of fresh clinical research evidence during the 1970s proving that they improved cardiac surgery outcome.[177] Cardiac anesthesiologists embraced the new synthetic narcotics because of their superior potency and relative lack of side effects compared to morphine, but they continued to routinely maintain patients on mechanical ventilation and wean them slowly over several hours.

Around 1990, insurance companies suddenly imposed new fast tracking discharge requirements. This resulted in drastic changes in cardiac anesthesia practice. Cardiac anesthesiologists replaced the synthetic narcotics with Propofol infusions to comply with the new reimbursement requirements. Soon thereafter, a new problem of severe post-bypass problem called *SIRS* (Systemic Inflammatory Response Syndrome) appeared in medical literature. The cardiac anesthesiologists were baffled by this problem because they had already forgotten the lessons of Crile and Lowenstein. Since then, they have resumed narcotic supplementation, but in the meantime insurance reimbursement had focused on angioplasty as a preferable treatment. Angioplasty is performed by cardiologists using local analgesia, which drastically reduced the need for cardiac surgeons and cardiac anesthesiologists. Even more recently, gathering evidence indicates that angioplasty is no better than coronary artery bypass surgery.

Stress theory suggests that the safest and most effective treatment for heart disease is Carbogen, chelation therapy, and powerful anticoagulants (see Chapter 16: The Unified Theory of Medicine, on page 175.

NON-CARDIAC ANESTHESIA, NARCOTICS, AND THE HOAX

During the 1980s, the Janssen Corporation began to promote its new synthetic narcotics for non-cardiac anesthesia supplementation, with the expectation that their short half-lives would mitigate the postoperative respiratory depression that plagued long-acting morphine. This caused yet another mysterious problem. Because of the hoax, patients were routinely hyperventilated, which paralyzed respiratory chemoreceptors. Patients would emerge from anesthesia and appear to be awake, alert, pain free, and breathing normally, but they would unexpectedly fall asleep and stop breathing. Nobody realized that hyperventilation was the underlying cause of the problem. Instead, the problem was blamed on the new narcotics. The problem was called *re-narcotization syndrome* or *narcotic hypersensitivity.*[178, 179] Nevertheless, the improved outcome afforded by supplementation with the new narcotics was recognized, and non-cardiac anesthesiologists began to imitate the cardiac anesthesia practice of prolonged postoperative "weaning" from mechanical ventilation. This necessitated armies of respiratory therapists to manage the weaning process while the anesthesiologists returned to the operating room. As we stated, around 1990 the insurance companies imposed their new fast tracking reimbursement requirements, and non-cardiac anesthesiologists resumed the classic Waters technique that avoided narcotic treatment and its lengthy weaning. The experience left all parties convinced that narcotics cause respiratory depression, and it discouraged beneficial narcotic treatment needed to minimize surgical morbidity and mortality.[180, 181] Nobody realized that mechanical hyperventilation was the underlying cause of the problem.

They still don't.

RESIDUAL PARALYSIS AND THE HOAX

The hoax emphasizes inhalation agent "overpressure" that abolishes cortical inhibition of spinal cord nociception pathways, and it discourages narcotics that inhibit spinal cord nociception. The exaggerated nociception causes spinal cord "windup" syndrome that increases the need for paralysis to control muscle tension and untoward muscle movements. The high doses of paralyzing medications exaggerate the necessity of toxic reversal agents such as neostigmine, which is a close chemical relative of toxic insecticides (parathion, malathion). Their side effects (bradycardia, salivation) require still more antidotes (atropine, glycopyrrolate). Furthermore, there is no reliable way to assess and eliminate residual postoperative paralysis, which is distressing to patients, not to mention a cause of inadequate breathing.[182, 183] Meanwhile, paralysis necessitates mechanical ventilation... that invites CO_2 depletion, lung damage, prolonged narcotic effects, respiratory depression, nausea, vomiting, cardiac inefficiency, and reduced tissue perfusion and oxygenation in the brain, bowel, and other vital organs and tissues.[184, 185, 186, 187, 188, 189, 190, 191, 192, 193, 194, 195, 196, 197]

"ANESTHESIA AWARENESS" AND THE HOAX

The hoax emphasizes paralysis and discourages narcotic dosage. Patients sometimes regain consciousness in the middle of surgery due to exhaustion of inhalation agent from anesthesia vaporizers. This might be compared to running out of gas while traveling down a freeway because the driver wasn't paying attention to the gas gauge. What happens is the patient regains consciousness during surgery and experiences excruciating pain due to inadequate narcotic treatment. He cannot move or talk to alert the anesthesiologist of his plight, because he is paralyzed and intubated.[198] Such intensely stressful experiences exaggerate surgical morbidity and mortality.[199]

STRESS RESEARCH AND THE HOAX

I hypothesize that the hoax derailed stress research and delayed the discovery of the mammalian stress mechanism. Because of the hoax, the important contributions of Crile and Henderson were erased from medical memory long before the era of stress research. Were it not for the hoax, the principles of Crile and Henderson would have guided stress researchers toward successful discovery of the stress mechanism.

RETROLENTAL FIBROPLASIA (NOW CALLED "RETINOPATHY OF PREMATURITY") AND THE HOAX

I hypothesize that the anesthesia hoax muddled medical reasoning, which caused medical mismanagement that produced the epidemic of retrolental fibroplasia in premature babies in the 1950s. Consider these basic principles:

- Premature babies are born while embryological development remains incomplete and still underway.
- Embryological development requires sustained oxygen supplies, especially for maturation of nervous tissues.
- High concentrations of oxygen are toxic to exposed lung tissues. Exposure to 100% oxygen at sea-level pressure for more than 24 hours causes pulmonary inflammation that is analogous to ARDS (Acute Respiratory Distress Syndrome). It causes exudate formation that disrupts gas exchange in the lung.
- The mechanism of oxygen transport and delivery is limited by 100% hemoglobin saturation. This protects internal cells and tissues from oxygen toxicity.

- Hyaline lung disease is a manifestation of immature lung tissues in premature babies.

- Immature lung tissues in premature babies are more susceptible to oxygen toxicity than mature lung tissues in full term babies.

- Treatment of premature babies with 100% oxygen causes pulmonary inflammation that disrupts oxygen uptake by hemoglobin in the lungs, causing hypoxia in peripheral tissues and organs. The hypoxia activates respiratory chemoreceptors that stimulate hyperventilation.

- Spontaneous hyperventilation depletes CO_2 tissue reserves and undermines oxygen release from hemoglobin into organs and tissues. The resulting oxygen starvation disrupts retinal tissue maturation, and results in retrolental fibroplasia.

The hoax caused this tragic epidemic of retrolental fibroplasia by promoting the notion that carbon dioxide is toxic and oxygen is not, when in fact the reverse is true. It condemns carbon dioxide as "toxic waste, like urine" so that helpful CO_2 supplementation is assiduously avoided. It further promotes mechanical hyperventilation to optimize oxygen transport and delivery, but this counterproductive approach cannot improve oxygen transport and delivery beyond 100% hemoglobin saturation. Instead, it depletes CO_2 tissue reserves and undermines oxygen delivery. Even today there is suspicion that carbon dioxide toxicity contributes to retrolental fibroplasia, when CO_2 supplementation should enhance tissue oxygenation.

In the 1950s, if not today, the *de-facto* management of hyaline membrane disease would have been elective intubation and mechanical hyperventilation with 100% oxygen. This explains the close association between hyaline membrane disease and retrolental fibroplasia. Even simple mask treatment with 100% oxygen is harmful, because it causes pulmonary oxygen toxicity that stimulates hyperventilation, depletes CO_2 reserves, and undermines oxygen delivery to developing retinal tissues.

The correct approach to preventing pulmonary oxygen toxicity and retrolental fibroplasia in premature babies should be as follows:

- Minimize pulmonary oxygen toxicity by restricting inhaled oxygen concentrations to levels necessary to maintain pulse oximeter saturation readings no higher than 90%.

- Avoid mechanical hyperventilation that depletes CO_2 tissue reserves.

- Use 5% carbogen instead of pure oxygen to optimize cardiorespiratory function and tissue oxygenation.

NEWBORN RESPIRATORY ARREST AND THE HOAX

Hyperventilation of pregnant women during Cesarean section surgery depletes carbon dioxide tissue reserves in babies as well as their mothers. As a result, the babies sometimes suffer lethal respiratory arrest when they fall asleep after they are delivered. Because of this, spinal or epidural analgesia is now preferred for Cesarean section. This problem could be eliminated by maintaining permissive hypercarbia during general anesthesia for cesarean section surgery.

ECMO AND THE HOAX

ECMO (Extra-Corporeal Membrane Oxygenation) is similar to the heart-lung bypass machine used for open heart surgery. It pumps and oxygenates a patient's blood outside the body. It is used to salvage critically ill patients suffering acute heart failure, lung failure, and critical illnesses. Because of the hoax, the machines are adjusted to maintain hypocarbia, which depletes CO_2 tissue reserves and exaggerates morbidity and mortality. ECMO outcome could be optimized by incorporating permissive hypercarbia and narcotic supplementation.[200]

ANESTHESIA JOURNALS AND THE HOAX

The hoax inspires the harmful habit of hyperventilation, which is so universal in anesthesia practice that it is seldom mentioned by authors or questioned by editors and reviewers. This is astonishing because CO_2 management is one of the most important elements of anesthetic management. On those rare occasions when it is mentioned, it is to confirm hyperventilation, as this is assumed to reflect careful practice.[201]

A blatant example of this imbalanced treatment of carbon dioxide is the November-December 1960 issue of Anesthesiology journal, which was devoted to a symposium entitled "Carbon Dioxide and Man" that consisted of some 17 pompous papers that purported to review all pertinent properties of CO_2.[202] These papers were vague, misleading, and uninformative. They promoted hyperventilation and discouraged hypercarbia via implication. There was no mention of CO_2 benefits, hyperventilation hazards, Crile, Henderson, Lundy, balanced anesthesia, or Anoci-Association. Thereafter, editors, researchers and practitioners alike have assiduously avoided mention of mechanical hyperventilation or CO_2 management in published studies as well as any and all evidence that the Waters technique exaggerates morbidity and mortality, to the point that all published anesthesia research since 1950 requires reappraisal.

ANIMAL RESEARCH AND THE HOAX

As the first chairman of a university anesthesiology department, Dr. Waters wielded enormous influence. His published claims that CO_2 has anesthetic properties that are suitable for animals are the source of the pervasive practice among animal researchers of using CO_2 for both anesthesia and euthanasia for their research animals. CO_2 is cheap, simple to use, and safe for the researchers. It would be ideal if it had genuine anesthetic properties, but it doesn't. Its use for these purposes is cruel and inhumane.[203, 204] It's another malignant manifestation of "monkey see, monkey do" that pervades all forms of human activity, with science being no exception.

THE SURGICAL STRESS SYNDROME (SSS) AND THE HOAX

The Waters hoax exaggerates surgical morbidity and mortality in the following ways:

- It mandates mechanical hyperventilation, which is especially dangerous during surgery. It depletes CO_2 tissue reserves, undermines oxygen transport and delivery, and causes oxygen starvation in tissues and organs throughout the body as explained in chapter 14.[205] The brain, lung, heart, and bowel tissues are particularly vulnerable because of their high oxygen requirements. This explains why dementia, strokes, heart attacks, dysrhythmias, bowel ileus, nausea, vomiting, atelectasis, and pneumonia commonly occur in the aftermath of surgical procedures.

- It mandates the use of toxic inhalation agent "overpressure" to control blood pressure. This abolishes cortical inhibition of spinal cord nociception pathways via descending inhibitory pathways from the brain to the spinal cord, which has the effect of exaggerating harmful spinal cord nociception pathway activity.

- It discourages helpful narcotics that inhibit harmful spinal cord nociception caused by surgery.

In contrast, the Crile/Henderson technique embraced by the nurse-anesthetists utilized generous morphine medication to control spinal cord nociception and CO_2 supplementation to optimize cardiorespiratory function and oxygen transport and delivery to tissues. This explains why the Beecher study revealed substantially lower rates of morbidity and mortality in cases managed by the nurses.

BELIEFS, HABITS, PRACTICES, AND THE HOAX

With the Crile/Henderson technique, practitioners routinely utilized morphine to control surgical stress, so that mild hypotension was regarded as a reflection of beneficial surgical stress control, and tachycardia and hypertension were attributed to inadequate analgesia. They regarded respiratory acidosis as a harmless effect of beneficial CO_2 supplementation, and they understood that respiratory alkalosis resulted from harmful hyperventilation.

With the Waters technique mechanical hyperventilation was assumed to be essential to eliminate the supposedly toxic and narcotic effects of CO_2. Morphine was replaced by toxic inhalation agent overpressure, which was incorrectly assumed to possess analgesic properties. Thus, respiratory alkalosis came to be regarded as beneficial, and respiratory acidosis as harmful.

Hypertension came to be regarded as a manifestation of healthy cardiac reserve (which doesn't really exist; the heart better resembles a 90-lb weakling than the Charles Atlas of the body), and hypotension came to be regarded as evidence of impending disaster.

Narcotics came to be regarded as unpredictable and dangerous because they are incompatible with hyperventilation.

Intravenous muscle relaxants and hypnotic agents came to be regarded as essential to prevent dental and vocal cord damage during intubation, despite the dearth of evidence of such danger.[206, 207, 208]

ANESTHESIA THEORY AND THE HOAX

The Waters technique assumes that the brain mediates all nervous activity induced by surgical stress, so that anesthesia is all that is needed to control surgical stress. The fallacy of this notion is detailed throughout this book, particularly in Chapter 15: Three Stress Mechanism Activation Pathways, on page 155. As a result, most attempts to investigate the cause of surgical stress have sought to identify brain mechanisms that mediate pain, while ignoring spinal cord nociception pathways that remain active regardless of brain function but that are inhibited by narcotics. This has paralyzed anesthesia theory development.[209] Conventional anesthesia theory also overlooks the independent effects of tissue disruption, which harmfully releases tissue factor into systemic circulation.[210, 211, 212]

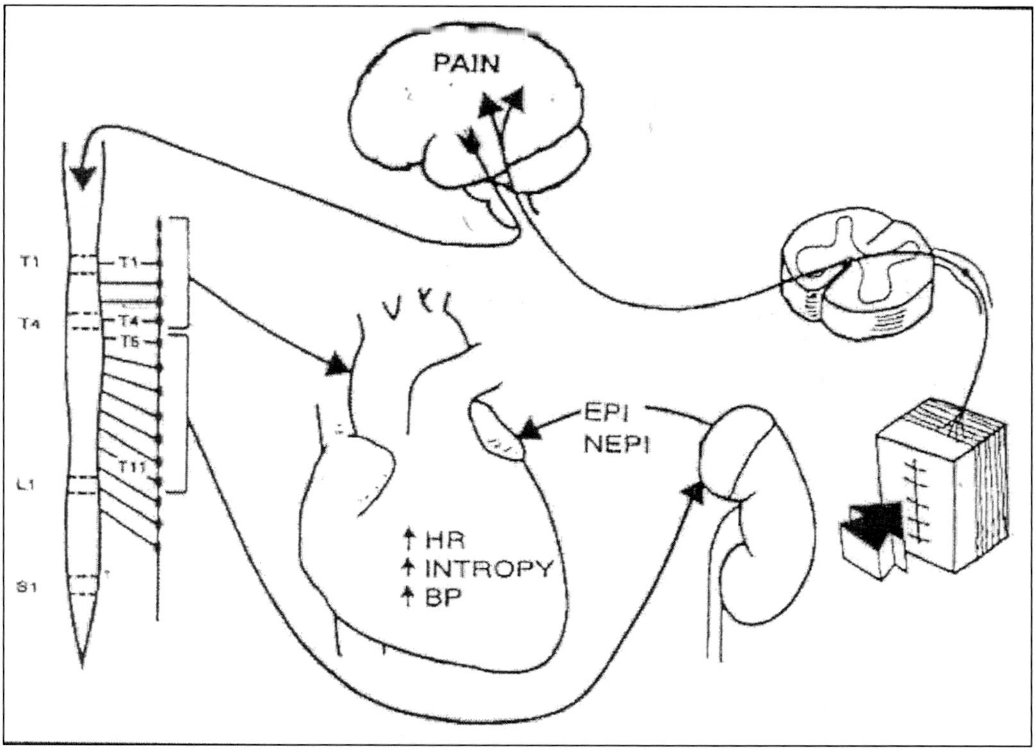

Figure 18.14: Incorrect notion of nociception.

This drawing illustrates the *incorrect* assumption that all sensory information is processed in the brain before it activates spinal cord nociception pathways, so that general anesthesia possesses analgesic properties, and narcotics are not needed to prevent harmful spinal cord nociception. This notion pervades anesthesia literature and frustrates theoretical advance.[213] Source: S. Liu, R. L. Carpenter, J. M. Neal, Epidural anesthesia and analgesia. Their role in postoperative outcome. Anesthesiology 82, 1474-1506 (1995).

CARBOGEN AND THE HOAX

Carbogen is arguably the most powerful, safe, useful, and effective medical treatment ever discovered. It is a mixture of carbon dioxide and oxygen, typically 5% CO_2 mixed with 95% O_2, that can optimize oxygen transport and delivery and enhance tissue oxygenation in a wide variety of medical conditions. Yandell Henderson's research inspired its installation on fire trucks in New York, Chicago, and other major cities. Physicians and firemen used it to treat myocardial infarction, stroke, pneumonia, asthma, newborn babies, drowning victims, carbon monoxide poisoning, and other maladies.[214] It saved many lives, but it was quickly abandoned and forgotten after Dr. Waters publicized the purported toxicity of carbon dioxide. The abandonment of Carbogen is especially unfortunate considering that modern research has demonstrated that inhaling small concentrations of CO_2 increases microvascular perfusion via a mechanism that involves nitric oxide.[215, 216, 217] This explains why inhaling CO_2 reduces blood viscosity and enhances cardiac output, cardiac efficiency, tissue perfusion, tissue oxygenation, and respiratory drive.[218, 219] CO_2 also offsets the respiratory depression of opioids. If all this were properly understood, Carbogen therapy could be restored as a useful, effective, inexpensive, and convenient treatment for acute myocardial infarction, stroke, and perioperative problems. It would be a perfect emergency treatment for suspected strokes because it does not affect coagulation. It could be used to enhance the potency and penetration of antibiotics to enable safe and efficient eradication of stubborn infections such as gas gangrene, tuberculosis, MRSA, Lyme Disease, and gram-negative facultative anaerobes that thrive in poorly perfused and oxygenated tissues. It would provide a powerful means to promote bone and wound healing. How many people must suffer and die before this simple, inexpensive treatment is restored?

CRILE AND THE HOAX

Solid science cannot be forgotten forever. In 1959 Dr. George De Castro described neuroleptanalgesia (combinations of fentanyl and droperidol that optimized surgical outcome),[220, 221] and in 1969 Lowenstein demonstrated that morphine supplementation optimizes cardiac bypass outcome.[222] In the 1980s anesthesiologists embraced the high potency, low toxicity, and short half-life of a new generation of synthetic opioids introduced by the Janssen Corporation as a means to optimize surgical outcome.[223, 224, 225] Unfortunately, the incompatibility of opioids and hyperventilation remained unappreciated, so that even short-acting synthetic narcotics necessitated prolonged, treacherous, impractical, and expensive postoperative "weaning" to restore CO_2 tissue reserves and reliable respiratory drive. Today, narcotics are persistently disparaged in anesthesia literature. If the relationship between hypercarbia and narcotic analgesia were properly understood, medical treatments would be revolutionized. For example, Crile demonstrated that narcotics can be used to treat dangerous infections such as sepsis and peritonitis, and he supplemented general anesthesia with morphine to optimize surgical outcome.[226] If Crile could do this 100 years ago, then why can't we utilize his principle today?

CRITICAL CARE AND THE HOAX

In the late 1980s critical care specialists were astonished when they inadvertently re-discovered the therapeutic properties of CO_2 supplementation while searching for better ways to manage ARDS.[227] Because of the hoax, they had no knowledge of the prior publications of Yandell Henderson, and they assumed that CO_2 was inherently dangerous. As a result, they invented the strange term *permissive hypercarbia* to describe their discovery. CO_2 supplementation soon became accepted to optimize cardiac output, minimize lung damage, and optimize respiratory drive in critically ill ICU patients. Both critical care and anesthesia research confirmed its merits, but no mention of permissive hypercarbia appeared in anesthesiology publications or practice guidelines and standards, even though capnography had become ubiquitous in operating rooms. Combinations of permissive hypercarbia and narcotics would have revolutionized anesthetic management and surgical outcome if their properties had been appreciated. Instead, the hoax prevailed, and anesthesiologists continued to hyperventilate their patients. If critical care specialists can employ hypercarbia in the ICU, then why can't anesthesiologists employ hypercarbia in the operating room to optimize narcotic dosage, cardiorespiratory function, and surgical outcome??

AIRWAY MANAGEMENT AND THE HOAX

Dr. Waters emphasized the principle of "secure airway," which was perhaps his only worthwhile contribution to the specialty he created. Even that laudable concept is being abandoned, mainly because the Waters technique promotes paralyzing agents that prolong emergence, recovery, and discharge. As a result, many practitioners have abandoned elective intubation in favor of "open airway" techniques that facilitate fast tracking but invite aspiration, airway obstruction, fire, laryngospasm, and hypoventilation.[228] Such techniques are inherently dangerous, and have caused countless disasters, most of which are concealed in the name of "privacy." A good example is the comedienne Joan Rivers, who suffered hypoxic brain damage at the age of 80 when she was subjected to open airway anesthesia for a minor vocal cord procedure in a doctor's office in New York City, and died a week later.

Also forgotten is the superior safety of mask induction. Sevoflurane has revolutionized its utility, and it deserves a fresh look.[229] Many practitioners presume that intravenous hypnosis and paralysis are essential to enable intubation, prevent dental and vocal cord damage, and assure patient comfort, but these assumptions lack substance.[230] Combinations of surprise and paralysis invariably precede induction airway disasters.[231] Fatigue, hunger, and hubris readily frustrate pre-operative detection and management of anatomical airway hazards.[232, 233] Routine mask induction and intubation without paralysis enables safe retreat from unexpected airway difficulties because it eliminates irreversible paralysis of airway muscles that can unexpectedly obstruct the airway, causing "can't intubate, can't ventilate" predicaments. Mask induction automatically incorporates "pre-oxygenation," and eliminates potentially lethal Propofol bolus toxicity.[234, 235] Its effects take place gradually, so that induction can be terminated at the first sign of trouble. Elective endotracheal intubation can be accomplished efficiently with the help of modern monitors, advanced inhalation agents, and small supplements of synthetic narcotics.[236, 237, 238] With CO_2 supplementation, nitrous oxide, and modern inhalation agents, mask induction can be comparably efficient and more pleasant than intravenous induction, which requires painful intravenous access.[239]

THE "SURGICAL HOME" AND THE HOAX

In recent years the hoax has caused persistent disparagement of narcotics in current literature, combined with fast tracking requirements. These considerations, plus the advent of improved ultrasound imaging technology, have encouraged anesthesiologists to embrace analgesic block techniques as an alternative to supplementing general anesthesia with narcotics. This approach is called the *Surgical Home*. Such techniques can produce excellent outcomes, but their shortcomings and complications are generally overlooked. Analgesic block installation requires expensive ultrasound equipment and nursing assistance, is time-consuming, and less reliable than narcotic analgesia. Often several nerves must be blocked simultaneously to optimize surgical outcome. For example, total knee replacement surgery requires simultaneous blockade of the femoral nerve, the lateral femoral cutaneous nerve, and the sciatic nerve. If any of these blocks fails or wears off prematurely, the analgesic benefits are largely negated. Blocks also invite nerve damage and toxic effects of the analgesic agents, and they often cause patient discomfort or distress. By comparison, supplementation of general anesthesia with hypercarbia and narcotics is safer, simpler, more efficient, and more reliable.

THE RED CROSS, CPR, AND THE HOAX

The American Heart Association provides a convenient summary of the history of cardiopulmonary resuscitation from the 1700s to the present at this website: https://cpr.heart.org/en/resources/history-of-cpr. What is most interesting about this summary is its complete omission of the utility of Carbogen, which is undoubtedly the single most effective emergency means to save lives yet discovered.

Life-threatening medical emergencies typically disrupt oxygen transport and delivery to brain cells, which may die in as few as five minutes if they are deprived of oxygen. This may be caused by a wide variety of circumstances, including heart attack (wherein the heart stops pumping oxygenated blood from the lungs to the brain and other tissues), stroke (which disrupts brain perfusion with oxygenated blood), drowning, electric shock, trauma, drug overdose, smoke inhalation, carbon monoxide poisoning, and so forth. Thus, the priority is *restoration of oxygen transport and delivery*, and time is of the essence. Such situations necessitate treatments that are simple, safe, effective, available, and easy to use.

Unwitnessed medical emergencies have negligible chance of survival, and even witnessed emergencies have very poor outcome without effective treatments. George Washington Crile proved the limited utility of chest compressions, epinephrine, and atropine, but these measures alone are seldom effective, plus epinephrine and atropine are seldom available outside hospitals. Yandell Henderson proved that Carbogen is by far the best single treatment for CPR:

1. It stimulates breathing to replenish lung oxygen.
2. It opens the capillary gate, which reduces microvascular flow resistance and optimizes cardiac output, tissue perfusion, and tissue oxygenation.
3. It releases oxygen from the hemoglobin molecule to enable cellular oxygen uptake.
4. It counteracts narcotic and toxic sedative respiratory depression.
5. It is safe and harmless.
6. It is inexpensive, so that it can be made available wherever it is needed.
7. It is simple and easy to use.
8. It complements defibrillation, atropine, and epinephrine.

One hundred years ago, Carbogen was commonly present in hospitals, doctors' offices, and fire trucks. It saved countless lives (see Figure 18.16 through Figure 18.35, following), but the very memory of this powerful treatment was extinguished by the CO_2 hoax. This situation cries for reform. Ideally IEDs, epinephrine, atropine, and Carbogen should be legally required wherever fire extinguishers are present, and it should be required equipment for fire trucks, police cars, city buses, taxicabs, airports, restaurants, stores, hospitals, clinics, golf courses, and anywhere else a cardiorespiratory arrest might be witnessed. Such a mandatory re-introduction of the therapeutic benefits of carbon dioxide would go far to reform medicine and research.

Unfortunately, in recent years the Red Cross has introduced counterproductive emergency treatment protocols including BLS (basic life support), ACLS (advanced cardiac life support), PALS (pediatric advanced life support), and ATLS (advanced trauma life support). Even the Red Cross admits that their ability to save lives is negligible. Where simplicity is in order, they promote confusing, complicated regimens that require intravenous access, advocate polypharmacy with unproven and even counterproductive drug combinations, excessive intravenous fluids, hyperventilation, and so forth. Such treatments are confusing and counterproductive even in hospitals where the drugs and equipment are immediately available. Worse yet, hospitals and clinics have embraced these training programs as a sort of "band-aid" that mimics effective measures, so that they are almost universally required for employment. They have become a time-wasting and onerously expensive burden for health care workers and a racket for the Red Cross. Their expensive re-certification is required every few years, and the Red Cross deliberately changes their protocols at regular intervals using manipulated statistics to force trainees to purchase the latest versions of their worthless books. In short, these programs do not serve public health. They should be abolished, and the Red Cross should be permanently banned from any such mischief.

More recently, computer technology has enabled the development of portable automated external defibrillator (AED) devices that automatically detect and treat ventricular fibrillation (disorganized heart contraction that disrupts effective heart function). Thus, the availability of Carbogen, AEDs, epinephrine, and atropine would greatly improve chances of positive outcome in medical emergencies. Unlike the Red Cross protocols, these treatments are simple, safe, complimentary, and proven effective, can be used successfully with minimal training, and are inexpensive enough to be made available wherever they might be needed.

NEUROSURGERY AND THE HOAX

The hoax appeals to neurosurgeons, because mechanical hyperventilation shrinks brain volume, lowers intracranial pressure, and reduces brain bleeding. Surgeons mistakenly regard this as beneficial because the brain is vulnerable to increased intracranial pressure, and brain tissue is extremely vulnerable to bleeding. However, recent evidence has revealed that hyperventilation somehow undermines brain surgery outcome.[240, 241, 242, 243, 244] The reason remains unclear to both surgeons and anesthesiologists, but it would be obvious were it not for the hoax.[245, 246]

Hyperventilation undermines tissue oxygenation throughout the body,[247] but the brain is particularly susceptible, because exaggerated brain cell metabolism demands continuous delivery of oxygen and nutrients. Brain tissue is accordingly equipped with three powerful mechanisms that maintain and regulate perfusion:

1. Carbon dioxide is the primary regulator of the capillary gate in all tissues throughout the body. It directly releases nitric oxide (NO) from capillary walls to cause nitrergic neurogenic vasodilation. The NO binds to Ca+, inhibits thrombin, and halts the generation and stabilization of insoluble fibrin to maintain capillary patency and flow (i.e. open the capillary gate). Compared to other tissues, the brain is exquisitely sensitive to CO_2 levels. Hypercarbia dramatically exaggerates brain blood flow, and hypocarbia drastically restricts it.[248]
2. The brain is richly endowed with both sympathetic and parasympathetic nerve endings, so that autonomic balance regulates microvascular flow resistance and brain blood flow distribution.[249] As in other tissues, parasympathetic nervous activity releases NO to open the capillary gate, and sympathetic nervous activity releases von Willebrand Factor to close the capillary gate.[250]
3. The brain is equipped with astrocytes that produce nitric oxide to open the capillary gate and maintain brain perfusion regardless of stress. Astrocytes explain why ischemic brain damage seldom occurs despite massive doses of epinephrine during cardiac arrest.[251, 252]

Astrocytes, Brain Bleeding, and Tissue Factor

Astrocytes maintain microvascular perfusion, but they cripple capillary hemostasis and render the brain vulnerable to bleeding. To counteract bleeding, brain tissue is rich in tissue factor, which activates blood enzyme factor VII to enhance coagulation and tissue repair. The tissue factor introduces yet another problem. Normally, the tissue factor is confined within the brain by a specialized brain vascular endothelium (this is called the *blood-brain barrier*), but brain surgery and brain trauma release the tissue factor into systemic circulation, which induces harmful systemic stress mechanism hyperactivity, and causes systemic inflammation and blood hypercoagulability. This, in turn, explains why brain surgery and brain trauma exaggerates morbidity and mortality more than do other types of surgery and injury.

Because brain tissue is so vulnerable to bleeding and brain cells themselves are fragile, the brain requires protection from trauma. This explains why the brain is enclosed within the protective skull. This invites yet another problem unique to the brain: the limited skull volume renders the brain vulnerable to swelling, which dangerously increases intracranial pressure. Severe increases in intracranial pressure can cause cerebellar tissue to “herniate” into the brain stem, causing death.

Mechanical Hyperventilation and Dementia

Mechanical hyperventilation during surgery invites hypoxic brain damage because it abnormally depletes CO_2 tissue reserves, closes the capillary gate, and drastically reduces brain perfusion and oxygenation. Surgical nociception activates sympathetic activity in brain tissue that closes the capillary gate and exaggerates the harmful effects of hyperventilation. The combination of CO_2 depletion and surgical stress is especially dangerous in geriatric patients who suffer accelerated capillary senescence, and causes hypoxic brain damage that manifests as postoperative dementia. Combinations of hypercarbia and narcotic supplementation protect the brain by synergistically opening the capillary gate, optimizing brain tissue oxygenation, and preventing postoperative dementia.

Rebound Syndrome

Hyperventilation inhibits brain perfusion, which shrinks brain volume, lowers intracranial pressure, and reduces the risk of bleeding during brain surgery.[253] However, when the hyperventilation is halted, the oxygen-starved brain suddenly swells with blood. This is called *rebound syndrome*. This can suddenly increase intracranial pressure and can cause fatal herniation of cerebellar tissue into the brain stem.

POLIO AND THE HOAX

In the mid-1950s the Hewlett-Packard Corporation tested the early prototype of their capnograph machine at Rancho Los Amigos Hospital, where polio victims were maintained in Drinker respirators (iron lungs). They discovered that the patients were being hyperventilated and adjusted the ventilators to maintain normocarbia in the patients. The rate of cure doubled and the death rate fell in half. It seems that hyperventilation may have been deadlier than polio.[254]

CORPORATE INFLUENCE AND THE HOAX

The hoax could never have survived so long without the support of medical corporations. Many understand that corporations reap government-subsidized profits, but few realize how they profit at the price of public health.[255, 256, 257] Corporations routinely subsidize fraudulent research to promote their products. The researchers who fabricate such frauds sometimes serve time in prison, but those who pay them and those who publish them go Scot-free.[258, 259, 260, 261, 262] The dead hand of corporate orthodoxy strangles fresh ideas to preserve the profitable status quo. Corporate influence is so thorough that no cogent description of the mechanism of oxygen transport and delivery (see Chapter 14: Oxygen Transport and Delivery, on page 137) can be found in medical textbooks today, or even the Internet, and major contributors to medical advance, such as Crile and Henderson, are forgotten along with their contributions.[263] The "International Standards for a Safe Practice of Anesthesia" exemplifies this untoward influence.[264] The "Standards" mandate pulse oximeters, oxygen monitors, and anesthesia machine modifications to prevent hypoxia, *but ignores guidelines for CO_2 management despite the dangers of hyperventilation, the known benefits of hypercarbia, and the universal availability of modern machines and monitors that enable the safe use of hypercarbia.*[265, 266, 267]

This is because the hoax has declared CO_2 depletion "not a problem" and "the less CO_2, the better." Considering the critical importance of CO_2 management during anesthesia, the lack of CO_2 standards is inexcusable because:

- CO_2 is necessary for oxygen transport and delivery, which is essential for vertebrate life.
- CO_2 enrichment of inhaled air enhances tissue oxygenation even without oxygen supplementation.[268]
- Hypocarbia harmfully inhibits tissue oxygenation, while hypercarbia improves oxygen transport and delivery to organs and tissues.
- Hypoxia provides warning in the form of visible cyanosis, while harmful hypocarbia reveals no distinguishing clinical manifestations during anesthesia.
- Capnography and pulse oximetry are now universally available to monitor and measure signs of impending CO_2 asphyxiation.
- Modern open circuit anesthesia machines have abolished the danger of CO_2 asphyxiation.

Yet another example of corporate influence on otherwise good medical practice is the manipulative reimbursements of medical insurance companies. Around 1995 insurance companies suddenly and arbitrarily altered their reimbursement policies to impose fast tracking (same-day surgery), which abolished the proven policy of retaining recovering patients for at least a day after surgery to protect them from postoperative problems.[269] This change occurred at a critical juncture, without debate or discussion, soon after capnography became available to monitor and measure carbon dioxide levels, soon after the therapeutic benefits of narcotics were re-discovered by anesthesiologists, and soon after the benefits of hypercarbia were re-discovered by critical care specialists. The change had the disastrous effect of preventing anesthesiologists from recognizing the benefits of the complimentary combination of narcotics and hypercarbia, and so comprehending the canard of the hoax.[270, 271, 272, 273, 274, 275, 276, 277] Worse yet, fast tracking thrust impossible responsibility upon anesthetists, nurses, and surgeons to detect pre-existing medical problems, because fast-tracked patients are dangerously discharged into the hands of their friends and families who have no medical training or facilities.[278, 279, 280, 281, 282, 283, 284] Given the timing and circumstances of these destructive changes, one cannot help but wonder if the underlying motive was to sustain the CO_2 hoax. If so, it worked: anesthesiologists promptly abandoned narcotics to facilitate fast tracking. Hospitals saved money and spared themselves from legal responsibility for the care of recovering patients. Soon thereafter, reports of previously unknown problems such as *Obstructive Sleep Apnea* (OSA) and *Systemic Inflammatory Response Syndrome* (SIRS) began to appear in medical literature.[285, 286, 287, 288, 289, 290, 291, 292, 293, 294, 295, 296, 297, 298,]

[299, 300, 301, 302, 303, 304] Increases in cancer, heart disease, and chronic illnesses no doubt also occurred, but remain overlooked — because nobody *has* looked.[305, 306]

Hospital corporations also influence medical practice for the sake of profit. In the late 1980s hospitals suddenly sought to control in-house physicians including anesthesiologists and radiologists. This involved legal maneuvers to create "closed" medical staffs. In the case of anesthesiology, this meant that the hospital administrator would issue a "master contract" to the anesthesiologist of his choice, whereupon that anesthesiologist would take responsibility for issuing "subcontracts" to subordinate anesthesiologists. This created an abusive environment wherein medical staff privileges were rendered meaningless, and the holder of the "master contract" could dictate practice standards in accord with his own political and financial considerations. Suffice it to say this created chaos at the expense of practice standards and patient care.

WHICH FORK IN THE YELLOW BRICK ROAD OF ANESTHESIA PROGRESS?

The hoax has devastated anesthesia progress, but the future of anesthesia is nevertheless pregnant with potential. Early anesthetists possessed superior science, but they lacked modern monitors, machines, drugs, and equipment. Safety, efficiency, and outcome can be optimized by combining their science with modern machines, medications, and monitors.[307, 308] Things have changed. New synthetic opioids are nearly devoid of side effects. Isoforane offers superior chemical stability. Sevoflurane is the best agent for mask induction yet developed. The dangers of explosion and CO_2 asphyxiation are gone, and soda lime serves no purpose.[309, 310, 311, 312, 313, 314] Pulse oximetry and capnography have revolutionized anesthetic management, but their ability to optimize opioid dosage and maximize surgical safety remains unappreciated. The anesthesiology profession is poised at the threshold of revolutionary advance if the full potential of modern machines, monitors, agents, and research information is embraced.

A UNIVERSAL TECHNIQUE TO OPTIMIZE SURGICAL OUTCOME

There are many ways to skin a cat, but there is also a best way. Although risk-free anesthesia is probably impossible, stress theory can be utilized to alter anesthetic technique to optimize surgical outcome using presently available medications and equipment. The hoax has prevented such progress. Routine mask induction and elective intubation followed by maintenance of general endotracheal anesthesia with ½ MAC anesthetic agent complemented by judicious opioid analgesia, minimal paralysis, and permissive hypercarbia combines the advantages of the Waters technique with those of the Crile/Henderson technique, and eliminates the shortcomings of both.[315, 316, 317, 318, 319, 320] It offers a practical, universally applicable means to minimize inhalation agent toxicity and optimize safety, simplicity, efficiency, cost, predictability, patient comfort, and outcome using presently available machines, monitors, and medications. Safe and beneficial hypercarbia is achieved with $FICO_2$ 10% (80 psi), which approaches the upper limit enabled by modern anesthesia machines.[321, 322, 323, 324, 325] The technique is pleasant for patients, safer than alternate techniques, and equally efficient. Narcotic supplementation is safer, simpler, faster, and less problematic than block techniques, and it enables effective control of pain and stress. It minimizes morbidity and mortality during surgery and there is every reason to believe that it can reduce cancer and heart disease by 30% or more in the distant aftermath of surgery.[326, 327, 328, 329, 330] The results of a clinical trial using these principles are presented in Appendix J: A Universal Anesthetic Approach that Optimizes Surgical Outcome, on page 453.

CONCLUSION

I value privacy, science, and common sense. I never desired to become a "reformer" or "whistleblower," but neither did I pursue a medical career to disseminate disease and death. I am appalled by the realization that my chosen profession was founded on false science by perverse purveyors of power, politics, privilege, persuasion, salesmanship, showmanship, cynicism, and deceit — the characteristics I least admire. Ralph Waters and Chauncey Leake were moral and mental midgets who contrived to corrupt medical science with their cunning confabulations and specious experiments for nothing more than shallow power and prestige. Seldom have so few inflicted so much harm upon so many for such trivial gain. They lacked the intelligence to anticipate, the courage to confront, and the decency to care about the consequences they caused. They contributed nothing but a fearsome example of the danger of polluting science with politics. Their mischief flung a "monkey wrench" into the delicate mechanism of scientific advance that has reversed medical progress for nearly four generations and killed and maimed countless patients. Their hoax must be revealed and reformed if medical progress is to have any hope of headway.

Medicine is of vital interest to every individual, rich and poor alike. Health, longevity, and productivity are essential to the survival and progress of civilization, the blessings of which are all too fragile. The only legitimate purpose of any medical profession is to provide the best medical care that science can offer.

The anesthesiology profession is a disaster. It has ignored and suppressed the proven benefits of anesthesia science and reversed medical progress for more than 60 years. It has failed to implement standards and guidelines for the management of carbon dioxide, narcotics, polypharmacy, acid/base balance, blood pressure, and intravenous fluids. Its policies are venal, self-serving, self-perpetuating, and immune to reform by its powerless, dues-paying members. Small wonder that its membership is declining.

The abandoned principles and practices of the previous generation merit reappraisal. With capnography, pulse oximetry, modern anesthesia machines, synthetic opioids, and improved inhalation agents it is now practical and safe to realize the full advantages of the Crile/Henderson technique and eliminate its shortcomings. Such advance can minimize surgical morbidity and mortality, pave the path for profitable and progressive pharmaceutical development, and re-revolutionize surgery. It promises priceless prestige for the profession, but reform won't be easy. Inspired leadership is lacking, and the hour is late. However, there are benefits for all concerned. In the opinion of this author, the complexity of anesthesia science and practice merits in-depth medical education and training, and the anesthesiology profession merits preservation — provided it can reform.

Must progress await our great-grandchildren? Why not us? Why not now?

Figure 18.15: Aqualumni Tree

This Aqualumni Tree cartoon illustrates the success of Dr. Waters' strategy of placing his residents in positions of academic influence, where they effectively promoted his disastrous anesthetic techniques and beliefs.[331]

MODERN MECHANIX
YESTERDAY'S *Tomorrow* TODAY

Mechanics And Handicraft
Issue: Oct, 1938
Posted in: Medical
Posted: 09/13/2007

1 Comment On A Dry-Ice Oxygen Tent For Infants (Oct, 1938)

A Dry-Ice Oxygen Tent for Infants (Oct, 1938)

I'm not exactly a medical expert, but this seems like a bad idea. CO2 stimulates respiration because you can't breathe it. Though this tent would probably work great for a baby tree.

A Dry-Ice Oxygen Tent for Infants

IN the constant battle to cut down the infant mortality rate due to diseases easily contracted when vitality is relatively low, physicians have developed a new type of oxygen tent, which uses dry-ice in combination with the oxygen. Of course, it is a well-known fact that carbon dioxide stimulates respiration; dry-ice is pure carbon dioxide.

The new oxygen tent is easily arranged behind the infant's crib and a transparent envelope is let down to lead the gases to the youngster. How this apparatus works was demonstrated at the annual convention of the American Medical Association, attended by 7,000 physicians. However, instead of using an infant, a life-size doll was substituted for the patient. So simple and convenient is the apparatus that the physician can transport it to the bedside of his patient. It is a safety factor for childbirths taking place at home.

Miss Helen Price demonstrating the newest oxygen tent to 7,000 physicians.

A Dry-Ice Oxygen Tent for Infants

IN the constant battle to cut down the infant mortality rate due to diseases easily contracted when vitality is relatively low, physicians have developed a new type of oxygen tent, which uses dry-ice in combination with the oxygen. Of course, it is a well-known fact that carbon dioxide stimulates respiration; dry-ice is pure carbon dioxide.

The new oxygen tent is easily arranged behind the infant's crib and a transparent envelope is let down to lead the gases to the youngster. How this apparatus works was demonstrated at the annual convention of the American Medical Association, attended by 7,000 physicians. However, instead of using an infant, a life-size doll was substituted for the patient. So simple and convenient is the apparatus that the physician can transport it to the bedside of his patient. It is a safety factor for childbirths taking place at home.

Related posts:

1. INSTANT OXYGEN (Oct, 1962)
2. SWIMMERS PEPPED UP BY WHIFF OF PURE OXYGEN (Oct, 1934)
3. Japanese Rollerblades in 1938 (Jul, 1936)
4. Auto Seat Gives Infant Comfort – AKA "Kiddie Catapult" (May, 1936)
5. Rotating Shelter Aids Sun-Bathers (May, 1938)

1 comment

Sorcerer Mickey says:
September 14, 2007
2:08 pm

And when Baby's looking a bit pale, there the Acme Carbon MOnoxide Tent to add a little color to his cheeks!

Submit comment

You must be logged in to post a comment.

ADVERTISEMENTS

SEARCH:

Search

CATEGORIES:

Categories: Select Category

ARCHIVES

Archives Select Month

SUBSCRIBE

Follow @ModernMechanix

Like Page

Be the first of your friends to like this

RSS - Posts
RSS - Comments

RECENT COMMENTS

Maritime Monday For September 19th, 2016: Disjecta Membra – GCaptain [[...] Big Dam to Water Sahara (Jul, 1933) – Yeah, well... what else are you going to do with all that leftover water, genius! The...]

10 Curious Automobile-Related Inventions From Bygone Days - JustTopTens.com [[...] Glare-proof glasses eliminated headlight glare from approaching cars. When meeting an oncoming car, the driver would simply turn his head slightly to the right,...]

11 Old-Timey Beauty Ingredients That Will Freak You Out | Backcap.org [[...] blog.modernmechanix.com [...]]

Is Earth Actually Flat? « YouTubeMonster [[...] http://blog.modernmechanix.com/5000-for-proving-the-earth-is-a-globe/ http://stevelyons.ca/projects/flat-earth-

Figure 18.16: Continuing use of CO_2

This article shows that physicians continued to utilize CO_2 supplementation at least as late as 1938.

FOREGGER SALES BROCHURES

The Foregger company, which was based in Manhattan, New York, became the most respected manufacturer of anesthesia equipment during the early days of modern anesthesia. The following copies of two of its sales brochures were kindly provided by the Wood Library and Museum of Anesthesia History. They reflect both the fear of CO_2 (as reflected by their advertisement of the Waters canister, which contained soda lime for the purpose of removing CO_2 from the gas mixture) and its use as a beneficial anesthesia supplement (as reflected by the carbogen equipment for sale). The first brochure is dated 1926. The second is undated, but was probably published in the early 1930s, shortly before Dr. Waters began his campaign to vilify carbon dioxide. It is interesting to note that the earlier brochure offers various versions of the Waters Canister for sale, but the newer brochure makes no mention of the Waters canister, and instead documents the enthusiastic acceptance of CO_2 supplementation by physicians in general as well as anesthetists, both in its text and in its presentation of Carbogen inhalation equipment.

The Waters Canister was a simple metal cylinder that was filled with soda lime. It was intended to absorb carbon dioxide, which Dr. Waters (a medical physician) alleged to possess dangerous toxic and narcotic properties, while Dr. Yandell Henderson (a PhD researcher) performed and published extensive research that demonstrated the powerful therapeutic properties of carbon dioxide and Carbogen (mixtures of CO_2 and O_2 in pressurized tanks). Henderson's research threatened the sales of the Waters Canister. Dr. Waters emerged as the clear winner in this contest between a medical doctor and a PhD researcher, as reflected by the continued presence of soda lime in anesthesia machines to this day, and the absence of Carbogen and CO_2 treatments from current medical practice.

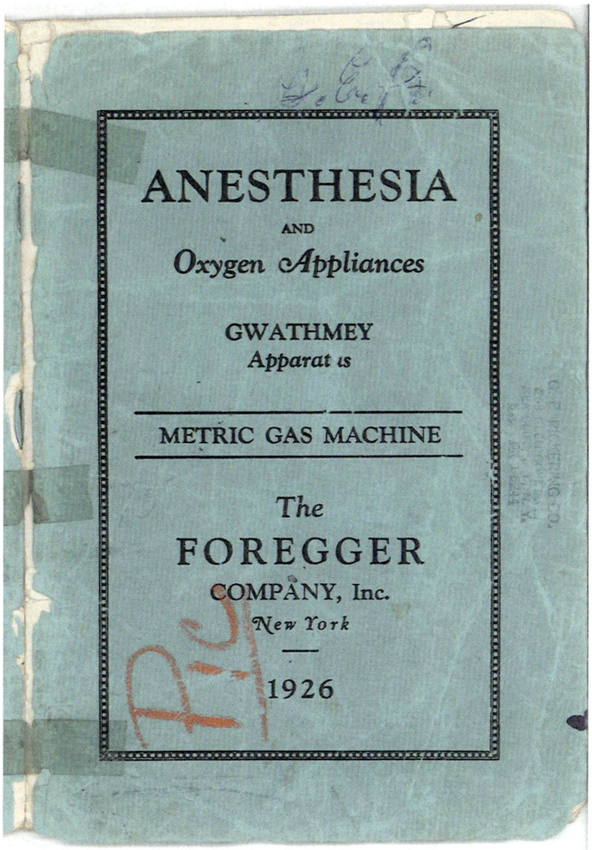

Figure 18.17: Foregger sales brochure, 1926

This is a Foregger anesthesia equipment sales brochure from the early days of modern anesthesia. It was generously shared by the Wood Library Museum of Anesthesiology (https://www.woodlibrarymuseum.org/). Note the date: 1926. This was when Dr. Ralph Waters was chosen to the Chairman of the Department of Anesthesia at Wisconsin University. He thus became the first chair of a university department of anesthesiology, which conferred considerable prestige. The Foregger company was already offering the patented Waters canister for sale.

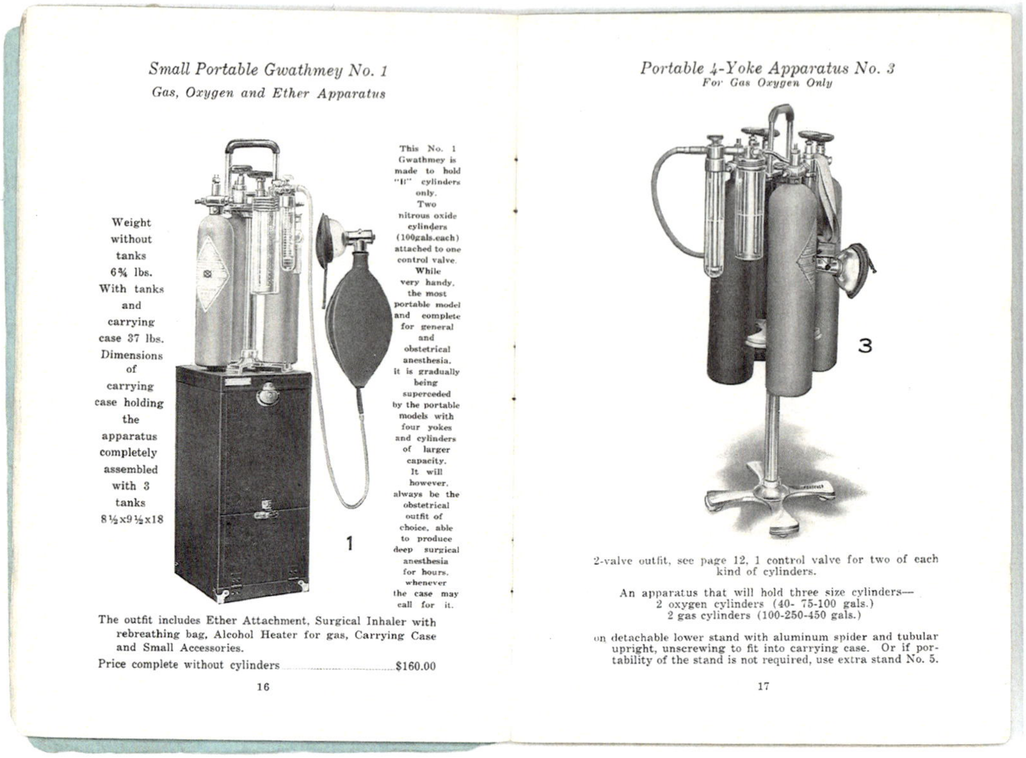

Small Portable Gwathmey No. 1
Gas, Oxygen and Ether Apparatus

Weight without tanks 6¾ lbs. With tanks and carrying case 37 lbs. Dimensions of carrying case holding the apparatus completely assembled with 3 tanks 8½x9½x18

This No. 1 Gwathmey is made to hold "B" cylinders only. Two nitrous oxide cylinders (100gals.each) attached to one control valve. While very handy, the most portable model and complete for general and obstetrical anesthesia, it is gradually being superceded by the portable models with four yokes and cylinders of larger capacity. It will however, always be the obstetrical outfit of choice, able to produce deep surgical anesthesia for hours, whenever the case may call for it.

1

The outfit includes Ether Attachment, Surgical Inhaler with rebreathing bag, Alcohol Heater for gas, Carrying Case and Small Accessories.

Price complete without cylinders $160.00

16

Portable 4-Yoke Apparatus No. 3
For Gas Oxygen Only

3

2-valve outfit, see page 12, 1 control valve for two of each kind of cylinders.

An apparatus that will hold three size cylinders—
2 oxygen cylinders (40- 75-100 gals.)
2 gas cylinders (100-250-450 gals.)

on detachable lower stand with aluminum spider and tubular upright, unscrewing to fit into carrying case. Or if portability of the stand is not required, use extra stand No. 5.

17

Figure 18.18: Portable tank systems.

These portable tank systems were apparently intended to be used for "home delivery" of babies. They delivered mixtures of oxygen and nitrous oxide.

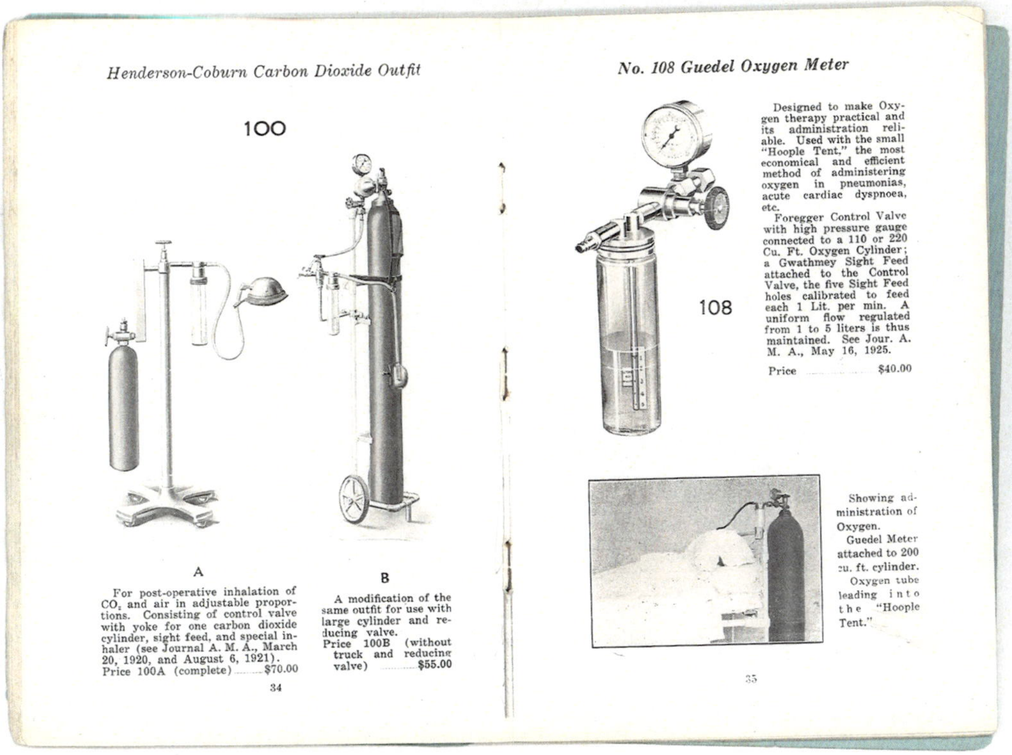

Henderson-Coburn Carbon Dioxide Outfit

100

A

For post-operative inhalation of CO_2 and air in adjustable proportions. Consisting of control valve with yoke for one carbon dioxide cylinder, sight feed, and special inhaler (see Journal A. M. A., March 20, 1920, and August 6, 1921).
Price 100A (complete) $70.00

B

A modification of the same outfit for use with large cylinder and reducing valve.
Price 100B (without truck and reducing valve) $55.00

34

No. 108 Guedel Oxygen Meter

108

Designed to make Oxygen therapy practical and its administration reliable. Used with the small "Hoople Tent," the most economical and efficient method of administering oxygen in pneumonias, acute cardiac dyspnoea, etc.

Foregger Control Valve with high pressure gauge connected to a 110 or 220 Cu. Ft. Oxygen Cylinder; a Gwathmey Sight Feed attached to the Control Valve, the five Sight Feed holes calibrated to feed each 1 Lit. per min. A uniform flow regulated from 1 to 5 liters is thus maintained. See Jour. A. M. A., May 16, 1925.

Price $40.00

Showing administration of Oxygen.
Guedel Meter attached to 200 cu. ft. cylinder.
Oxygen tube leading into the "Hoople Tent."

35

Figure 18.19: Device to add carbon dioxide

The apparatus on page 34 was designed to add carbon dioxide to atmospheric air to enhance cardiopulmonary function. Thus, carbon dioxide was understood as a treatment unto itself, with or without oxygen. Medical insurance was uncommon in that era, and CO_2 was far less expensive than other medical gases, as it was a "waste product" of oil processing. The "Gwathmey Sight Feed" apparatus on page 35 was a practical device that measured gas flows using calibrated holes that bubbled the gas through a small jar of water. It has been replaced by more expensive and less reliable pressure gauges.

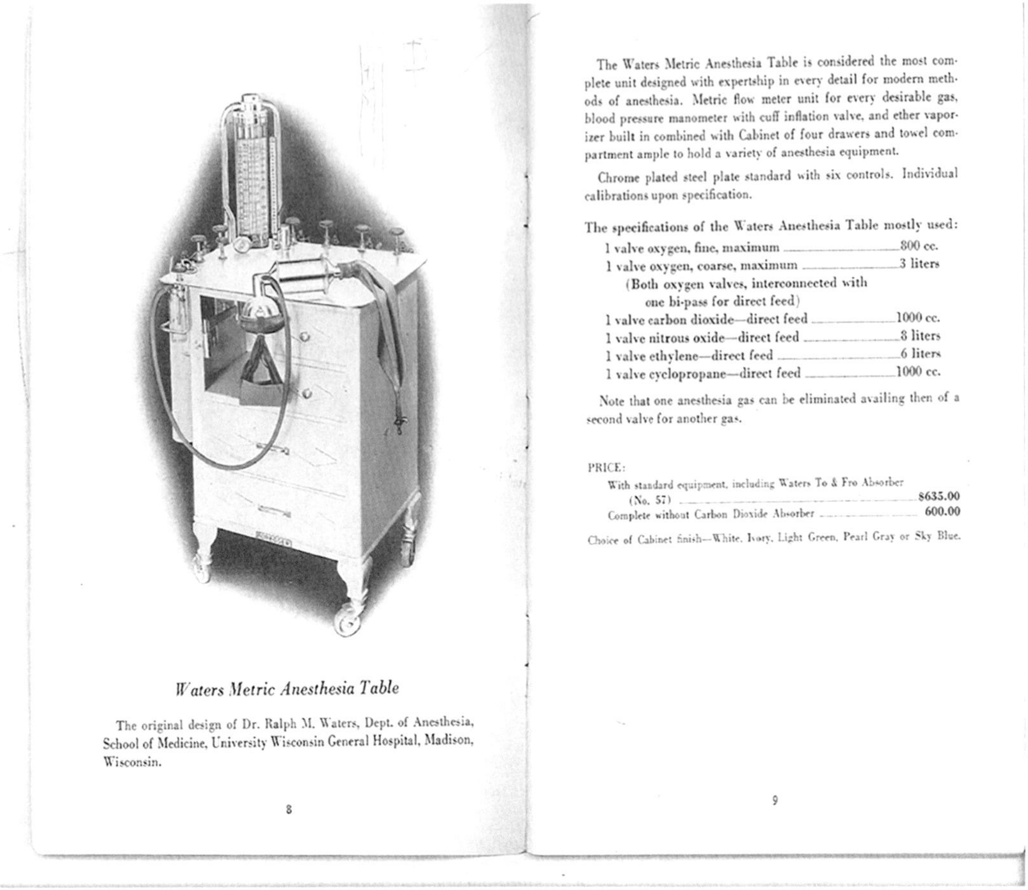

Waters Metric Anesthesia Table

The original design of Dr. Ralph M. Waters, Dept. of Anesthesia, School of Medicine, University Wisconsin General Hospital, Madison, Wisconsin.

8

The Waters Metric Anesthesia Table is considered the most complete unit designed with expertship in every detail for modern methods of anesthesia. Metric flow meter unit for every desirable gas, blood pressure manometer with cuff inflation valve, and ether vaporizer built in combined with Cabinet of four drawers and towel compartment ample to hold a variety of anesthesia equipment.

Chrome plated steel plate standard with six controls. Individual calibrations upon specification.

The specifications of the Waters Anesthesia Table mostly used:

1 valve oxygen, fine, maximum	800 cc.
1 valve oxygen, coarse, maximum	3 liters
(Both oxygen valves, interconnected with one bi-pass for direct feed)	
1 valve carbon dioxide—direct feed	1000 cc.
1 valve nitrous oxide—direct feed	8 liters
1 valve ethylene—direct feed	6 liters
1 valve cyclopropane—direct feed	1000 cc.

Note that one anesthesia gas can be eliminated availing then of a second valve for another gas.

PRICE:

With standard equipment, including Waters To & Fro Absorber (No. 57)	$635.00
Complete without Carbon Dioxide Absorber	600.00

Choice of Cabinet finish—White, Ivory, Light Green, Pearl Gray or Sky Blue.

9

Figure 18.20: Anesthesia machine attributed to Dr. Waters

Note how this early anesthesia machine design was attributed to Dr. Waters, and how it incorporated his Waters canister. It delivered mixtures of oxygen, nitrous oxide, ethylene, cyclopropane, and *pure carbon dioxide* to the Waters canister. The use of soda lime to absorb CO_2 from the breathing circuit was called the *absorption technique*. This illustrates the CO_2 insanity that has persisted throughout the history of anesthesia and medical practice. It makes no sense to combine CO_2 delivery with CO_2 absorption!

Soda lime consists of small rocks coated with sodium hydroxide that absorb carbon dioxide. Then and now its effects are mitigated by "channeling" so that most of the soda lime does not interact with the gas flowing through the canister.[332, 333, 334] Ignorance is bliss; then as now, patients survived, so everybody was happy. Gas flows were adjusted in accord with clinical signs and symptoms because practitioners of that era lacked modern monitors capable of measuring medical gas concentrations. However, modern studies have demonstrated that soda lime is harmful and counterproductive. It serves no useful purpose whatsoever.

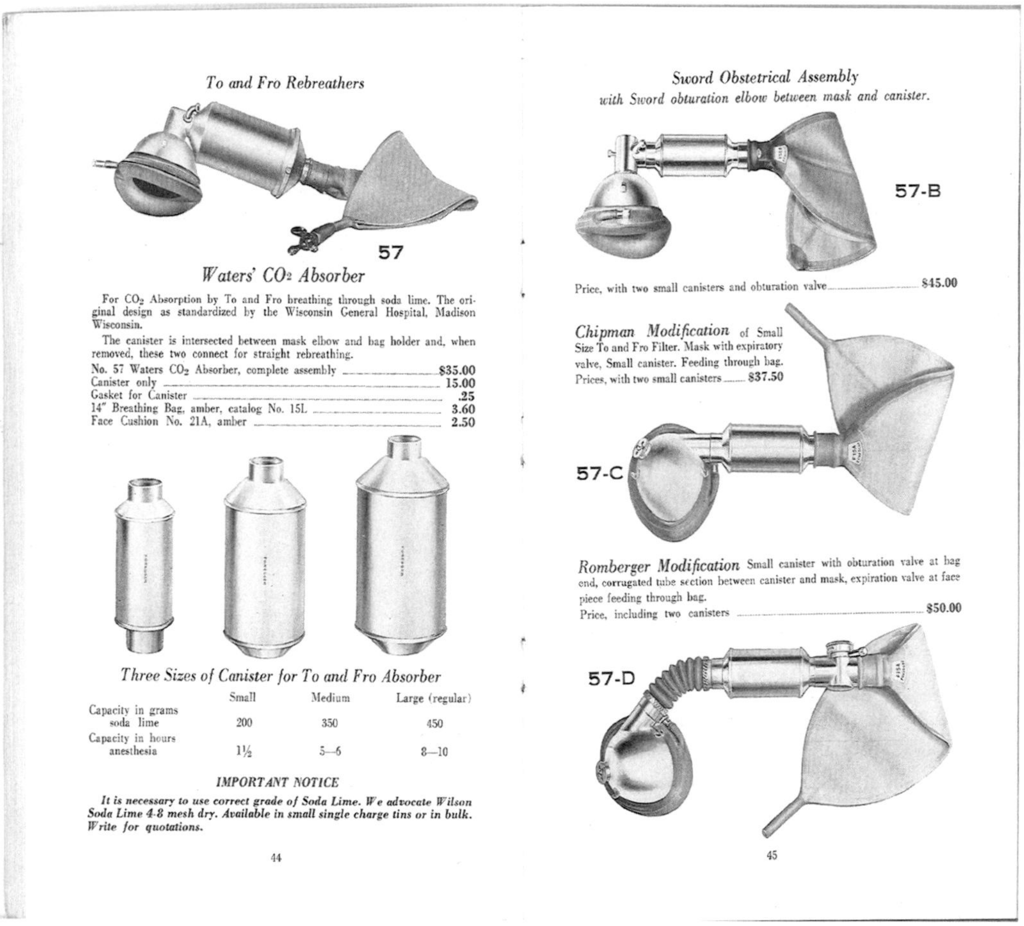

To and Fro Rebreathers

57

Waters' CO_2 Absorber

For CO_2 Absorption by To and Fro breathing through soda lime. The original design as standardized by the Wisconsin General Hospital, Madison Wisconsin.

The canister is intersected between mask elbow and bag holder and, when removed, these two connect for straight rebreathing.

No. 57 Waters CO_2 Absorber, complete assembly $35.00
Canister only 15.00
Gasket for Canister .... .25
14" Breathing Bag, amber, catalog No. 15L 3.60
Face Cushion No. 21A, amber 2.50

Three Sizes of Canister for To and Fro Absorber

	Small	Medium	Large (regular)
Capacity in grams soda lime	200	350	450
Capacity in hours anesthesia	1½	5—6	8—10

IMPORTANT NOTICE

It is necessary to use correct grade of Soda Lime. We advocate Wilson Soda Lime 4-8 mesh dry. Available in small single charge tins or in bulk. Write for quotations.

44

Sword Obstetrical Assembly
with Sword obturation elbow between mask and canister.

57-B

Price, with two small canisters and obturation valve $45.00

Chipman Modification of Small Size To and Fro Filter. Mask with expiratory valve, Small canister. Feeding through bag.
Prices, with two small canisters $37.50

57-C

Romberger Modification Small canister with obturation valve at bag end, corrugated tube section between canister and mask, expiration valve at face piece feeding through bag.
Price, including two canisters $50.00

57-D

45

Figure 18.21: Waters CO_2 absorber

Note these versions of the patented Waters CO_2 absorber, which was a simple to-and-fro breathing device with a canister containing soda lime that was intended to remove supposedly dangerous carbon dioxide from the gas mixture inhaled by the patient. This reflects the widespread belief that CO_2 is toxic "waste gas" that must be "rid from the body" by breathing that had prevailed from the earliest days of gas research. Dr. Waters capitalized on this belief by patenting and selling this device. Soda lime releases heat as it chemically reacts with carbon dioxide, so the practitioner would replace the canister when it was no longer warm to the touch.

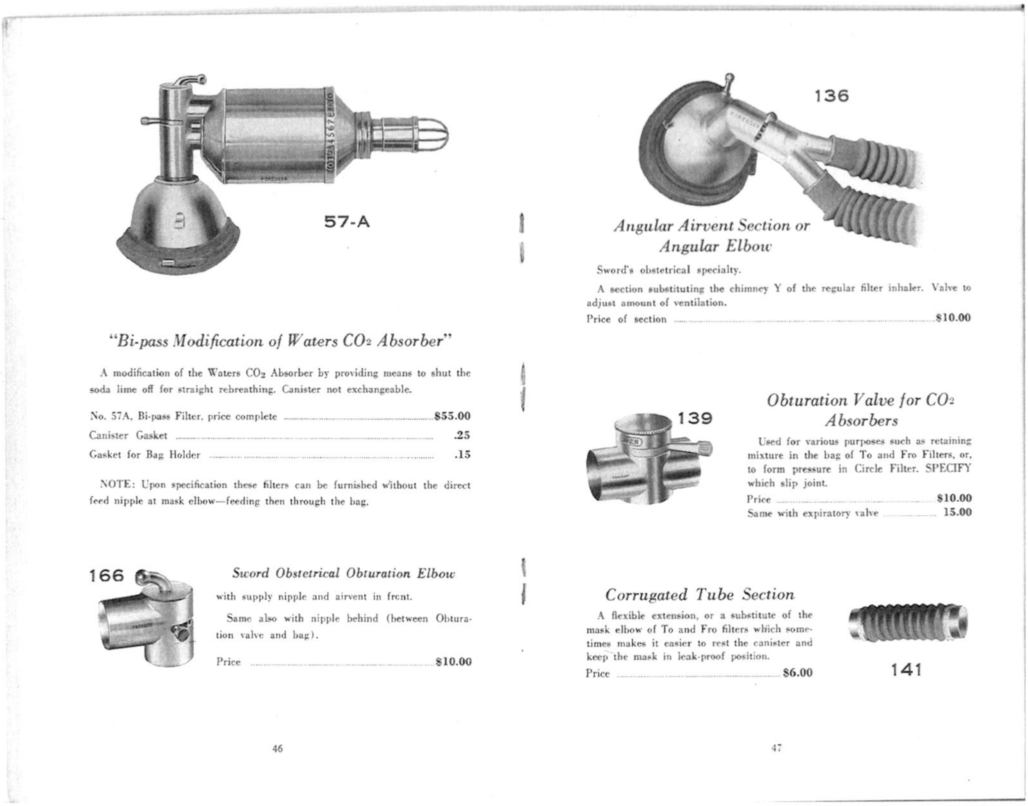

57-A

"Bi-pass Modification of Waters CO_2 Absorber"

A modification of the Waters CO_2 Absorber by providing means to shut the soda lime off for straight rebreathing. Canister not exchangeable.

No. 57A, Bi-pass Filter, price complete	$55.00
Canister Gasket	.25
Gasket for Bag Holder	.15

NOTE: Upon specification these filters can be furnished without the direct feed nipple at mask elbow—feeding then through the bag.

166

Sword Obstetrical Obturation Elbow

with supply nipple and airvent in front.

Same also with nipple behind (between Obturation valve and bag).

Price $10.00

46

136

Angular Airvent Section or Angular Elbow

Sword's obstetrical specialty.

A section substituting the chimney Y of the regular filter inhaler. Valve to adjust amount of ventilation.

Price of section $10.00

139

Obturation Valve for CO_2 Absorbers

Used for various purposes such as retaining mixture in the bag of To and Fro Filters, or, to form pressure in Circle Filter. SPECIFY which slip joint.

Price	$10.00
Same with expiratory valve	15.00

Corrugated Tube Section

A flexible extension, or a substitute of the mask elbow of To and Fro filters which sometimes makes it easier to rest the canister and keep the mask in leak-proof position.

Price $6.00

141

47

Figure 18.22: Diversion valve

Note the diversion valve that could be used to divert gas flow away from the Waters Canister while the soda lime was being refreshed.

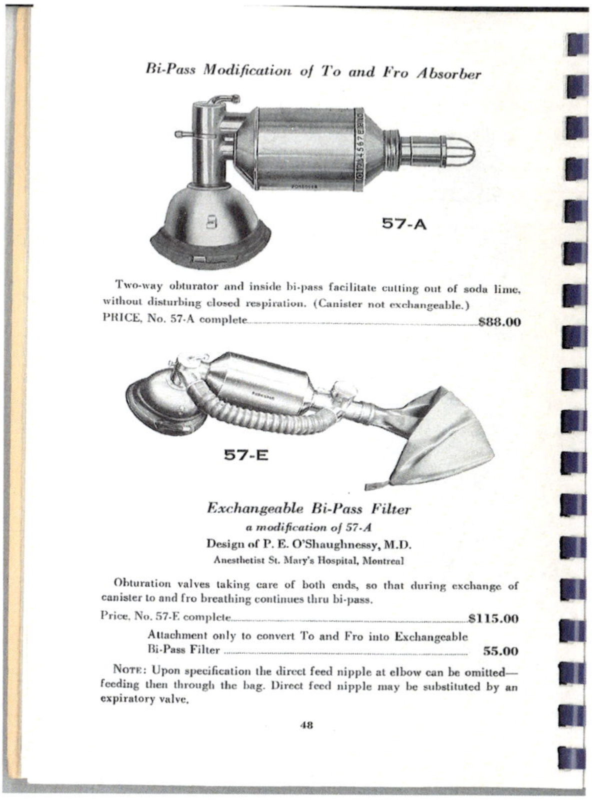

Bi-Pass Modification of To and Fro Absorber

57-A

Two-way obturator and inside bi-pass facilitate cutting out of soda lime, without disturbing closed respiration. (Canister not exchangeable.)

PRICE, No. 57-A complete..$88.00

57-E

Exchangeable Bi-Pass Filter

a modification of 57-A

Design of P. E. O'Shaughnessy, M.D.

Anesthetist St. Mary's Hospital, Montreal

Obturation valves taking care of both ends, so that during exchange of canister to and fro breathing continues thru bi-pass.

Price, No. 57-E complete..$115.00

Attachment only to convert To and Fro into Exchangeable Bi-Pass Filter 55.00

NOTE: Upon specification the direct feed nipple at elbow can be omitted—feeding then through the bag. Direct feed nipple may be substituted by an expiratory valve.

48

Figure 18.23: Bypass valve.
Note the bypass valve combined with the Waters canister.

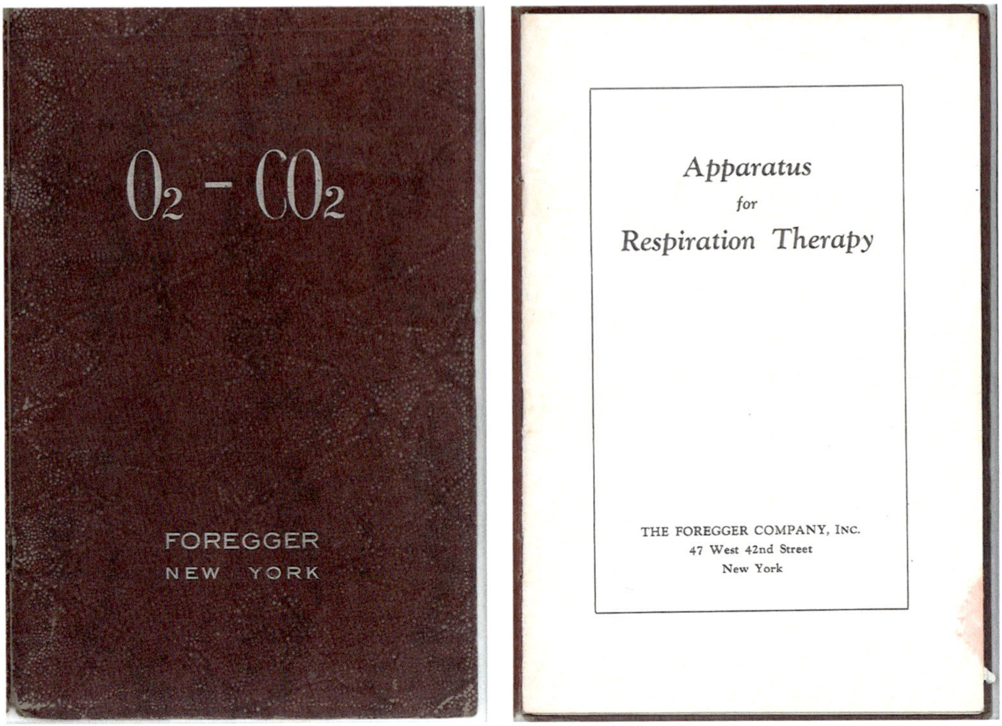

Figure 18.24: Foregger brochure from the 1930s

This is a copy of a more recent Foregger sales brochure that was generously provided by the Wood Library and Anesthesiology Museum. It is undated, but was probably produced in the early 1930s after CO_2 had been widely embraced as a medical treatment, and Carbogen (a pressurized mixture of oxygen and carbon dioxide) was commonly present on fire trucks and in hospitals and physicians' offices.

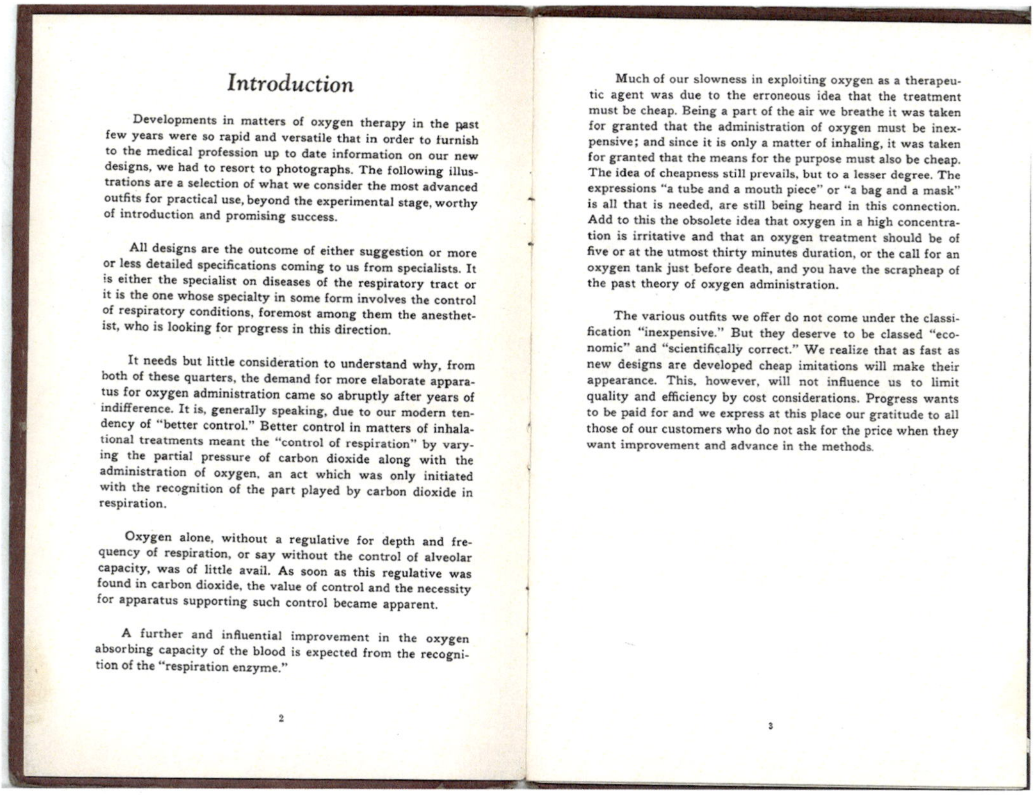

Introduction

Developments in matters of oxygen therapy in the past few years were so rapid and versatile that in order to furnish to the medical profession up to date information on our new designs, we had to resort to photographs. The following illustrations are a selection of what we consider the most advanced outfits for practical use, beyond the experimental stage, worthy of introduction and promising success.

All designs are the outcome of either suggestion or more or less detailed specifications coming to us from specialists. It is either the specialist on diseases of the respiratory tract or it is the one whose specialty in some form involves the control of respiratory conditions, foremost among them the anesthetist, who is looking for progress in this direction.

It needs but little consideration to understand why, from both of these quarters, the demand for more elaborate apparatus for oxygen administration came so abruptly after years of indifference. It is, generally speaking, due to our modern tendency of "better control." Better control in matters of inhalational treatments meant the "control of respiration" by varying the partial pressure of carbon dioxide along with the administration of oxygen, an act which was only initiated with the recognition of the part played by carbon dioxide in respiration.

Oxygen alone, without a regulative for depth and frequency of respiration, or say without the control of alveolar capacity, was of little avail. As soon as this regulative was found in carbon dioxide, the value of control and the necessity for apparatus supporting such control became apparent.

A further and influential improvement in the oxygen absorbing capacity of the blood is expected from the recognition of the "respiration enzyme."

2

Much of our slowness in exploiting oxygen as a therapeutic agent was due to the erroneous idea that the treatment must be cheap. Being a part of the air we breathe it was taken for granted that the administration of oxygen must be inexpensive; and since it is only a matter of inhaling, it was taken for granted that the means for the purpose must also be cheap. The idea of cheapness still prevails, but to a lesser degree. The expressions "a tube and a mouth piece" or "a bag and a mask" is all that is needed, are still being heard in this connection. Add to this the obsolete idea that oxygen in a high concentration is irritative and that an oxygen treatment should be of five or at the utmost thirty minutes duration, or the call for an oxygen tank just before death, and you have the scrapheap of the past theory of oxygen administration.

The various outfits we offer do not come under the classification "inexpensive." But they deserve to be classed "economic" and "scientifically correct." We realize that as fast as new designs are developed cheap imitations will make their appearance. This, however, will not influence us to limit quality and efficiency by cost considerations. Progress wants to be paid for and we express at this place our gratitude to all those of our customers who do not ask for the price when they want improvement and advance in the methods.

3

Figure 18.25: Carbogen supplementation gains acceptance

This introduction reflects the remarkable speed and enthusiasm with which CO_2 supplementation during anesthesia gained acceptance among the nurse-anesthetists who dominated anesthesia practice, and physicians subsequently embraced Carbogen as a powerful treatment. They clearly understood that carbon dioxide stimulates breathing that oxygenates blood, and that oxygen therapy confers limited therapeutic benefit unless it is combined with carbon dioxide. Note the reference to the cost of medical gases and the irritative effect of pure oxygen. They understood that breathing pure oxygen is harmful. Unlike today, medical gases including oxygen were expensive, while CO_2 was cheap. Combining oxygen and carbon dioxide synergistically enhanced therapeutic efficiency, and Carbogen was widely accepted as an emergency treatment for heart attacks, strokes, drowning, smoke inhalation, carbon monoxide poisoning, asthma, pneumonia, and breathing problems in newborn babies. Anesthetists used CO_2 supplementation to speed ether induction and emergence, stimulate breathing, optimize oxygen uptake and delivery, reduce the risk of ether fires and explosions, and prevent postoperative nausea, vomiting, respiratory depression, atelectasis, and pneumonia.

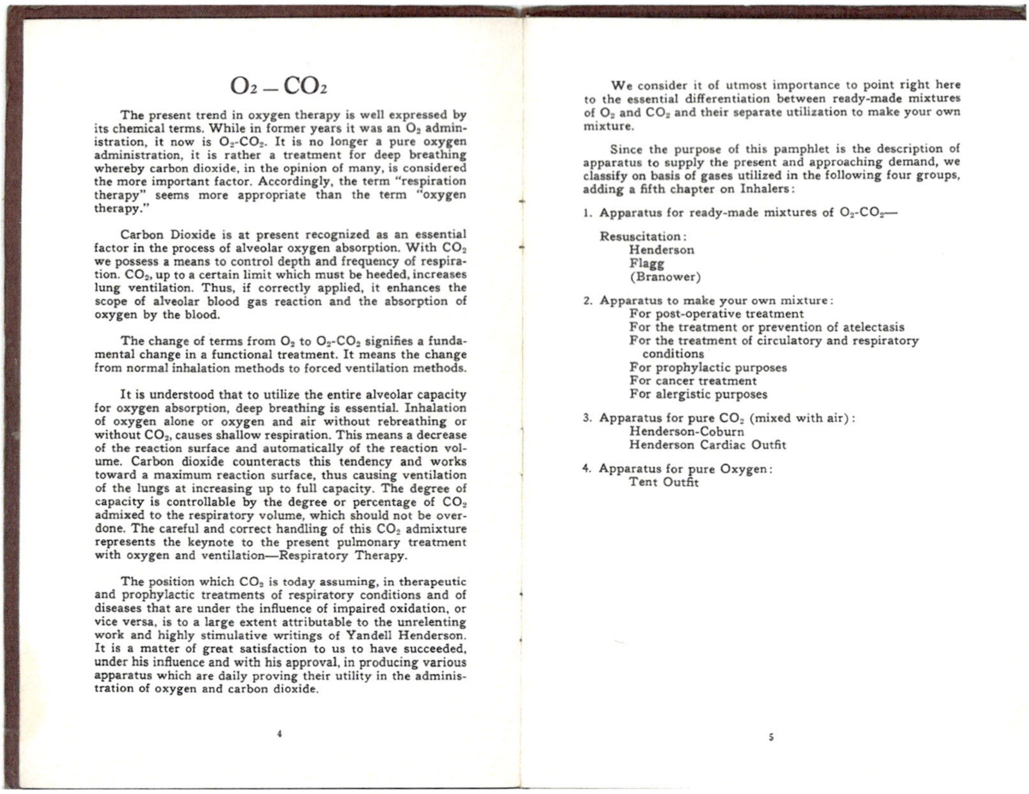

$O_2 - CO_2$

The present trend in oxygen therapy is well expressed by its chemical terms. While in former years it was an O_2 administration, it now is O_2-CO_2. It is no longer a pure oxygen administration, it is rather a treatment for deep breathing whereby carbon dioxide, in the opinion of many, is considered the more important factor. Accordingly, the term "respiration therapy" seems more appropriate than the term "oxygen therapy."

Carbon Dioxide is at present recognized as an essential factor in the process of alveolar oxygen absorption. With CO_2 we possess a means to control depth and frequency of respiration. CO_2, up to a certain limit which must be heeded, increases lung ventilation. Thus, if correctly applied, it enhances the scope of alveolar blood gas reaction and the absorption of oxygen by the blood.

The change of terms from O_2 to O_2-CO_2 signifies a fundamental change in a functional treatment. It means the change from normal inhalation methods to forced ventilation methods.

It is understood that to utilize the entire alveolar capacity for oxygen absorption, deep breathing is essential. Inhalation of oxygen alone or oxygen and air without rebreathing or without CO_2, causes shallow respiration. This means a decrease of the reaction surface and automatically of the reaction volume. Carbon dioxide counteracts this tendency and works toward a maximum reaction surface, thus causing ventilation of the lungs at increasing up to full capacity. The degree of capacity is controllable by the degree or percentage of CO_2 admixed to the respiratory volume, which should not be overdone. The careful and correct handling of this CO_2 admixture represents the keynote to the present pulmonary treatment with oxygen and ventilation—Respiratory Therapy.

The position which CO_2 is today assuming, in therapeutic and prophylactic treatments of respiratory conditions and of diseases that are under the influence of impaired oxidation, or vice versa, is to a large extent attributable to the unrelenting work and highly stimulative writings of Yandell Henderson. It is a matter of great satisfaction to us to have succeeded, under his influence and with his approval, in producing various apparatus which are daily proving their utility in the administration of oxygen and carbon dioxide.

4

We consider it of utmost importance to point right here to the essential differentiation between ready-made mixtures of O_2 and CO_2 and their separate utilization to make your own mixture.

Since the purpose of this pamphlet is the description of apparatus to supply the present and approaching demand, we classify on basis of gases utilized in the following four groups, adding a fifth chapter on Inhalers:

1. Apparatus for ready-made mixtures of O_2-CO_2—

   Resuscitation:
   - Henderson
   - Flagg
   - (Branower)

2. Apparatus to make your own mixture:
   - For post-operative treatment
   - For the treatment or prevention of atelectasis
   - For the treatment of circulatory and respiratory conditions
   - For prophylactic purposes
   - For cancer treatment
   - For alergistic purposes

3. Apparatus for pure CO_2 (mixed with air):
   - Henderson-Coburn
   - Henderson Cardiac Outfit

4. Apparatus for pure Oxygen:
   - Tent Outfit

5

Figure 18.26: Carbon dioxide and oxygen transport

Note how this brochure reflects the understanding that carbon dioxide is essential for oxygen transport and delivery, and thus more important than oxygen alone. It also reflects the remarkable speed with which the therapeutic properties of CO_2 were embraced, and how the research and writings of Dr. Yandell Henderson inspired this change. It is shocking to realize that these invaluable medical treatments and principles are now abandoned, forgotten, and banished from medical textbooks and literature.

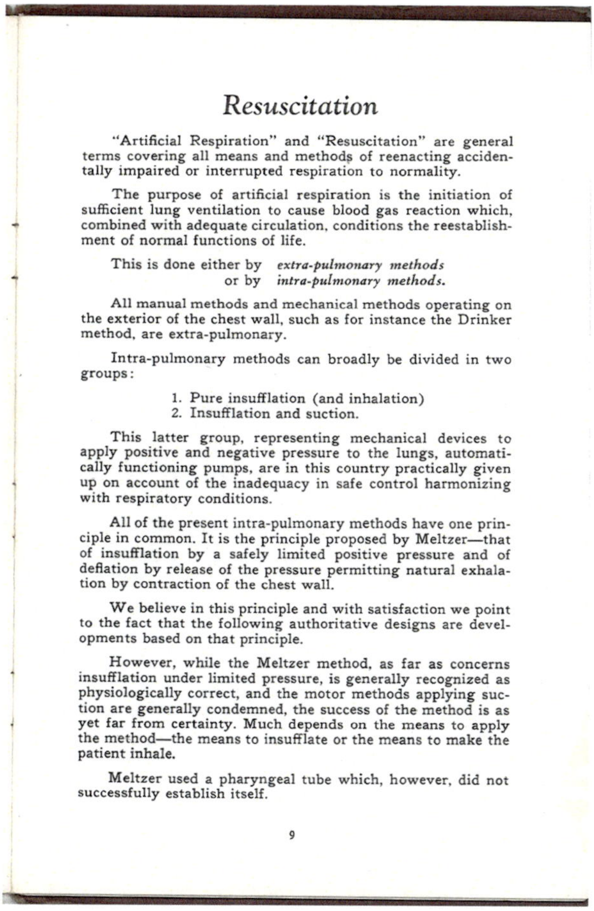

Resuscitation

"Artificial Respiration" and "Resuscitation" are general terms covering all means and methods of reenacting accidentally impaired or interrupted respiration to normality.

The purpose of artificial respiration is the initiation of sufficient lung ventilation to cause blood gas reaction which, combined with adequate circulation, conditions the reestablishment of normal functions of life.

This is done either by *extra-pulmonary methods*
or by *intra-pulmonary methods.*

All manual methods and mechanical methods operating on the exterior of the chest wall, such as for instance the Drinker method, are extra-pulmonary.

Intra-pulmonary methods can broadly be divided in two groups:

1. Pure insufflation (and inhalation)
2. Insufflation and suction.

This latter group, representing mechanical devices to apply positive and negative pressure to the lungs, automatically functioning pumps, are in this country practically given up on account of the inadequacy in safe control harmonizing with respiratory conditions.

All of the present intra-pulmonary methods have one principle in common. It is the principle proposed by Meltzer—that of insufflation by a safely limited positive pressure and of deflation by release of the pressure permitting natural exhalation by contraction of the chest wall.

We believe in this principle and with satisfaction we point to the fact that the following authoritative designs are developments based on that principle.

However, while the Meltzer method, as far as concerns insufflation under limited pressure, is generally recognized as physiologically correct, and the motor methods applying suction are generally condemned, the success of the method is as yet far from certainty. Much depends on the means to apply the method—the means to insufflate or the means to make the patient inhale.

Meltzer used a pharyngeal tube which, however, did not successfully establish itself.

9

Figure 18.27: Drinker Method

The *Drinker Method* refers to the "iron lung" that used the "extra-pulmonary" method of artificial breathing (subjecting the chest to negative pressure). It was superior for long-term breathing support and was commonly used to treat polio patients.[335, 336] It is nearly forgotten since the polio vaccine has eradicated polio. The Meltzer method refers to positive pressure ventilation, which necessitates endotracheal intubation and positive pressure insufflation of the lungs.[337] This method now prevails in medical practice, largely because the iron lung requires vast storage space. It is interesting to note that elective endotracheal intubation and mechanical ventilation was already familiar practice at the turn of the previous century, but intubation remained rare until the 1930s when advancing needle technology enabled intravenous access to administer drugs that relaxed the vocal cords and pharyngeal muscles.

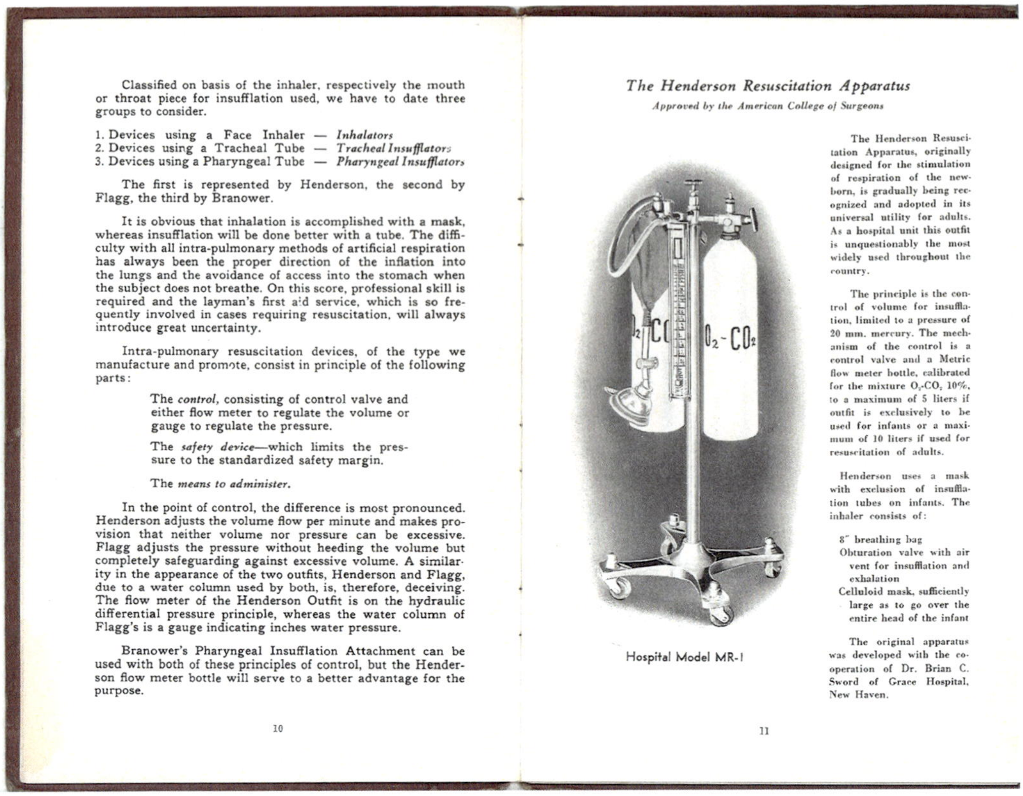

Classified on basis of the inhaler, respectively the mouth or throat piece for insufflation used, we have to date three groups to consider.

1. Devices using a Face Inhaler — *Inhalators*
2. Devices using a Tracheal Tube — *Tracheal Insufflators*
3. Devices using a Pharyngeal Tube — *Pharyngeal Insufflators*

The first is represented by Henderson, the second by Flagg, the third by Branower.

It is obvious that inhalation is accomplished with a mask, whereas insufflation will be done better with a tube. The difficulty with all intra-pulmonary methods of artificial respiration has always been the proper direction of the inflation into the lungs and the avoidance of access into the stomach when the subject does not breathe. On this score, professional skill is required and the layman's first aid service, which is so frequently involved in cases requiring resuscitation, will always introduce great uncertainty.

Intra-pulmonary resuscitation devices, of the type we manufacture and promote, consist in principle of the following parts:

The *control*, consisting of control valve and either flow meter to regulate the volume or gauge to regulate the pressure.

The *safety device*—which limits the pressure to the standardized safety margin.

The *means to administer*.

In the point of control, the difference is most pronounced. Henderson adjusts the volume flow per minute and makes provision that neither volume nor pressure can be excessive. Flagg adjusts the pressure without heeding the volume but completely safeguarding against excessive volume. A similarity in the appearance of the two outfits, Henderson and Flagg, due to a water column used by both, is, therefore, deceiving. The flow meter of the Henderson Outfit is on the hydraulic differential pressure principle, whereas the water column of Flagg's is a gauge indicating inches water pressure.

Branower's Pharyngeal Insufflation Attachment can be used with both of these principles of control, but the Henderson flow meter bottle will serve to a better advantage for the purpose.

10

The Henderson Resuscitation Apparatus

Approved by the American College of Surgeons

Hospital Model MR-I

The Henderson Resuscitation Apparatus, originally designed for the stimulation of respiration of the new-born, is gradually being recognized and adopted in its universal utility for adults. As a hospital unit this outfit is unquestionably the most widely used throughout the country.

The principle is the control of volume for insufflation, limited to a pressure of 20 mm. mercury. The mechanism of the control is a control valve and a Metric flow meter bottle, calibrated for the mixture O_2-CO_2 10%, to a maximum of 5 liters if outfit is exclusively to be used for infants or a maximum of 10 liters if used for resuscitation of adults.

Henderson uses a mask with exclusion of insufflation tubes on infants. The inhaler consists of:

8" breathing bag
Obturation valve with air vent for insufflation and exhalation
Celluloid mask, sufficiently large as to go over the entire head of the infant

The original apparatus was developed with the cooperation of Dr. Brian C. Sword of Grace Hospital, New Haven.

11

Figure 18.28: Henderson's Outfit

Note the mention that Henderson's Outfit was designed to avoid excessive pressure and volume that might deplete CO_2 or cause lung damage.

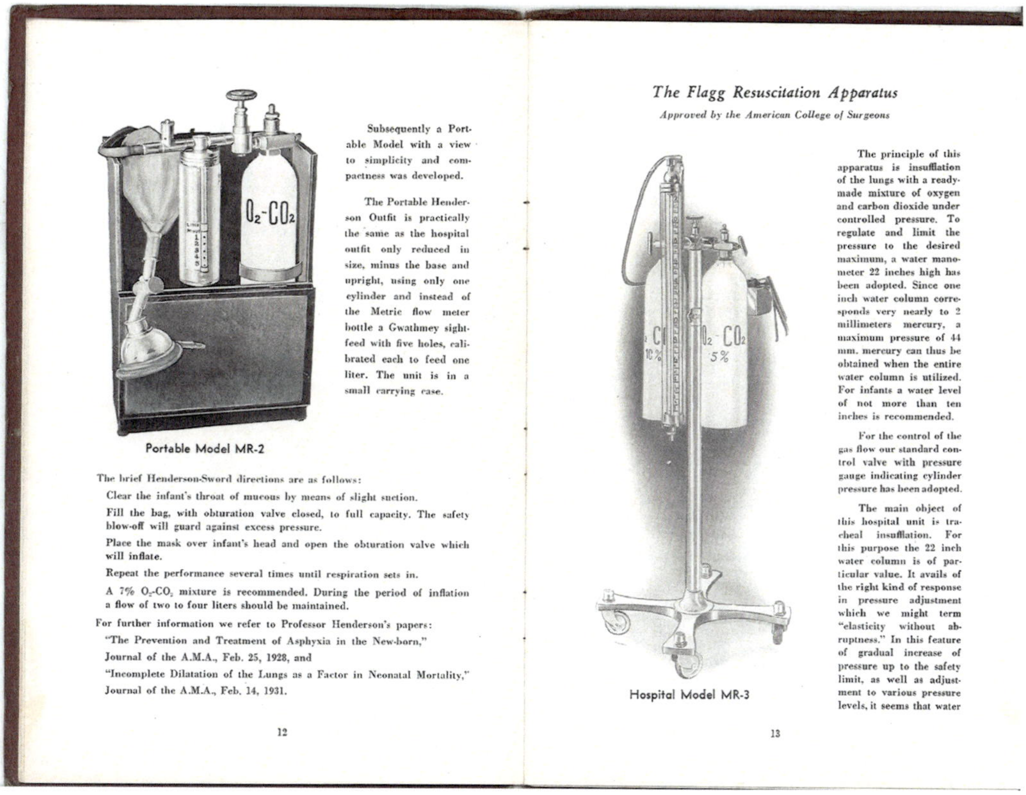

Portable Model MR-2

Subsequently a Portable Model with a view to simplicity and compactness was developed.

The Portable Henderson Outfit is practically the same as the hospital outfit only reduced in size, minus the base and upright, using only one cylinder and instead of the Metric flow meter bottle a Gwathmey sight-feed with five holes, calibrated each to feed one liter. The unit is in a small carrying case.

The brief Henderson-Sword directions are as follows:

Clear the infant's throat of mucous by means of slight suction.

Fill the bag, with obturation valve closed, to full capacity. The safety blow-off will guard against excess pressure.

Place the mask over infant's head and open the obturation valve which will inflate.

Repeat the performance several times until respiration sets in.

A 7% O_2-CO_2 mixture is recommended. During the period of inflation a flow of two to four liters should be maintained.

For further information we refer to Professor Henderson's papers:

"The Prevention and Treatment of Asphyxia in the New-born,"
Journal of the A.M.A., Feb. 25, 1928, and
"Incomplete Dilatation of the Lungs as a Factor in Neonatal Mortality,"
Journal of the A.M.A., Feb. 14, 1931.

12

The Flagg Resuscitation Apparatus

Approved by the American College of Surgeons

Hospital Model MR-3

The principle of this apparatus is insufflation of the lungs with a ready-made mixture of oxygen and carbon dioxide under controlled pressure. To regulate and limit the pressure to the desired maximum, a water manometer 22 inches high has been adopted. Since one inch water column corresponds very nearly to 2 millimeters mercury, a maximum pressure of 44 mm. mercury can thus be obtained when the entire water column is utilized. For infants a water level of not more than ten inches is recommended.

For the control of the gas flow our standard control valve with pressure gauge indicating cylinder pressure has been adopted.

The main object of this hospital unit is tracheal insufflation. For this purpose the 22 inch water column is of particular value. It avails of the right kind of response in pressure adjustment which we might term "elasticity without abruptness." In this feature of gradual increase of pressure up to the safety limit, as well as adjustment to various pressure levels, it seems that water

13

Figure 18.29: Carbogen

The O_2-CO_2 tanks designate Carbogen, which was typically a pressurized mixture of 5% carbon dioxide and 95% oxygen that was superior to either oxygen or carbon dioxide alone. The portable equipment on page 12 of the brochure appears to have been used primarily to promote breathing in newborn babies who were delivered at home. The equipment on page 13 of the brochure J30was designed to prevent excessive positive pressure that might cause lung damage. It is interesting to note that medical equipment and practices were already advanced 100 years ago, and sometimes superior to presently prevailing equipment and concepts.

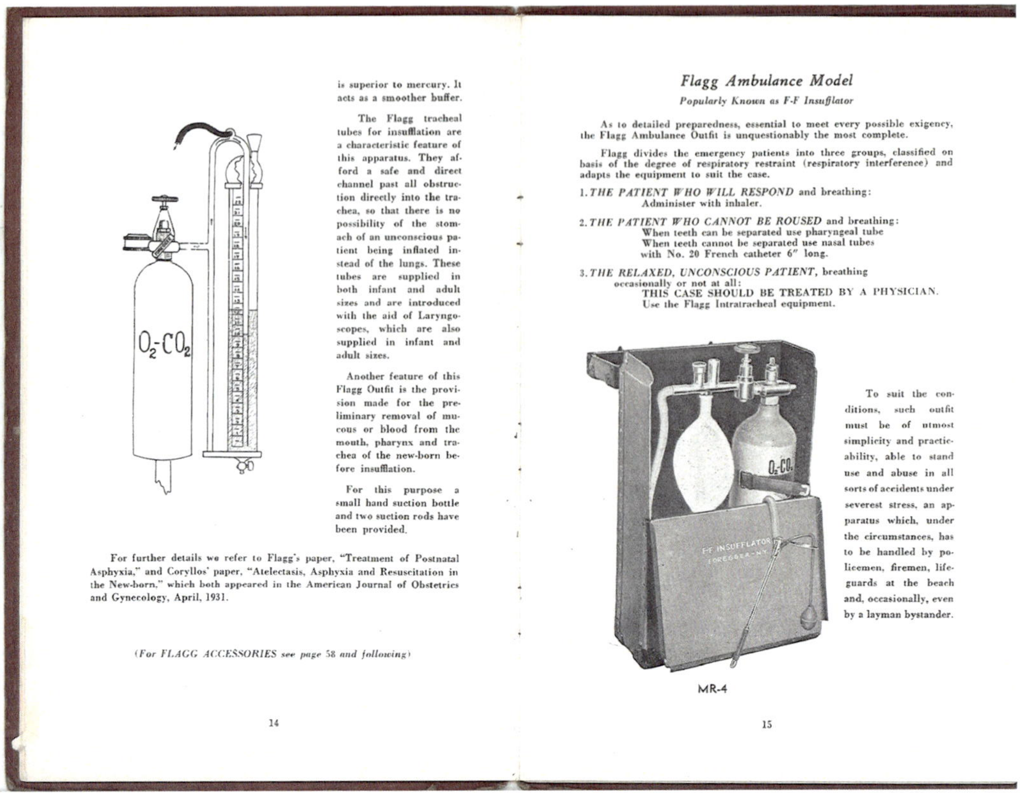

is superior to mercury. It acts as a smoother buffer.

The Flagg tracheal tubes for insufflation are a characteristic feature of this apparatus. They afford a safe and direct channel past all obstruction directly into the trachea, so that there is no possibility of the stomach of an unconscious patient being inflated instead of the lungs. These tubes are supplied in both infant and adult sizes and are introduced with the aid of Laryngoscopes, which are also supplied in infant and adult sizes.

Another feature of this Flagg Outfit is the provision made for the preliminary removal of mucous or blood from the mouth, pharynx and trachea of the new-born before insufflation.

For this purpose a small hand suction bottle and two suction rods have been provided.

For further details we refer to Flagg's paper, "Treatment of Postnatal Asphyxia," and Coryllos' paper, "Atelectasis, Asphyxia and Resuscitation in the New-born," which both appeared in the American Journal of Obstetrics and Gynecology, April, 1931.

(For FLAGG ACCESSORIES see page 58 and following)

14

Flagg Ambulance Model

Popularly Known as F-F Insufflator

As to detailed preparedness, essential to meet every possible exigency, the Flagg Ambulance Outfit is unquestionably the most complete.

Flagg divides the emergency patients into three groups, classified on basis of the degree of respiratory restraint (respiratory interference) and adapts the equipment to suit the case.

1. *THE PATIENT WHO WILL RESPOND* and breathing:
   Administer with inhaler.

2. *THE PATIENT WHO CANNOT BE ROUSED* and breathing:
   When teeth can be separated use pharyngeal tube
   When teeth cannot be separated use nasal tubes with No. 20 French catheter 6" long.

3. *THE RELAXED, UNCONSCIOUS PATIENT*, breathing occasionally or not at all:
   THIS CASE SHOULD BE TREATED BY A PHYSICIAN.
   Use the Flagg Intratracheal equipment.

To suit the conditions, such outfit must be of utmost simplicity and practicability, able to stand use and abuse in all sorts of accidents under severest stress, an apparatus which, under the circumstances, has to be handled by policemen, firemen, lifeguards at the beach and, occasionally, even by a layman bystander.

15

Figure 18.30: Carbogen apparatus

Note the reference on page 14 of the brochure to both pediatric and adult laryngoscopes that facilitated installation of airway devices which prevented dangerous stomach insufflation that could cause vomiting and aspiration. The Carbogen apparatus on page 15 of the brochure was intended to be rugged, practical, and simple so that it could be used by firemen, policemen, and even bystanders. This would have been superior to presently prevailing "Advanced Cardiac Life Support" (ACLS) dogma that promotes 100% oxygen and intravenous medications that are less practical and effective.

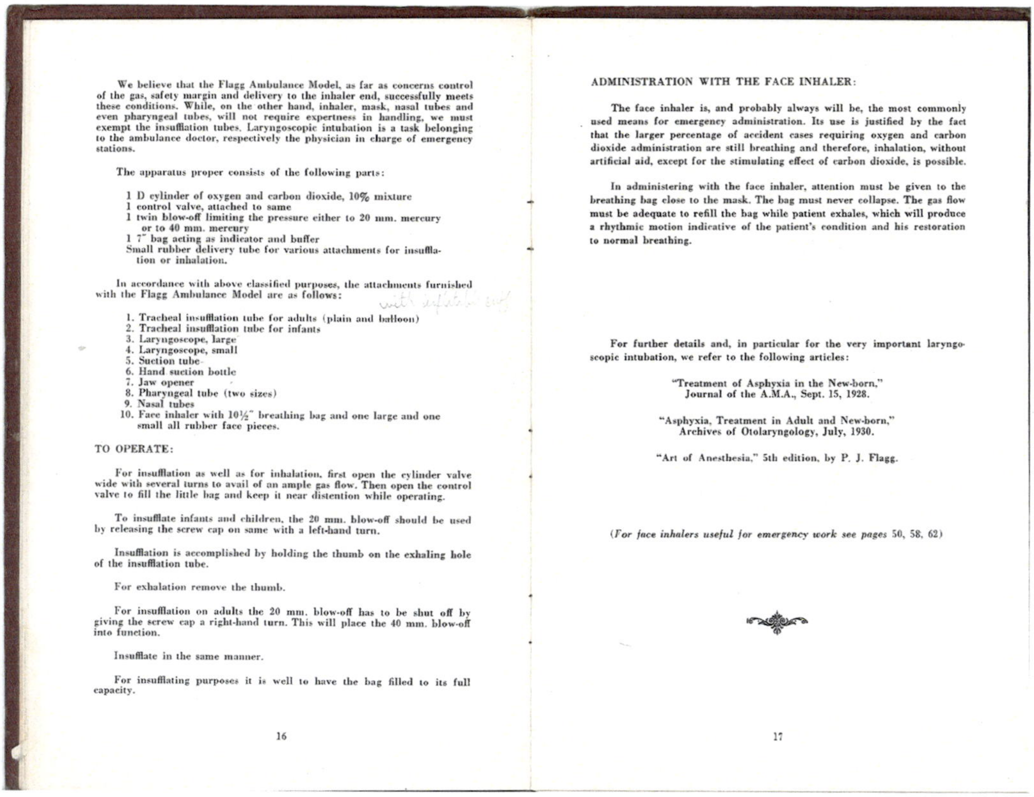

We believe that the Flagg Ambulance Model, as far as concerns control of the gas, safety margin and delivery to the inhaler end, successfully meets these conditions. While, on the other hand, inhaler, mask, nasal tubes and even pharyngeal tubes, will not require expertness in handling, we must exempt the insufflation tubes. Laryngoscopic intubation is a task belonging to the ambulance doctor, respectively the physician in charge of emergency stations.

The apparatus proper consists of the following parts:

1 D cylinder of oxygen and carbon dioxide, 10% mixture
1 control valve, attached to same
1 twin blow-off limiting the pressure either to 20 mm. mercury or to 40 mm. mercury
1 7" bag acting as indicator and buffer
Small rubber delivery tube for various attachments for insufflation or inhalation.

In accordance with above classified purposes, the attachments furnished with the Flagg Ambulance Model are as follows:

1. Tracheal insufflation tube for adults (plain and balloon)
2. Tracheal insufflation tube for infants
3. Laryngoscope, large
4. Laryngoscope, small
5. Suction tube
6. Hand suction bottle
7. Jaw opener
8. Pharyngeal tube (two sizes)
9. Nasal tubes
10. Face inhaler with 10½" breathing bag and one large and one small all rubber face pieces.

TO OPERATE:

For insufflation as well as for inhalation, first open the cylinder valve wide with several turns to avail of an ample gas flow. Then open the control valve to fill the little bag and keep it near distention while operating.

To insufflate infants and children, the 20 mm. blow-off should be used by releasing the screw cap on same with a left-hand turn.

Insufflation is accomplished by holding the thumb on the exhaling hole of the insufflation tube.

For exhalation remove the thumb.

For insufflation on adults the 20 mm. blow-off has to be shut off by giving the screw cap a right-hand turn. This will place the 40 mm. blow-off into function.

Insufflate in the same manner.

For insufflating purposes it is well to have the bag filled to its full capacity.

16

ADMINISTRATION WITH THE FACE INHALER:

The face inhaler is, and probably always will be, the most commonly used means for emergency administration. Its use is justified by the fact that the larger percentage of accident cases requiring oxygen and carbon dioxide administration are still breathing and therefore, inhalation, without artificial aid, except for the stimulating effect of carbon dioxide, is possible.

In administering with the face inhaler, attention must be given to the breathing bag close to the mask. The bag must never collapse. The gas flow must be adequate to refill the bag while patient exhales, which will produce a rhythmic motion indicative of the patient's condition and his restoration to normal breathing.

For further details and, in particular for the very important laryngoscopic intubation, we refer to the following articles:

"Treatment of Asphyxia in the New-born," Journal of the A.M.A., Sept. 15, 1928.

"Asphyxia, Treatment in Adult and New-born," Archives of Otolaryngology, July, 1930.

"Art of Anesthesia," 5th edition, by P. J. Flagg.

(For face inhalers useful for emergency work see pages 50, 58, 62)

17

Figure 18.31: Carbogen mixture

Again, note the emphasis on simplicity, practicality, and cost. Also note that the Carbogen mixture recommended for emergency resuscitation consisted of 10% carbon dioxide and 90% oxygen as opposed to the more common 5% carbon dioxide and 95% oxygen. Also note the reference to scientific literature of the time.[338, 339, 340]

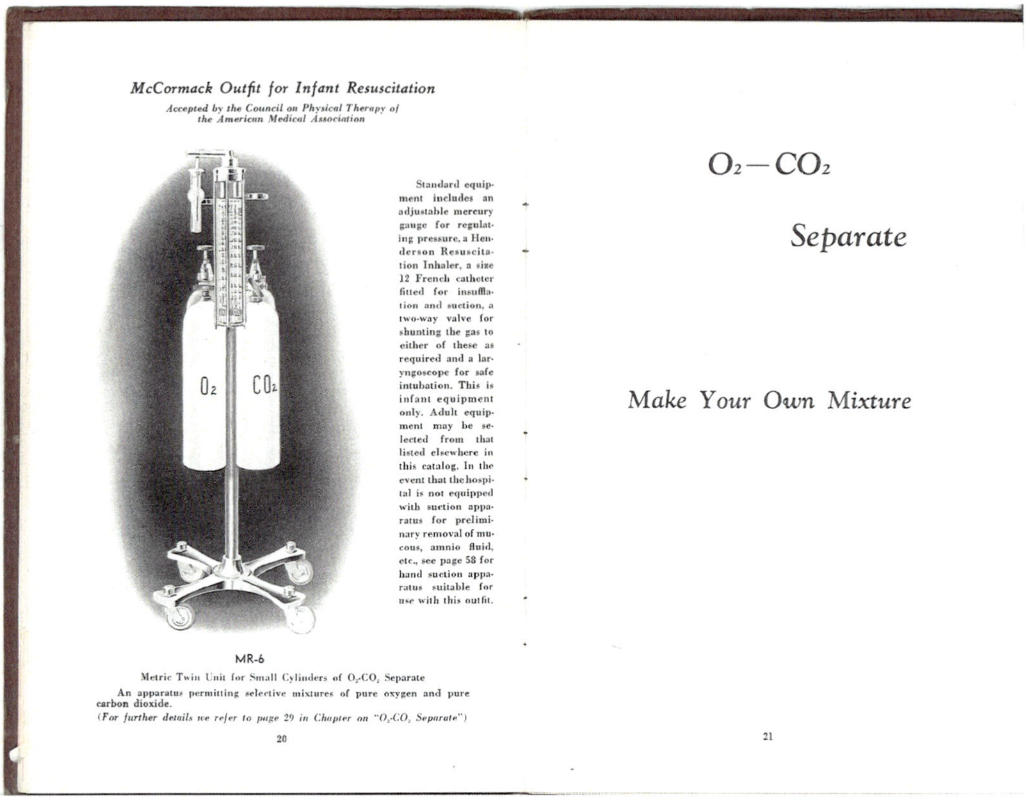

McCormack Outfit for Infant Resuscitation

Accepted by the Council on Physical Therapy of the American Medical Association

Standard equipment includes an adjustable mercury gauge for regulating pressure, a Henderson Resuscitation Inhaler, a size 12 French catheter fitted for insufflation and suction, a two-way valve for shunting the gas to either of these as required and a laryngoscope for safe intubation. This is infant equipment only. Adult equipment may be selected from that listed elsewhere in this catalog. In the event that the hospital is not equipped with suction apparatus for preliminary removal of mucous, amnio fluid, etc., see page 58 for hand suction apparatus suitable for use with this outfit.

MR-6

Metric Twin Unit for Small Cylinders of O_2-CO_2 Separate

An apparatus permitting selective mixtures of pure oxygen and pure carbon dioxide.

(For further details we refer to page 29 in Chapter on "O_2-CO_2 Separate")

20

O_2—CO_2

Separate

Make Your Own Mixture

21

Figure 18.32: Carbogen equipment for hospitals

This Carbogen equipment was not portable and was obviously intended for use in hospital delivery suites. The CO_2 concentration could be tailored according to the preference of the operator.

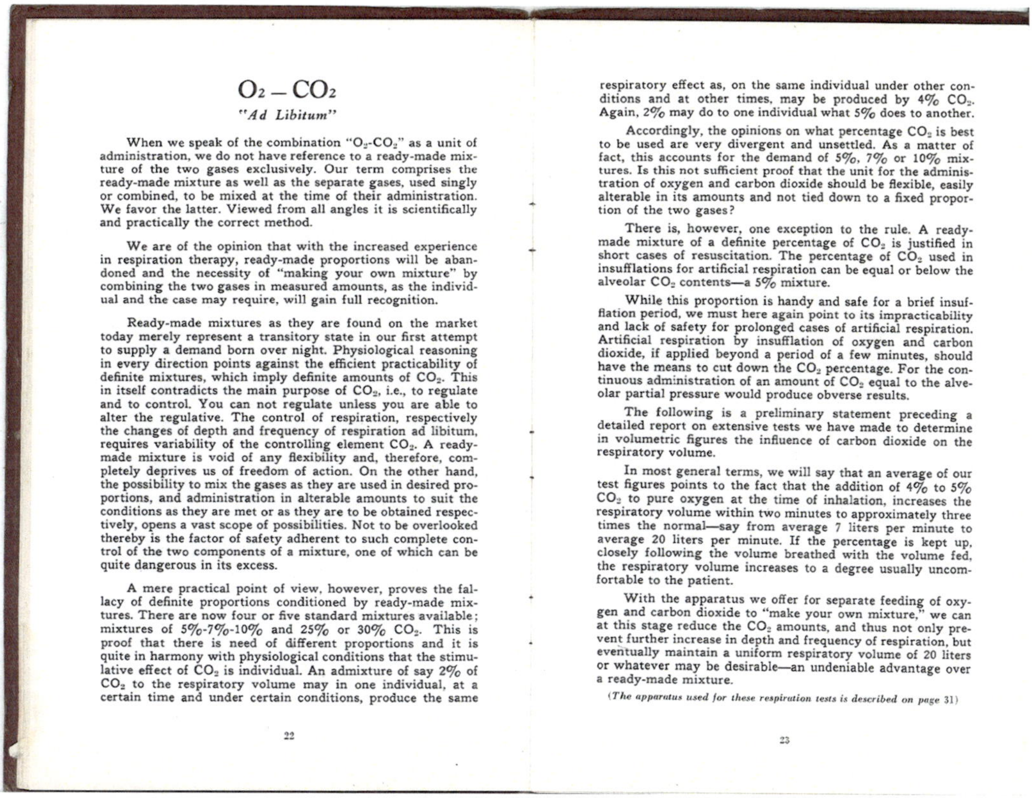

$O_2 - CO_2$

"Ad Libitum"

When we speak of the combination "O_2-CO_2" as a unit of administration, we do not have reference to a ready-made mixture of the two gases exclusively. Our term comprises the ready-made mixture as well as the separate gases, used singly or combined, to be mixed at the time of their administration. We favor the latter. Viewed from all angles it is scientifically and practically the correct method.

We are of the opinion that with the increased experience in respiration therapy, ready-made proportions will be abandoned and the necessity of "making your own mixture" by combining the two gases in measured amounts, as the individual and the case may require, will gain full recognition.

Ready-made mixtures as they are found on the market today merely represent a transitory state in our first attempt to supply a demand born over night. Physiological reasoning in every direction points against the efficient practicability of definite mixtures, which imply definite amounts of CO_2. This in itself contradicts the main purpose of CO_2, i.e., to regulate and to control. You can not regulate unless you are able to alter the regulative. The control of respiration, respectively the changes of depth and frequency of respiration ad libitum, requires variability of the controlling element CO_2. A ready-made mixture is void of any flexibility and, therefore, completely deprives us of freedom of action. On the other hand, the possibility to mix the gases as they are used in desired proportions, and administration in alterable amounts to suit the conditions as they are met or as they are to be obtained respectively, opens a vast scope of possibilities. Not to be overlooked thereby is the factor of safety adherent to such complete control of the two components of a mixture, one of which can be quite dangerous in its excess.

A mere practical point of view, however, proves the fallacy of definite proportions conditioned by ready-made mixtures. There are now four or five standard mixtures available; mixtures of 5%-7%-10% and 25% or 30% CO_2. This is proof that there is need of different proportions and it is quite in harmony with physiological conditions that the stimulative effect of CO_2 is individual. An admixture of say 2% of CO_2 to the respiratory volume may in one individual, at a certain time and under certain conditions, produce the same respiratory effect as, on the same individual under other conditions and at other times, may be produced by 4% CO_2. Again, 2% may do to one individual what 5% does to another.

Accordingly, the opinions on what percentage CO_2 is best to be used are very divergent and unsettled. As a matter of fact, this accounts for the demand of 5%, 7% or 10% mixtures. Is this not sufficient proof that the unit for the administration of oxygen and carbon dioxide should be flexible, easily alterable in its amounts and not tied down to a fixed proportion of the two gases?

There is, however, one exception to the rule. A ready-made mixture of a definite percentage of CO_2 is justified in short cases of resuscitation. The percentage of CO_2 used in insufflations for artificial respiration can be equal or below the alveolar CO_2 contents—a 5% mixture.

While this proportion is handy and safe for a brief insufflation period, we must here again point to its impracticability and lack of safety for prolonged cases of artificial respiration. Artificial respiration by insufflation of oxygen and carbon dioxide, if applied beyond a period of a few minutes, should have the means to cut down the CO_2 percentage. For the continuous administration of an amount of CO_2 equal to the alveolar partial pressure would produce obverse results.

The following is a preliminary statement preceding a detailed report on extensive tests we have made to determine in volumetric figures the influence of carbon dioxide on the respiratory volume.

In most general terms, we will say that an average of our test figures points to the fact that the addition of 4% to 5% CO_2 to pure oxygen at the time of inhalation, increases the respiratory volume within two minutes to approximately three times the normal—say from average 7 liters per minute to average 20 liters per minute. If the percentage is kept up, closely following the volume breathed with the volume fed, the respiratory volume increases to a degree usually uncomfortable to the patient.

With the apparatus we offer for separate feeding of oxygen and carbon dioxide to "make your own mixture," we can at this stage reduce the CO_2 amounts, and thus not only prevent further increase in depth and frequency of respiration, but eventually maintain a uniform respiratory volume of 20 liters or whatever may be desirable—an undeniable advantage over a ready-made mixture.

(*The apparatus used for these respiration tests is described on page 31*)

22

23

Figure 18.33: Ready-made Carbogen

Note the statement "Ready-made (Carbogen) mixtures as they are found on the market today merely represent a transitory state in our first attempt to supply a demand born overnight." This reflects the remarkable speed with which the therapeutic properties of carbon dioxide were widely accepted in the absence of counterproductive government regulatory restrictions. Note also the discussion of the advantages of equipment that allows adjustment of the relative concentrations of oxygen and carbon dioxide to suit circumstances.

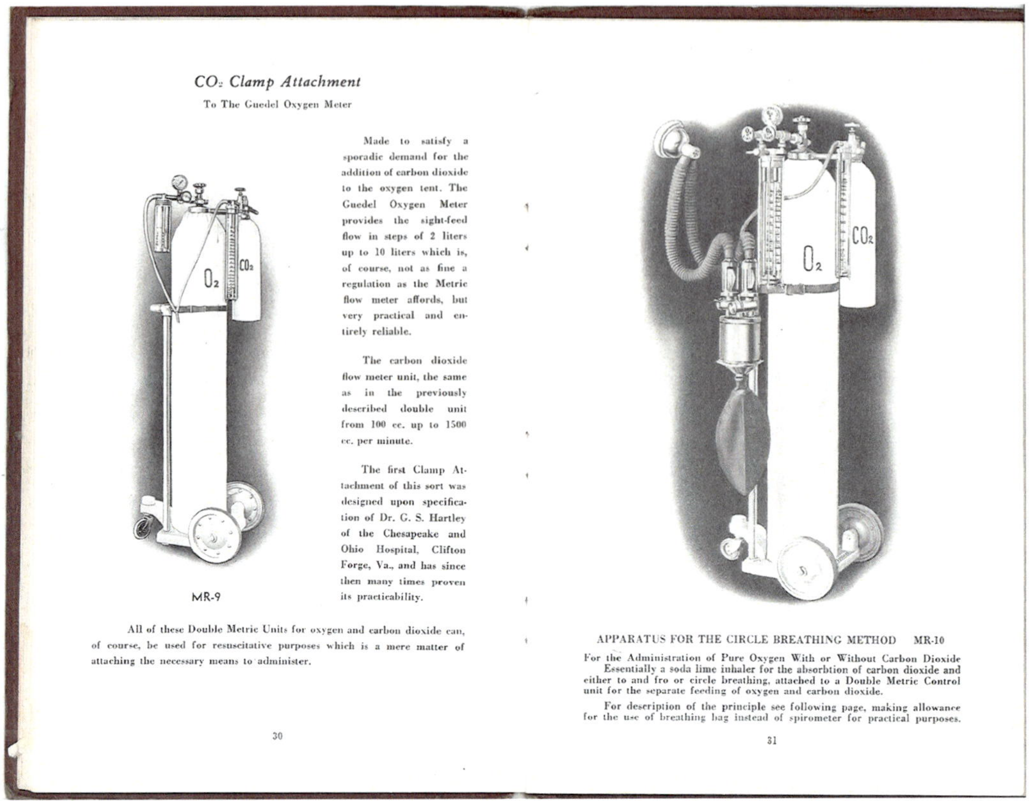

CO_2 Clamp Attachment

To The Guedel Oxygen Meter

MR-9

Made to satisfy a sporadic demand for the addition of carbon dioxide to the oxygen tent. The Guedel Oxygen Meter provides the sight-feed flow in steps of 2 liters up to 10 liters which is, of course, not as fine a regulation as the Metric flow meter affords, but very practical and entirely reliable.

The carbon dioxide flow meter unit, the same as in the previously described double unit from 100 cc. up to 1500 cc. per minute.

The first Clamp Attachment of this sort was designed upon specification of Dr. G. S. Hartley of the Chesapeake and Ohio Hospital, Clifton Forge, Va., and has since then many times proven its practicability.

All of these Double Metric Units for oxygen and carbon dioxide can, of course, be used for resuscitative purposes which is a mere matter of attaching the necessary means to administer.

30

APPARATUS FOR THE CIRCLE BREATHING METHOD MR-10

For the Administration of Pure Oxygen With or Without Carbon Dioxide

Essentially a soda lime inhaler for the absorbtion of carbon dioxide and either to and fro or circle breathing, attached to a Double Metric Control unit for the separate feeding of oxygen and carbon dioxide.

For description of the principle see following page, making allowance for the use of breathing bag instead of spirometer for practical purposes.

31

Figure 18.34: Supplementation and absorbtion

Despite the rapid acceptance of the therapeutic benefits of carbon dioxide, there remained a stubborn fear of mythical carbon dioxide toxicity, as reflected by the equipment displayed on page 31 of the brochure. The incompatible combination of CO_2 supplementation and CO_2 absorption by soda lime reflects the schizophrenic attitude that endured despite scientific advance.

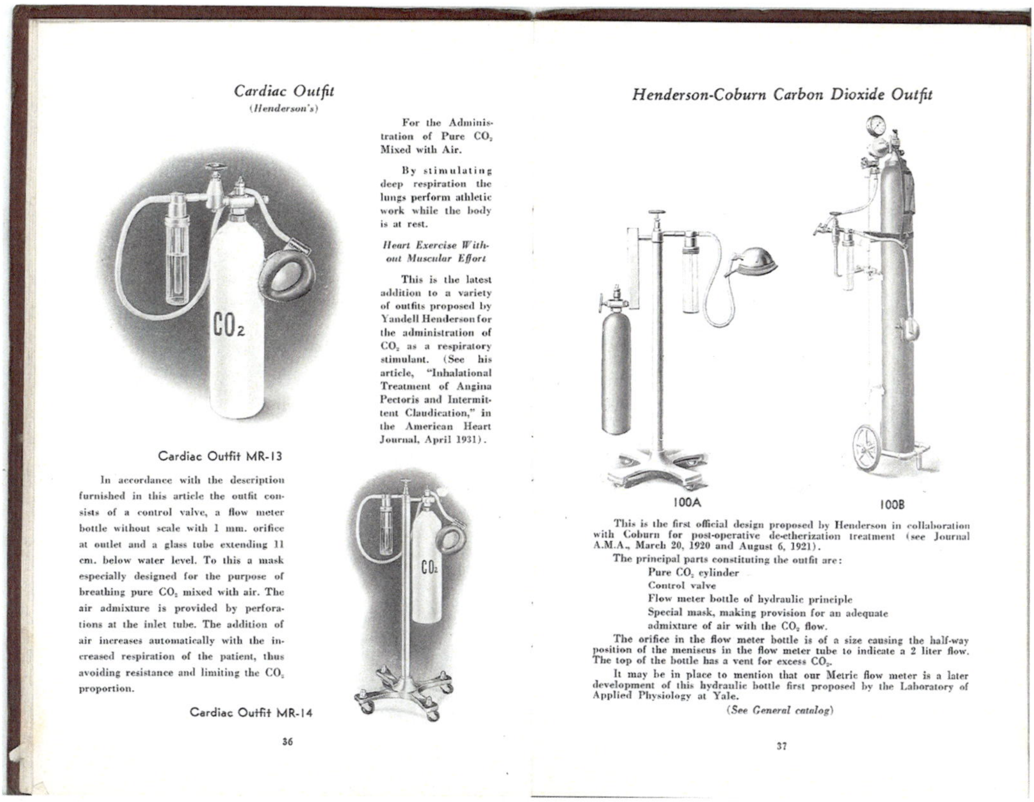

Cardiac Outfit
(Henderson's)

For the Administration of Pure CO_2 Mixed with Air.

By stimulating deep respiration the lungs perform athletic work while the body is at rest.

Heart Exercise Without Muscular Effort

This is the latest addition to a variety of outfits proposed by Yandell Henderson for the administration of CO_2 as a respiratory stimulant. (See his article, "Inhalational Treatment of Angina Pectoris and Intermittent Claudication," in the American Heart Journal, April 1931).

Cardiac Outfit MR-13

In accordance with the description furnished in this article the outfit consists of a control valve, a flow meter bottle without scale with 1 mm. orifice at outlet and a glass tube extending 11 cm. below water level. To this a mask especially designed for the purpose of breathing pure CO_2 mixed with air. The air admixture is provided by perforations at the inlet tube. The addition of air increases automatically with the increased respiration of the patient, thus avoiding resistance and limiting the CO_2 proportion.

Cardiac Outfit MR-14

36

Henderson-Coburn Carbon Dioxide Outfit

100A 100B

This is the first official design proposed by Henderson in collaboration with Coburn for post-operative de-etherization treatment (see Journal A.M.A., March 20, 1920 and August 6, 1921).

The principal parts constituting the outfit are:

Pure CO_2 cylinder
Control valve
Flow meter bottle of hydraulic principle
Special mask, making provision for an adequate admixture of air with the CO_2 flow.

The orifice in the flow meter bottle is of a size causing the half-way position of the meniscus in the flow meter tube to indicate a 2 liter flow. The top of the bottle has a vent for excess CO_2.

It may be in place to mention that our Metric flow meter is a later development of this hydraulic bottle first proposed by the Laboratory of Applied Physiology at Yale.

(*See General catalog*)

37

Figure 18.35: CO_2 supplementation of atmospheric air

Pages 36 and 37 of the brochure are particularly noteworthy because they demonstrate that CO_2 supplementation of atmospheric air, which contains only 21% oxygen, offers an effective treatment of myocardial infarction as well as an inexpensive and effective means to accelerate emergence from ether anesthesia. It is truly shocking that current practice has abandoned the valuable therapeutic properties of carbon dioxide. Carbogen is a perfect emergency treatment for stroke, because it has no anticoagulant effect and can thus be used to immediately treat both hemorrhagic and ischemic types of stroke without fear of worsening the hemorrhagic form of stroke. Carbogen has the added advantage of immediately diagnosing the type of stroke at hand, because it will immediately restore perfusion and oxygenation to ischemic brain tissue and restore brain function, while having no effect on hemorrhagic stroke. Once ischemic stroke has been confirmed, the Carbogen treatment can be supplemented with powerful anticoagulants.

NOTES

1 Z. Zhang, Q. Guo, E. Wang, Hyperventilation in neurological patients: from physiology to outcome evidence. *Current opinion in anaesthesiology* 32, 568-573 (2019).

2 D. W. Marion, A. Firlik, M. R. McLaughlin, Hyperventilation therapy for severe traumatic brain injury. *New Horiz* 3, 439-447 (1995).

3 F. Plum, Hyperpnea, hyperventilation, and brain dysfunction. *Annals of internal medicine* 76, 328 (1972).

4 F. Plum, J. B. Posner, W. W. Smith, Effect of hyperbaric-hyperoxic hyperventilation on blood, brain, and CSF lactate. *The American journal of physiology* 215, 1240-1244 (1968).

5 K. M. Conlee, M. L. Stephens, A. N. Rowan, L. A. King, Carbon dioxide for euthanasia: concerns regarding pain and distress, with special reference to mice and rats. *Lab Anim* 39, 137-161 (2005).

6 G. P. Boivin, D. L. Hickman, M. A. Creamer-Hente, K. R. Pritchett-Corning, N. A. Bratcher, Review of CO(2) as a Euthanasia Agent for Laboratory Rats and Mice. *J Am Assoc Lab Anim Sci* 56, 491-499 (2017).

7 P. J. Danneman, S. Stein, S. O. Walshaw, Humane and practical implications of using carbon dioxide mixed with oxygen for anesthesia or euthanasia of rats. *Lab Anim Sci* 47, 376-385 (1997).

8 P. Hawkins *et al.*, A Good Death? Report of the Second Newcastle Meeting on Laboratory Animal Euthanasia. *Animals (Basel)* 6, (2016).

9 J. Makowska, H. Golledge, N. Marquardt, D. M. Weary, Sedation or inhalant anesthesia before euthanasia with CO_2 does not reduce behavioral or physiologic signs of pain and stress in mice. *J Am Assoc Lab Anim Sci* 51, 396-397; author reply 397-399 (2012).

10 L. S. Coleman, Should soda lime be abolished? *Anesth Analg* 102, 1290-1291 (2006).

11 L. S. Coleman, Soda lime is a dangerous relic of a forgotten past. *Anaesthesia* 70, (2016).

12 C. Prys-Roberts, W. D. Smith, J. F. Nunn, Accidental severe hypercapnia during anaesthesia. A case report and review of some physiological effects. *Br J Anaesth* 39, 257-267 (1967).

13 M. J. Frumin, R. M. Epstein, G. Cohen, Apneic oxygenation in man. *Anesthesiology* 20, 789-798 (1959).

14 W. H. Forrest, Jr., J. W. Bellville, The Effect of Sleep Plus Morphine on the Respiratory Response to Carbon Dioxide. *Anesthesiology* 25, 137-141 (1964).

15 E. L. Bernays, *Propaganda*. (H. Liveright, New York,, 1928), pp. 159 p.

16 https://en.wikipedia.org/wiki/Cardiff_Giant

17 I. A. Murdoch, S. A. Qureshi, I. C. Huggon, Perioperative haemodynamic effects of an intravenous infusion of calcium chloride in children following cardiac surgery. *Acta Paediatr* 83, 658-661 (1994).

18 L. S. Coleman, Four Forgotten Giants of Anesthesia History. *Journal of Anesthesia and Surgery* 3, 1-17 (2015).

19 C. Lynch, 3rd, M. J. Frazer, Depressant effects of volatile anesthetics upon rat and amphibian ventricular myocardium: insights into anesthetic mechanisms of action. *Anesthesiology* 70, 511-522 (1989).
20 L. S. Coleman, Four Forgotten Giants of Anesthesia History. *Journal of Anesthesia and Surgery* 3, 1-17 (2015).
21 L. S. Coleman, in *apsf Newsletter*. (Anesthesia Patient Safety Foundation, Administrator, Deanna Walker Anesthesia Patient Safety Foundation Building One, Suite Two 8007 South Meridian Street Indianapolis, IN 46217-2922 e-mail address: walker@apsf.org FAX: (317) 888-1482, 2010), vol. Winter 2009-2020.
22 L. S. Coleman, A call for standards on perioperative CO(2) regulation. *Can J Anaesth*, (2011).
23 L. S. Coleman, 30 Years Lost in Anesthesia Theory. *Cardiovasc Hematol Agents Med Chem* 10, 31-49 (2012).
24 L. S. Coleman, Routine Mask Induction and Intubation without Paralysis may be Safer. *J Anesth Surg* 2, 1-2 (2015).
25 L. W. Crile GW, *Anoci-association*. (Saunders, Philadelphia, 1914), pp. 223-225.
26 J. H. Eisele, E. I. Eger, 2nd, M. Muallem, Narcotic properties of carbon dioxide in the dog. *Anesthesiology* 28, 856-865 (1967).
27 T. H. Stanley, T. D. Egan, H. Van Aken, A tribute to Dr. Paul A. J. Janssen: entrepreneur extraordinaire, innovative scientist, and significant contributor to anesthesiology. *Anesth Analg* 106, 451-462, table of contents (2008).
28 L. S. Coleman, in *apsf Newsletter*. (Anesthesia Patient Safety Foundation, Administrator, Deanna Walker Anesthesia Patient Safety Foundation Building One, Suite Two 8007 South Meridian Street Indianapolis, IN 46217-2922 e-mail address: walker@apsf.org FAX: (317) 888-1482, 2010), vol. Winter 2009-2020.
29 L. S. Coleman, A call for standards on perioperative CO(2) regulation. *Can J Anaesth*, (2011).
30 P. Cartwright *et al.*, Ventilatory depression related to plasma fentanyl concentrations during and after anesthesia in humans. *Anesth Analg* 62, 966-974 (1983).
31 V. J. Cardenas, Jr. *et al.*, Correction of blood pH attenuates changes in hemodynamics and organ blood flow during permissive hypercapnia. *Critical care medicine* 24, 827-834 (1996).
32 C. R. Carvalho *et al.*, Temporal hemodynamic effects of permissive hypercapnia associated with ideal PEEP in ARDS. *American journal of respiratory and critical care medicine* 156, 1458-1466 (1997).
33 K. G. Hickling, Permissive hypercapnia. *Respir Care Clin N Am* 8, 155-169, v (2002).
34 J. G. Laffey, D. O'Croinin, P. McLoughlin, B. P. Kavanagh, Permissive hypercapnia--role in protective lung ventilatory strategies. *Intensive Care Med* 30, 347-356 (2004).
35 H. Morisaki, R. Serita, Y. Innami, Y. Kotake, J. Takeda, Permissive hypercapnia during thoracic anaesthesia. *Acta anaesthesiologica Scandinavica* 43, 845-849 (1999).

36 A. Ohmura, M. Sha, J. Katagiri, How far can we go with permissive hypercapnia? A case presentation and some biased comments with emphasis on maintaining normal haemoglobin level. *Acta Anaesthesiol Scand Suppl* 107, 209-213 (1995).

37 U. H. Thome, W. A. Carlo, Permissive hypercapnia. *Semin Neonatol* 7, 409-419 (2002).

38 Q. Zhou *et al.*, Effects of permissive hypercapnia on transient global cerebral ischemia-reperfusion injury in rats. *Anesthesiology* 112, 288-297 (2010).

39 K. G. Hickling, S. J. Henderson, R. Jackson, Low mortality associated with low volume pressure limited ventilation with permissive hypercapnia in severe adult respiratory distress syndrome. *Intensive Care Med* 16, 372-377 (1990).

40 O. Akca, Hypercapnia: what is the upper limit? *Paediatric anaesthesia* 15, 80-81 (2005).

41 O. Akca, Optimizing the intraoperative management of carbon dioxide concentration. *Current opinion in anaesthesiology* 19, 19-25 (2006).

42 O. Akca, Carbon dioxide and tissue oxygenation: is there sufficient evidence to support application of hypercapnia for hemodynamic stability and better tissue perfusion in sepsis? *Intensive Care Med* 34, 1752-1754 (2008).

43 O. Akca *et al.*, Hypercapnia improves tissue oxygenation. *Anesthesiology* 97, 801-806 (2002).

44 O. Akca *et al.*, Effect of intra-operative end-tidal carbon dioxide partial pressure on tissue oxygenation. *Anaesthesia* 58, 536-542 (2003).

45 O. Akca *et al.*, Tissue oxygenation response to mild hypercapnia during cardiopulmonary bypass with constant pump output. *Br J Anaesth* 96, 708-714 (2006).

46 W. H. Forrest, Jr., J. W. Bellville, The Effect of Sleep Plus Morphine on the Respiratory Response to Carbon Dioxide. *Anesthesiology* 25, 137-141 (1964).

47 J. W. S. Bellville, John C., The Effect of Drugs on the Respiratory Response to Carbon Dioxide. *Anesthesiology* 21, 727 (1960).

48 J. W. Bellville, W. S. Howland, J. C. Seed, R. W. Houde, The effect of sleep on the respiratory response to carbon dioxide. *Anesthesiology* 20, 628-634 (1959).

49 K. J. Tuman *et al.*, Does choice of anesthetic agent significantly affect outcome after coronary artery surgery? *Anesthesiology* 70, 189-198 (1989).

50 J. Marquez, J. Magovern, P. Kaplan, T. Sakert, G. P. Gravlee, Cardiac surgery "fast tracking" in an academic hospital. *Journal of cardiothoracic and vascular anesthesia* 9, 34-36 (1995).

51 B. Bissonnette, H. Swan, P. Ravussin, V. Un, Neuroleptanesthesia: current status. *Can J Anaesth* 46, 154-168 (1999).

52 E. Lowenstein *et al.*, Cardiovascular response to large doses of intravenous morphine in man. *N Engl J Med* 281, 1389-1393 (1969).

53 S. N. Raja, E. Lowenstein, The birth of opioid anesthesia. *Anesthesiology* 100, 1013-1015 (2004).

54 M. G. Davies, P. O. Hagen, Systemic inflammatory response syndrome. *Br J Surg* 84, 920-935 (1997).

55 C. Adembri *et al.*, Pulmonary injury follows systemic inflammatory reaction in infrarenal aortic surgery. *Critical care medicine* 32, 1170-1177 (2004).

56 Y. G. Liu, C. S. Li, [Coagulation dysfunctions in patients with systemic inflammatory response syndrome]. *Zhonghua Nei Ke Za Zhi* 43, 94-97 (2004).
57 K. Takenaka, E. Ogawa, H. Wada, T. Hirata, Systemic inflammatory response syndrome and surgical stress in thoracic surgery. *J Crit Care* 21, 48-53; discussion 53-45 (2006).
58 T. G. Monk, V. Saini, B. C. Weldon, J. C. Sigl, Anesthetic management and one-year mortality after noncardiac surgery. *Anesth Analg* 100, 4-10 (2005).
59 Y. Shavit, S. Ben-Eliyahu, A. Zeidel, B. Beilin, Effects of fentanyl on natural killer cell activity and on resistance to tumor metastasis in rats. Dose and timing study. *Neuroimmunomodulation* 11, 255-260 (2004).
60 D. I. Sessler, Does regional analgesia reduce the risk of cancer recurrence? A hypothesis. *European journal of cancer prevention : the official journal of the European Cancer Prevention Organisation* 17, 269-272 (2008).
61 F. J. Overdyk, Postoperative opioids remain a serious patient safety threat. *Anesthesiology* 113, 259-260; author reply 260-251 (2010).
62 http://www.apsf.org/newsletters/html/2010/winter/11_opioids.htm
63 http://www.apsf.org/newsletters/html/2010/winter/12_hypervent.htm
64 V. Bockris, The life and death of Andy Warhol. (Bantam Books, New York, 1989), pp. 392 p.
65 L. S. Coleman, 30 Years Lost in Anesthesia Theory. *Cardiovasc Hematol Agents Med Chem* 10, 31-49 (2012).
66 N. Riegels, M. J. Richards, Humphry Davy: his life, works, and contribution to anesthesiology. *Anesthesiology* 114, 1282-1288 (2011).
67 Y. Henderson, Resuscitation with Carbon Dioxide. *Science* 83, 399-402 (1936).
68 L. W. Crile GW, *Anoci-association*. (Saunders, Philadelphia, 1914), pp. 223-225.
69 Y. Henderson, Physiological Observations on Pike's Peak, Colorado, made in the Summer of 1911. *Trans Am Climatol Assoc* 28, 11-15 (1912).
70 Y. Henderson, Public Service as an Element in the Life of the American Scientist. *Science* 77, 584-585 (1933).
71 Y. Henderson, Resuscitation with Carbon Dioxide. *Science* 83, 399-402 (1936).
72 Y. Henderson, in *Cyclopedia of Medicine*. (1940).
73 Y. Henderson, Resuscitation with Carbon Dioxide. *Science* 83, 399-402 (1936).
74 J. S. Lundy, Carbon Dioxide as an Aid In Anesthesia. *J.A.M.A* 85:1953, 1925, (1925).
75 Y. Henderson, Resuscitation with Carbon Dioxide. *Science* 83, 399-402 (1936).
76 Y. Henderson, in *Cyclopedia of Medicine*. (1940).
77 Y. Henderson, The Wesley M. Carpenter Lecture: Atelectasis, Massive Collapse and Related Post-Operative Conditions. *Bull N Y Acad Med* 11, 639-656 (1935).
78 Y. Henderson, How Breathing Begins at Birth. *Science* 85, 89-91 (1937).
79 http://www.asevet.com/resources/circuits/ayres.htm
80 D. E. Jackson, Anesthesia Equipment From 1914 to 1954 and Experiments Leading To Its Development. *Anesthesiology* 16, 953-969 (1955).
81 https://patents.google.com/patent/US9032954B2/en
82 D. E. Jackson, Anesthesia Equipment From 1914 to 1954 and Experiments Leading To Its Development. *Anesthesiology* 16, 953-969 (1955).
83 L. S. Coleman, Routine Mask Induction and Intubation without Paralysis may be Safer. *J Anesth Surg* 2, 1-2 (2015).

84 https://www.howequipmentworks.com/circle_breathing_system/
85 J. S. Lundy, Convulsions Associated with General Anesthesia. *Mayo Clinic Health Letter*, (1936).
86 M. C. Seevers, WH; et al, The Role of Hypercapnia in the Production of "Ether Convulsions". *J Pharmacol Exp Ther* 63, (1938).
87 W. H. B. Cassels, T.J., Seevers, M.H., Convulsions During Anesthesia. *American Society for Pharmacology and Experimental Therapeutics, Baltimore*, (1938).
88 M. C. Seevers, WH; et al, The Role of Hypercapnia in the Production of "Ether Convulsions". *J Pharmacol Exp Ther* 63, (1938).
89 https://en.wikipedia.org/wiki/Achilles%27_heel
90 D. R. Bacon, Special article: Francis Hoeffer McMechan, MD: creator of modern anesthesiology? *Anesth Analg* 115, 1393-1400 (2012).
91 D. R. Bacon, Special article: Francis Hoeffer McMechan, MD: creator of modern anesthesiology? *Anesth Analg* 115, 1393-1400 (2012).
92 D. R. Bacon, Special article: Francis Hoeffer McMechan, MD: creator of modern anesthesiology? *Anesth Analg* 115, 1393-1400 (2012).
93 https://www.annualreviews.org/doi/pdf/10.1146/annurev.pa.16.040176.000245
94 B. A. Drew, Arthur Guedel and the Ascendance of Anesthesia: A Teacher, Tinkerer, and Transformer. *J Anesth Hist* 5, 85-92 (2019).
95 B. A. Drew, Arthur Guedel and the Ascendance of Anesthesia: A Teacher, Tinkerer, and Transformer. *J Anesth Hist* 5, 85-92 (2019).
96 R. M. Waters, *Holding Court With the Ghost of Gilman Terrace: Selected Writings of Ralph Milton Waters, M.D.* M. David C. Lai, Ed., In Commemoration of the 75th anniversary of Academic Anesthesiology (Wood Library-Museum of Anesthesiology, 520 N. Northwest Highway, Park Ridge, Illinois, 60068 USA, 2002).
97 L. E. Morris, M. E. Schroeder, M. E. Warner, Wood Library-Museum of Anesthesiology (Park Ridge Ill.), *Ralph Milton Waters, M.D., mentor to a profession : proceedings, the Ralph M. Waters International Symposium on Professionalism in Anesthesia, Madison, Wisconsin, 2002.* (Wood Library-Museum of Anesthesiology, Park Ridge, Ill., 2004), pp. xviii, 260 p.
98 C. D. W. Leake, R.M., The Anesthetic Properties of Carbon Dioxid. *J. Pharmacol. Exp. Ther.* 33, (1928).
99 C. D. W. Leake, R.M., The Anesthetic Properties of Carbon Dioxid. *J. Pharmacol. Exp. Ther.* 33, (1928).
100 R. M. Waters, Carbon Dioxide: Its Place in Anaesthesia. *Can Med Assoc J* 17, 1510-1513 (1927).
101 R. M. Waters, Carbon Dioxide Absorption from Anaesthetic Atmospheres : (Section of Anaesthetics). *Proc R Soc Med* 30, 11-22 (1936).
102 R. M. Waters, Absorption of carbon dioxide from anaesthetic atmosphere; historical aspects. *Anesthesiology* 8, 339-347 (1947).
103 R. M. Waters, Clinical Scope and Utility of Carbon Dioxid Filtration in Inhalation Anesthesia. *Anesthesia and Analgesia*, 20-22 (1924).
104 R. M. Waters, Toxic Effects of Carbon Dioxide. *J.A.M.A* 100:519, 1933, 219-224 (1933).
105 R. M. Waters, Carbon Dioxide Absorption Technic in Anesthesia. *Ann Surg* 103, 38-45 (1936).

106 R. M. Waters, Carbon Dioxide. *Can Med Assoc J* 38, 240-243 (1938).

107 R. M. Waters, Carbon Dioxide: Its Place in Anaesthesia. *Can Med Assoc J* 17, 1510-1513 (1927).

108 C. Thompson, Henry Hill Hickman. A forgotten pioneer of anesthesia. *BMJ* 1, 843-845 (1912).

109 F. G. Page.

110 W. D. A. Smith, *Henry Hill Hickman.* (History of Anesthesia Society, Sheffield).

111 G. Crile, D. H. Dolley, An Experimental Research into the Resuscitation of Dogs Killed by Anesthetics and Asphyxia. *The Journal of experimental medicine* 8, 713-725 (1906).

112 C. D. W. Leake, R.M., The Anesthetic Properties of Carbon Dioxid. *J. Pharmacol. Exp. Ther.* 33, (1928).

113 C. D. W. Leake, R.M., The Anesthetic Properties of Carbon Dioxid. *J. Pharmacol. Exp. Ther.* 33, (1928).

114 C. D. Leake, The Historical Development of Surgical Anesthesia. Scientific Monthly 20, 304-328 (1925).

115 E. Brown, Physiological effects of high concentrations of ca. *United States naval medical bulletin*, (1930).

116 A. Campbell, E. P. Poulton, *Oxygen and carbon dioxide therapy.* Oxford medical publications (Oxford University Press, H. Milford, London,, 1934), pp. xv, 179 p.

117 L. E. Morris, M. E. Schroeder, M. E. Warner, Wood Library-Museum of Anesthesiology (Park Ridge Ill.), *Ralph Milton Waters, M.D., mentor to a profession : proceedings, the Ralph M. Waters International Symposium on Professionalism in Anesthesia, Madison, Wisconsin, 2002.* (Wood Library-Museum of Anesthesiology, Park Ridge, Ill., 2004), pp. xviii, 260 p.

118 L. E. Morris, M. E. Schroeder, M. E. Warner, Wood Library-Museum of Anesthesiology (Park Ridge Ill.), *Ralph Milton Waters, M.D., mentor to a profession : proceedings, the Ralph M. Waters International Symposium on Professionalism in Anesthesia, Madison, Wisconsin, 2002.* (Wood Library-Museum of Anesthesiology, Park Ridge, Ill., 2004), pp. xviii, 260 p.

119 https://www.annualreviews.org/doi/pdf/10.1146/annurev.pa.16.040176.000245

120 R. M. Waters, Toxic Effects of Carbon Dioxide. *J.A.M.A* 100:519, 1933, 219-224 (1933).

121 J. S. Lundy, Carbon Dioxide as an Aid In Anesthesia. *J.A.M.A* 85:1953, 1925, (1925).

122 Y. Henderson, A Lecture ON RESPIRATION IN ANAESTHESIA: CONTROL BY CARBON DIOXIDE. *Br Med J* 2, 1170-1175 (1925).

123 R. D. Freidman, Carbon Dioxide in General Anesthesia. *Dental Outlook* June, (1928).

124 J. S. Lundy, Convulsions Associated with General Anesthesia. *Mayo Clinic Health Letter*, (1936).

125 M. C. Seevers, WH; et al, The Role of Hypercapnia in the Production of “Ether Convulsions”. *J Pharmacol Exp Ther* 63, (1938).

126 W. H. B. Cassels, T.J., Seevers, M.H., Convulsions During Anesthesia. *American Society for Pharmacology and Experimental Therapeutics, Baltimore*, (1938).

127 https://www.showcaves.com/english/it/caves/Cane.html

128 Y. Henderson, H. W. Haggard, *Noxious gases and the principles of respiration influencing their action*. (Chemical Catalog Co., New York,, 1927), pp. 220 p.
129 https://www.osha.gov/dts/hib/hib_data/hib19960605.html
130 C. D. W. Leake, R.M., The Anesthetic Properties of Carbon Dioxid. *J. Pharmacol. Exp. Ther.* 33, (1928).
131 R. M. Waters, Carbon Dioxide. *Can Med Assoc J* 38, 240-243 (1938).
132 R. M. Waters, Carbon Dioxide. *Can Med Assoc J* 38, 240-243 (1938).
133 R. M. Waters, Carbon Dioxide. *Can Med Assoc J* 38, 240-243 (1938).
134 C. D. W. Leake, R.M., The Anesthetic Properties of Carbon Dioxid. *J. Pharmacol. Exp. Ther.* 33, (1928).
135 R. F. Albrecht, D. J. Miletich, M. Ruttle, Cerebral effects of extended hyperventilation in unanesthetized goats. *Stroke* 18, 649-655 (1987).
136 M. Komori *et al.*, Permissive range of hypercapnia for improved peripheral microcirculation and cardiac output in rabbits. *Critical care medicine* 35, 2171-2175 (2007).
137 A. W. Sielenkamper, K. Eicker, H. Van Aken, Thoracic epidural anesthesia increases mucosal perfusion in ileum of rats. *Anesthesiology* 93, 844-851 (2000).
138 Y. Henderson, in *Cyclopedia of Medicine*. (1940).
139 R. M. Waters, Carbon Dioxide. *Can Med Assoc J* 38, 240-243 (1938).
140 Carbon Copy of a letter from Ralph Waters to Lincoln Sise, May 5, 1933, Ralph Waters Papers, University of Wisconsin-Madison Archives, Madison, WI. 163.
141 L. E. Morris, M. E. Schroeder, M. E. Warner, Wood Library-Museum of Anesthesiology (Park Ridge Ill.), *Ralph Milton Waters, M.D., mentor to a profession : proceedings, the Ralph M. Waters International Symposium on Professionalism in Anesthesia, Madison, Wisconsin, 2002*. (Wood Library-Museum of Anesthesiology, Park Ridge, Ill., 2004), pp. xviii, 260 p.
142 K. M. Conlee, M. L. Stephens, A. N. Rowan, L. A. King, Carbon dioxide for euthanasia: concerns regarding pain and distress, with special reference to mice and rats. *Lab Anim* 39, 137-161 (2005).
143 P. J. Danneman, S. Stein, S. O. Walshaw, Humane and practical implications of using carbon dioxide mixed with oxygen for anesthesia or euthanasia of rats. *Lab Anim Sci* 47, 376-385 (1997).
144 C. D. W. Leake, R.M., The Anesthetic Properties of Carbon Dioxid. *J. Pharmacol. Exp. Ther.* 33, (1928).
145 A. Campbell, E. P. Poulton, *Oxygen and carbon dioxide therapy*. Oxford medical publications (Oxford University Press, H. Milford, London,, 1934), pp. xv, 179 p.
146 L. S. Coleman, in *apsf Newsletter*. (Anesthesia Patient Safety Foundation, Administrator, Deanna Walker Anesthesia Patient Safety Foundation Building One, Suite Two 8007 South Meridian Street Indianapolis, IN 46217-2922 e-mail address: walker@apsf.org FAX: (317) 888-1482, 2010), vol. Winter 2009-2020.
147 L. S. Coleman, A call for standards on perioperative CO_2 regulation. *Can J Anaesth*, (2011).
148 F. J. Overdyk, Postoperative opioids remain a serious patient safety threat. *Anesthesiology* 113, 259-260; author reply 260-251 (2010).

149 F. J. Overdyk, postoperative Opioids Need System-Wide Overhaul. *Anesthesia Patient Safety Foundation Newsletter*, (2010).

150 L. E. Morris, M. E. Schroeder, M. E. Warner, Wood Library-Museum of Anesthesiology (Park Ridge Ill.), *Ralph Milton Waters, M.D., mentor to a profession : proceedings, the Ralph M. Waters International Symposium on Professionalism in Anesthesia, Madison, Wisconsin, 2002.* (Wood Library-Museum of Anesthesiology, Park Ridge, Ill., 2004), pp. xviii, 260 p.

151 W. N. Kemp, The overuse of carbon dioxide absorption in general anesthesia. *American journal of surgery* 87, 864-868 (1954).

152 H. K. Beecher, D. P. Todd, A study of the deaths associated with anesthesia and surgery: based on a study of 599, 548 anesthesias in ten institutions 1948-1952, inclusive. *Ann Surg* 140, 2-35 (1954).

153 H. K. Beecher, D. P. Todd, A study of the deaths associated with anesthesia and surgery: based on a study of 599, 548 anesthesias in ten institutions 1948-1952, inclusive. *Ann Surg* 140, 2-35 (1954).

154 B. Kavanagh, Normocapnia vs hypercapnia. *Minerva Anestesiol* 68, 346-350 (2002).

155 L. S. Coleman, in *apsf Newsletter*. (Anesthesia Patient Safety Foundation, Administrator, Deanna Walker Anesthesia Patient Safety Foundation Building One, Suite Two 8007 South Meridian Street Indianapolis, IN 46217-2922 e-mail address: walker@apsf.org FAX: (317) 888-1482, 2010), vol. Winter 2009-2020.

156 L. S. Coleman, A call for standards on perioperative CO(2) regulation. *Can J Anaesth*, (2011).

157 M. Brattwall, M. Warren-Stomberg, F. Hesselvik, J. Jakobsson, Brief review: Theory and practice of minimal fresh gas flow anesthesia. *Can J Anaesth* 59, 785-797 (2012).

158 H. J. Ladegaard-Pedersen, A Circle System Without Carbon Dioxide Absorption. *Acta Anaesth. Scand.* 22, 281-286 (1978).

159 K. J. Boniface, J. M. Brown, Effect of carbon dioxide excess on contractile force of heart, in situ. *The American journal of physiology* 172, 752-756 (1953).

160 K. J. Boniface, J. M. Brown, Effect of carbon dioxide excess on contractile force of heart, in situ. *The American journal of physiology* 172, 752-756 (1953).

161 J. H. Eisele, E. I. Eger, 2nd, M. Muallem, Narcotic properties of carbon dioxide in the dog. *Anesthesiology* 28, 856-865 (1967).

162 J. H. Eisele, E. I. Eger, 2nd, M. Muallem, Narcotic properties of carbon dioxide in the dog. *Anesthesiology* 28, 856-865 (1967).

163 http://en.wikipedia.org/wiki/Arterial_blood_gas

164 D. J. Cullen, E. I. Eger, 2nd, The effect of extreme hypocapnia on the anaesthetic requirement (MAC) of dogs. *Br J Anaesth* 43, 339-343 (1971).

165 M. Komori *et al.*, Permissive range of hypercapnia for improved peripheral microcirculation and cardiac output in rabbits. *Critical care medicine* 35, 2171-2175 (2007).

166 R. J. Brosnan, E. I. Eger, 2nd, M. J. Laster, J. M. Sonner, Anesthetic properties of carbon dioxide in the rat. *Anesth Analg* 105, 103-106 (2007).

167 C. J. S. a. H. H. H. Thompson, A Forgotten pioneer of anesthesia. *Brit. Med. J.1:843-845, 1912,* (1912).

168 E. Lowenstein et al., Cardiovascular response to large doses of intravenous morphine in man. *N Engl J Med* 281, 1389-1393 (1969).
169 L. W. Crile GW, *Anoci-association.* (Saunders, Philadelphia, 1914), pp. 223-225.
170 Y. Henderson, Resuscitation with Carbon Dioxide. *Science* 83, 399-402 (1936).
171 Y. Henderson, Public Service as an Element in the Life of the American Scientist. *Science* 77, 584-585 (1933).
172 Y. Henderson, Physiological Observations on Pike's Peak, Colorado, made in the Summer of 1911. *Trans Am Climatol Assoc* 28, 11-15 (1912).
173 Y. Henderson, in *Cyclopedia of Medicine*. (1940).
174 J. S. Lundy, Carbon Dioxide as an Aid In Anesthesia. *J.A.M.A* 85:1953, 1925, (1925).
175 Y. Henderson, The Wesley M. Carpenter Lecture: Atelectasis, Massive Collapse and Related Post-Operative Conditions. *Bull N Y Acad Med* 11, 639-656 (1935).
176 Y. Henderson, How Breathing Begins at Birth. *Science* 85, 89-91 (1937).
177 O. Akca, Optimizing the intraoperative management of carbon dioxide concentration. *Current opinion in anaesthesiology* 19, 19-25 (2006).
178 L. S. Coleman, in *apsf Newsletter*. (Anesthesia Patient Safety Foundation, Administrator, Deanna Walker Anesthesia Patient Safety Foundation Building One, Suite Two 8007 South Meridian Street Indianapolis, IN 46217-2922 e-mail address: walker@apsf.org FAX: (317) 888-1482, 2010), vol. Winter 2009-2020.
179 L. S. Coleman, A call for standards on perioperative CO(2) regulation. *Can J Anaesth*, (2011).
180 T. G. Monk, V. Saini, B. C. Weldon, J. C. Sigl, Anesthetic management and one-year mortality after noncardiac surgery. *Anesth Analg* 100, 4-10 (2005).
181 L. S. Coleman, A testable hypothesis that may explain the morbidity and mortality caused by surgical stress. *Anesth Analg* 103, 1589; author reply 1589-1590 (2006).
182 H. K. Beecher, D. P. Todd, A study of the deaths associated with anesthesia and surgery: based on a study of 599, 548 anesthesias in ten institutions 1948-1952, inclusive. *Ann Surg* 140, 2-35 (1954).
183 http://www.anesthesiologynews.com/ViewArticle.aspx?d=Clinical+Anesthesiology&d_id=1&i=July+2012&i_id=864&a_id=21217
184 R. F. Albrecht, D. J. Miletich, M. Ruttle, Cerebral effects of extended hyperventilation in unanesthetized goats. *Stroke* 18, 649-655 (1987).
185 B. Kavanagh, Normocapnia vs hypercapnia. *Minerva Anestesiol* 68, 346-350 (2002).
186 J. H. Havill, Prolonged hyperventilation and intracranial pressure. *Critical care medicine* 12, 72-74 (1984).
187 J. Huttunen *et al.*, Effects of voluntary hyperventilation on cortical sensory responses. Electroencephalographic and magnetoencephalographic studies. *Exp Brain Res* 125, 248-254 (1999).
188 J. G. Laffey *et al.*, Carbon dioxide attenuates pulmonary impairment resulting from hyperventilation. *Critical care medicine* 31, 2634-2640 (2003).
189 D. W. Marion, A. Firlik, M. R. McLaughlin, Hyperventilation therapy for severe traumatic brain injury. *New Horiz* 3, 439-447 (1995).

190 K. Nishitateno, S. H. Ngai, A. D. Finck, B. A. Berkowitz, Pharmacokinetics of morphine: concentrations in the serum and brain of the dog during hyperventilation. *Anesthesiology* 50, 520-523 (1979).
191 M. E. Raichle, F. Plum, Hyperventilation and cerebral blood flow. *Stroke* 3, 566-575 (1972).
192 A. J. Salvatore, S. F. Sullivan, E. M. Papper, Postoperative hypoventilation and hypoxemia in man after hyperventilation. *N Engl J Med* 280, 467-470 (1969).
193 A. Taniguchi, S. Kitamura, R. Kawahara, L. Maeda, K. Hitomi, [Neonatal apnea after general anesthesia--effects of intraoperative hyperventilation and serum ionized Ca concentration]. *Masui* 44, 499-502 (1995).
194 R. A. van Hulst, J. J. Haitsma, T. W. Lameris, J. Klein, B. Lachmann, Hyperventilation impairs brain function in acute cerebral air embolism in pigs. *Intensive Care Med* 30, 944-950 (2004).
195 M. Weckesser *et al.*, Functional imaging of the visual cortex with bold-contrast MRI: hyperventilation decreases signal response. *Magn Reson Med* 41, 213-216 (1999).
196 S. G. Ainslie, J. H. Eisele, Jr., G. Corkill, Fentanyl concentrations in brain and serum during respiratory acid--base changes in the dog. *Anesthesiology* 51, 293-297 (1979).
197 S. Corne, K. Webster, M. Younes, Hypoxic respiratory response during acute stable hypocapnia. *American journal of respiratory and critical care medicine* 167, 1193-1199 (2003).
198 Anesthesiology News ISSUE: JANUARY 2015 | VOLUME: 41:1
199 L. S. Coleman, 30 Years Lost in Anesthesia Theory. *Cardiovasc Hematol Agents Med Chem* 10, 31-49 (2012).
200 J. Hanawald, Rapid Gas Change During ECMO Linked to Mortality Risk. *Anesthesiology News* 37:8, (2011).
201 M. M. S. a. J. Wang, Central Nervous System Toxicity during the Induction of General Anesthesia After Intravenous Bolus of Lidocaine. *Journal of Clinical Research and Anesthesia* 2(1)-10004 www.auctoresonline.org, 4 (2019).
202 J. E. Eckenhoff, Carbon Dioxide and Man. *Anesthesiology* 21, 585-586 (1960).
203 K. M. Conlee, M. L. Stephens, A. N. Rowan, L. A. King, Carbon dioxide for euthanasia: concerns regarding pain and distress, with special reference to mice and rats. *Lab Anim* 39, 137-161 (2005).
204 P. J. Danneman, S. Stein, S. O. Walshaw, Humane and practical implications of using carbon dioxide mixed with oxygen for anesthesia or euthanasia of rats. *Lab Anim Sci* 47, 376-385 (1997).
205 D. J. Cullen, E. I. Eger, 2nd, The effect of extreme hypocapnia on the anaesthetic requirement (MAC) of dogs. *Br J Anaesth* 43, 339-343 (1971).
206 L. S. Coleman, Routine Mask Induction and Intubation without Paralysis may be Safer. *J Anesth Surg* 2, 1-2 (2015).
207 T. E. Smith, W. G. Elliott, Routine inhaled induction in adults: a safe practice? *Anesth Analg* 102, 646-647; author reply 647 (2006).
208 A. A. van den Berg, D. A. Chitty, R. D. Jones, M. S. Sohel, A. Shahen, Intravenous or inhaled induction of anesthesia in adults? An audit of preoperative patient preferences. *Anesth Analg* 100, 1422-1424, table of contents (2005).

209 E. R. John, L. S. Prichep, The anesthetic cascade: a theory of how anesthesia suppresses consciousness. *Anesthesiology* 102, 447-471 (2005).

210 L. S. Coleman, 30 Years Lost in Anesthesia Theory. *Cardiovasc Hematol Agents Med Chem* 10, 31-49 (2012).

211 L. Coleman, A Testable Stress Mechanism Explains Both Unified Stress Theory and Capillary Gate Theory. *FDA Anesthesia & Life Support Drugs Advisory Committee pp239-243 3/29/07*, (2007).

212 K. E. Welty-Wolf *et al.*, Coagulation blockade prevents sepsis-induced respiratory and renal failure in baboons. *American journal of respiratory and critical care medicine* 164, 1988-1996 (2001).

213 S. Liu, R. L. Carpenter, J. M. Neal, Epidural anesthesia and analgesia. Their role in postoperative outcome. *Anesthesiology* 82, 1474-1506 (1995).

214 A. Campbell, E. P. Poulton, *Oxygen and carbon dioxide therapy.* Oxford medical publications (Oxford University Press, H. Milford, London,, 1934), pp. xv, 179 p.

215 M. Komori *et al.*, Permissive range of hypercapnia for improved peripheral microcirculation and cardiac output in rabbits. *Critical care medicine* 35, 2171-2175 (2007).

216 O. J. Kirkeby, S. Kutzsche, C. Risoe, I. R. Rise, Cerebral nitric oxide concentration and microcirculation during hypercapnia, hypoxia, and high intracranial pressure in pigs. *J Clin Neurosci* 7, 531-538 (2000).

217 J. J. Smith *et al.*, The role of nitric oxide in the cerebrovascular response to hypercapnia. *Anesth Analg* 84, 363-369 (1997).

218 O. Akca, Optimizing the intraoperative management of carbon dioxide concentration. *Current opinion in anaesthesiology* 19, 19-25 (2006).

219 O. Akca *et al.*, Hypercapnia improves tissue oxygenation. *Anesthesiology* 97, 801-806 (2002).

220 G. De Castro, P. Mundeleer, [Anesthesia without sleep: "neuroleptanalgesia".]. *Acta Chir Belg* 58, 689-693 (1959).

221 J. de Castro *et al.*, Comparative study of cardiovascular, neurological and metabolic side effects of 8 narcotics in dogs. Pethidine, piritramide, morphine, phenoperidine, fentanyl, R 39 209, sufentanil, R 34 995. III. Comparative study of the acute metabolic toxicity of the narcotics used in high and massive doses in curarised and mechanically ventilated dogs. *Acta anaesthesiologica Belgica* 30, 71-90 (1979).

222 E. Lowenstein *et al.*, Cardiovascular response to large doses of intravenous morphine in man. *N Engl J Med* 281, 1389-1393 (1969).

223 T. H. Stanley, T. D. Egan, H. Van Aken, A tribute to Dr. Paul A. J. Janssen: entrepreneur extraordinaire, innovative scientist, and significant contributor to anesthesiology. *Anesth Analg* 106, 451-462, table of contents (2008).

224 D. T. Mangano *et al.*, Postoperative myocardial ischemia. Therapeutic trials using intensive analgesia following surgery. The Study of Perioperative Ischemia (SPI) Research Group. *Anesthesiology* 76, 342-353 (1992).

225 P. S. Myles *et al.*, Remifentanil, fentanyl, and cardiac surgery: a double-blinded, randomized, controlled trial of costs and outcomes. *Anesth Analg* 95, 805-812, table of contents (2002).

226 L. W. Crile GW, *Anoci-association.* (Saunders, Philadelphia, 1914), pp. 223-225.

227 B. Kavanagh, Normocapnia vs hypercapnia. *Minerva Anestesiol* 68, 346-350 (2002)
228 L. S. Coleman, Routine Mask Induction and Intubation without Paralysis may be Safer. *J Anesth Surg* 2, 1-2 (2015).
229 A. A. van den Berg, D. A. Chitty, R. D. Jones, M. S. Sohel, A. Shahen, Intravenous or inhaled induction of anesthesia in adults? An audit of preoperative patient preferences. *Anesth Analg* 100, 1422-1424, table of contents (2005).
230 T. E. Smith, W. G. Elliott, Routine inhaled induction in adults: a safe practice? *Anesth Analg* 102, 646-647; author reply 647 (2006).
231 H. K. Beecher, D. P. Todd, A study of the deaths associated with anesthesia and surgery: based on a study of 599, 548 anesthesias in ten institutions 1948-1952, inclusive. *Ann Surg* 140, 2-35 (1954).
232 J. L. Benumof, Management of the difficult adult airway. With special emphasis on awake tracheal intubation. *Anesthesiology* 75, 1087-1110 (1991).
233 J. L. Benumof, Awake intubations are alive and well. *Can J Anaesth* 62, 723-726 (2015).
234 B. Vasile, F. Rasulo, A. Candiani, N. Latronico, The pathophysiology of propofol infusion syndrome: a simple name for a complex syndrome. *Intensive Care Med* 29, 1417-1425 (2003).
235 M. Weisenberg *et al.*, Dose-dependent hemodynamic effects of propofol induction following brotizolam premedication in hypertensive patients taking angiotensin-converting enzyme inhibitors. *J Clin Anesth* 22, 190-195 (2010).
236 T. E. Smith, W. G. Elliott, Routine inhaled induction in adults: a safe practice? *Anesth Analg* 102, 646-647; author reply 647 (2006).
237 A. A. van den Berg, D. A. Chitty, R. D. Jones, M. S. Sohel, A. Shahen, Intravenous or inhaled induction of anesthesia in adults? An audit of preoperative patient preferences. *Anesth Analg* 100, 1422-1424, table of contents (2005).
238 S. A. Coleman, J. W. McCrory, C. J. Vallis, R. J. Boys, Inhalation induction of anaesthesia with isoflurane: effect of added carbon dioxide. *Br J Anaesth* 67, 257-261 (1991).
239 S. A. Coleman, J. W. McCrory, C. J. Vallis, R. J. Boys, Inhalation induction of anaesthesia with isoflurane: effect of added carbon dioxide. *Br J Anaesth* 67, 257-261 (1991).
240 R. J. Ede, A. E. Gimson, D. Bihari, R. Williams, Controlled hyperventilation in the prevention of cerebral oedema in fulminant hepatic failure. *J Hepatol* 2, 43-51 (1986).
241 D. A. Godoy, A. Seifi, D. Garza, S. Lubillo-Montenegro, F. Murillo-Cabezas, Hyperventilation Therapy for Control of Posttraumatic Intracranial Hypertension. *Front Neurol* 8, 250 (2017).
242 D. W. Marion, A. Firlik, M. R. McLaughlin, Hyperventilation therapy for severe traumatic brain injury. *New Horiz* 3, 439-447 (1995).
243 M. E. Raichle, F. Plum, Hyperventilation and cerebral blood flow. *Stroke* 3, 566-575 (1972).

244 Z. Zhang, Q. Guo, E. Wang, Hyperventilation in neurological patients: from physiology to outcome evidence. *Current opinion in anaesthesiology* 32, 568-573 (2019).
245 J. E. Heffner, S. A. Sahn, Controlled hyperventilation in patients with intracranial hypertension. Application and management. *Arch Intern Med* 143, 765-769 (1983).
246 J. Huttunen et al., Effects of voluntary hyperventilation on cortical sensory responses. Electroencephalographic and magnetoencephalographic studies. *Exp Brain Res* 125, 248-254 (1999).
247 K. Nakao et al., Hyperventilation as a specific test for diagnosis of coronary artery spasm. *Am J Cardiol* 80, 545-549 (1997).
248 R. A. van Hulst, J. J. Haitsma, T. W. Lameris, J. Klein, B. Lachmann, Hyperventilation impairs brain function in acute cerebral air embolism in pigs. *Intensive Care Med* 30, 944-950 (2004).
249 T. Itakura, K. Yamamoto, M. Tohyama, N. Shimizu, Central dual innervation of arterioles and capillaries in the brain. *Stroke* 8, 360-365 (1977).
250 Y. Itoh, N. Suzuki, Control of brain capillary blood flow. *J Cereb Blood Flow Metab* 32, 1167-1176 (2012).
251 Y. Itoh, N. Suzuki, Control of brain capillary blood flow. J *Cereb Blood Flow Metab* 32, 1167-1176 (2012).
252 J. A. dKim, N. D. Tran, S. J. Wang, M. J. Fisher, Astrocyte regulation of human brain capillary endothelial fibrinolysis. *Thromb Res* 112, 159-165 (2003).
253 R. F. Albrecht, D. J. Miletich, M. Ruttle, Cerebral effects of extended hyperventilation in unanesthetized goats. *Stroke* 18, 649-655 (1987).
254 M. B. Jaffe, Infrared measurement of carbon dioxide in the human breath: "breathe-through" devices from Tyndall to the present day. *Anesth Analg* 107, 890-904 (2008).
255 A. Marcus, in *Anesthesiology News* (McMahon, New York, NY, October 8, 2010).
256 P. Bufano, in *Anesthesiology News*. (McMahon, New York, NY, 2014).
257 A. Marcus, in *Anesthesiology News*. (McMahon, New York, 2011).
258 A. Marcus, in *Anesthesiology News*. (McMahon, New York, 2011).
259 A. Marcus, in *Anesthesiology News*. (McMahon, New York, NY, October 8, 2010).
260 A. Marcus, in *Anesthesiology News* (McMahon, New York, NY, July 2, 2012).
261 A. Schapiro, in *Anesthesiology News*. (McMahon, New York, NY, June 9, 2017).
262 J. Virapen, *Side Effects (of Prozac): Death* (Money Tree Publishing, 6256 Bullet Drive, Crestview, Fl. 32536, 2015), pp. 189.
263 L. S. Coleman, Four Forgotten Giants of Anesthesia History. *Journal of Anesthesia and Surgery* 3, 1-17 (2015).
264 A. W. Gelb et al., World Health Organization-World Federation of Societies of Anaesthesiologists (WHO-WFSA) International Standards for a Safe Practice of Anesthesia. *Anesth Analg* 126, 2047-2055 (2018).
265 A. W. Gelb et al., World Health Organization-World Federation of Societies of Anaesthesiologists (WHO-WFSA) International Standards for a Safe Practice of Anesthesia. *Anesth Analg* 126, 2047-2055 (2018).
266 L. S. Coleman, A call for standards on perioperative CO(2) regulation. *Can J Anaesth*, (2011).

267 R. N. Merchant, Special announcement: Guidelines to the practice of anesthesia — revised edition 2015. *Can J Anaesth* 62, 8-9 (2015).

268 T. C. Harvey et al., Effect of carbon dioxide in acute mountain sickness: a rediscovery. *Lancet* 2, 639-641 (1988).

269 J. Marquez, J. Magovern, P. Kaplan, T. Sakert, G. P. Gravlee, Cardiac surgery "fast tracking" in an academic hospital. *Journal of cardiothoracic and vascular anesthesia* 9, 34-36 (1995).

270 T. H. E. Stanley, Talmage D., *Anesthesia for the New Millennium*. (Springer-Science + Business Media, 1999), vol. 34, pp. 288.

271 K. G. Hickling, Permissive hypercapnia. *Respir Care Clin N Am* 8, 155-169, v (2002).

272 J. G. Laffey, D. O'Croinin, P. McLoughlin, B. P. Kavanagh, Permissive hypercapnia--role in protective lung ventilatory strategies. *Intensive Care Med* 30, 347-356 (2004).

273 H. Morisaki, R. Serita, Y. Innami, Y. Kotake, J. Takeda, Permissive hypercapnia during thoracic anaesthesia. *Acta anaesthesiologica Scandinavica* 43, 845-849 (1999).

274 A. Ohmura, M. Sha, J. Katagiri, How far can we go with permissive hypercapnia? A case presentation and some biased comments with emphasis on maintaining normal haemoglobin level. *Acta Anaesthesiol Scand Suppl* 107, 209-213 (1995).

275 U. H. Thome, W. A. Carlo, Permissive hypercapnia. *Semin Neonatol* 7, 409-419 (2002).

276 K. G. Hickling, S. J. Henderson, R. Jackson, Low mortality associated with low volume pressure limited ventilation with permissive hypercapnia in severe adult respiratory distress syndrome. *Intensive Care Med* 16, 372-377 (1990).

277 M. Komori et al., Permissive range of hypercapnia for improved peripheral microcirculation and cardiac output in rabbits. *Critical care medicine* 35, 2171-2175 (2007).

278 M. C. Geoffrey D. Muller, FFICM, FRCA David W. Healy, MD, MRCP, FRCA, in *Anesthesiology News Airway Management* 2015. (McMahon, 2015).

279 M. D. John Doyle, in *Anesthesiology News*. (McMahon, New York, NY).

280 J. L. Benumof, Obesity, sleep apnea, the airway and anesthesia. *Current opinion in anaesthesiology* 17, 21-30 (2004).

281 F. J. Overdyk, Q. Ahmed, Postoperative monitoring of obese patients with obstructive sleep apnea. *Anesth Analg* 108, 1044-1045; author reply 1045 (2009).

282 J. L. Benumof, The elephant in the room is bigger than you think: finding obstructive sleep apnea patients dead in bed postoperatively. *Anesth Analg* 120, 491 (2015).

283 M. e. a. Mahesh Nagappa, Postoperative Outcomes in Obstructive Sleep Apnea Patients Undergoing Cardiac Surgery: A Systematic Review and Meta-analysis of Comparative Studies. Anesth Analg 125, (2017).

284 B. Kronemyer, in *Anesthesiology News*. (McMahon, New York, NY, October 12, 2016).

285 M. D. John Doyle, in *Anesthesiology News*. (McMahon, New York, NY).

286 J. L. Benumof, Obesity, sleep apnea, the airway and anesthesia. *Current opinion in anaesthesiology* 17, 21-30 (2004).
287 F. J. Overdyk, Q. Ahmed, Postoperative monitoring of obese patients with obstructive sleep apnea. *Anesth Analg* 108, 1044-1045; author reply 1045 (2009).
288 J. L. Benumof, The elephant in the room is bigger than you think: finding obstructive sleep apnea patients dead in bed postoperatively. *Anesth Analg* 120, 491 (2015).
289 M. e. a. Mahesh Nagappa, Postoperative Outcomes in Obstructive Sleep Apnea Patients Undergoing Cardiac Surgery: A Systematic Review and Meta-analysis of Comparative Studies. *Anesth Analg* 125, (2017).
290 B. Kronemyer, in *Anesthesiology News*. (McMahon, New York, NY, October 12, 2016).
291 W. H. Reinhart, J. Oswald, R. Walter, M. Kuhn, Blood viscosity and platelet function in patients with obstructive sleep apnea syndrome treated with nasal continuous positive airway pressure. *Clin Hemorheol Microcirc* 27, 201-207 (2002).
292 J. L. Benumof, Obstructive sleep apnea in the adult obese patient: implications for airway management. *J Clin Anesth* 13, 144-156 (2001).
293 B. Bhushan, B. Ayub, D. M. Thompson, F. Abdullah, K. R. Billings, Impact of short sleep on metabolic variables in obese children with obstructive sleep apnea. *Laryngoscope* 127, 2176-2181 (2017).
294 J. B. Gross et al., Practice guidelines for the perioperative management of patients with obstructive sleep apnea: a report by the American Society of Anesthesiologists Task Force on Perioperative Management of patients with obstructive sleep apnea. *Anesthesiology* 104, 1081-1093; quiz 1117-1088 (2006).
295 M. B. Manne, G. Rutecki, Obstructive sleep apnea: Who should be tested, and how? *Cleve Clin J Med* 83, 25-27 (2016).
296 R. Mehra, Sleep apnea ABCs: airway, breathing, circulation. *Cleve Clin J Med* 81, 479-489 (2014).
297 R. Mehra, N. Foldvary-Schaefer, The USPSTF and screening for obstructive sleep apnea: Dispelling misconceptions. *Cleve Clin J Med* 84, 429-431 (2017).
298 M. G. Davies, P. O. Hagen, Systemic inflammatory response syndrome. *Br J Surg* 84, 920-935 (1997).
299 Y. G. Liu, C. S. Li, [Coagulation dysfunctions in patients with systemic inflammatory response syndrome]. *Zhonghua Nei Ke Za Zhi* 43, 94-97 (2004).
300 K. Takenaka, E. Ogawa, H. Wada, T. Hirata, Systemic inflammatory response syndrome and surgical stress in thoracic surgery. *J Crit Care* 21, 48-53; discussion 53-45 (2006).
301 E. Deflandre, S. Lacroix, OSA and adverse perioperative outcomes: we should distinguish them. *Reg Anesth Pain Med* 44, 754 (2019).
302 K. M. Sundar, S. E. Daly, Chronic cough and OSA: an underappreciated relationship. *Lung* 192, 21-25 (2014).
303 F. Chung, H. R. Abdullah, P. Liao, STOP-Bang Questionnaire: A Practical Approach to Screen for Obstructive Sleep Apnea. *Chest* 149, 631-638 (2016).
304 J. L. Benumof, Mismanagement of obstructive sleep apnea may result in finding these patients dead in bed. *Can J Anaesth* 63, 3-7 (2016).

305 T. G. Monk, V. Saini, B. C. Weldon, J. C. Sigl, Anesthetic management and one-year mortality after noncardiac surgery. *Anesth Analg* 100, 4-10 (2005).
306 J. H. Dreyfuss, in *Anesthesiology News*. (May 5, 2016).
307 L. S. Coleman, 30 Years Lost in Anesthesia Theory. *Cardiovasc Hematol Agents Med Chem* 10, 31-49 (2012).
308 L. S. Coleman, A testable hypothesis that may explain the morbidity and mortality caused by surgical stress. *Anesth Analg* 103, 1589; author reply 1589-1590 (2006).
309 L. S. Coleman, Should soda lime be abolished? *Anesth Analg* 102, 1290-1291 (2006).
310 H. J. Ladegaard-Pedersen, A Circle System Without Carbon Dioxide Absorption. *Acta Anaesth. Scand.* 22, 281-286 (1978).
311 M. J. Laster, E. I. Eger, 2nd, Temperatures in soda lime during degradation of desflurane, isoflurane, and sevoflurane by desiccated soda lime. *Anesth Analg* 101, 753-757, table of contents (2005).
312 C. Keijzer, R. S. Perez, J. J. De Lange, Carbon monoxide production from five volatile anesthetics in dry sodalime in a patient model: Halothane and sevoflurane do produce carbon monoxide; temperature is a poor predictor of carbon monoxide production. *BMC Anesthesiol* 5, 6 (2005).
313 G. Moriwaki, H. Bito, K. Ikeda, Partly exhausted soda lime or soda lime with water added, inhibits the increase in compound A concentration in the circle system during low-flow sevoflurane anaesthesia. *Br J Anaesth* 79, 782-786 (1997).
314 C. G. Schoonbee, C. M. Conway, Factors affecting carbon dioxide homeostasis during controlled ventilation with circle systems. *Br J Anaesth* 53, 471-477 (1981).
315 A. A. van den Berg, D. A. Chitty, R. D. Jones, M. S. Sohel, A. Shahen, Intravenous or inhaled induction of anesthesia in adults? An audit of preoperative patient preferences. *Anesth Analg* 100, 1422-1424, table of contents (2005).
316 S. A. Coleman, J. W. McCrory, C. J. Vallis, R. J. Boys, Inhalation induction of anaesthesia with isoflurane: effect of added carbon dioxide. *Br J Anaesth* 67, 257-261 (1991).
317 I. Ben-Shlomo *et al.*, Propofol and fentanyl act additively for induction of anaesthesia. *Anaesthesia* 48, 111-113 (1993).
318 M. D. Brunner *et al.*, MAC reduction of isoflurane by sufentanil. *Br J Anaesth* 72, 42-46 (1994).
319 A. I. McEwan *et al.*, Isoflurane minimum alveolar concentration reduction by fentanyl. *Anesthesiology* 78, 864-869 (1993).
320 C. L. Westmoreland, P. S. Sebel, A. Gropper, Fentanyl or alfentanil decreases the minimum alveolar anesthetic concentration of isoflurane in surgical patients. *Anesth Analg* 78, 23-28 (1994).
321 A. Ohmura, M. Sha, J. Katagiri, How far can we go with permissive hypercapnia? A case presentation and some biased comments with emphasis on maintaining normal haemoglobin level. *Acta Anaesthesiol Scand Suppl* 107, 209-213 (1995).

322 U. H. Thome, W. A. Carlo, Permissive hypercapnia. *Semin Neonatol* 7, 409-419 (2002).
323 Q. Zhou *et al.*, Effects of permissive hypercapnia on transient global cerebral ischemia-reperfusion injury in rats. *Anesthesiology* 112, 288-297 (2010).
324 O. Akca, Hypercapnia: what is the upper limit? *Paediatric anaesthesia* 15, 80-81 (2005).
325 M. Komori *et al.*, Permissive range of hypercapnia for improved peripheral microcirculation and cardiac output in rabbits. *Critical care medicine* 35, 2171-2175 (2007).
326 T. G. Monk, V. Saini, B. C. Weldon, J. C. Sigl, Anesthetic management and one-year mortality after noncardiac surgery. *Anesth Analg* 100, 4-10 (2005).
327 D. T. Mangano *et al.*, Postoperative myocardial ischemia. Therapeutic trials using intensive analgesia following surgery. The Study of Perioperative Ischemia (SPI) Research Group. *Anesthesiology* 76, 342-353 (1992).
328 P. S. Myles *et al.*, Remifentanil, fentanyl, and cardiac surgery: a double-blinded, randomized, controlled trial of costs and outcomes. *Anesth Analg* 95, 805-812, table of contents (2002).
329 A. K. Exadaktylos, D. J. Buggy, D. C. Moriarty, E. Mascha, D. I. Sessler, Can anesthetic technique for primary breast cancer surgery affect recurrence or metastasis? *Anesthesiology* 105, 660-664 (2006).
330 D. T. Mangano *et al.*, Perioperative myocardial ischemia in patients undergoing noncardiac surgery--I: Incidence and severity during the 4 day perioperative period. The Study of Perioperative Ischemia (SPI) Research Group. *J Am Coll Cardiol* 17, 843-850 (1991).
331 L. E. Morris, M. E. Schroeder, M. E. Warner, Wood Library-Museum of Anesthesiology (Park Ridge Ill.), *Ralph Milton Waters, M.D., mentor to a profession : proceedings, the Ralph M. Waters International Symposium on Professionalism in Anesthesia, Madison, Wisconsin, 2002.* (Wood Library-Museum of Anesthesiology, Park Ridge, Ill., 2004), pp. xviii, 260 p.
332 L. S. Coleman, Should soda lime be abolished? *Anesth Analg* 102, 1290-1291 (2006).
333 L. S. Coleman, Soda lime is a dangerous relic of a forgotten past. *Anaesthesia* 70, (2016).
334 E. S. Brown, The activity and surface area of fresh soda lime. *Anesthesiology* 19, 208-212 (1958).
335 J. Rossman, Drinker Respirator Patents Held Invalid. *Science* 82, 221-222 (1935).
336 E. B. Shaw, H. E. Thelander, M. A. Limper, Respiratory Failure in Poliomyelitis-Its Treatment with the Drinker Respirator: Report of Cases. *Cal West Med* 35, 5-7 (1931).
337 C. A. Elsberg, III. Clinical Experiences with Intratracheal Insufflation (Meltzer), with Remarks upon the Value of the Method for Thoracic Surgery. *Ann Surg* 52, 23-29 (1910).
338 Y. Henderson, Observations on RESUSCITATION FROM ASPHYXIA AND PREVENTION AND TREATMENT OF SECONDARY PNEUMONIA BY INHALATION OF CO(2). *Br Med J* 2, 687-689 (1931).

339 Y. Henderson, Applications of the Physiology of Respiration to Resuscitation from Asphyxia and Drowning and to the Prevention and Treatment of Secondary Pneumonia. *Yale J Biol Med* 4, 429-436 (1932).

340 P. J. Flagg, Asphyxia Treatment in the Adult and in the Newborn. *JAMA* Arch Otolaryngol. 1930;12(1):23-30. doi:10.1001/archotol.1930.03570010033004, 23-30 (1930).

341 http://en.wikipedia.org/wiki/Yoshitaka_Fujii

342 http://www.anesthesiologynews.com/ViewArticle.aspx?d=Web%2bExclusives&d_id=175&i=October+2010&i_id=669&a_id=16095

343 J. H. Eisele, E. I. Eger, 2nd, M. Muallem, Narcotic properties of carbon dioxide in the dog. *Anesthesiology* 28, 856-865 (1967).

344 R. J. Brosnan, E. I. Eger, 2nd, M. J. Laster, J. M. Sonner, Anesthetic properties of carbon dioxide in the rat. *Anesth Analg* 105, 103-106 (2007).

345 J. E. Eckenhoff, Carbon Dioxide and Man. *Anesthesiology* 21, 585-586 (1960).

Chapter 19

What Needs to Happen Next

"Using mathematics when it's not needed is not science but scientism — at the time of this writing, science has been taken over by vendors using it to sell products (like margarine or genetically modified solutions) and, ironically, the skeptical enterprise is being used to silence skeptics."
— Nassim Nicholas Taleb, Skin in the Game

The great discoveries of science often consist in the uncovering of a truth buried under the rubble of traditional prejudice, in getting out of the cul-de-sacs into which formal reasoning divorced from reality leads; in liberating the mind trapped between the iron teeth of dogma.
— Arthur Koestler "The Sleepwalkers: A History of Man's Changing Vision of the Universe," page 214

The stress mechanism is a *testable mechanism,* and it needs to be tested.

The term *theory* is often used loosely. Strictly speaking, stress theory (the idea that a single mechanism explains all types of disease) is not a theory. It is a powerful hypothesis (a proposed explanation) that potentially explains the confusing manifestations of physiology, pathology, and stress. The concept is so simple, logical, appealing, and compatible with known science that the best minds in medicine *assumed* that it was valid and spoke of it as a theory. Nevertheless, it has always remained a humble hypothesis for lack of a testable mechanism that explains how it works and can be tested to confirm it. Without such a mechanism, the hypothesis is useless, like a car without an engine.

The importance of a testable mechanism is exemplified by the story of DNA. Johannes Friedrich Miescher discovered DNA in 1869 and hypothesized that it was the medium of genetic inheritance, but this could not be confirmed until 1953, when Watson and Crick described the testable DNA mechanism that explains how DNA stores and replicates genetic information and enabled its testing and confirmation. This famously elevated Miescher's hypothesis to the status of confirmed theory.

Were it not for Watson and Crick, there would be no DNA theory, and Miescher's hypothesis would be forgotten. Today Miescher, who originated the DNA hypothesis, is nearly forgotten, but Watson and Crick, who discovered its mechanism, are famous. Life isn't fair. This is why the objective of stress research was to find a testable mechanism that could either confirm or refute Selye's hypothesis, and explain how it works. The effort was finally abandoned for failure to find such a testable stress mechanism, and Selye's ideas were largely forgotten, even though his hypothesis was never refuted, and it remained as virile as ever. Since then, many have argued that no single mechanism could possibly explain the bewildering blizzard of disease manifestations. They forget the very purpose of science (or perhaps never understood it to begin with):

"It is the function of science to discover the existence of a general reign of order in nature and to find the causes governing this order. And this refers in equal measure to the relations of man-social and political-and to the entire universe as a whole."
—Dmitri Mendeleev

Things have changed. The testable stress mechanism described in this book explains how Selye's hypothesis works, and enables it to be tested and either confirmed or refuted. This is mandatory to further the acceptance of stress theory. If confirmed, stress theory will revolutionize medicine and introduce a new era of human existence, free from the eternal curse of disease and premature death. If refuted, it will be relegated to the dustbin of scientific ideas.

Before such testing can happen, the stress mechanism must gain recognition. That won't be easy, because the stress mechanism is complex, counterintuitive, and at odds with conventional theory. It is no accident that the stress mechanism has eluded detection until now, and it won't gain acceptance quickly. Furthermore, peacetime progress is excruciatingly slow, and acceptance must overcome political opposition from economic interests.

Coincidentally, as this book was being compiled, the CBS-TV news show *60 Minutes* (US, 7 March 2021) included a program that featured Dr. David Seftel, the physician for 500 employees of the racetrack Golden Gate Fields in California. Dr. Seftel used the drug fluvoxamine (intended for obsessive/compulsive disorder) to treat 65 COVID-19 sufferers at Golden Gate Fields. He based his use on papers he had read about the drug. The drug already had undergone a small-scale trial treating sepsis in mice and was long-approved for human use, although not specifically for treating COVID-19.

Among the 65 people who accepted the drug, none needed hospitalization. Among the 48 who declined the drug, 12.5% of the cases deteriorated — this is evidently about par for COVID-19. Dr. Seftel and other physicians interviewed did not suggest an underlying mechanism for why the drug worked; only that it did. They stated that it demands further testing.[1] Testing, however, will yield nothing but unconvincing statistics *unless it is accompanied by a testable mechanism that explains how and why the treatment works.* The stress mechanism explains how fluvoxamine and other sedatives improve COVID outcome: their anti-anxiety effects inhibit harmful sympathetic nervous system hyperactivity, mitigate harmful stress mechanism hyperactivity, and allow the body to heal itself. Furthermore, the stress mechanism confers far more powerful and effective treatments than mere sedation. It indicates powerful treatments that can halt stress mechanism hyperactivity, and restore organ function, including normal lung function. COVID kills by disrupting lung function. Restoring normal lung function should cure COVID and other forms of viral and bacterial pneumonia in as little as 24 hours.

The implications of stress theory exceed the bounds of medicine and strain the limits of imagination. Life has evolved countless stress mechanisms that optimize survival in alternate environments, so that life is ubiquitous throughout the earth, including the volcanic environment deep beneath the earth's surface. Stress theory may pave the path to alter evolution, with ramifications that reside in the realm of science fiction.

INDUSTRIAL POLLUTION AND STRESS

Cancer, heart disease, asthma, emphysema, hypertension, diabetes, and rheumatoid diseases were rare before the modern age of industrialization, but they are now increasing in epidemic fashion. This cannot be accounted for by the increasing life span conferred by adequate food, clothing, and shelter. Stress theory explains how toxic combinations of automobile exhaust, chlorinated water, pesticides, herbicides, food additives, and industrial waste, plus rich diets and inadequate exercise that promote obesity, induce subclinical stress mechanism hyperactivity that causes chronic illnesses. The problem now threatens the very survival of civilization.

How we choose to fix these problems may be open to debate. *That* we need to fix them is obvious. The stress mechanism confers fresh ways to detect and eliminate environmental stresses, treat the diseases they cause, and assess treatment success to eliminate this threat.

CARBOGEN

One thing that can be done quickly and at minimal cost is to require Carbogen in all public places and on fire trucks, police cars, hospitals, apartment buildings, and medical offices — much as we do fire extinguishers and first aid kits. Its synergistic combination of oxygen and carbon dioxide optimizes all aspects of oxygen transport and delivery, and provides a simple, safe, inexpensive, proven treatment that can be administered by nearly anyone with minimal training, and is perfect for nearly all medical emergencies including strokes, heart attacks, trauma, asthma, atelectasis, pneumonia, influenza, drowning, smoke inhalation, carbon monoxide poisoning, drug overdose, alcohol inebriation, electrical shock, and breathing problems in newborn babies. It can be used safely to treat both hemorrhagic and ischemic forms of stroke, because it does not affect bleeding. 100 years ago, Carbogen was everywhere. *It needs to be everywhere again.*

DIRECTED RESEARCH

Specific aspects of the stress mechanism demand further research to confirm and clarify how they work. These include the relationship of the stress mechanism to the human genome; the role of PAR receptors on the surface of cells; the mechanism of malignancy; the nature of amyloidosis; and fresh pharmaceuticals that alter specific aspects of stress mechanism activity.

Mapping the human genome is an outstanding achievement in human research. Other genomes also have been mapped — plant, animal, extinct animal, bacteria, and virus. However, the bacterial *one gene = one protein* dictate of the *universal dogma of molecular biology* has failed to tell us how the eukaryotic cell genome encodes the genetic characteristics of complex multicellular plants and animals. Anatomical shapes, sizes, behavioral characteristics (and the unknown role of "junk DNA") are likely encoded on multiple genes that express chaotic equations, as opposed to today's prevailing notion of "one gene = one characteristic." As a trivial analogy, in computing, 255 different values can be expressed by only the eight on-off bits of a single byte, but the human genome contains approximately 30,000 genes.

Nor do we know the mechanism that communicates genetic information through the nuclear membrane into the cytoplasm (and beyond the individual cell to coordinate with other cells and form them into structures). The RNA and mRNA models apparently do not work well with the interposition of the nuclear membrane of the eukaryotic cell. Pores in the nuclear membrane seem to exist; but what they permit to pass — and what that does — is unknown.

While mapping the human genome and understanding the mechanism that communicates through the nuclear membrane are not strictly part of the stress mechanism, they are important factors that affect our understanding of how the stress mechanism works with the other mechanisms of the body, such as embryological development.

PAR RECEPTORS

The stress mechanism suggests that PAR receptors on the surface of a cell govern cell activities and the cell's coordination with neighboring cells in forming structures. How these PAR receptors are regulated is unknown. This may be the key to understanding the eukaryotic genome and the mechanism of cell differentiation and mammalian embryological development. Research must follow up on this lead.

A CURE FOR CANCER

The stress mechanism confers a fresh explanation of malignancy, and predicts that a combination of analgesia, anesthesia, hypercarbia, and thrombin suppression can disrupt the malignant mechanism and cure cancer. It likewise predicts that similar treatments can improve surgical outcome, optimize antibiotic potency and penetration, control critical illness, and revolutionize health care. These ideas must be tested, preferably beginning with animal models.

Confirming and clarifying this prediction using animal and human clinical studies could easily produce Nobel Prize winners.

AMYLOIDOSIS AND DISEASE

Amyloidosis is the abnormal accumulation of fibrillar amyloid protein in blood vessels, organs, and tissues. It is difficult to diagnose. The stress mechanism suggests that chronic, subacute stress mechanism hyperactivity accelerates amyloid protein production and causes chronic illnesses.

Where does amyloid protein come from? The stress mechanism suggests that the coagulation/capillary gate mechanism — common to all vertebrates — continually generates and disintegrates insoluble fibrin into multiple forms of fibrin split products (FSP). Interaction with factor X causes FSP to mis-fold, ultimately producing the amyloid proteins, the root cause of many chronic diseases such as Type II diabetes.

However, while this is a reasonable hypothesis, it must be tested and tested again. This single prediction, if correct, could lead to curing many diseases.

PHARMACEUTICAL DEVELOPMENT

The stress mechanism offers entirely new drug development strategies and new modes of treatment. Based on the stress mechanism, here are some suggestions for pharmaceutical development to find "new worlds to conquer."

- Investigate the relationship between insoluble fibrin and amyloid protein. If confirmed, this relationship would likely confer profitable pharmacological development of treatments for chronic illnesses.

- Develop safe treatments to neutralize tissue factor in blood circulation. This would provide a powerful means to control and cure critical illnesses.

- Investigate the effects of existing drugs on stress mechanism activity to clarify their therapeutic properties.

- Develop pharmaceuticals that inhibit harmful stress mechanism hyperactivity.

- Develop hypnotic agents that lack toxic and addictive properties. Hypnotic agents exert their benefits by controlling consciousness and preventing harmful fear and anxiety, but their beneficial effects are plagued by tissue toxicity, especially when they are used for prolonged periods of time.

MAKE LIFE EASIER FOR RESEARCHERS

The Internet has proven that the free exchange of information helps research, and yet important research information — even research funded by governments through tax dollars — is locked behind paywalls and subscriptions to learned journals. Creating this book would not have been possible without easy access to information, and it could not have been created ten years ago. This by itself has helped lose us the "50 years" in medical research of the title.

This plea is not dependent on the stress mechanism, but creating this book about the stress mechanism illustrates the plight of many researchers. The author would like to see:

- Prohibit scientific and technical journals from charging fees for access to papers documenting government-financed research. Make electronic versions of such reports available to the public free of charge within one year of original publication. The availability of such information would greatly facilitate scientific advance.

- Research financed by private corporations should be published in journals openly financed by and controlled by the corporations themselves. This would eliminate any taint of bias — and the corporations should be justly proud of their successful research.

- Require any journal calling itself independent to be free of financial influence from corporations and foundations, including advertising. If a journal is independent and valuable, it should exist on the revenue of its subscriptions.

Additionally, we would like to see a concerted effort to digitize all STEM (science-technology-engineering-math) publications from the distant past through today and provide free access. This project would preserve knowledge — such as the use of Carbogen, which has almost been forgotten.

DISCOURAGE ABUSIVE PRACTICES

Professional medical and scientific organizations must be *required by law* to allow their dues-paying members to elect their leaders and ratify all policies. Such organizations must be required to pay *all* expenses using membership dues and must not be allowed to accept funds from drug companies, manufacturers, hospitals, or private donors. For example, the American Society of Anesthesiologists (ASA) chooses its leadership from within, and realizes more income from private manufacturers and drug companies than from membership dues. As a result, the ASA is isolated from the interests of its powerless but dues-paying membership.

Hospitals, clinics, and medical boards must be *prohibited by law* from imposing routine re-certification requirements and continuing medical education (CME) requirements as conditions for licensing, medical staff privileges, or employment — unless they pay for all costs of such training including compensation of physicians, nurses, and employees for their time. Practical experience has demonstrated that such requirements provide the means to brainwash health care workers with counterproductive concepts. For example, hospitals have embraced confusing, complex, counterproductive, and onerously expensive Red Cross certification programs including BLS, ACLS, PALS, and ATLS that offer no redeeming value.

CONCLUSION

"It must be considered that there is nothing more difficult to carry out nor more doubtful of success nor more dangerous to handle than to initiate a new order of things; for the reformer has enemies in all those who profit by the old order, and only lukewarm defenders in all those who would profit by the new order; this lukewarmness arising partly from the incredulity of mankind who does not truly believe in anything new until they actually have experience of it"
—Niccolo Machiavelli

David Rockefeller confirmed that money cannot (yet) buy life when he died after his seventh heart transplant at the age of 101. The Hayflick limit remains the limit of life span, and our understanding of the genetic code remains incomplete. However, stress theory promises to pave the path to understanding the genetic code and explaining how the genetic blueprint is translated into embryological development, which would confer the ability to alter evolution and extend health and longevity beyond the Hayflick limit. Plant and animal characteristics could be altered to revolutionize food production. Environmental stress resistance could be augmented. Genetic defects could be corrected. Intelligence and exercise capacity could be optimized. Criminal behavior could be constrained to enable true civilization.[2] Physiology could be adapted to hibernation, space travel, and survival in foreign planetary environments. The brain could be altered to merge consciousness with computing.[3]

Obviously, much work remains to be done, but the mere possibility of such prospects merits the resurrection of stress theory as the prevailing paradigm of medical and biological research, and the restoration of wartime scientific rigor, vigor, and integrity.

NOTES

1 https://sfist.com/2021/03/08/golden-gate-fields-physician-featured-on-60-minutes-story-about-new-drug-for-covid-patients/
2 L. S. Coleman, Take a Walk on the Wild Side. *Voluntaryist*. 1995.
3 R. Kurzwell, *The Singularity is Near*.

Appendix A
Medical History

"Man errs as long as he strives."
— Johann Wolfgang von Goethe

A brief review of medical history illustrates how the excruciatingly slow pace of "trial and error" medical progress has evolved presently prevailing practices and beliefs, and set the stage for the resurrection of stress theory as the prevailing paradigm of medical research.

ANCIENT ORIENTAL MEDICINE

Ancient Chinese civilization evolved independently due to geographical isolation. Chinese medicine theorized that vital energy called *Chi* consists of opposing elements (*Yin* and *Yang*) that circulate through meridians, and emphasized acupuncture, herbs, massage, exercise, and diet. Western science has dismissed this approach for lack of detectable mechanisms, but the recent recognition of subtle electrical and magnetic forces in the body may confer a scientific basis for acupuncture.[1, 2, 3, 4] The history of smallpox suggests that Chinese medical sophistication deteriorated as their civilization declined even as Western Civilization recovered from the collapse of the Roman Empire.[5] Chinese physicians practiced smallpox vaccination (aka *variolation*) by blowing powdered smallpox scabs into the noses of healthy individuals.[6] The Persians learned variolation from the Chinese, and Lady Mary Wortley Montagu discovered the practice while living in the Ottoman Empire with her husband, the British ambassador. She had survived smallpox, but it killed her brother, so she arranged for her children to be vaccinated. She promoted variolation enthusiastically after her return to London, but confronted conflict with the medical profession. Edward Jenner, who was a boy when she died, developed a safer vaccination technique using cowpox instead of smallpox, and revolutionized smallpox prevention.

ANCIENT OCCIDENTAL MEDICINE

Ancient Western physicians embraced the concept of *Vis medicatrix naturae* (the healing power of nature), which acknowledged the ability of animals and people to recover without assistance. Progress was slow, because ancient civilizations prohibited human dissection on penalty of death; technology was primitive; and the most conservative of all professions eternally opposes change.

Humorism was the first documented medical theory. It attributed disease to excesses and deficiencies of humors of black bile, yellow bile, phlegm, and blood that moved through the body. It rationalized bloodletting, emetics, purges, and herbs. Its origin remains unknown, but Fahraeus (1796-1884) speculated that it was based on the observation that blood left undisturbed in a clear container forms a dark clot at the bottom (black bile). Above the clot is a layer of red cells (blood). Above the red cells is a layer of white cells (phlegm). The top layer is a clear yellow serum (yellow bile).

Hippocrates (460–370 BC) is considered the Founder of Western medicine. He was a contemporary of Pericles during the pinnacle of Athenian power and progress and is credited for the famous *Hippocratic Oath* (do no harm...) that is traditionally sworn by graduating medical students. He embraced humorism, but he believed that diseases have natural causes. His quotes are revealing:

"Disease is not an entity, but a fluctuating condition of the patient's body, a battle between the substance of disease and the natural self-healing tendency of the body."

"Illnesses do not come upon us out of the blue. They are developed from small daily sins against Nature. When enough sins have accumulated, illnesses will suddenly appear."

Greek physicians viewed disease as a process that leads to a *crisis* whereupon the disease resolves, or death ensues. Their efforts thus concentrated on *prognosis* rather than *diagnosis*. They relied on observation, and prepared detailed case histories, including records of disease progression, family history, and environmental circumstances that predicted disease progression. Their tradition of careful documentation endures. They identified diseases and conditions, and sought to facilitate the natural healing process by gentle measures such as soothing balms, cleanliness, and pouring clean water or wine on wounds. They used traction to stretch muscles and re-align broken bones. They burned, strangled, and excised hemorrhoids. Hippocrates advocated diet and exercise to treat diabetes. He recognized the relationship of clubbed fingers to lung disease, lung cancer, and cyanotic heart disease. He used lead pipes to drain chest abscesses. He identified the *Hippocratic face* that characterizes prolonged illness and impending death ("eyes sunken, nose sharp, skin stretched hard and dry, complexion pale and dusky"). He categorized illnesses as acute, chronic, endemic, and epidemic, and introduced the concepts of exacerbation, relapse, resolution, crisis, paroxysm, peak, and convalescence.

Aristotle (384–322 BC) was the most famous philosopher in ancient history. His books encompassed physics, biology, metaphysics, money, medicine, logic, poetry, theatre, and music. He embraced empiricism (experimentation to develop hypothetical ideas) but he advocated simple observation and logic as a reliable means to understand observed phenomena. He entered Plato's Academy at the age of 18 and remained there until Plato's death 20 years later, but after Theophrastus was chosen as Plato's successor Aristotle returned to his native Macedonia, where Phillip of Macedon chose him to tutor his son, Alexander the Great. Alexander subsequently united the known world into the greatest empire then known, whereupon Ptolemy Soter, Alexander's Egyptian successor, founded the famous library of Alexandria that methodically accumulated human knowledge. This largely accounts for Aristotle's influence through the ages. Much of Aristotle's legacy was lost when the library was burned during the Roman era, but many of his books survived in the Muslim world. His entrenched teleological reasoning restrained scientific and medical progress until the Renaissance.

Theophrastus (371–287 BC) refined humorism. He alleged that those with too much blood were sanguine; those with too much phlegm were phlegmatic; those with too much yellow bile were choleric; and those with too much black bile were melancholic.

Galen (AD 129–216) was a Roman physician who dissected animals, because human dissection was illegal. He identified the larynx as the source of the voice and hypothesized that blood produced by the right side of the heart penetrates through invisible pores in the cardiac septum into the left side of the heart, where it "gains spirit" (turns bright red). He thus distinguished arterial blood from venous blood, but he mistakenly assumed that arterial blood originates in the heart and is absorbed by tissues, while venous blood originates in the liver and is likewise absorbed by tissues. His humorism and anatomical misinterpretations frustrated medical progress for centuries.

Ibn Sina (Avicenna) (AD 980–1037) was a prominent Persian polymath during the Islamic Golden Age. He is considered one of the greatest scholars of all time. He produced some 450 books that explored astronomy, alchemy, geography, geology, psychology, theology, logic, mathematics, physics, and poetry. He studied medicine at the age of 16, discovered new treatments, and became a qualified physician at age 18. His fame spread quickly. His Canon of Medicine reconciled the contributions of Aristotle and Galen, and became a standard medical text in medieval universities. It remained in use in the 18th Century and was recently re-printed.

Ibn al-Nafis (AD 1213–1288) was a Persian polymath, like Avicenna, who wrote some 110 books on jurisprudence and medicine, including original works on diet, health, and ophthalmology. He was born in Damascus, where he studied medicine before moving to Egypt, whereupon he became a personal physician for prominent politicians. This enabled him to perform human dissections, and he correctly concluded that blood flows through the pulmonary artery from the right chamber of the heart to the lungs before reaching the left side of the heart. His work paved the path for the discoveries of William Harvey.

THE RISE OF MODERN MEDICINE

The tale of money wags the dogs of history, science, and medicine. The final Crusade sacked the gold reserves of Constantinople, released Western civilization from the strangulating Byzantine gold standard, and freed the kings of Europe to introduce fresh money systems that initiated the Renaissance, liberated science and medicine from dark age dogma, and undermined humorism. Medical reformers invariably encountered dangerous opposition as they challenged the dogmas of Galen and Avicenna. The invention of the printing press provided paper money that enabled the industrial revolution and ignited a violent struggle for political and economic supremacy that culminated with WWI and WWII. As sick soldiers are useless, and disease is deadlier than bullets, government subsidies accelerated medical progress, and physicians reluctantly abandoned the Greek emphasis on prognosis and embraced accurate diagnosis and specialized treatment.

William of Occam (1287–1347) was an English Franciscan friar who is remembered for *Occam's Razor*, which is the problem-solving paradigm that promotes the simplest explanation as the one most likely to be correct ("Among competing hypotheses, the one with the fewest assumptions should be selected"). His critical reasoning exposed the teleological shortcomings of Aristotle's philosophy, which brought him into conflict with Pope John XXII and led to his excommunication. He became the leader of a small band of dissident Franciscans living in exile from Avignon, and spent the remainder of his life writing about political issues. The Razor became a fundamental precept of modern science after his death, and he was he was officially rehabilitated by Innocent VI.

Paracelsus (Philip von Hohenhelm) (1493–1541) is remembered as the founder of pharmacology (i.e., toxicology). He was a Renaissance contemporary of Martin Luther, Copernicus, Galileo, and Leonardo da Vinci. He paved the path for modern medical research and practice by utilizing objective observations to challenge the traditional medical authority of Avicenna and Galen. He rejected humorism, and believed that beneficial medical substances could be found in alchemical combinations of roots, herbs, and minerals. He recognized the importance of nutrition and advocated cleanliness during an era when physicians routinely applied feces to open wounds to promote pus. He understood that syphilis can only be transmitted by contact, and that it can be treated using carefully regulated doses of mercury. He thus anticipated antisepsis, germ theory, the importance of nutrition, and the modern concept of accurate clinical diagnosis and specialized treatments. His defiance of medical tradition engendered the wrath of medical authorities who banned his books and forced him to flee from Basel and wander through central Europe. When compared to Luther he said, "I leave it to Luther to defend what he says and I will be responsible for what I say. That which you wish to Luther, you wish also to me: you wish us both in the fire."

Andreas Vesalius (1514–1564) (Andriese van Wesel) is remembered as the founder of modern anatomy. He studied the theories of Galen at the University of Paris, where he developed an interest in anatomy. He obtained his medical degree at the University of Padua in 1537, whereupon he was offered the chair of surgery and anatomy. He discovered that Galen's classical texts were based on the dissections of Barbary Macaques, which had been overlooked for some 1400 years. This led him to abandon medieval tradition and utilize human dissection and direct hands-on observation, whereupon he discovered revolutionary discrepancies in Galen's works. Generations of anatomists had claimed to observe connecting passages between the cardiac ventricles that confirmed Galen's faulty theories, but Vesalius admitted that he was unable to find such openings. In 1543 he published "*De humani corporis* *fabrica*" (the human body). Despite collegial adversity, his work established anatomy as a modern descriptive science.

Michael Servitus (1509–1553) was a religious idealist, linguist, and polymath versed in mathematics, astronomy, meteorology, geography, jurisprudence, translation, biblical studies, human anatomy, medicine, and pharmacology. He discovered that pulmonary circulation flows from the heart to the lungs, rather than air in the lungs flowing to the heart, as had previously been thought. His reasoning was based on the color of the blood, the size and location of the ventricles, and the large diameter of the pulmonary vein.

Servitus was repelled by the pomp and wealth he witnessed at the coronation of Pope Charles and decided to follow the path of the Reformation. He began his medical studies in Paris in 1536. His teachers considered his knowledge and dissection skill comparable to that of Andreas Vesalius, who was his contemporary. However, he annoyed his teachers by teaching mathematics and astrology and predicting the eclipse of Mars by the Moon, and the Dean, Jean Tagault, suspended this teaching, whereupon Servitus wrote, "Apologetic Discourse of Michel de Villeneuve in Favor of Astrology and against a Certain Physician." This inspired Tagault to argue for the death penalty. Servitus then completed his medical studies in Montpellier and became a Doctor of Medicine in 1539. Soon thereafter he was imprisoned for several days after he injured a jealous physician who attacked him in a sword fight. He began a correspondence with Calvin in which he repudiated predestination and the trinity, which offended Catholics as well as Calvin and Luther, and Calvin convicted Servitus to be burned at the stake atop a pyre of his own books. His last words were, "Jesus, Son of the Eternal God, have mercy on me." Calvin was harshly denounced for the execution, and the fact that Servitus was dead allowed his writings to be distributed widely, whereupon they influenced the Unitarian movement. It is said that Servitus died so that freedom of conscience could become a civil right in modern society.

Hieronymus Fabricius (1537–1619) is remembered as the Founder of embryology. He dissected animals, and studied the formation of the esophagus, stomach, intestines, eyes, ears, and larynx. He was the first to identify venous valves, but he failed to perceive that they prevent retrograde flow. His work facilitated the evolution of surgical techniques and set the stage for Harvey's discoveries.

William Harvey (1578–1657) was an English physician who first correctly described blood circulation. He met Fabricius when he studied at the University of Padua, and read his book that described venous valves (*De Venarum Ostiolis,* The Doors of the Arteries). This provided him with the crucial evidence for his argument for systemic blood circulation. He refuted Galen's theory by demonstrating that the heart pumps large volumes of blood. He recognized the necessity of a connection between arterial and venous circulation, but did not detect capillaries. Collegial derogation damaged his practice after he published his completed treatise on blood circulation, *De Motu Cortis*, but he persevered and was re-elected to the College of Physicians and subsequently elected Treasurer.

Francis Bacon (1561–1626) was an English philosopher, statesman, scientist, jurist, orator, author, and aristocrat who proposed the modern scientific method that revolutionized science. The objective of science is to identify a simple truth. The scientific method begins with unexplained observations that inspire a testable hypothesis, followed by controlled experimentation to test the hypothesis. If the hypothesis survives experimental testing, it becomes accepted as theory — though it remains open to challenge, and may be rejected in favor of a better theory. His biographer observed that "Bacon's influence in the modern world is so great that every man who rides in a train, sends a telegram, follows a steam plough, sits in an easy chair, crosses the Atlantic, eats a good dinner, enjoys a beautiful garden, or undergoes a painless surgical operation, owes him something."

Marcello Malpighi (1628–1694) is remembered as the founder of microscopic anatomy, histology (study of tissues), physiology, and embryology. He pioneered the use of microscopy to study animal and plant tissues, and studied plants, animals, and insects. Several physiological structures bear his name. He was among the first to view and describe red blood cells. In 1661 he identified and described capillary networks that join small arteries to small veins, which complemented the observations of Harvey and Ibn al-Nafis, and completed the understanding of blood circulation. This discovery engendered animosity from adherents to Galen's traditional views. His epitaph states, "SUMMUM INGENIUM/INTEGERRIMAM VITAM/ FORTEM STRENUAMQUE MENTEM/AUDACEM SALUTARIS ARTIS AMOREM" (great genius, honest life, strong and tough mind, daring love for the medical art).

Marie Francois Xavier Bichat (1771–1802) is remembered as the founder of histology. He identified 21 types of elementary tissues that compose organs in the human body, including connective tissue, nerve tissue, and muscle. He distrusted the microscope, and thus overlooked cellular structure. He theorized that the ganglionic nervous system regulates internal organ activity.

Rudolf Virchow (1821–1902) is remembered as the founder of modern pathology. His work devastated the theory of humorism. He described and named leukemia, chordoma, ochronosis, embolism, and thrombosis. He introduced the terms chromatin, agenesis, parenchyma, osteoid, amyloid degeneration, and spina bifida. He rejected the notion that organisms could arise from nonliving matter, and recognized that cells arise from pre-existing cells. He observed that chronic tissue inflammation causes cancers to arise from otherwise normal cells, and that anti-inflammatory medications minimize cancer risk and inhibit tumor development. He demonstrated how emboli are formed in the legs and transported to the lungs. He encouraged medical students to "think microscopically." He described the transmission cycle of the roundworm, and established the link between animal and human infectious diseases. He demonstrated that infected meat transmits parasites, and that cooking meat prevents parasite transmission. He developed a systematic method of autopsy that remains standard.

Though he was a contemporary of Pasteur, Virchow rejected germ theory. He believed instead that social inequality is the primary cause of disease, and his passionate promotion of political reform caused conflict with Bismarck. He famously stated: *"Medicine is a social science, and politics is nothing else but medicine on a large scale — The physicians are the natural attorneys of the poor, and the social problems should largely be solved by them."* He also rejected Darwin's theory of evolution, and opposed social Darwinism and the idea of German cultural superiority. The concept of *Virchow's Triad* was named in his honor after his death.

Louis Pasteur (1822–1895) is remembered as the founder of microbiology. He discovered the principles of vaccination, microbial fermentation, pasteurization, and molecular isomerism. His medical experiments disproved the theory of spontaneous generation, confirmed the germ theory of disease, and guided its application in disease prevention and treatment. He solved the mysteries of rabies, anthrax, chicken cholera, and silkworm disease. He famously stated: "In the field of observation, chance favors only the prepared mind." He became a French national hero, but he admonished his family never to expose his laboratory notebooks. They became available to public inspection only in 1985 after his last surviving male descendant donated them to the French National Library. They revealed that his most important discoveries involved deceptions that may have been intended to embellish his reputation, promote profitable products, or obtain research funds.

Antoine Bechamp (1816–1908) was Pasteur's bitter scientific opponent. Like Bernard and Virchow, he did not believe that bacteria could thrive in healthy animal tissue. He claimed priority over Pasteur in discovering that microbes cause fermentation. He founded the synthetic dye industry by discovering an inexpensive method to produce aniline dye. He developed the first therapeutic chemical by synthesizing arsanilic acid. He died in obscurity, but his ideas are still promoted by alternative medicine advocates who dismiss Pasteur's germ theory, and believe that Pasteur plagiarized and suppressed Bechamp's work.

Claude Bernard (1813–1878) is remembered as the founder of physiology. He demonstrated the role of the pancreas in digestion, the processing of blood sugar into glycogen by the liver, the existence of vasomotor nerves that affect cardiac function, and the effects of curare and carbon monoxide. He originated the concept of *milieu interieur* (interior conditions) wherein the internal stability of the animal body is maintained independent of environmental circumstances to facilitate cellular survival and function. He was Pasteur's friend and contemporary, but he disagreed with Pasteur's germ theory. He believed that organisms normally present in the body cause most microbial diseases, and that the body becomes susceptible to infectious agents only if the milieu interieur becomes disturbed. His viewpoint accords with the ancient medical beliefs of Greece, Rome, India, and China. Though Pasteur's viewpoint presently prevails, the debate remains unresolved.

Bernard was known as the "prince of vivisection." He wrote: "The physiologist is no ordinary man. He is a learned man, a man possessed and absorbed by a scientific idea. He does not hear the animals' cries of pain. He is blind to the blood that flows. He sees nothing but his idea, and organisms which conceal from him the secrets he is resolved to discover." His arranged marriage disintegrated and his wife became an opponent of vivisection after she returned home with her daughters to discover that Bernard had dissected their dog. His vivisection experiments were so gruesome that Dr. George Hoggan, who spent four months observing Bernard's experiments, was moved to state that he was "prepared to see not only science, but even mankind, perish rather than have recourse to such means of saving it."

Bernard believed that the same physical laws apply to both living creatures and inanimate matter. His personal ambition was to introduce the scientific method to medicine, and his classic book "An Introduction to the Study of Experimental Medicine" personified his scientific idealism. He stated: "Art is I; science is we. The experimenter who does not know what he is looking for will not understand what he finds. It is what we know already that often prevents us from learning. A fact in itself is nothing. It is valuable only for the idea attached to it, or for the proof that it furnishes." He may have detected Pasteur's deceptions, for he regarded those who ignore results that support their rivals as "despisers of their fellows."

Joseph Lister (1827–1912) is remembered as the founder of modern surgery. He was inspired by Pasteur's research to utilize carbolic acid to sterilize surgical instruments and wounds, which drastically reduced wound infections and revolutionized surgery.

Robert Koch (1843–1910) is remembered as the founder of modern bacteriology. He identified the microbes that cause tuberculosis, cholera, and anthrax. He refined and published four criteria that establish a causative relationship between a microbe and a disease. These are known as "Koch's postulates":

1. The microorganism must be found in abundance in all organisms suffering from the disease, but should not be found in healthy organisms.
2. The microorganism must be isolated from a diseased organism and grown in pure culture.
3. The cultured microorganism should cause disease when introduced into a healthy organism.
4. The microorganism must be re-isolated from the inoculated, diseased experimental host and identified as being identical to the original specific causative agent.

Koch abandoned his first postulate after he discovered asymptomatic cholera carriers, and subclinical carriers have subsequently been found to be a common feature of communicable diseases. The second postulate must be suspended in the case of prions and viruses that cannot grow in pure cultures. The third postulate is also qualified, because infective agents often fail to induce disease.

George Washington Crile (1864–1943) was an American trauma surgeon who demonstrated that nociception and nervous fear undermine surgical outcome, and that analgesic agents can treat surgical shock and sepsis by inhibiting harmful nociception. He performed the first successful blood transfusion, demonstrated that blood transfusion treats shock, invented cardiopulmonary resuscitation, and demonstrated the resuscitation utility of epinephrine and atropine. He abhorred war, but served as a physician in both the Spanish American War and WWI.

Yandell Henderson (1873–1944) was a PhD researcher and director of the Yale Laboratory of Applied Physiology who was the foremost authority on the pharmacology and toxicology of gases in his time. He demonstrated the therapeutic benefits of carbon dioxide and led a famous expedition of international experts who studied respiratory physiology at the top of Pike's Peak. He believed that science should serve the public interest, and testified before Congress about the dangers of tetraethyl lead in gasoline. He opposed American involvement in WWI.

Walter B. Cannon (1871–1945) was a Harvard physiologist who extended Bernard's *milieu interieur* concept and coined the terms *fight or flight* and *homeostasis*. He combined newly discovered X-rays with heavy metal salts to study digestion and peristalsis. He studied wound shock, the effects of emotional excitement, the chemical mediation of nervous signals, and the effects of chemicals on denervated organs.

Cannon's famous book "The Wisdom of the Body" provides a fascinating glimpse of the medical advance attained in the WWI era. Blood and lymph proteins were known to be necessary for blood coagulation and categorized as fibrinogen, albumen, and globulin. Today they are known as fibrinogen, soluble fibrin, and insoluble fibrin. Cannon developed a practical plasma substitute that employed freeze-dried fibrinogen to treat acute blood loss. The product remains available in France but is inexplicably absent in North America. It was clear in Cannon's time that blood coagulability increases in accord with surgical stress, trauma, blood loss, increased sympathetic tone, and adrenalin; that adrenalin and sympathetic nervous activity produce identical effects; that water entering the GI tract restores blood volume more effectively than salt solutions injected into a vein or under the skin; that calcium is controlled within narrow limits by the parathyroid glands and is essential for brain and muscle function, milk production, blood coagulation, bone formation, and bone repair; that glucose is stored in the liver as glycogen and released by sympathetic nervous activity; that insulin enables cellular glucose uptake and lowers serum blood sugar; that blood enzymes destroy bacteria; that estrogen and testosterone determine sexual differentiation; that thyroid activity regulates body metabolism; and that fear and anxiety are adverse to health and surgical outcome.

Cannon's persuasive presentation of hemodynamic physiology pervades modern physiology, but is riddled with shortcomings. Earlier researchers had noted the shortcomings of blood pressure measurement long before Cannon wrote his book, as exemplified by these quotes:

"Each time a change can be observed in the height of a manometer applied to an artery, the experimenter must ask himself which of the two factors of the arterial pressure has varied, or whether the two factors, power and resistance, have been modified at the same time. In the absence of a criterion which would allow one to decide with complete certainty this litigious question, often enough physiologists have chosen the hypothesis which best accorded with their preconceived ideas."
— Etienne Jules Marey 1881

"The development of our knowledge of the circulation has been bedeviled by the fact that the measurement of blood flow is so complicated, whereas that of blood pressure is so easy: hence the blood pressure manometer has exerted an almost hypnotic influence, though bodily organs don't need pressure but flow."
— Adolf Jarisch 1928

Cannon ignored such inconvenient observations, and embraced the intuitively appealing idea that blood pressure is the driving force of blood flow and tissue perfusion. His persuasive presentation of hemodynamic physiology persists to the present day despite numerous inconsistencies and unfounded assumptions. In Cannon's view, the heart is "richly endowed with reserves of force and can increase its contractile force and volume to accommodate physiological requirements"; "the mechanical work of the powerfully contracting heart is spent in overcoming frictional resistance in the blood vessels"; "sympathetic tone determines vascular contractility and shifts blood flow from one region to another"; and "exercise is accompanied by a six-fold increase in pulmonary ventilation, doubling of heart rate, and raising of arterial pressure, all of which presumably increase oxygen delivery." Cannon's concept implies that energy-dependent muscular vasoconstriction opposes energy-dependent cardiac work, so that hemodynamic function is inherently inefficient. He noted that autonomic balance governs heart rate, which he attributed to direct autonomic innervation of vascular and cardiac muscle, even though he acknowledged that adrenalin effects are identical to those of direct sympathetic innervation, and the Starling effect remains intact in the absence of innervation.

Cannon can be forgiven for his shortcomings, given the limited information available in his time, but it is difficult to understand why his ideas remain enshrined while the repeatedly reaffirmed discoveries of Crile and Henderson, who were Cannon's contemporaries, are abandoned and forgotten. This is addressed in the final portion of this book. Meanwhile patient research toil has provided fresh evidence that enables an improved understanding of physiology that is the focus of this book.

NOTES

1 S. Zarlenga, The lost science of money : the mythology of money, the story of power. (*American Monetary Institute*, Valatie, NY, 2002), pp. xii, 724 p.

2 W. H. Dixon, George Fabyan Collection (Library of Congress), Personal history of Lord Bacon. From unpublished papers. (Ticknor and Fields, Boston,, 1861), pp. xix, 424 p.

3 E. Williams, The forgotten giants behind Louis Pasteur: contributions by the veterinarians Toussaint and Galtier. *Vet Herit* 33, 33-39 (2010)

4 E. D. Hume, M. R. Leverson, Béchamp or Pasteur? A lost chapter in the history of biology. (*The C.W. Daniel Company*, London, 1932), pp. 301 p. 301 l.

5 https://en.wikipedia.org/wiki/Virchow%27s_triad

6 https://en.wikipedia.org/wiki/Marie_Françoise_Bernard

Appendix B
Hemostasis History

"Science like life feeds on its own decay. New facts burn old rules; then newly developed concepts bind old and new together into a reconciling law."
— William James

Conventional coagulation theory is as lacking as conventional physiology theory. Blood is a living substance that coagulates even after it is removed from the body, but the nature of coagulation remains elusive despite decades of research. It is generally assumed that the sole purpose of coagulation is to halt blood loss in the face of tissue damage. Physiologists call this *hemostasis* (i.e. [1] Arrest of the escape of blood by either natural means such as clot formation or vessel spasm, [2] Interruption of blood flow to a part).

Rudolf Virchow (1821-1902, see Appendix A: Medical History, on page 395), the famous German pathologist, identified tissue crushing, stasis (slowing) of blood flow, and hypercoagulability as precursors of coagulation. A committee named these three factors *Virchow's Triad* in his honor after his death, but the triad confers more questions than answers. It doesn't explain fluctuating blood viscosity (flow resistance) and coagulability, or their close relationship. Stasis of blood flow is attributed typically to platelet activation that induces *vasoconstriction* in the immediate vicinity of tissue damage, but the cause of platelet activation remains unknown, and how platelets might cause the stasis that contributes to clot formation remains unclear. However, it is now clear that disruption of the vascular endothelium activates the coagulation cascade by exposing tissue factor to factor VII, but even this fails to explain platelet activation and blood flow stasis.

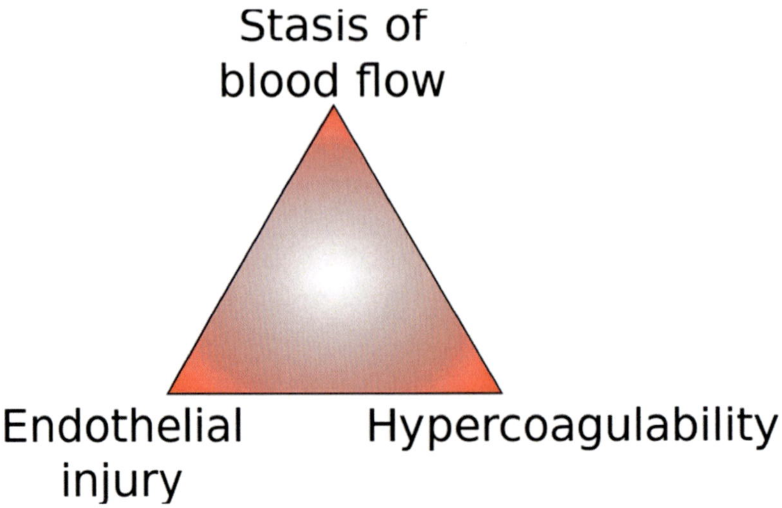

Figure B.1: Virchow's Triad.
Virchow's Triad identifies circumstances associated with clot formation but leaves many questions unanswered.

THE COAGULATION CASCADE

In the 1950s, another committee introduced the confusing system of Roman numerals, as opposed to systemic names, to identify the enzymes and substances involved in coagulation. Unfortunately, this nomenclature frustrates understanding the coagulation process. Appendix I: Summary of Coagulation Enzymes and Substances, on page 449 is intended to serve as a convenient reference source for readers. It summarizes the names and characteristics of the various enzymes and substances that are identified by Roman numerals. This presentation will limit the use of the Roman numeral system to enzymatic factors VII, VIII, IX, and X that are the focus of mammalian stress mechanism function — if only because they are known by no other names.

During the 1960s stress researchers described the *coagulation cascade* (see Figure B.2) that consists of two pathways of enzyme interactions that culminate in coagulation. Like Virchow's triad, the coagulation cascade raises more questions than it answers. It identifies *fibrin* as the end product of the enzymatic interactions, but fails to explain how the enzymatically inert fibrin induces clot formation. It fails to explain the relationships between nervous activity, clot formation, blood viscosity, clot dissolution, tissue repair, atherosclerosis, blood pressure, and tissue repair.

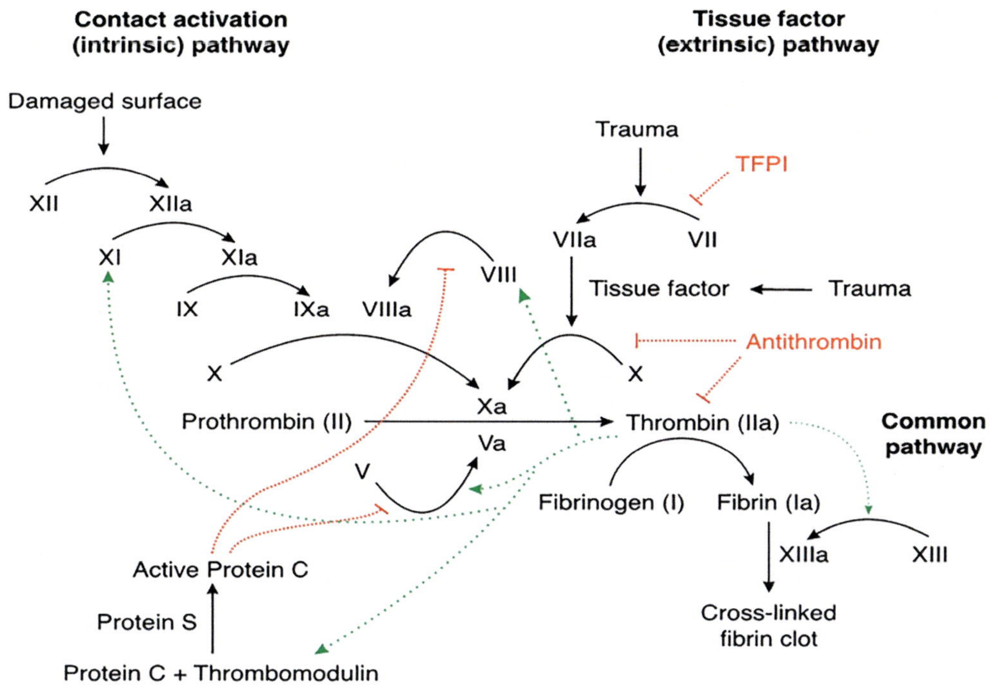

Figure B.2: Coagulation cascade.

The coagulation cascade is founded on laboratory (*in vitro*) studies. It identifies two pathways of enzyme interactions that culminate in the production of *fibrin* (it lumps soluble fibrin and insoluble fibrin into the category of *fibrin*.) that enables clot formation. The intrinsic pathway involves the interaction of factors VIII and IX, while the extrinsic pathway involves the interaction of *tissue factor* (TF) with factor VII. Both pathways interact with factor X to form a final common pathway that converts prothrombin to thrombin, and fibrinogen to fibrin, to enable coagulation. The era of stress research produced evidence that nervous activity exaggerates blood coagulability, and that coagulation is related to tissue repair, but there was inadequate evidence available to explain these observations, and the cascade was viewed only as a mechanism that prevents blood loss after injury. Since stress research was abandoned, evidence has accumulated that explains the relationships of nervous activity, hemodynamic physiology, tissue disruption and tissue repair, and revealed that the coagulation cascade was the first crude description of the mammalian stress mechanism.

PLATELETS

Until recently, platelets were assumed to produce tissue factor, but recent research has dispelled this notion.[1] Like red cells, platelets lack a nucleus, do not produce tissue factor, and are unique to mammals. DNA regulation cannot explain platelet activities, because platelets lack a nucleus. Nevertheless, many medical textbooks assume that platelets initiate and orchestrate blood coagulation. Platelets somehow become activated in the presence of tissue damage, whereupon they change shape and become "sticky." Platelet activation is accompanied by unexplained flood flow stasis, and activated platelets bind together to form a white clot before the appearance of the red clot in damaged tissues, so that platelets appear to initiate and orchestrate coagulation. However, recent research indicates that although platelets play a pivotal role in coagulation, they do not initiate or control it. Instead, coagulation is initiated by damage to the vascular endothelium which exposes tissue factor in extravascular tissues to factor VII in flowing blood. This initiates the enzymatic interaction of factors VII, VIII, IX, and X that generates thrombin, whereupon thrombin energizes both platelet activation and insoluble fibrin generation to enable coagulation.[2]

SUMMARY

Though researchers have long suspected that blood coagulation is somehow involved in tissue repair, the *coagulation cascade* remains regarded merely as a mechanism that prevents blood loss after injury. Fresh research since stress theory was abandoned has revealed that *fibrin* consists of a mixture of closely-related fibrinogen, soluble fibrin, and insoluble fibrin, each of which has unique properties that explain not only coagulation, but also tissue repair and blood flow regulation. In addition, fresh research has revealed how combinations of tissue disruption and nervous activity alters the activities of blood enzymes to regulate tissue repair and hemodynamic physiology. Thus, in retrospect, the coagulation cascade can now be recognized as the first crude description of the mammalian stress mechanism.

NOTES

1 B. A. Bouchard, K. G. Mann, S. Butenas, No evidence for tissue factor on platelets. *Blood* 116, 854-855 (2010).

2 L. S. Coleman, Insoluble fibrin may reduce turbulence and bind blood components into clots. *Med Hypotheses* 65, 820-821 (2005).

Appendix C
Conventional Theory

"The key to wisdom is this — constant and frequent questioning, for by doubting we are led to question and by questioning we arrive at the truth."
— Peter Abelard

An effective medical theory must explain the nature of disease. Since disease disturbs physiological homeostasis, the theory should explain physiology. Since humans are mammals, the theory should explain the physiology and pathology of all mammals. Since microbial invasion causes disease, the theory should clarify the nature of microbiology. In addition to microbial invasion, numerous forms of stress induce or aggravate disease, including radiation, toxic chemicals, and emotion, so the theory should explain the nature of stress. Since mammals are animals, the theory should explain all aspects of zoology, including embryology, evolution, ethology, intelligence, immunity, inflammation, hemostasis, chemotaxis, chemokines, cytokines, prostaglandins, apoptosis, sex, parasitology, longevity, speciation, environmental adaptability,[1] anatomy, taxonomy, paleontology, cell differentiation, cell de-differentiation, and cell re-differentiation. Since animals and plants presumably share a common origin, the theory should be compatible with botany as well. In other words, a truly effective theory of medicine must be analogous to a Unified Theory of Biology that explains the very nature of life. Unfortunately, life is so overwhelmingly complex, confusing, and counterintuitive, that the prospect of an effective theory of biology seems unimaginable, and the nature of life remains in the realm of religion. The following examples illustrate the difficulty.

Mammals evolved directly from reptiles, but they are nevertheless distinctly different. Reptiles thrive in warm climates, are covered with scales, and have minimal food requirements, limited exercise tolerance, inferior intelligence, and poor cold tolerance. Mammals suffer exaggerated food requirements and poor heat tolerance, but they enjoy superior intelligence and exercise capacity, are covered with fur, and thrive in cold climates. The temperature sensitive metabolism of eukaryotic cells provides only a partial explanation, because the exercise capacity of reptiles remains inferior to that of mammals regardless of temperature.[2] Disease further frustrates theoretical advance by perturbing physiology and producing mutable manifestations. Since both reptiles and mammals are composed of essentially identical eukaryotic cells, it would seem logical to expect that their differences are determined by extracellular mechanisms, but such mechanisms have eluded detection.

Paleontology provides mute testimony of the extinction of most known species, but reveals few clues about behavior, intelligence, anatomy, physiology, and environmental circumstances that might explain their origin and extinction.

Taxonomy crudely categorizes animals and plants into kingdom, phylum, class, order, family, genus, species and even sub-species on the basis of observable anatomy and behavior but cannot identify the underlying physiological mechanisms that presumably determine such relationships.

The DNA mechanism explains how genetic information is retained, replicated, and transmitted to subsequent generations, but how the chromosomal DNA of eukaryotic cells communicates outside the nucleus to enable complex multicellular life forms remains unknown. Cellular mitosis, metabolism, differentiation, and apoptosis proceed at higher levels during embryological development than at any other time of life, but the nature of such development remains unknown. The specialized cells of complex multicellular creatures retain the ability to re-organize themselves to restore organ tissues when they are separated from one another, and they retain the ability to de-differentiate and then re-differentiate into totally different specialized cell types,[3] but the nature of these feats remains unknown. The nature and relationships of anatomy, physiology, ethology, and intelligence in complex multicellular life forms remains unknown.

Darwin's theory is the presently prevailing explanation of evolution, but it is plagued by shortcomings. It cannot explain the Cambrian Explosion where multiple complex animal forms suddenly appeared at the bottom of the oceans. It cannot explain Lamarck's observation that life grows more complex with time. Darwin insisted that evolution proceeds slowly via small gradual changes, so his theory cannot explain the phenomenon of saltation, wherein drastically different offspring suddenly and inexplicably appear in both animals and plants. In Darwin's time, the term *mutation* referred to such sudden changes, but since the discovery of DNA the term has come to be defined as "a sudden departure from the parent type in one or more heritable characteristics, caused by a change in a gene or chromosome." This idea is problematic, because there is considerable evidence that sudden, drastic environmental change precipitates extinction and the sudden appearance of new species. Such environmentally induced change cannot be explained by DNA damage, because Max Delbruck and Salvador Luria, the mentors of James Watson, famously devised elegant bacterial phage (virus) experiments that proved that mutations occur at random and are not affected by environmental factors.[4] Lastly, Darwin's theory fails to explain the origin of life. Explanations of life's origin have thus far failed to gain universal acceptance, perhaps because human egocentricity mistakenly assumes that life originated on the earth's surface.[5]

Disease adds another layer of confusion by disturbing homeostatic mechanisms, but despite decades of intense medical research, modern medical science cannot explain the relationships between and among seemingly unrelated diseases, immune activity, inflammation, tissue repair, hemostasis, hemodynamic physiology, and cellular hormones such as chemokines, cytokines and prostaglandins. The innate ability of the body to mend itself without assistance remains so inexplicable as to be seldom discussed. Medicine remains riddled with *syndromes* (groups of associated symptoms) that are unexplained by definition, such as SIRS, ARDS, MOFS, eclampsia, and rheumatoid disease. The medical missive "not uncommon" means "we don't know."

Cannon's lingering legacy likens hemodynamic physiology to the static hydraulic brake system of a car, where brake pedal force transmits instantaneously and equally throughout all inner surfaces of the brake system. This implies that the blood pressure force should appear instantly with the onset of cardiac systole, and remain equal throughout the arterial tree until the onset of diastole. Instead, blood pressure appears during diastole, after the heart muscle is at rest, in the form of a burst of turbulence that speeds throughout the arterial tree, generating turbulent lateral forces that press against the inner walls of arteries as it propagates towards the periphery. These lateral forces manifest as the palpable pulse and enable blood pressure measurement, but they are perturbed by temperature, arterial length, arterial diameter, arterial curvature, arterial bifurcations, arterial strictures, wave reflections, heart rate, blood viscosity, and small variations in cardiac contractility. Thus, their intensity varies from location to location, so that blood pressure measurements obtained at different locations are invariably different, and "accurate" blood pressure is a misnomer. Invasive monitoring is hardly more "accurate," for it must be calibrated to cuff pressure. Meanwhile, cardiac contractility varies little from beat to beat, and flow resistance (viscosity) is the primary determinant of cardiac output, cardiac efficiency, tissue perfusion, and tissue oxygenation. Counter intuitively, blood pressure and atherosclerosis resistance increase in concert as arterial diameter declines toward the periphery. Efforts to derive cardiac output from conventional blood pressure and pulse wave measurements have proved futile.

The terms *vasoconstriction*, *vasodilation* and *vasoactive* are misnomers that reflect Cannon's concept that opposing muscular forces of cardiac contractility and vascular contraction regulate cardiac output, and regulate tissue perfusion in accord with autonomic balance. This implies that hemodynamic physiology is inherently inefficient, because muscle contraction is an energy-expensive event. Cannon accordingly regarded the heart as the "Charles Atlas" of the mammalian body and supposed that it possesses enormous reserves of power and force that enable cardiac output to increase by a factor of ten during athletic endeavor. The available evidence, however, indicates that animal physiology is extremely efficient, and that the mammalian heart is a "90 pound weakling" that consumes less energy than a 40-watt light bulb and generates the same force with each heartbeat, even during intense exercise. Cannon's concept implies that elephants should possess gigantic hearts that generate high blood pressure to propel blood to distant parts of their gigantic body, but mammalian hearts consistently weigh 1% of total body weight and are so weak that they cannot pump blood more than a few inches above their height.[6] Mammalian brains are accordingly located only inches above the heart to facilitate perfusion. The giraffe is an obvious exception, but it requires a disproportionately gigantic heart that compromises other organs.

Were blood pressure the driving force of blood flow then hypertension should be accompanied by increased cardiac output and enhanced tissue perfusion; instead, hypertension is a prelude to congestive heart failure that is characterized by low cardiac output, poor tissue perfusion, impaired exercise capacity, and limited longevity. Cannon's concept implies that exercise conditioning should exaggerate the size and weight of the heart in order to generate higher blood pressure to sustain exercise requirements. Instead, exercise conditioning induces negligible increases in heart size and weight, and trained athletes exhibit abnormally *low* blood pressure at rest, normal blood pressure during intense exercise, and enhanced longevity.

Blood pressure and pulse are undetectable in patients with left ventricular turbine pumps that function as artificial hearts in the presence of complete heart failure. In terms of conventional belief, such patients should be dead. Nevertheless, these pumps sustain tissue perfusion and oxygenation that enables moderate exercise. These observations indicate that conventional instruments do not measure the driving force of blood flow, and conventional blood pressure is not the driving force of tissue perfusion.

Muscular vasoconstriction is energy-extravagant and soon followed by muscular exhaustion and obligatory vasodilation, so arteriolar vasoconstriction cannot explain essential hypertension. Instead, essential hypertension is characterized by accelerated microvascular deterioration that undermines tissue perfusion, exaggerates flow resistance, and increases cardiac work. The increased cardiac work induces myocyte hypertrophy that is soon followed by myocyte apoptosis and pathological collagen deposition that progressively exaggerates cardiac wall thickness and undermines cardiac function until heart failure ensues.[7] In contrast, exercise conditioning proliferates capillaries in affected muscles. The resulting reduction in flow resistance causes hypotension in resting athletes, and it enables 10-fold increases in cardiac output during intense exercise with negligible increases in cardiac size, muscle mass, or blood pressure. Such observations indicate that the heart functions on the basis of extreme efficiency rather than power and force.

The shortcomings of blood pressure were recognized from the beginning, but it nevertheless remains a basic standard for medical monitoring and treatment. It is a counterproductive guide to anesthetic management. Researchers and clinicians intuitively equate hypertension with effective cardiac reserve, and regard hypotension as a harbinger of death and disaster. However, tachycardia and hypertension invariably reflect cardiac distress, reduced cardiac output, and tissue ischemia, while bradycardia and hypotension characterize enhanced cardiac efficiency and tissue perfusion, as in normal sleep and resting athletes. Vasodilator drugs such as nitroprusside lower blood pressure even as they increase cardiac output and tissue perfusion, and alleviate congestive heart failure.[8] Vasopressor drugs such as Ca^+ and epinephrine exaggerate blood pressure even as they reduce cardiac output and tissue perfusion.[9] Streptokinase anticoagulation has no effect on muscular contractility, but it restores perfusion to infarcted tissues even as it inexplicably reduces blood pressure.

Familiar *Newtonian fluids* such as water, oil, steam, and atmospheric gases exhibit exponential increases in flow resistance (viscosity) when they are accelerated in pipes, but non-Newtonian blood exhibits exponential decreases in flow resistance when it is accelerated in arteries. The significance of this observation is ignored in medical literature.

The purpose of red cells remains obscure. Medical textbooks routinely assume that the sole purpose of red cells is to transport oxygen to peripheral tissues, but hemoglobin and its congeners circulate freely in the blood of invertebrates, and hemoglobin encapsulation does not enhance oxygen transport and delivery. Nevertheless, hemoglobin encapsulation distinguishes vertebrates from invertebrates, and distinct red cell morphology characterizes each class of vertebrates. Mammalian red cell mass is far greater than necessary for effective oxygen transport; approximately one-quarter of all cells in the adult human body are blood cells, and stressful nervous activity exaggerates red cell production.[10] Most vertebrates have large, nucleated red cells that retain the Krebs cycle, but mammals are characterized by small, enucleated, biconcave red cells that lack the Krebs cycle, have abbreviated life spans, and are so closely similar as to be nearly indistinguishable among species.

Camels, llamas, and alpacas are presently classed as mammals, but their red cells are oval and retain a nucleus. Compared to other mammals they suffer notoriously inferior exercise capacity. A satisfactory explanation is presently lacking.

Atherosclerosis defies explanation. Down's Syndrome, distance from the heart, anemia, exercise, coagulopathies, anticoagulants, and moderate red wine consumption all inhibit atherosclerosis. Advancing age, increased blood viscosity, emotional stress, sedentary life style, hyperthyroidism, hypothyroidism, hypertension, polycythemia, congestive heart failure, obesity, and chronic diseases all accelerate atherosclerosis. Earlier researchers attributed atherosclerosis to inadequate shear stress,[11] but this fails to explain why atherosclerosis decreases with distance from the heart. Presently prevailing theories that attribute atherosclerosis to lipoproteins or primary lesions of the vascular endothelium are similarly lacking.[12]

The nature of hemostasis remains unclear. Capillary hemostasis occurs within minutes even though capillaries cannot contract. Virchow's Triad identifies hypercoagulability of blood, stasis of blood flow, and endothelial injury as essential elements of clot formation, but does not explain how these factors initiate or enable hemostasis. The triad is named after the eminent German researcher Rudolf Virchow (1821-1902, see Appendix A: Medical History, on page 395), but Virchow himself never suggested the idea of a triad.[13] Instead the Triad represents a consensus that was achieved decades after Virchow's death. Modern research has identified the coagulation cascade, which describes the sequence of blood enzyme interactions that enable clot formation. Insoluble fibrin, the end product of the cascade, is essential for both clot formation and capillary hemostasis, but it has no enzymatic properties that might explain the nature of hemostasis. There is considerable evidence that coagulation cascade enzymes influence tissue repair, but the relationship remains unclear.

The nature of tissue repair remains unclear. Modern studies have demonstrated that tissue injury somehow initiates clot formation that is followed by an orderly progression of inflammation in surrounding tissues, repair cell migration into the damaged tissues from undamaged surrounding tissues, repair cell proliferation and differentiation in the damaged tissues, immune activity in the healing tissues, and so forth, but no mechanism has been identified that governs the tissue repair process and explains its nature.

Conventional medical theory cannot explain the nature of disease and disease relationships. Most diseases produce strikingly similar symptoms and manifestations, such as fever, rashes, exudates, bowel disturbance, dementia, tachycardia, hypertension, malaise, fatigue, edema, catabolism, and inflammation.

Seemingly unrelated diseases have unexplained relationships. Hypertension is typically accompanied by diabetes, and vice-versa. Obesity exaggerates the risk of hypertension, diabetes, and cancer. Cancer increases the risk of infarction. Chronic infections and rheumatoid diseases increase the risk of cancer, diabetes, and hypertension.

Exaggerated immune activity occurs in association with surgery, trauma, tissue repair, rheumatoid disease, and malignancy. Rheumatoid disease is often categorized as an *autoimmune disease* simply because immune activity can be detected, but there is no evidence that immune activity attacks body tissues. Cancer is sometimes attributed to excessive immune activity and alternatively attributed to inadequate immune activity that fails to prevent it, but there is no convincing evidence of either notion.

Anesthesia suffers for lack of an effective theory that enables the alteration of anesthetic technique to optimize surgical outcome.[14] The nature of anesthesia, analgesia, allostasis, nociception, pain, emotion, and their relationships remains unclear.

Animal studies suggest that stress is a normal and even necessary part of life.[15] Animals exhibit physiological perturbations in response to numerous forms of environmental stress, including injury, disease, starvation, cold, heat, predation, parasitism, and confrontations over mates and food. They maintain and repair their structural integrity, and sustain their internal milieu despite such stresses. The nature of these phenomena remains unclear.

NOTES

1 S. C. Campbell-Staton *et al.*, Winter storms drive rapid phenotypic, regulatory, and genomic shifts in the green anole lizard. *Science* 357, 495-498 (2017).

2 A. F. Bennett, Gleeson, Todd T., Gorman, George C., Anaerobic metabolism in a lizard (Anolis bonairensis) under natural conditions. *Physiol. Zool.* 54, 237-241 (1981).

3 R. O. Becker, G. Selden, *The body electric : electromagnetism and the foundation of life*. (Quill, New York, ed. 1st Quill, 1985), pp. 364 p.

4 S. E. Luria, M. Delbruck, Mutations of Bacteria from Virus Sensitivity to Virus Resistance. *Genetics* 28, 491-511 (1943).

5 S. L. Miller, A production of amino acids under possible primitive earth conditions. *Science* 117, 528-529 (1953).

6 J. Prothero, Heart weight as a function of body weight in mammals. *Growth* 43, 139-150 (1979).

7 V. P. van Empel *et al.*, Myocyte apoptosis in heart failure. *Cardiovasc Res* 67, 21-29 (2005).

8 G. Tatu-Chitoiu *et al.*, Streptokinase-induced hypotension has no detrimental effect on patients with thrombolytic treatment for acute myocardial infarction. A substudy of the Romanian Study for Accelerated Streptokinase in Acute Myocardial Infarction (ASK-ROMANIA). *Rom J Intern Med* 42, 557-573 (2004).

9 L. S. Coleman, A Stress Repair Mechanism that Maintains Vertebrate Structure during Stress. *Cardiovasc Hematol Disord Drug Targets*, (2010).

10 M. A. Al-Drees *et al.*, Making Blood: The Haematopoietic Niche throughout Ontogeny. *Stem Cells Int* 2015, 571893 (2015).

11 M. Texon, The Hemodynamic Concept of Atherosclerosis. *Bull N Y Acad Med* 36, 263-274 (1960).

12 P. Libby, Atherosclerosis: the new view. *Sci Am* 286, 46-55 (2002).

13 C. N. Bagot, R. Arya, Virchow and his triad: a question of attribution. *British journal of haematology* 143, 180-190 (2008).

14 L. S. Coleman, 30 Years Lost in Anesthesia Theory. *Cardiovasc Hematol Agents Med Chem* 10, 31-49 (2012).

15 S. M. Korte, J. M. Koolhaas, J. C. Wingfield, B. S. McEwen, The Darwinian concept of stress: benefits of allostasis and costs of allostatic load and the trade-offs in health and disease. *Neurosci Biobehav Rev* 29, 3-38 (2005).

Appendix D
Classical Hemodynamic Physiology

"The trouble with the world is that the stupid are cocksure and the intelligent are full of doubt."
— Bertrand Russell

Claude Bernard (1813-1878, see Appendix A: Medical History, on page 395), the father of medical physiology, endeavored to prove that animal life functions in accord with the principles that apply to physical sciences. He demonstrated that nervous stimulation alters organ perfusion, which in turn regulates organ function. He contributed the concept that cells require a milieu interieur that sustains their survival and function regardless of environmental circumstances.[1]

Like many others, his personal life was chaotic. His arranged marriage to a physician's daughter provided a dowry that financed his research, but his wife divorced him and became an ardent opponent of vivisection after she returned home with her daughters to discover him vivisecting their pet dog. He is thus remembered as the "prince of vivisection" as well as for his scientific contributions.

Bernard's research was impressive, but like all good research, it produced more questions than answers. Bernard proved that nervous activity regulates tissue perfusion, which in turn regulates organ function, but the mechanism that explains this remains elusive. The presently accepted vasoconstriction/vasodilation explanation is sorely lacking (see Appendix B: Hemostasis History, on page 407).

To appreciate stress theory, it will be helpful to understand the shortcomings of conventional medical physiology and the origins of these shortcomings. Presently prevailing hemodynamic physiology theory (the explanation of blood flow and distribution) has evolved slowly over hundreds of years and reflects intuitively appealing — and misleading — assumptions. The following modern quote summarizes prevailing hemodynamic physiology dogma:

"It is generally accepted that the peripheral circulation of the blood is determined by the arteriovenous pressure gradient across the vascular bed and by the vascular resistance to blood flow. The resistance to flow depends on both the viscous properties of the blood and on the calibers of the resistance blood vessels. By far the most potent variable affecting resistance is the tone of the vascular smooth muscle cells. Although, basically, this tone appears to be an inherent and automatic characteristic of muscle, it may be strongly modified, and even dominated, by extrinsic influences such as nervous stimuli and chemical agents. These influences constitute the most powerful tools for the control for the vascular bed."[2]

On first glance, this explanation appears reasonable and intuitively appealing, but on closer scrutiny it is riddled with reservations, which I will review along with a little historical background.

The idea of blood circulation is not new. Greek and Egyptian physicians of antiquity surmised that the heart pumps blood to distant parts of the body, and that blood returns to the heart and passes through the lung where it releases waste products and absorbs life-giving substance,[3] but this insight was forgotten due to inadequate technology, the prohibition of human dissection, and the cyclical collapse of civilizations.

Galen (129-216), the legendary Roman physician, dissected Barbary apes (as opposed to humans), but this nuance was neglected for centuries. He presumed that the right side of the heart produces blood that penetrates to the left side of the heart, where it "gains spirit" (turns bright red), and is absorbed by tissues. He further assumed that venous blood originates in the liver and is likewise absorbed by tissues. His misinformation mangled medical progress for centuries.

Ibn al-Nafis (1213-1288), a Persian polymath, correctly concluded that blood flows from the right chamber of the heart through the pulmonary artery to the lungs before reaching the left side of the heart.

Hieronymus Fabricius (1537-1619), a professor of anatomy at the University of Padua, discovered membranous folds in the interior of veins that he called "valves." He did not appreciate their purpose (they account for one-way blood flow in veins), but his observations enabled his student William Harvey (1578-1657) to recognize the nature of blood circulation. Harvey refuted Galen's theory by demonstrating that the heart pumps large volumes of blood. He postulated the presence of capillaries that connect arteries to veins, but failed to find them. Marcello Malpighi (1628-1694) later discovered capillaries that completed the understanding of blood circulation.

Stephen Hales (1677-1761) measured blood pressure by inserting small tubes into arteries and measuring the height of the resulting blood column. He provided persuasive evidence that blood pressure is the driving force of blood flow by demonstrating that blood pressure declines as animals bleed to death. However, modern research at the turn of the previous century challenged this intuitively appealing assessment:

"Each time a change can be observed in the height of a manometer applied to an artery, the experimenter must ask himself which of the two factors of the arterial pressure has varied, or whether the two factors, power and resistance, have been modified at the same time. In the absence of a criterion which would allow one to decide with complete certainty this litigious question, often enough physiologists have chosen the hypothesis which best accorded with their preconceived ideas."

— Etienne Jules Marey, 1881

"The development of our knowledge of the circulation has been bedeviled by the fact that the measurement of blood flow is so complicated, whereas that of blood pressure is so easy: hence the blood pressure manometer has exerted an almost hypnotic influence, though bodily organs don't need pressure but flow."
— Adolf Jarisch 1928

Walter B. Cannon (1871-1945), a prominent Harvard professor during the WWI era, ignored such inconvenient reservations and embraced Hales' alluring concept that blood pressure is the driving force of tissue perfusion. He published a persuasive book called "The Wisdom of the Body" that has subsequently dominated medical beliefs and practices. He alleged that the heart is "richly endowed with reserves of force and can increase its contractile force and volume to accommodate physiological requirements." He assumed that the mechanical work of the powerfully contracting heart is spent in overcoming frictional resistance in the blood vessels; that sympathetic tone determines vascular contractility and shifts blood flow from one region to another.[4]

Neither Cannon's concept nor presently prevailing hemodynamic physiology theory can withstand modern evidence, critical scrutiny, and common sense. Here are some of its shortcomings:

1. Cannon assumed that what applies to water also applies to blood, but water is a Newtonian fluid that exhibits exponential increases in turbulent flow resistance when it is accelerated in pipes, whereas blood is a non-Newtonian fluid that exhibits exponential *decreases* in flow resistance when it is accelerated in arteries. This exponential decrease is due to the presence of biconcave red blood cells, which inhibit turbulent flow friction. Because of its non-Newtonian nature and laminar (friction-free) flow characteristics, blood accelerates as arterial diameter decreases. Thus, arterial vasoconstriction cannot effectively regulate blood flow. This principle remains overlooked in current medical literature. It is elaborated on in Chapter 13: The Turbulence Mechanism, on page 119 that explains the laminar and the turbulent nature of blood.
2. Most evidence suggests that nature is extremely efficient. Cannon argued that opposing muscular forces of cardiac contractility and vasoconstriction determine blood pressure and blood flow regulation, as if the opposing forces waste energy by "fighting" with each other. Furthermore, both muscular contraction *and* relaxation consume ATP, which further exaggerates the wasteful nature of Cannon's concept.[5]
3. Cannon ignored Tannenberg's capillary gate theory,[6] which was well known in his time. It postulates that capillaries are the most efficient site of blood flow regulation, because their surface area is far greater than the sum total of larger vessels', and blood flow rate, pressure, and turbulence are minimal in capillaries.

4. Cannon's theory fails to explain *capillary recruitment* wherein microvascular flow varies in accord with autonomic balance, which suggests that a sub-microscopic mechanism regulates blood flow at the capillary level.[7, 8]
5. Cannon ignored capillary hemostasis — the cessation of microvascular blood loss within 15 minutes — which cannot be explained by vasoconstriction because *capillaries lack musculature*.
6. Cannon's concept that the heart is the "Charles Atlas" of the body implies that cardiovascular conditioning should increase the size, weight, and muscular power of the heart, but exercise has negligible effect on the heart. Instead, exercise conditioning induces capillary proliferation in affected muscles, which reduces flow resistance and enhances cardiac output, tissue perfusion, and tissue oxygenation. This suggests that capillaries are the focal site of flow resistance.
7. Cannon assumes that blood pressure is the driving force of blood flow and that increased blood pressure explains the ten-fold increase in cardiac output that occurs during intense exercise. Instead, trained athletes exhibit abnormally *low* blood pressure at rest and normal blood pressure during intense exercise.
8. Blood, like water, is incompressible; therefore, Cannon's concept implies that cardiac contraction should instantaneously increase blood pressure equally throughout the arterial tree during systole, as in the static hydraulic brake system in a car. Instead, measured blood pressure varies from point-to-point due to changing arterial length, diameter, bifurcations, curvature, constrictions, and wave reflections, and blood pressure increases with distance from the heart as declining arterial diameter exaggerates diastolic blood turbulence. This is discussed in detail in Chapter 13: The Turbulence Mechanism, on page 119.
9. Cannon's concept cannot explain why low blood pressure characterizes normal healthy sleep and athletic conditioning, or why pathological essential hypertension invariably reflects microvascular damage that elevates flow resistance, inhibits cardiac output, undermines exercise tolerance, and causes Type II diabetes and congestive heart failure.
10. Cannon's concept cannot explain the confusing relationships of blood pressure, blood viscosity, blood coagulability, red cell mass, and atherosclerosis. For example, blood pressure increases and atherosclerosis decreases as arterial diameter declines toward the periphery. Atherosclerosis is associated with hypertension, but it appears first in large central arteries where blood pressure is low, and is rare in small peripheral arteries where blood pressure is exaggerated. Anemia elevates pulse pressure, undermines hemostasis, and inhibits atherosclerosis, but does not cause microvascular damage, while microvascular damage characterizes essential hypertension. These confusing observations are explained by the diastolic turbulent pulse wave, which is discussed in Chapter 13.

Cannon's crippled concepts endure despite their shortcomings, while the constructive contributions of his contemporaries — who were more famous at the time — are abandoned and forgotten.[9] This is better explained by politics than science. The mammalian stress mechanism explains all these baffling manifestations and relationships.

NOTES

1 Bernard C. An introduction to the study of experimental medicine. New York: Schuman; 1949. xix, 226 p. p.

2 Fishman AP, Richards DW. Circulation of the blood : men and ideas. Bethesda, Md. Baltimore, Md.: *American Physiological Society*; Williams & Wilkins Co. distributor; 1982. xx, 859 p. p.

3 Fishman AP, Richards DW. Circulation of the blood : men and ideas. Bethesda, Md. Baltimore, Md.: *American Physiological Society*; Williams & Wilkins Co. distributor; 1982. xx, 859 p. p.

4 Cannon WB. The wisdom of the body. New York,: W.W. Norton & Company; 1939. xviii, 19-333 p. p.

5 Coleman LS. The Mammalian Stress Mechanism (MSM) Explains Radial Artery Spasm (RAS). J Anes Surg. 2018;5(1):81-94.

6 Tannenberg J. Uber die Kapillartatigkeit. Verh dtsch path. 1925;Ges., 20:374.

7 Fishman AP, Richards DW. Circulation of the blood : men and ideas. Bethesda, Md. Baltimore, Md.: *American Physiological Society*; Williams & Wilkins Co. distributor; 1982. xx, 859 p. p.

8 Coleman LS. Four Forgotten Giants of Anesthesia History. Journal of Anesthesia and Surgery. 2015;3(2):1-17.

9 Coleman LS. Four Forgotten Giants of Anesthesia History. Journal of Anesthesia and Surgery. 2015;3(2):1-17.

Appendix E
Cell Biology

"The good life is one inspired by love and guided by knowledge."
— Bertrand Russell

Independent, free-living cells constitute a vast biomass of countless species that evolved eons before the appearance of complex multicellular animals and plants. Cellular life probably originated in the stable, hot, nutrient-rich volcanic environment miles beneath the earth's surface and subsequently adapted to the oceans, the terrestrial surface, and ultimately the cold, turbulent, nutrient-poor atmosphere. In so doing, cells have accumulated countless useful mutations, and evolved alternate metabolic pathways and intracellular stress mechanisms.

Microbes exist in mind-boggling variety, and microbe classification is a rapidly changing and contentious field. During the era of stress research, all cells were categorized as either *prokaryotes* or *eukaryotes*. More recently, a third category of cells has been identified, and cells are now classified into three domains: *Bacteria*, *Archaea*, and *Eukarya*. Our present knowledge is founded on organisms that thrive on or near the earth's surface. It seems likely that additional undiscovered microbial domains thrive deep beneath the earth's surface.

BACTERIA

Bacteria are prokaryotes. They are probably the original cell type, and they cause most non-viral infections in humans. Bacterial DNA floats free in the cytoplasm, where it *transcribes* (copies) its genetic information onto RNA, which then *translates* (converts) the information into proteins and enzymes that enable cell function.

Bacteria respire (generate energy) via their cell walls, which limits them to small size, unicellular existence, and a few shapes, (rods and spheres) that optimize their surface area relative to their internal volume.

ARCHAEA

Archaea are prokaryotes that are similar to Bacteria in that they lack visible internal structures, respire via their cell wall so that they are restricted to small size and single cell existence, and their DNA floats free in their cytoplasm. However, their DNA does not utilize the transcription/translation mechanism. For that reason, they are thought to be more closely related to the Eukarya than Bacteria. They exist in a greater variety of shapes than Bacteria, and many respire (generate energy) via *methanogenesis* metabolic pathways that are absent in Bacteria.

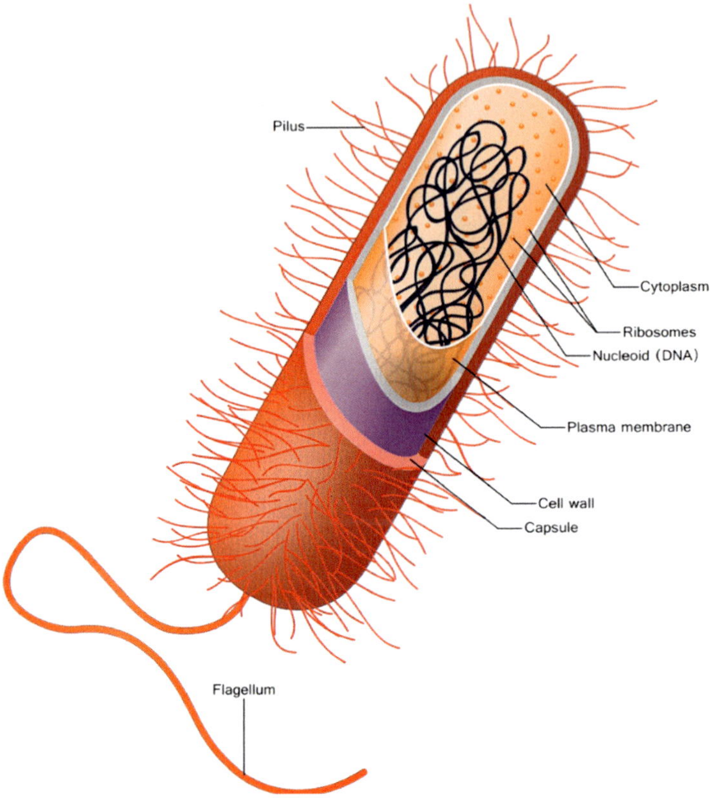

Figure E.1: Prokaryotic cell.
A depiction of a typical free-living prokaryotic cell. Prokaryotes typically propel themselves via a tail-like flagellum. Both Bacteria and Archaea respire (generate ATP) via their cell wall, and their DNA floats free in the cytoplasm in the form of strands and circles, but the DNA of Archaea does not utilize the transcription/translation mechanism. This vector image is completely made by Ali Zifan - Own work; used information from Biology 10e Textbook (chapter 4, Page: 63) by: Peter Raven, Kenneth Mason, Jonathan Losos, Susan Singer · McGraw-Hill Education., CC BY-SA 4.0, https://commons.wikimedia.org/w/index.php?curid=44194140

EUKARYA

Eukarya is the only cell domain that is pertinent to medicine. All complex multicellular plants and animals are composed exclusively of eukaryotic cells that are far larger, more complex, and more capable than prokaryotes. For example, they are capable of differentiation, de-differentiation, and re-differentiation into specialized cell types. Eukaryotes lack the flagellum found in prokaryotes, but they can alter their shape to move via *chemotaxis* using tiny muscle fibers that are lacking in prokaryotes. Some eukaryotes also possess cilia that enable movement.

They are thought to share a common ancestor with Archaea, because neither Archaea nor Eukarya mediate their genetic information via the bacterial transcription/translation mechanism.

Eukaryotic cytoplasm (the substance contained within the cell walls) is highly structured. It contains organelles that perform specialized functions and can be visualized via microscopy. Some of these, such as chloroplasts (in plant cells) and mitochondria (in all eukaryotic cells) contain their own DNA that utilizes the bacterial transcription/translation mechanism. These are thought to represent previously independent, free-living bacterial cells that were incorporated into eukaryotic cells via *endosymbiosis*,[1] which suggests that eukaryotes may have originated as "predator" cells that engulfed and digested Bacteria, much as immune cells engulf and digest bacteria that invade the body and cause infections.

Eukaryotic cells contain two different types of DNA. This causes confusion. The primary DNA that determines inheritance is contained in the form of chromosomes. These consist of tightly wound strands of DNA that retain vast quantities of genetic information. The chromosomes are confined within the thick walls of a nucleus and thereby isolated from the secondary DNA found in chloroplasts and mitochondria. A complex mechanism called *mitosis* maintains chromosomal isolation from cytoplasmic DNA during eukaryotic cell division. Organelle DNA thus replicates independently of nuclear DNA and does not mediate animal and plant inheritance.

During the era of stress research it was known that eukaryotic cells can move via chemotaxis, respond to electricity and magnetism, communicate with one another by releasing chemical signals, spontaneously re-organize themselves into complex organ tissues, and differentiate, de-differentiate, and re-differentiate into alternate cell types.[2, 3] Since then, it has been discovered that they can grow as large as a grape, connect to one another via microtubules, assume astonishing shapes (see Acetabularia in Figure E.4 on page 431), detect and respond to light, discern the difference between left and right, attack and digest other cells while preserving their organelles, negotiate mazes, and propel themselves by multiple means.[4] It now seems likely that science has only begun to understand their capabilities.

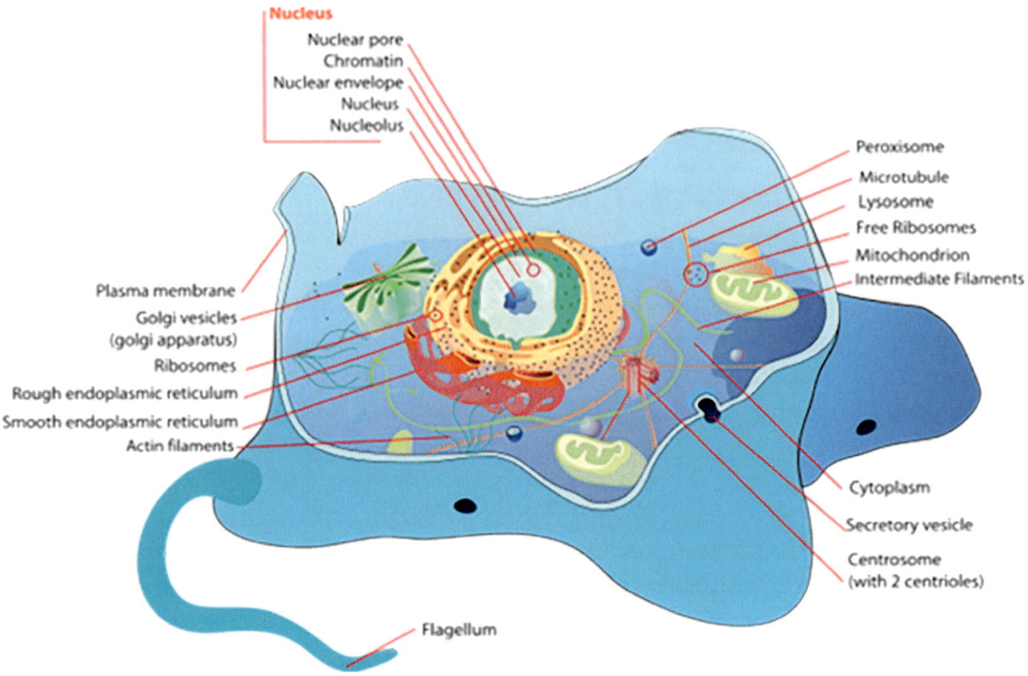

Figure E.2: Eukaryotic animal cell.

A depiction of a typical eukaryotic animal cell. Animal cells move by activating muscle fibers within the cell to alter their shape. Unlike prokaryotes and Archaea, eukaryotic cytoplasm contains organelles that perform specific functions, including mitochondria, chloroplasts, Golgi apparatus, vacuoles, endoplasmic reticulum, and reticular endothelium. Unlike prokaryotes, eukaryotic DNA is tightly coiled in the form of chromosomes that are confined within the thick walls of the nucleus, which isolates them from the cytoplasm. NB: the artist who prepared this drawing mistakenly depicted the eukaryotic cell as having a flagellum, which is exclusive to prokaryotes. By LadyofHats (Mariana Ruiz) - Own work using Adobe Illustrator. Image renamed from Image: Animal cell structure.svg, Public Domain, https://commons.wikimedia.org/w/index.php?curid=4266142

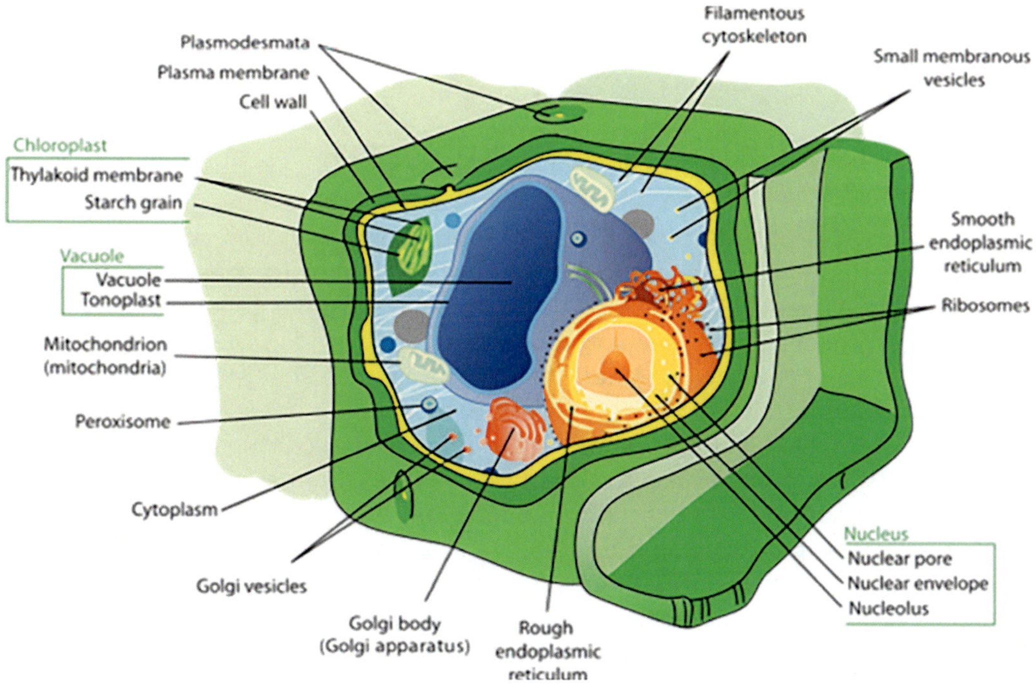

Figure E.3: Eukaryotic plant cell.
A prototypical eukaryotic plant cell. Most plant cells are surrounded by a thick, protective cellulose wall that renders them immobile. Courtesy of LadyofHats - Self-made using Adobe Illustrator

Figure E.4: Acetabularia.
This drawing of Acetabularia cells demonstrates the ability of individual eukaryotic cells

to create exotic shapes. [[File:Haeckel Acetabularia acetabulum.jpg|Haeckel Acetabularia acetabulum]]

CELLULAR RESPIRATION AND ATP

The food we eat is digested into molecular fats, carbohydrates, and proteins that are transported to our cells via blood circulation. Cells convert these substances into a chemical called ATP (adenosine tri-phosphate). All cells generate ATP and utilize it to energize their activities, much as gasoline provides the energy that propels cars. Thus, ATP is the *universal medium of cellular energy*. ATP is often referred to as *phosphate* in scientific literature.

Biologists refer to ATP generation as *respiration*. This is confusing, because the term respiration commonly refers to breathing — air exchange in the lungs that facilitates oxygen capture from the atmosphere and transport to cells via blood circulation, where it is utilized to manufacture ATP. From a biologist's viewpoint, breathing and blood circulation are accessory mechanisms that capture and transport oxygen to eukaryotic cells to enable cellular respiration that generates ATP. Therefore, the term respiration must be understood in context.

Prokaryotes, eukaryotes and Archaea are also distinguished by their differing metabolic pathways that generate ATP. In multicellular animals and plants, which are composed exclusively of eukaryotic cells, ATP energizes embryological development, hemodynamic physiology, and tissue repair. Disease wastefully exaggerates ATP consumption.

PROKARYOTIC ATP GENERATION

Prokaryotes respire (generate ATP) via their cell walls, and are thus limited to small size, single-cell existence, and a few shapes (mostly spheres and rods) that optimize ATP generation by maximizing their surface area relative to their internal volume. Prokaryotes have evolved numerous respiratory pathways that enable them to thrive in vast abundance and variety in diverse environments. Numerous prokaryotes respire and thrive in the absence of oxygen. In contrast to Bacteria, the Archaea possess methanogenesis pathways that enable them to metabolize (digest) sewage in the absence of oxygen. One wonders what the world would be like in their absence.

EUKARYOTIC ATP GENERATION

In contrast to prokaryotes, eukaryotic cells do not respire via their cell walls. Instead they generate ATP internally via an efficient respiratory pathway called the *Krebs cycle* that consumes glucose and oxygen and produces ATP, water, and carbon dioxide. The Krebs cycle operates in mitochondria, which are organelles located in the cytoplasm. The ability to generate ATP from within the cell enables eukaryotes to grow larger, move about by changing their shape, and sub-specialize to form organs, tissues, and anatomical structures in complex multicellular animals and plants.

In addition to the Krebs cycle, eukaryotes possess *glycolysis* pathways that generate ATP in the absence of oxygen, but these pathways are far less efficient than the Krebs Cycle. Their by-product is lactic acid, which painfully accumulates in muscles and tissues. If oxygen is restored in a timely fashion, the Krebs Cycle converts lactic acid into ATP.

ATPASE AND INTRACELLULAR ATP UTILIZATION

ATPase enzymes utilize ATP to energize intracellular activities in both prokaryotes and eukaryotes. There are several closely-related types of intracellular ATPase enzymes, all of which are calcium-dependent. In eukaryote muscle cells, a calcium pump sequesters Ca+ within an intracellular organelle called the *sarcoplasmic reticulum.* The sarcoplasmic reticulum releases Ca+ into the cell cytoplasm to activate ATPase.

Oxygen starvation undermines the Krebs cycle and causes eukaryotic animal cells to revert to inefficient anaerobic ATP generation, causing painful lactic acid accumulation and muscle fatigue. This is called *metabolic acidosis*. Exercise conditioning (in mammals) induces angiogenesis (capillary proliferation) that enhances oxygen delivery, increases mitochondrial ATP generation, and improves exercise tolerance.

THROMBIN AND EXTRACELLULAR ATP UTILIZATION

Outside eukaryotic cells, in the blood and extracellular tissues of complex multicellular mammals, calcium-dependent thrombin converts ATP energy into cell and extracellular enzyme activities. All animal cells possess thrombin receptors (Protease-Activated Receptors, or PAR) that determine how individual cell types respond to thrombin elevations.

Like ATPase, thrombin requires calcium and ATP. This suggests that thrombin is closely related to intracellular ATPase enzymes. Until recently, thrombin was distinguished from other coagulation cascade enzymes only by its location in the sequence of enzymatic interactions that produced *fibrin*. It is now apparent that thrombin is the primary product of the interaction of factors VII, VIII, IX, and X, and that thrombin utilizes ATP to energize extracellular enzymes and cell activities. In mammals, extracellular Ca+ levels are maintained within a narrow range to optimize thrombin activity, and the mammalian stress mechanism governs the rate, location, and magnitude of thrombin generation.

THROMBIN VS. AUXIN

Yet another variation of ATP physiology has evolved in plants, which generate the extracellular enzyme auxin to energize growth and development. Like thrombin, auxin requires ATP and Ca+, and functions as a universal enzyme of extracellular energy transformation. Plants obtain Ca+ from the soil, which explains why plants thrive in calcium-rich soil. The strikingly similar molecular structures and functions of thrombin and auxin probably reflect their common origin.

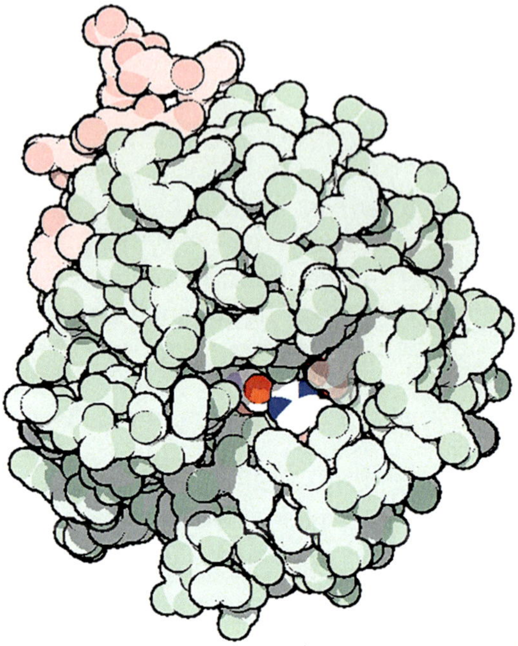

Figure E.5: Thrombin molecule.
Thrombin (courtesy of "Molecule of the Month") Its active site is shown in red.

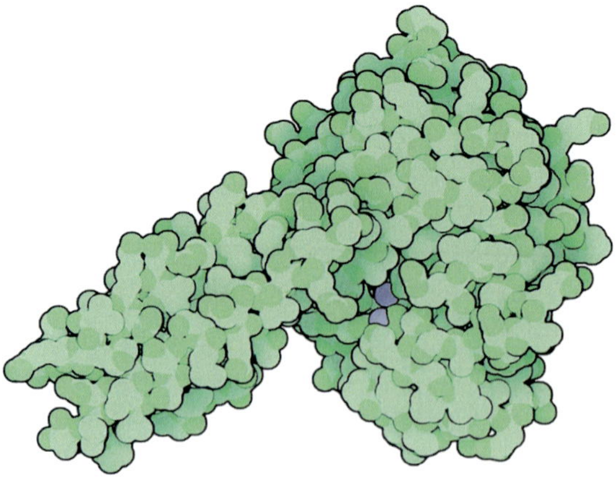

Figure E.6: Auxin molecule.
Auxin, shown here conjugated with amidohydrolase ILL2 (courtesy of "Molecule of the Month") Plants bind auxin to amino acids or sugars that can be clipped off to expose its active site, which is shown in blue. Note the structural similarity of auxin and thrombin. I hypothesize that they are essentially the same.

NOTES

1 https://en.wikipedia.org/wiki/Symbiogenesis

2 R. O. Becker, J. A. Spadaro, Electrical stimulation of partial limb regeneration in mammals. *Bull N Y Acad Med* 48, 627-641 (1972).

3 P. J. Rosch, Bioelectromagnetic and subtle energy medicine: the interface between mind and matter. *Annals of the New York Academy of Sciences* 1172, 297-311 (2009).

4 https://www.youtube.com/watch?v=ooA0J6DWWTM

Appendix F

The Structure of Scientific Revolutions

"Most people say that it is the intellect which makes a great scientist. They are wrong: it is character."
— Albert Einstein

During the era of stress research Thomas Kuhn (1922-1996) authored an influential book called "The Structure of Scientific Revolutions."[1] As a physicist, Kuhn was keenly aware that professors and researchers can only advance their careers within the context of orthodox beliefs, and emphasized that amateurs working outside the field contribute most revolutionary scientific advances. The DNA discovery is an example. Both Watson and Crick were youthful amateurs who were unfettered by vulnerable reputations. Watson was an American PhD candidate in ornithology — the study of birds. Crick was a British mathematician who helped develop magnetic mines that devastated German shipping during WWII. They met by chance when they were assigned to share the same office at Cambridge University. Neither had any experience with cell research, but both were fascinated by DNA, and they combined their talents to achieve the most astonishing advance in biological history.

Kuhn's book places the era of stress research in historical perspective. He observed that scientific progress proceeds in four phases:

1. THE PRE-PARADIGM PHASE

During the pre-paradigm phase researchers pursue ideas independently, embrace disparate techniques, and cannot communicate effectively with one another for lack of common technology and terminology that enables them to compare results. The pre-paradigm phase of stress theory evolved during WWII when war-related subsidies fueled diverse and seemingly unrelated research endeavors. Vaccines and sulfa drugs were the focus of pharmaceutical development. Oswald Avery's meticulous bacterial experiments proved that DNA mediates inheritance, and refuted the prevailing belief that proteins mediate inheritance. Max Delbruck and Salvador Luria proved that mutations occur at random and are not affected by environmental stress. Goodman and Gilman discovered that toxic chemicals can shrink malignant tumors. Salk developed the first viral vaccine (polio), which was previously believed to be impossible. Fleming discovered the first true antibiotic (penicillin).

2. PARADIGM FORMATION AND NORMAL SCIENCE

Kuhn argued that disparate research devolves into a common set of ideas, assumptions, terms, and technologies that he called a *paradigm*. This produces a prolonged period of normal science wherein researchers produce useful information within the context of the paradigm. The path-breaking discovery and description of the DNA mechanism by Watson and Crick galvanized divergent medical research into a stress research paradigm that assumed that an *extracellular stress mechanism* converts chromosomal DNA genetic information into embryological development, and that it maintains and repairs the resulting structures. Thus, research became focused on a search for the hypothetical stress mechanism. This produced a productive period of normal science (stress research) that lasted 30 years.

3. THE SCIENTIFIC CRISIS

Kuhn noted that scientists are not as critical and objective as commonly assumed. They are reluctant to question the shortcomings of the prevailing paradigm during periods of normal science until they become discouraged by its failure to attain the objective posed by the paradigm. This leads to a crisis, wherein the prevailing paradigm is challenged, and fresh ideas are entertained. Stress researchers became discouraged after 30 years of fruitless and frustrating research failed to find any clue of Selye's hypothetical stress mechanism, whereupon they began to question its existence.

4. PARADIGM SHIFT VERSUS REVOLUTION

Kuhn describes two possible outcomes of a scientific crisis. Most commonly, there is a paradigm shift that alters the paradigm so as to restore normal science, and productive research resumes. Less commonly, a revolutionary idea is contributed by an amateur working outside the field, and the world suddenly looks different. Unfortunately, when stress research failed to find any clue of the existence of a stress mechanism, and subsidies for stress research declined, scientists were left without a viable research goal. For lack of a better alternative, they embraced the only available alternative: the *Central Dogma of Molecular Biology* — the assumption that DNA activity directly explains physiology, pathology, and stress via the transcription/translation mechanism. This idea became known as *genomics*. Unfortunately, the shortcomings of genomics were what inspired stress theory to begin with, and the crippled genomic paradigm has failed to produce progress. As a result, medical research is again stuck in a crisis, where no viable alternative to genomics is yet available.[2] This book argues that fresh information from unrelated research has belatedly discovered Selye's stress mechanism, and stress theory is now poised to resume its rightful role as the prevailing paradigm of medical and biological research, and revolutionize medicine.

NOTES

1 Kuhn TS, Hacking I. The structure of scientific revolutions. Fourth edition. ed. Chicago ; London: The University of Chicago Press; 2012. xlvi, 217 p. p.

2 Pontin J. A Decade of Genomics. *Technology Review.* 2010:1.

Appendix G

The Central Dogma of Molecular Biology

"It has been said that man is a rational animal. All my life I have been searching for evidence which could support this."
— Bertrand Russell

Have you ever wondered why a cure for cancer never materializes despite decades of expensive research? Why there has been no significant theoretical advance in medicine and biology since the DNA discovery? Why Americans now seldom win Nobel prizes? Why the cost of health care keeps inexplicably climbing? These problems can be traced to the defective logic of the presently prevailing paradigm of medical research, which suits the interests of powerful corporations that dominate health care using specious drug development, ineffective treatments, misleading advertisements, and government largess that sustains their bloated profits.

Almost all of our DNA knowledge is founded on bacterial cell research, but bacteria are far smaller, simpler, and less capable than the eukaryotic cells that compose our bodies. They employ different metabolic pathways and internal mechanisms than those of human cells, and vice-versa (see Appendix E: Cell Biology, on page 427). As a result, most of what we know about DNA is irrelevant to medicine.

At the time Watson and Crick discovered the DNA mechanism, they assumed that the bacterial *transcription/translation* mechanism converts genetic information into proteins and enzymes that account for all the internal activities of all types of cells. This simple assumption seemed reasonable at that time, when little was known about eukaryotic cells. Today their assumption is known as the *central dogma of molecular biology* and research based on this assumption is called *genomics*. A *dogma* is a principle or set of principles laid down by an authority as incontrovertibly true, such as the edict of the Catholic church during the middle ages that the Earth is the center of the universe. Dogma is incompatible with science, which questions everything. *Thus, genomics is inherently unscientific.* To understand the central dogma, I suggest these simple video presentations by the "Khan Institute":

- https://www.google.com/search?client=safari&rls=en&q=dogma+of+molecular+biology+diagram&ie=UTF-8&oe=UTF-8

The *transcription/translation* mechanism has been confirmed in bacteria, mitochondria, and chloroplasts (the latter of which are thought to have originated as independent, free-living bacteria that were engulfed and incorporated into eukaryotic cells) and it may suffice to explain the inner workings of bacteria, but it doesn't explain how the chromosomal DNA blueprint is communicated outside the nucleus of eukaryotic cells, let alone outside the cell wall, so as to enable the development of complex multicellular structures during embryological development, not to mention the manifestations of tissue maintenance, tissue repair, blood flow regulation, or disease in mature animals and plants. These shortcomings were understood and openly acknowledged during the era of stress research, and are the reason that so many scientists embraced stress theory after the DNA discovery. *Today these shortcomings are tacitly overlooked and ignored.* Furthermore, research has refuted the original Watson and Crick assumption that the transcription/translation mechanism is irreversible:

- https://en.wikipedia.org/wiki/Central_dogma_of_molecular_biology

Thus, the means by which the chromosomal DNA of eukaryotic cells transmits its information outside the nucleus remains a complete mystery, as does the means of tissue repair and blood flow regulation.

Medical research is complex and expensive. Researchers require years of expensive education, plus elaborate equipment and sustained financial support to pursue their ideas. During the era of worldwide warfare that enabled medical advances, medical researchers were largely funded by private sources such as the Rockefeller Institute that demanded scientific integrity and valid results. Since then, medical research has come to be funded by government grants or private pharmaceutical corporations whose interests are often at odds with public health. Sickness sustains their profits, and their research seeks to rationalize profitable patents, and often overlooks dangerous side effects.[1] Researchers supported by pharmaceutical companies often manipulate statistics to justify worthless or dangerous products. Worse yet, they sometimes fabricate false data that is published without proper scrutiny and review. In rare cases where such chicanery is detected, the individual researchers sometimes spend time in prison, but the journals that publish their false information and the pharmaceutical corporations that pay for it seldom suffer consequences.[2]

Governments around the world withdrew funding for stress research more than forty years ago after it failed to find any clue of any testable mechanism that could confirm any aspect of Selye's theory. With the abandonment of stress research, the transcription/translation idea became the prevailing paradigm of medical research by default, for there were no available alternatives that researchers could use to apply for research grants. Applications for government research support must now be framed in the context of genomics if they are to have any hope of acceptance. It should be of no surprise that the transcription/translation mechanism has failed to explain the biology of complex multicellular animals and plants, or that research framed in its context has failed to produce theoretical progress, because genomics fails to explain how chromosomal DNA communicates not only outside the thick walls of the nucleus, but also outside the cell wall to enable embryological development, which entails rapid cell multiplication, specialization, and organization into complex multicellular organs and tissues. There is no evidence of DNA/RNA signaling that might explain such communication.[3]

The second shortcoming of genomics is that it fails to explain tissue maintenance, tissue repair, and hemodynamic physiology. Genomics implies that this would entail a continuous flood of RNA emanating not only from the nucleus, but also outside cells, to regulate tissue repair, organ function, blood flow, and so forth. There is no such evidence.

The third shortcoming of genomics is that it fails to explain the nature of disease. Except for rare illnesses such as Cystic fibrosis, congenital dysautonomia, or hemophilia — all of which entail genetic defects — *there is no evidence that disease is inherited.*

There must be undiscovered mechanisms that explain how chromosomal DNA communicates outside the nucleus and outside the cell to enable embryological development, but no evidence of such mechanisms has been found. Presently available evidence indicates that at least some chromosomal DNA becomes active only during embryological development, but most chromosomal DNA remains quiescent once embryological development is complete.[4, 5, 6]

The previous generation of stress theorists hypothesized that every cell retains and replicates the genetic blueprint that determines embryological development, and that this genetic blueprint directs the appearance of the extracellular stress mechanism at the very outset of embryological development, whereupon the stress mechanism works closely with DNA to enable embryological development. At the conclusion of embryological development, DNA resumes quiescence, while the stress mechanism remains active for the remainder of life to maintain and repair mature structures.

The genomics paradigm has utterly failed to explain any aspect of the biology of complex multicellular plants and animals. The transcription/translation mechanism exists only in bacteria, chloroplasts, and mitochondria. Chloroplasts and mitochondria are found in eukaryotic cells, but they are isolated from the chromosomal DNA that determines inheritance in multicellular animals and plants (see Appendix E: Cell Biology, on page 427). There is no convincing evidence or hypothesis to support the idea that this simple bacterial mechanism can explain genetic inheritance in complex multicellular animals and plants. Nevertheless, the political context of medical research requires that applications for research subsidies, reports submitted to medical journals, and professional research careers must be confined within the context of this fatally flawed idea. Thus, derailed by dogma that is enforced by straitjacket of government policy at the behest of commercial interests, it is hardly surprising that medical research has resembled a massive "snipe hunt" for the past 50 years. Small wonder that there is growing frustration with the failure of genomics to enable theoretical progress.[7, 8] Thomas Kuhn would describe this as a scientific "crisis" that precedes the appearance of a fresh paradigm (see Appendix F: The Structure of Scientific Revolutions, on page 437).

Meanwhile, despite being crippled by these constraints, the patient toil of "normal science" has yielded critical information that has belatedly enabled the first crude description of Selye's stress mechanism, which explains embryological development, physiology, pathology, and stress. It is as if Selye's ideas, like the mythological Phoenix,[9] are being reborn after they were burned alive, and are now poised to resume their role as the prevailing paradigm of medical research, and revolutionize medicine and biology.

NOTES

1 J. Virapen, *Side Effects: Death*. (Money Tree Publishing 6256 Bullet Drive, Crestview, Fl. 32536, 2015), pp. 189.
2 http://www.anesthesiologynews.com/Article/PrintArticle?articleID=12634
3 J. Pontin, in *Technology Review,* J. Pontin, Ed. (MIT, MIT, 2010), vol. 114, chap. 8, pp. 1.
4 J. S. Mattick, Challenging the dogma: the hidden layer of non-protein-coding RNAs in complex organisms. *Bioessays* 25, 930-939 (2003).
5 J. S. Mattick, The hidden genetic program of complex organisms. *Sci Am* 291, 60-67 (2004).
6 J. S. Mattick, The regulatory architecture of the human genome. *Asia Pac J Clin Nutr* 13, S14 (2004).
7 J. Pontin, in *Technology Review,* J. Pontin, Ed. (MIT, MIT, 2010), vol. 114, chap. 8, pp. 1.
8 T. S. Kuhn, I. Hacking, *The structure of scientific revolutions*. (The University of Chicago Press, Chicago ; London, ed. Fourth edition., 2012), pp. xlvi, 217 p.
9 https://en.wikipedia.org/wiki/Phoenix_(mythology)

Appendix H

Hypotheses, Theories, and Mechanisms

"Extreme positions are not succeeded by moderate ones, but by contrary extreme positions."
— Friedrich Nietzsche

The terms *hypothesis* and *theory* are often confused.

A hypothesis is a proposed explanation of observed phenomena. For scientific purposes, the hypothesis is worthless unless it can be verified by testing, and testing requires a *testable mechanism*.

A theory is a hypothesis that is complemented by a testable mechanism that confirms its validity, so that it has gained general acceptance among scientists. Thus, a hypothesis is useless unless it is complemented by a testable mechanism that confirms its validity. Even so, many theories are overturned after fresh evidence reveals their shortcomings.

Examples illustrate these subtleties:

The competing hypotheses of *geocentrism* (the earth is the center of the universe) and *heliocentrism* (the sun is the center of the universe) waxed and waned through history until Copernicus and Kepler paved the path for Newton's mathematical mechanism that explains the elliptical orbits of the earth and other planets around the sun (heliocentrism). Geocentrism was thus discarded, and heliocentrism became accepted theory. However, accumulating evidence has revealed shortcomings in Newton's mathematical mechanism, and Einstein's theory of relativity now prevails as the preferred explanation of astronomical observations. Thus, even accepted theories are open to challenge and vulnerable to replacement.

Before the discovery of DNA, many biologists hypothesized that proteins were the medium of inheritance, while others hypothesized that DNA was the medium of inheritance. Thus, *DNA theory* existed before Watson and Crick described the testable DNA mechanism. Their mechanism enabled the verification of the DNA hypothesis, which was subsequently accepted as theory, while the protein hypothesis was generally discarded. Even so, a few biologists still cling to the protein hypothesis.

Appendix I

Summary of Coagulation Enzymes and Substances

"Ignorance is the curse of God; knowledge is the wing wherewith we fly to heaven."
— William Shakespeare

NUMBER AND/OR NAME	FUNCTION	ASSOCIATED GENETIC DISORDERS (NOTES IN ITALIC)
Factor I (fibrinogen)	Precursor of soluble fibrin (tissue repair) and insoluble fibrin (coagulation, capillary hemostasis).	Congenital afibrinogenemia, Familial renal amyloidosis *Abnormalities in fibrinogen metabolism cause amyloidosis, which causes rheumatoid diseases!*
Factor II (prothrombin)	Precursor of thrombin, the Universal Enzyme of Extracellular Energy Transformation.	Prothrombin G20210A, Thrombophilia *Prothrombin is the precursor of THROMBIN.*
Factor III (tissue factor or tissue thromboplastin)	Found in extravascular tissues. Stabilized labile hepatic factor VII.	*Essential for life. Tissues rich in tissue factors are targets for stress.*
Factor IV (calcium)	Essential for thrombin activity.	*EDTA, trisodium citrate, $MgSO_2$, nitric oxide (NO), nitroprusside (NTP), and* calcium channel blocker *drugs bind to Ca+ and inactivate thrombin.*
Factor V (proaccelerin, labile factor)	Co-factor of X with which it forms the prothrombinase complex.	*Activated protein C resistance.*
Factor VI	Unassigned — old name for Factor Va.	

NUMBER AND/OR NAME	FUNCTION	ASSOCIATED GENETIC DISORDERS (NOTES IN ITALIC)
Factor VII (stable factor, proconvertin)	Interacts with factor X and tissue factor to generate thrombin for tissue repair. Also activates factor VIII for hemostasis.	Congenital factor VII deficiency (also activates VIII and IX). *Essential for life. Extremely labile and must be stabilized by tissue factor (TF) to exert its enzymatic effects.*
Factor VIII (antihemophilic factor A)	Chimeric combination of hepatic enzyme factor VIIIC and von Willebrand Factor. Links nervous activity to blood enzyme activity to enable capillary gate mechanism.	Defects in VIIIC cause true hemophilia, a severe clotting defect; defects in VWF cause von Willebrand Disease, which is a relatively mild coagulation defect.
Factor IX (antihemophilic factor B or Christmas factor)	Enhances factor VIII activity but has no effect on other enzymes.	Haemophilia B.
Factor X (Stuart-Prower factor)	Interacts with factor VII and tissue factor to enable tissue repair; interacts with factor VIII to enable coagulation.	*Essential for tissue repair and therefore essential for life.*
Factor XI (plasma thromboplastin antecedent)	Activates IX.	Haemophilia C
Factor XII (Hageman factor)	Activates factor XI, VII and prekallikrein.	Hereditary angioedema type III.
Factor XIII (fibrin-stabilizing factor)	Crosslinks fibrin.	Congenital Factor XIIIa/b deficiency (rare).
von Willebrand Factor	Binds to VIIIC to enable its enzymatic activity; also maintains capillary structure.	von Willebrand disease, angiodysplasia, angioneurotic edema.
prekallikrein (Fletcher factor)	Activates XII and prekallikrein; cleaves HMWK.	Prekallikrein/Fletcher Factor deficiency.
High-molecular-weight kininogen (HMWK) (Fitzgerald factor).	Supports reciprocal activation of XII, XI, and prekallikrein.	Kininogen deficiency.
Fibronectin (produced by vascular endothelium).	Mediates cell adhesion (also binds to strands of soluble fibrin to generate insoluble fibrin).	Glomerulopathy with fibronectin deposits. *In other words, it causes kidney damage when generated in excess.*

NUMBER AND/OR NAME	FUNCTION	ASSOCIATED GENETIC DISORDERS (NOTES IN ITALIC)
Antithrombin III (ATIII, produced by vascular endothelium).	Inhibits IIa, Xa, and other proteases to regulate capillary gate.	Antithrombin III deficiency.
Heparin cofactor II	Inhibits IIa, cofactor for heparin and dermatan sulfate (*minor antithrombin*).	Heparin cofactor II deficiency *This inhibits thrombin generation.*
Protein C (produced by the vascular endothelium).	Inactivates Va and VIIIa.	Protein C deficiency.
Protein S	Cofactor for activated protein C (APC, inactive when bound to C4b-binding protein).	Protein S deficiency.
Protein Z	Mediates thrombin adhesion to phospholipids and stimulates degradation of factor X by ZPI.	Protein Z deficiency. *Protein Z appears to degrade factor X, probably to inhibit excessive factor X generation; its function remains unclear.*
Protein Z-related protease inhibitor (ZPI)	Degrades factors X (in presence of protein Z) and XI (independently).	
Plasminogen	Converts to plasmin, lyses fibrin and other proteins (plasminogen is inherently unstable and must be stabilized by *Thrombin-Activated Fibrinolysis Inhibitor* (TAFI)).	Plasminogen deficiency, type I (ligneous conjunctivitis). *Plasminogen forms crosslinks that connect strands of soluble fibrin together to form insoluble fibrin; its instability explains why insoluble fibrin self-destructs.*
Alpha 2-antiplasmin	Inhibits plasmin.	Antiplasmin deficiency.
Tissue Plasminogen Activator (TPA)	Activates plasminogen.	Familial hyperfibrinolysis and thrombophilia.
Urokinase	Activates plasminogen.	Quebec platelet disorder. *Urokinase is a powerful anticoagulant that disintegrates insoluble fibrin and restores microvascular perfusion in the presence of acute infarction.*
Plasminogen activator inhibitor-1 (PAI1)	Inactivates tPA & urokinase (endothelial PAI).	Plasminogen activator inhibitor-1 deficiency.

NUMBER AND/OR NAME	FUNCTION	ASSOCIATED GENETIC DISORDERS (NOTES IN ITALIC)
Plasminogen activator inhibitor-2 (PAI2)	Inactivates tPA & urokinase (placental PAI).	
Cancer procoagulant	Pathological factor X activator linked to thrombosis in cancer.	*Note that mammalian stress mechanism hyperactivity generates excessive quantities of thrombin that is the underlying cause of cancer. Conventional literature and theory overlook the significance. It's the key to a cure for cancer.*

Appendix J

A Universal Anesthetic Approach that Optimizes Surgical Outcome

Lewis S. Coleman, MD,
Francisco Jimenez Riera, MD

INTRODUCTION

There are many ways to skin a cat, but there is also a best way. This appendix reports the results of 1500 adult and pediatric dental surgery patients who were anesthetized with a method guided by stress theory to minimize toxicity and optimize safety, simplicity, predictability, efficiency, economy, and outcome using presently available machines, monitors, and medications. The technique combined mask induction with Sevoflurane and nitrous oxide, plus front-loading with Fentanyl that enabled elective endotracheal intubation without paralysis. This was followed by 5-15 minutes of apneic oxygenation, until metabolic hypercarbia restored respiratory drive. General anesthesia was subsequently maintained using ½ MAC Isoflurane supplemented with Fentanyl to sustain beneficial hypercarbia and prevent untoward movements and harmful sympathetic nervous system hyperactivity by limiting respiratory rate within 8-12 breaths/minute in adults or 9-16 breaths/minute in children. Noninvasive transcutaneous O_2/CO_2 monitoring confirmed that supplementation of general anesthesia with hypercarbia and narcotic analgesia optimizes organ oxygenation, and refutes the destructive Leake-Waters hoax described in Chapter 18 on page 277.

WHAT INSPIRED THE STUDY

My discovery of the mammalian stress mechanism and recognition of the havoc wreaked by the Leake-Waters hoax described in Chapter 18, plus my observations of inexorably deteriorating safety standards and practices has convinced me of the need for reform. When I began my career in private anesthesia practice, there were ubiquitous “safety standards” that were established long before I began my training. These included the following:

- All diagnostic and surgical procedures requiring anesthesia care were performed in hospitals as opposed to physicians’ offices, dental suites, and outpatient surgical centers.

- Patients were admitted to the hospital at least 24 hours before surgery.

- Pre-operative blood, urine, and X-ray tests were required for all surgery patients.

- Anesthesiologists performed pre-operative visits to evaluate medical condition, explain anesthesia procedures, and order pre-operative sedation to allay harmful pain, fear, and anxiety.

- Patients remained hospitalized for at least 24 hours after surgery, and longer after major procedures. This enabled post-operative observation by trained nurses to detect occult problems such as OSA (Obstructive Sleep Apnea) and care for pain and other postoperative needs.

- Anesthesiologists performed post-operative visits to assess anesthetic and surgical problems.

- Surgeons screened patients carefully and ordered additional diagnostic procedures as necessary before hospital admission to avoid oversight or last-minute delays.

Few interrupt when money talks. During my anesthesia career, I observed how corporate influence manipulates anesthesiologists like puppets on strings:

- In the early 1980s medical insurance paid generously for hazardous “invasive monitoring” techniques including central venous pressure catheters, arterial cannulas, and Swan-Ganz catheters. The reimbursements were rescinded several years later, whereupon these techniques were relegated to medical memory.

- Anesthesiologists learned to supplement general anesthesia with the new generation of short-acting synthetic narcotics (Fentanyl, Sufentanil, etc.) introduced by the Janssen Corporation in the early 1980s. This improved surgical outcome, but necessitated prolonged postoperative “weaning” from mechanical hyperventilation entailed by the Leake-Waters hoax that was detailed in Chapter 18. The onerous weaning process was managed by armies of respiratory therapists.

- Later in the 1980s medical insurance companies eliminated reimbursement for routine preoperative laboratory and X-ray screening because it wasn't cost effective. There was no protest from professional organizations.

- Capnography became ubiquitous in hospitals around 1990 due to the efforts of the Anesthesia Patient Safety Foundation (APSF), a private organization founded by manufacturing interests. Capnography eliminated accidental esophageal intubation, and slashed malpractice insurance costs, but its ability to monitor and measure exhaled CO_2 levels was ignored, and practitioners continued to hyperventilate in accord with the Leake-Waters hoax.

- Contiguous with capnography, critical care specialists re-discovered the therapeutic benefits of permissive hypercarbia, *but anesthesiologists continued to hyperventilate their patients as before.*

- Around 1995, insurance companies arbitrarily imposed fast tracking (same day surgery) reimbursement requirements that necessitated admitting patients on the day of surgery and discharging them immediately after surgery. This annihilated long-established safety principles of early admission and late discharge. There was no debate, discussion, or protest from "organized medicine." The respiratory therapists disappeared.

- Doctors accommodated fast tracking with unseemly haste. Cardiac anesthesiologists promptly replaced their customary high dose narcotic technique with intravenous Propofol infusions, as if they had no appreciation of the benefits of narcotic supplementation.[1, 2] Within a year there were reports of a new form of post-bypass morbidity and mortality called SIRS (Systemic Inflammatory Response Syndrome).[3, 4, 5] In reality, this was just another manifestation of the lethal post-bypass syndrome that Lowenstein had controlled with narcotics.[6, 7]

- Non-cardiac anesthesiologists likewise abandoned narcotics and resumed the Leake-Waters technique with paralysis and hyperventilation to satisfy the fast tracking requirements. Their anesthetic outcomes were soon bedeviled by unexplained postoperative respiratory arrest that was attributed to *narcotic hypersensitivity*.[8, 9, 10]

- Next, medical insurance reduced anesthesia reimbursement for "minor" procedures such as cataract extractions, limb surgery, dental surgery, and endoscopies.

- Insurance allowed reimbursement for "monitored anesthesia care" (open airway techniques) introduced by dental schools. Anesthesiologists promptly forgot the importance of airway security (general endotracheal anesthesia) and embraced open airway techniques as a substitute for general anesthesia that could be used for rapid discharge after "minor" diagnostic and surgical procedures. Hospitals emasculated their policies and procedures to permit these hazardous techniques.

- Hospitals designated portions of their buildings to serve as outpatient surgery centers to accommodate the insurance edicts. The distinction was entirely artificial and legal, because the outpatient facilities shared the same surgeons, anesthesiologists, and buildings as the hospitals, except that nursing staffs were kept separate, and the outpatients were discharged home in the care of medically naïve civilian caregivers immediately after recovering from anesthesia.

- Reimbursements suddenly supported angioplasty in favor of cardiac bypass surgery. As cardiac bypass cases declined, cardiac anesthesiologists suddenly demanded to be included among the non-cardiac anesthesiologists.

- Declining reimbursements caused anesthesiologists to petition hospitals for subsidies. This enabled hospital administrators to dictate anesthesia practice. The administrators issued "exclusive contracts" to a "Director of Anesthesia" who hired and fired members of the anesthesia team and skimmed from their income. Numerous anesthesia teams disintegrated as their members quit in disgust.

- Next, medical insurance elevated reimbursements for nurse-anesthetists, which inspired a new "team" approach where anesthesiologists supervised the nurse-anesthetists. This confused responsibility and undermined safety.

- The altered reimbursements inspired the surgical home concept where anesthesiologists installed local analgesic blocks before surgery, and nurses managed the subsequent anesthetics. The blocks invited nerve damage, local analgesic systemic toxicity (LAST), and block failure, and required expensive ultrasound equipment and nursing support.

- The analgesic blocks and open airway techniques exaggerated malpractice lawsuits.[11]

- Hospitals competed with one another for declining insurance reimbursements, which further exaggerated the chaos.

Amid this mayhem, I sought better ways to cope with the increasingly sick and fragile surgical patient population. In the process I re-discovered the principles and practices of the forgotten first generation of modern anesthesia practitioners. This improved my outcomes and enabled me to adapt to the fast tracking requirements without abandoning airway safety and narcotics, and ultimately inspired this study.[12, 13, 14, 15, 16, 17, 18, 19, 20]

DENTAL ANESTHESIA AND THE OPEN AIRWAY TECHNIQUE

The increasing acceptance of hazardous "open airway" techniques also inspired this study. It is no accident that a dentist devised the first practical anesthesia device to deal with dental pain and fear, but since then dentistry has diverged from medicine. Dentists are seldom prepared to pay for the expensive machines, equipment, anesthetists, and nurses needed for hospital-style general endotracheal anesthesia. Instead, they rely primarily on relatively safe analgesic blocks and sedatives. However, these are often inadequate to deal with dental pain and fear.

During the 1980s, the introduction of short-acting Propofol and synthetic narcotics (Fentanyl, Alfentanil, Remifentanil) enabled dental schools to devise "open airway" techniques that can be managed in dental offices and clinics as a substitute for general endotracheal anesthesia. The technique replaces anesthesia machines with portable pumps that infuse intravenous mixtures of Propofol and narcotics, and relegates endotracheal intubation to the status of an emergency procedure. This seemingly convenient technique has been embraced by hospital anesthesiologists to facilitate rapid discharge after minor surgical and diagnostic procedures such as endoscopies, but it is inherently dangerous and has resulted in numerous tragedies and malpractice lawsuits.[21, 22, 23, 24, 25, 26, 27]

- Without a secure airway (elective intubation), the synergistic mixtures of Propofol and narcotics are difficult to manage, like balancing on a board perched on a rubber ball. Excessive infusion depresses respiratory drive and invites hypoxia, but breathing assistance is impossible without intubation. Inadequate infusion invites laryngospasm, unpredictable movements, and awareness.

- Fluctuating infusions invite tachyphylaxis (decreasing drug effectiveness), as well as nausea and vomiting.

- Capnography cannot measure exhaled carbon dioxide without intubation, so that it cannot warn of respiratory depression short of respiratory arrest.

- Like a vacuum cleaner nozzle, the unprotected airway can aspirate vomit, foreign objects, and dental debris, or suffer laryngospasm and airway obstruction.

- Urgent endotracheal intubation must deal with the dentist, dental debris, dental equipment, blood, saliva, vomit, fire, and laryngospasm.

Open airway practitioners employ counterproductive solutions for these problems:

- Dental anesthesia training emphasizes meticulous pre-operative examination to detect enlarged tonsils and anatomical airway abnormalities, but 100% detection is impossible. Such conditions are exceedingly dangerous without a secure airway (elective intubation).

- Practitioners flood the pharynx with oxygen to assure 100% hemoglobin saturation, but this disguises dangerous tissue hypoxia and CO_2 accumulation until it reaches a critical threshold.

- Flooding the pharynx with oxygen invites fire.[28, 29]

- Practitioners place cotton sponges over the laryngeal orifice to prevent aspiration of dental debris, but this provides fuel for fire.

- To prevent fire, practitioners wet the sponge with water, but the wet sponge can unpredictably obstruct the tracheal orifice

- Practitioners sample exhaled CO_2 using small tubes placed in the pharynx, but exhaled CO_2 in the open pharynx is mixed with atmospheric air and infused oxygen so that capnograph machines can only detect CO_2 and cannot measure it to detect impending respiratory insufficiency.

- Mucus and saliva can clog the small CO_2 sampling tubes and unexpectedly disrupt capnograph function altogether.

The open airway technique seems so convenient that board-certified anesthesiologists and nurse-anesthetists have embraced it for brief but deceptively dangerous hospital procedures such as endoscopies, which are often performed in small, dark rooms crowded with equipment and personnel that render patient access difficult, and anesthesia machines are absent. Sick, obese, and elderly patients are often placed in prone or lateral positions for prolonged periods of time with their airway occluded by endoscopy equipment that frustrates emergency airway management. These circumstances invite catastrophe and have resulted in multiple malpractice lawsuits.[30, 31, 32, 33, 34]

The practice guidelines of the American Society of Anesthesiologists, the Canadian Society of Anesthesiologists, and the International Standards for a Safe Practice of Anesthesia have progressively abandoned the traditional safety principles to accommodate fast tracking and the open airway technique in hospitals.[35, 36, 37, 38, 39, 40, 41] Their compromised guidelines are often rationalized on the assumption that infusion techniques can maintain patients in a safe state of semi-consciousness that sustains respiratory drive. This is a foolish fiction because semi-consciousness invites untoward movements, muscle tension, and laryngospasm. Effective infusion of synergistic mixtures of opioids and Propofol that prevents such problems unavoidably threatens respiratory drive, laryngospasm, aspiration, and hypoxia. Without elective endotracheal intubation (airway security), laryngospasm and aspiration cannot be prevented; CO_2 cannot be measured; and respiratory support cannot be rendered. The results have been disastrous. Open airway techniques are the underlying cause of numerous anesthetic fires, deaths, and malpractice lawsuits, most of which remain unpublicized for the sake of legal "privacy." One well-known example of this presently prevailing insanity the indefensible death of prominent comedienne Joan Rivers at the age of 80 while undergoing endoscopy and vocal cord examination in a Manhattan doctor's office.[42, 43] Numerous healthy children and teenagers have been killed under similar circumstances in both hospitals and dental offices.[44, 45, 46, 47, 48, 49, 50, 51] These tragedies are avoidable, and therefore inexcusable.[52]

Some dental anesthetists have recognized the dangers of the unsecured airway and devised clever ways to combine elective endotracheal intubation with intravenous infusions. This enables respiratory support and capnography but entails other compromises.

The technique of general endotracheal anesthesia utilized in this study can be adapted to dental offices using small, portable anesthesia machines and oxygen tanks to enable superior safety and equivalent convenience, efficiency, economy, and outcome as compared to open airway techniques. Indeed, I have used this technique myself in a dental office to provide general endotracheal anesthesia for a close acquaintance. *There is no advantage to the open airway technique than cannot be matched or exceeded by properly managed general endotracheal anesthesia.*

THE STUDY

Anesthesia is synonymous with stress control. Its purpose is to prevent stress mechanism hyperactivity induced by surgery, and thereby mitigate the surgical stress syndrome (SSS) that manifests as fever, laryngospasm, tachycardia, hypertension, dysrhythmias, atelectasis, pneumonia, delirium, dementia, heart attacks, strokes, tachypnea, respiratory arrest, and exaggerated pain (allodynia) in the immediate aftermath of surgery as well as cancer, heart disease, and chronic illnesses in the distant aftermath of surgery. It is generally agreed that tissue oxygenation is the gold standard of effective surgical stress control. This study kept track of induction and emergence times and measured the transcutaneous partial pressures of oxygen and carbon dioxide (TcO_2 & $TcCO_2$) in adult and pediatric dental surgery patients to confirm that supplementing general anesthesia with hypercarbia and narcotic analgesia improves tissue oxygenation and confers a method of general anesthesia that is safer, simpler, and at least as efficient as general anesthesia with mechanical hyperventilation and/or "open airway" techniques.

Furthermore, stress theory postulates that critical illnesses are likewise caused by excessive stress mechanism hyperactivity, so that measures that mitigate the SSS should also provide useful treatments for all forms of critical illness, including SIRS, SARS, COVID, MERS, ARDS, MOFS, eclampsia, asthma, pneumonia, and so forth. By confirming that the combination of general anesthesia, hypercarbia, and narcotic analgesia controls surgical stress and optimizes tissue oxygenation, the study establishes these treatments as a fundamental means to control stress mechanism hyperactivity, and pave the path for improved treatments and cures for all forms of disease.

HOW STRESS THEORY AND SIMPLICITY GUIDED STUDY DESIGN

The purposes of the study were as follows:

1. Refute the Leake-Waters hoax by demonstrating the safety and benefits of hypercarbia.
2. Demonstrate that eliminating intravenous fluids improves safety and saves money.
3. Demonstrate that properly managed narcotic supplementation is safe and beneficial.
4. Demonstrate that mask induction and elective endotracheal intubation with narcotic supplementation is as efficient as intravenous induction and elective intubation with paralysis as well as safer.
5. Demonstrate that complimentary combinations of hypercarbia and narcotic supplementation offers a convenient and safe means to optimize tissue oxygenation and organ protection during surgery that is compatible with fast tracking, equally economical and efficient as well as safer than open airway techniques.
6. Demonstrate that monitoring with capnography and pulse oximetry enables the safe management of beneficial hypercarbia.
7. Demonstrate that the patient's normal metabolic CO_2 production is sufficient to provide adequate hypercarbia that optimizes tissue oxygenation during anesthesia without the need for external CO_2 sources.

Foolproof anesthesia has yet to be invented, but the anesthetic technique utilized during this study was intended to provide the safest, simplest, and best surgical outcome possible with presently available medications, monitors, and equipment. It embraced the engineering principle that simplicity promotes safety, economy, and efficiency. This is known as the "KISS" principle ("Keep It Simple, Stupid!"). For example, Charles Lindberg insisted on a single-engine design for the Spirit of St. Louis because he knew that extra engines would exaggerate fuel consumption (undermine efficiency) and invite engine failure (undermine safety). In the study design, the KISS principle manifests as elimination of problematic infusion pump devices, elimination of polypharmacy that undermines predictability, and the removal of unnecessary soda lime and intravenous fluids. The study technique departed from the KISS principle in one respect: Sevoflurane and nitrous oxide facilitated mask induction because of their relatively inoffensive smell, but they were replaced by Isoflurane during anesthetic maintenance because of its superior chemical stability, potency, and low cost. Theoretically, it would have been preferable to utilize a single inhalation agent.

HOW THE STUDY UTILIZED STRESS THEORY

In addition to the KISS principle, the study embraced the principles of stress theory that are elaborated in this book. Stress theory identifies three synergistic pathways that activate the stress mechanism and induce the stress mechanism hyperactivity that causes the surgical stress syndrome (tissue disruption pathway, cognitive pathway, and nociception pathway). It also identifies three synergistic treatment modalities (anesthesia, analgesia, and hypercarbia) that inhibit harmful stress mechanism hyperactivity induced by surgery and thereby optimize surgical outcome.[53, 54] The following table presents a summary of the principles and practices that guided the study technique:

PRINCIPLES AND PRACTICES USED BY THE STUDY	CONSIDERATIONS
Anesthesia inhibits the cognitive pathway (consciousness) that activates harmful sympathetic nervous activity in accord with fear and anxiety. Sympathetic nervous system hyperactivity closes the capillary gate, increases microvascular flow resistance, and undermines cardiac efficiency, cardiac output, tissue perfusion, tissue oxygenation, and organ protection. Only one anesthetic (hypnotic) agent is needed, which minimizes polypharmacy.	*All hypnotic agents are toxic, so it is best to minimize their dosage to mitigate their undesirable effects. There are no advantages to doses greater than necessary to obtund consciousness.* Narcotic supplementation synergistically reduces hypnotic requirements in half (1/2 MAC)[55] Presently available inhalation agents include Desflurane, Isoforane, and Sevoflurane. Desflurane is unsuitable because its toxicity causes pulmonary inflammation that threatens bronchospasm.[56] Sevoflurane offers the least offensive smell. Isoforane offers the lowest cost, best potency, and chemical stability.[57]
Analgesia inhibits the nociception pathway that activates harmful sympathetic nervous activity. Sympathetic nervous system hyperactivity closes the capillary gate, increases microvascular flow resistance, and undermines cardiac efficiency, cardiac output, tissue perfusion, tissue oxygenation, and organ protection. Only one analgesic agent or technique is needed, which minimizes polypharmacy. Fentanyl was selected for the study because its half-life is sufficient to provide postoperative pain control.	Analgesia reduces inhalation agent requirements, which speeds induction and emergence and minimizes toxicity. It prevents nociception that induces hyperventilation, depletes CO_2 tissue reserves, and paralyzes respiratory chemoreceptors. It opens the capillary gate and promotes cardiac output and tissue perfusion. Synthetic narcotics are the most practical analgesics because they are devoid of toxicity, and produce dependable, renewable, dose-dependent inhibition of spinal cord nociception pathways. They reduce toxic inhalation agent requirements by half. They obtund the smell of inhalation agents. NSAIDs are notoriously toxic to tissues, and they undermine coagulation and tissue repair. Peripheral nerve blocks invite nerve damage, are time-consuming, unreliable, uncomfortable for patients, often require blocking multiple nerves, necessitate special training and experience, require expensive ultrasound equipment, and invite local analgesic systemic toxicity (LAST).

PRINCIPLES AND PRACTICES USED BY THE STUDY	CONSIDERATIONS
No satisfactory means to control the **tissue disruption pathway** is currently available. Theoretically, it should be possible to develop an antidote for tissue factor released into systemic circulation by surgery. Such a treatment could be administered soon after surgery to prevent positive feedback that exaggerates harmful stress mechanism hyperactivity.	Stress mechanism hyperactivity elevates thrombin generation that causes the surgical stress syndrome, eclampsia, and other critical illnesses. $MgSO_4$ mitigates eclampsia by competing with Ca+ to inhibit thrombin activity. Therefore, it should be possible to optimize surgical outcome by treating surgery patients with $MgSO_4$ in the immediate aftermath of surgery once hemostasis is assured. This is less ideal than a specific tissue factor antidote, but it might offer a means to complement the benefits of anesthesia, analgesia, and hypercarbia. EDTA, and tri-sodium citrate may offer more potent alternatives to $MgSO_4$, but these agents are potentially dangerous and must be managed carefully. They are best suited for treating critically ill patients in the ICU where intense monitoring is available. Theoretically they would be most effective if they were administered in modest doses within a few hours after surgery to disrupt positive feedback that induces stress mechanism hyperactivity.
Hypercarbia optimizes all elements of oxygen transport and delivery. Carbon dioxide is the primary regulator of the capillary gate. It directly releases nitric oxide from the capillary endothelium, reduces microvascular flow resistance, and optimizes cardiac output, cardiac efficiency, tissue perfusion, tissue oxygenation, and organ safety.	Hypercarbia stimulates respiratory chemoreceptors, optimizes respiratory drive, and counteracts narcotic respiratory depression. It releases oxygen from hemoglobin to optimize cellular oxygenation and prevent organ damage. It accelerates the metabolism and clearance of Fentanyl and other opioids in a dose-related fashion.[58] It accelerates inhalation agent induction and emergence.[59, 60] It prevents intraoperative heart attacks, strokes, and postoperative nausea, vomiting, atelectasis, pneumonia, and dementia.
Mask induction was a universal approach during the Golden Era of Ether, but it is largely ignored in present anesthesia training. It is safer than intravenous induction because it introduces inhalation agent toxicity gradually so that it can be safely aborted in the event of unexpected anatomical airway problems or dangerous toxic effects. It is equally efficient because blood flow transports inhalation agents rapidly to the brain.	Mask induction and elective intubation can be accelerated by supplementation with carbon dioxide and/or intravenous narcotics or sedatives to make it still faster. It eliminates the need for painful and time-wasting preoperative intravenous catheter installation, which can be accomplished more easily when patients are anesthetized. It eliminates the need for pre-oxygenation.

STUDY METHOD

The anesthetic methods employed in this study were in no way experimental. Every aspect of the study technique, including apneic oxygenation,[61, 62] permissive hypercarbia,[63, 64, 65, 66, 67, 68, 69, 70] narcotic supplementation,[71, 72, 73, 74, 75, 76, 77, 78] mask induction, and intubation without paralysis,[79, 80, 81, 82] was based on established theory, published procedures, research evidence, documented clinical experience of others, and the author's own experience with narcotic supplementation and hypercarbia for more than 20 years. The author's experience is documented by his preserved AIMS anesthetic records from Enloe Hospital in Chico, CA that may be reviewed at http://www.stressmechanism.com/. On the web site, follow the link to *50 Years Downloads*, and click the link for the Enloe anesthetic records.

MASK INDUCTION AND INTUBATION WITHOUT PARALYSIS

Mask induction was the universal method of anesthetic induction during the early days of modern anesthesia when primitive needle technology precluded intravenous access. It is safer than intravenous induction because it introduces the toxic effects of hypnotic inhalation agents gradually, and can be safely aborted in the event of unexpected difficulties. Elective intubation without paralysis is easily accomplished using modern inhalation agents alone, and can be accelerated by narcotic, hypnotic, or CO_2 supplementation.[83, 84] Narcotics can be readily reversed if necessary.

Using paralysis to enable elective intubation is inherently dangerous. It can cause a sudden relaxation of pharyngeal muscles that obscures the airway and causes "can't intubate, can't ventilate" predicaments that are analogous to the airway obstruction that occurs during Obstructive Sleep Apnea (OSA) that sometimes threatens anesthetic outcome (see Chapter 19: Obstructive Sleep Apnea, on page 387). Mask management was still regarded as an essential skill when I performed my residency at UCLA but now it is nearly abandoned and forgotten.

NARCOTIC SUPPLEMENTATION

Narcotics beneficially inhibit harmful nociception in spinal cord pathways. They prevent the harmful release of von Willebrand hormone and spinal cord "windup" syndrome. Narcotics inhibit respiratory chemoreceptors, which inhibits breathing, and this causes CO_2 accumulation that stimulates respiratory chemoreceptors and counteracts the respiratory depression of the narcotics. The net result is a decrease in respiratory rate and an increase in respiratory volume that sustains effective gas exchange in the lung. At the same time, the increased CO_2 optimizes tissue perfusion and oxygen release from the hemoglobin molecule to sustain tissue oxygenation. The effect is like normal sleep, which also causes a mild decrease in respiratory drive that is compensated by CO_2 accumulation that sustains tissue oxygenation.

One hundred years ago, Dr. George Washington Crile determined that maintaining respiratory rate between 8-12 breaths per minute during anesthesia using morphine supplementation prevents both harmful nociception and hyperventilation, and minimizes surgical morbidity and mortality.[85] This is the oldest technique of modern anesthesia, but its benefits are forgotten on account of the Leake-Waters hoax[86] (see Chapter 18). Dr. Herman Schwartz, one of the older professors at UCLA, taught me to supplement Halothane (an inhalation agent no longer used) with Demerol (a narcotic that is also no longer used). I was impressed by the improved outcome, but the technique could only be used for patients breathing spontaneously. The study technique used newer agents (Isoflurane and Fentanyl) that offer improved characteristics.

The benefits of narcotic supplementation were re-discovered by Lowenstein in the 1960s,[87, 88] which drastically reduced the morbidity and mortality caused by the heart-lung bypass machine, and they were re-re-discovered for non-cardiac anesthesia in the 1980s after the Janssen Corporation introduced its new generation of short-acting synthetic narcotics.[89, 90, 91, 92, 93, 94]

APNEIC OXYGENATION

In 1959 Frumin & Epstein published classic apneic oxygenation studies by anesthetizing, paralyzing, and intubating anesthesiology resident volunteers and connecting their endotracheal tubes to anesthesia machines flushed with 100% oxygen. The paralyzed volunteers could not breathe, but their blood gas measurements documented that their arterial oxygen saturation remained at or near 100%, while CO_2 levels in their arterial blood harmlessly climbed as high as 250 mm Hg (which provided clear evidence of the lack of CO_2 toxicity).[95, 96] In 1961 Eger and Severinghaus duplicated these conditions to measure the rate of rise of CO_2.[97] Animal studies and human accident reports have documented that CO_2 elevations at least as high as 500 torr are harmless so long as hemoglobin oxygen saturation and tissue oxygen delivery are sustained.[98, 99, 100, 101, 102]

My study used the principle of apneic oxygenation to safely maintain oxygenation after patients were treated with bolus doses of Fentanyl to facilitate elective endotracheal intubation and eliminate the need for dangerous paralysis.

PERMISSIVE HYPERCARBIA

During anesthesia, hypercarbia supercharges the mechanism of oxygen transport and delivery during anesthesia when patients are breathing gas mixtures enriched with oxygen, which assures effective saturation of hemoglobin with oxygen as blood traverses the lung. Hypercarbia opens the capillary gate, reduces microvascular flow resistance, optimizes cardiac output and oxygen transport from the lungs to capillary beds, and releases oxygen from the hemoglobin molecule into tissues to optimize tissue oxygenation and prevent hypoxic organ damage due to surgical stress. The therapeutic benefits of carbon dioxide were first described by Dr. Yandell Henderson at the turn of the previous century (see Chapter 18) and were more recently re-discovered by critical care specialists and anesthesiology researchers in the 1990s.[103, 104, 105, 106, 107, 108, 109, 110, 111, 112, 113, 114, 115]

Historically, the first serious problem attributed to anesthesia was unexplained respiratory arrest after seemingly uneventful recovery from uneventful surgery. This problem was caused by spontaneous hyperventilation (that is, the patient naturally hyperventilated) induced by uncontrolled surgical nociception. The hyperventilation depleted CO_2 tissue reserves and undermined respiratory chemoreceptor function (see Chapter 14: Oxygen Transport and Delivery, on page 137). Hyperventilation and CO_2 depletion paralyzes respiratory chemoreceptors that are essential to sustain breathing when the primary respiratory drive of consciousness is extinguished by sleep. Thus, if hyperventilated patients fall asleep while they are recovering from surgery, they may stop breathing and die because their respiratory chemoreceptors remain paralyzed by CO_2 depletion.

Dr. Yandell Henderson investigated this phenomenon, determined its cause, and recommended that patients breathe 5% Carbogen (a mixture of 5% carbon dioxide and 95% oxygen in a pressurized tank) during general anesthesia to prevent lethal CO_2 depletion due to spontaneous hyperventilation. This not only abolished the postoperative respiratory arrests, but also reduced intraoperative morbidity and mortality due to CO_2 depletion.[116, 117, 118]

NURSE ANESTHESIA AND CO_2 MANAGEMENT

In the aftermath of WWI, when physicians were in short supply, Dr. George Washington Crile helped Agatha Hudgens, his nurse-anesthetist, to found a school of nurse-anesthesia in Cleveland, Ohio. The nurses combined Crile's narcotic supplementation with Henderson's and Lundy's CO_2 supplementation.[119] They supplemented ether anesthesia with intramuscular morphine and CO_2 from external tanks. They became famous for their excellent anesthetic outcomes. The morphine supplementation prevented harmful nociception, and the CO_2 supplementation counteracted morphine respiratory depression and optimized cardiorespiratory function, but overenthusiastic CO_2 supplementation from external tanks using Donald Jackson's original closed circuit anesthesia machines occasionally caused hypoxic brain convulsions due to CO_2 asphyxiation. This problem caused the modern misunderstanding that carbon dioxide has harmful toxic and narcotic properties (see Chapter 18).

Yandell Henderson described how the nurses influenced doctors to embrace the therapeutic properties of carbon dioxide:

"What I could not achieve by the first fifteen years of direct advocacy has now been won by a maneuver as old as the Trojan horse. This success is the result of another fifteen years in which, avoiding theoretical discussion, I have devoted my efforts mainly along practical lines. I stopped arguing with physiologists and made friends with the anesthetists. We first introduced, as part of the technique of anesthesia, the use of a rebreathing bag to conserve the body's store of carbon dioxide. Then gradually we introduced a mixture of oxygen and carbon dioxide as part of the equipment of the anesthetist. And now cylinders of carbon dioxide, either mixed or pure, are a feature of nearly all operating rooms. And in this way the acapnia

theory has captured the citadel of surgery, much as Ulysses and the Greeks captured Troy. They used a wooden horse. The acapnia theory won its entrance in those cylinders of carbon dioxide. And since this almost unnoticed victory, deaths from failure of respiration under anesthesia-once common have almost ceased to occur."[120]

CO_2 MANAGEMENT IN THE PRESENT STUDY

Unlike the nurse-anesthesia technique, the present study did not employ CO_2 supplementation from external tanks. Instead, beneficial hypercarbia was maintained using narcotic supplementation that prevented stress and reduced respiration to enable beneficial CO_2 accumulation that preserved respiratory drive and improved tissue perfusion and oxygenation. A similar effect occurs during normal sleep, which extinguishes the primary respiratory drive of consciousness. With this approach, there is no danger of CO_2 asphyxiation, because modern "open circuit" anesthesia machines continuously flush the breathing circuit with fresh gases.[121] In addition, capnography monitors end-tidal CO_2 and pulse oximetry monitors hemoglobin saturation.

I used a front-loading dose of Fentanyl which facilitated elective endotracheal intubation by inducing vocal cord relaxation. The Fentanyl halted respiration for 5-15 minutes, during which time tissue oxygenation was safely sustained by apneic oxygenation.[122, 123, 124] During this time, CO_2 accumulation from cellular metabolism stimulated respiratory chemoreceptors, counteracted the Fentanyl, and restored respiration, whereupon patients were maintained with ½ MAC Isoflurane (because Fentanyl reduces Isoflurane requirements by half). This minimized Isoflurane toxicity and accelerated anesthetic emergence at the conclusion of surgery.[125, 126, 127, 128]. The combined effects of anesthesia, analgesia and hypercarbia optimized cardiorespiratory function, tissue oxygenation, and organ protection. Additional Fentanyl dosage maintained respiratory rate between 8-12 breaths per minute in adults and 12-16 breaths per minute in children to optimize outcome.[129]

The notion that carbon dioxide has toxic and narcotic properties is an insipid myth, as explained in Chapter 18. The therapeutic benefits of carbon dioxide were discovered by Yandell Henderson 100 years ago.[130, 131, 132, 133, 134, 135] They were rediscovered by Lowenstein in the 1960s.[136, 137] They were re-rediscovered in the 1990s by critical care and anesthesia researchers.[138, 139, 140, 141, 142, 143, 144, 145, 146, 147, 148, 149, 150, 151, 152, 153]

Carbon dioxide is chemically inert and devoid of either toxicity or anesthetic effects.[154] When mixed with oxygen, it cannot cause asphyxiation unless its concentration exceeds 30% (more than 220 mm Hg at sea level where this study was conducted), which is far above levels approached during this study (116 mm Hg). Numerous reports and studies have demonstrated that hypercarbia is harmless as long as hemoglobin remains saturated with oxygen.[155, 156, 157, 158, 159] Asphyxiating elevations of carbon dioxide are impossible with modern anesthesia machines that continuously introduce fresh gases into the breathing circuit.[160] There have been no reports of CO_2 asphyxiation since these machines were introduced 70 years ago. In addition, patients were monitored with pulse oximetry and capnography to assure adequate saturation of hemoglobin with oxygen and to maintain CO_2 elevations within a therapeutic range (60-200 torr).

Capnography and pulse oximetry are now standard in operating rooms in North America and Europe. Pulse oximetry provides warning if hypercarbia disrupts hemoglobin saturation in the lung. Capnography provides warning of asphyxiating CO_2 elevations in the breathing circuit (which is impossible with open circuit anesthesia machines). There is no need for CO_2 supplementation from external sources; cellular metabolism generates adequate carbon dioxide to enable the benefits of therapeutic hypercarbia so long as mechanical hyperventilation is avoided.

TRANSCUTANEOUS OXYGEN AND CARBON DIOXIDE MONITORING

Transcutaneous oxygen and CO_2 (TcO_2 and $TcCO_2$) monitoring was used during the study to prove that the combined effects of hypercarbia and narcotic supplementation enhance tissue oxygenation during anesthesia. When I was a resident, pulse oximetry (SpO_2) and transcutaneous oxygen monitoring (TcO_2) were regarded as competing candidates for monitoring oxygen transport and delivery during anesthesia. Both have advantages and shortcomings.

Pulse oximetry monitors hemoglobin saturation with oxygen in red blood cells. It uses an algorithm that interprets light frequencies reflected from blood. Pulse oximetry became a standard of anesthesia practice because it is reliable and easy to use. However, few practitioners realize that pulse oximetry is "fooled" by the prevailing practice of mechanical hyperventilation, so that it provides false assurance of effective tissue oxygenation. The near-universal practice of mechanical hyperventilation supersaturates blood plasma with oxygen and causes the pulse oximeter to read 100% even while it depletes CO_2 tissue reserves.

CO_2 is necessary for the release of oxygen from hemoglobin to the tissues. Hyperventilation thus undermines the release of oxygen from hemoglobin and causes covert hypoxia in organs and tissues.[161] This misleading phenomenon supports and promotes the Leake-Waters hoax. (See Chapter 18: The Great Medical Hoax of the 20th Century, on page 277.)

Transcutaneous oxygen monitoring is theoretically superior to pulse oximetry because it measures the partial pressure of oxygen in tissues. This provides a direct assessment of tissue oxygenation during anesthesia. During my residency, the primitive transcutaneous oxygen monitors of the time could not be calibrated to measure oxygen partial pressure. Instead, they produced a numerical reading that reflected the combined effects of tissue perfusion and oxygen release from hemoglobin. As a result, they seemed relatively useless and were soon forgotten, while pulse oximetry became a standard monitor in the operating room. However, the transcutaneous technology has improved in subsequent years. It can now be calibrated so that it continuously monitors and measures the partial pressures of both carbon dioxide and oxygen in subcutaneous tissues. This is more meaningful than the hemoglobin saturation in blood that is measured by the pulse oximeter.

The Perimed representative who demonstrated the transcutaneous oxygen monitor also demonstrated a new Doppler device that detects and measures microvascular perfusion. It confirmed the prediction of stress theory that my anesthetic technique enhances microvascular flow as well as tissue oxygenation, but adding this feature would have entailed considerable expense. Even so, the used machine I purchased cost $13,000.00, not counting the expensive disposable supplies it required for each patient. I decided that the additional feature was not essential for my purposes. In retrospect, I regret that I didn't purchase the Doppler feature.

Although the transcutaneous monitoring technology has improved since my training, it remains "fussy" and impractical for routine use in the operating room. It required the following measures:

- Careful cleaning of the monitoring probe.
- Wetting the probe with special saline mixture.
- Covering the wetted probe with two thin membranes.
- Calibrating the probe three times using a standardized mixture of oxygen and carbon dioxide.
- Anesthetizing the patient. (The sensor could not be attached to pediatric patients while they were conscious.)
- Cleaning the patient's skin to remove oils. (I always installed the probe on the right arm.)
- Attaching a small disposable adhesive plastic cup on the skin. (They could only be used once.)
- Dripping saline solution into the plastic cup.
- Attaching the calibrated sensor to the plastic cup.
- Securing the plastic cup and monitoring cable.
- Initiating monitor operation.
- Securing the sensor to a special protective mount between cases to prevent drying of the membrane.
- Replacing the membrane and re-calibrating the machine at regular intervals.

Managing the transcutaneous monitor required practice, but I was eventually able to achieve consistent readings. The partial pressure of oxygen consistently climbed as oxygen accumulated in tissues faster than cells consumed it. This proved that the technique protects tissues and organs from surgical stress.

Other studies have used invasive monitors to demonstrate that either hypercarbia or analgesia improves tissue oxygenation, but no current studies have demonstrated the combined benefits of hypercarbia and narcotic analgesia.[162, 163, 164, 165] It should not be surprising that combinations of hypercarbia and narcotic analgesia provide a practical means to optimize tissue oxygenation and surgical outcome, because this capability has been understood for 100 years and was universally utilized for countless patients during the Golden Era of Ether.[166, 167]

Unfortunately, the TcO_2/$TcCO_2$ sensor cable proved to be fragile, and the CO_2 sensing capability failed during the study. Nevertheless, I was able to capture enough information to prove that the combination of hypercarbia and narcotic supplementation progressively increases the partial pressure of oxygen and carbon dioxide in subcutaneous tissues. This confirmed the ability of the anesthetic technique to prevent the tissue hypoxia caused by surgical stress. Copies of anesthetic records that display the TcO_2 and $TcCO_2$ data can be downloaded from:

http://www.stressmechanism.com/

On the web site, follow the link to *50 Years Downloads*, and click the link for TcO_2 and $TcCO_2$ anesthetic records. My published papers and stress theory slideshow presentations also are there.

SODA LIME

Soda lime was removed from the anesthesia machine during the study, because it confers no benefit, invites potential dangers, and introduces an unnecessary variable.[168, 169, 170, 171] It is a dangerous relic from a forgotten past, and should be banished from anesthesia machines.

INTRAVENOUS FLUIDS

During the study, intravenous access was established to enable drug administration but then "capped off" so that no intravenous fluids were administered. This eliminated tubing that can entangle feet, trip operating room staff, and yank the intravenous catheter out of its vein.

Intravenous fluids are potentially harmful, and are a waste of money except in rare conditions such as cholera. Thousands of invasive surgeries were accomplished without intravenous fluids during the Golden Era of Ether when primitive needle technology precluded intravenous access. A century ago, George Washington Crile performed extensive animal experiments and determined that intravenous fluids cause the bowel to swell like a sausage, disrupt surgical bowel repairs, and interfere with breathing by pressing on the diaphragm from below. They also undermine blood coagulability, exaggerate surgical blood loss, and disrupt the glycocalyx that lines the inner surface of blood vessels.[172]

SHORTCOMINGS OF THE STUDY

I was obliged to pay for all expenses of the study, and the small staff at the dental clinic was unavailable to help with the record keeping tasks. The clinic owner insisted that I maintain paper anesthetic records in addition to the AIMS records. This added to my ordinary anesthesia responsibilities and hampered my ability to collect study data using an Excel spreadsheet on my portable computer. Fortunately, Dr. Francisco Jimenez Riera, a Spanish anesthesiology resident, discovered my published papers and traveled from Spain to observe my practice in the clinic for a month during the TcO_2/$TcCO_2$ data collection. His assistance proved invaluable.

I recorded the total dose of Fentanyl received by each patient during the study, but lacked the time to record the loading dose at the beginning of the case separately from supplementary doses administered later. I didn't document the duration of apneic oxygenation. This crippled the clarification of the relationship between Fentanyl dosage and the duration of apneic oxygenation. Also, I didn't document the precise time when patients were prepared for surgery in addition to noting the incision time. By using the stopwatch function of my wristwatch, I did note that most patients were ready for surgery within five minutes after they entered the operating room, and the duration of apneic oxygenation varied from five to fifteen minutes; but I was unable to capture this data and report it because of multiple duties at the outset of each case. However, the duration of apneic oxygenation is reflected by the printed anesthetic records that I retained to illustrate the TcO_2 and $TcCO_2$ readings.

Anesthetic records showing TcO_2 and $TcCO_2$ are available for review at http://www.stressmechanism.com/. On the website, click the link for 50 Years Downloads, and then click the link for the TcO_2 and $TcCO_2$ anesthetic records.

STUDY DETAILS

With the approval of the Protection of Human Subjects and patient consent, approximately 1500 ASA class 1–3 pediatric and adult patients presented for ambulatory dental surgery under general anesthesia with elective nasotracheal intubation administered by the author.

The following measures applied to all cases:

- All patients were evaluated by the clinic nurse and the anesthesiologist prior to surgery, and the procedures and risks were explained.
- Each patient or caregiver completed a standard pre-operative questionnaire and instruction sheet that included a consent to publish data.
- Patients with BMI (body mass index) greater than 35 were referred to a hospital, and patients with class III or worse medical problems were referred to a hospital.

- All patients and their caregivers were specifically queried to determine if they persistently snored while sleeping or had any signs of obstructive sleep apnea. See Chapter 19: Obstructive Sleep Apnea, on page 387. All such patients were referred to Rades Hospital for anesthesia care.

- Patients who admitted eating or drinking were re-scheduled.

- The mask induction procedure was described to patients as pleasant and safe, and they were instructed to breathe normally during the induction. They were informed that painful intravenous access and intubation would take place while they were unconscious. None objected, and none complained afterward.

- Soda lime was removed from the anesthesia machine during the study because it confers no benefits, introduces unnecessary risk, adds unnecessary expense, produces toxic combustion products by chemically reacting with inhalation agents, adds an unnecessary variable that might perturb study results, and harmfully extracts carbon dioxide from the breathing circuit. It is a dangerous relic that has no place in modern anesthesia machines.

- All patients visited the bathroom immediately before surgery.

- All breeding age females were screened for pregnancy with urine tests.

- No pre-medication was used because it added an unnecessary variable to the study, and because combining benzodiazepines with narcotics causes respiratory depression and invites obstructive sleep apnea.[173, 174, 175]

- Prior to the study we had determined that Decadron has no effect on nausea and vomiting with this technique, so it was not used during the study. During the Golden Era of Ether hypercarbia was found to prevent nausea and vomiting.

- All able adult patients walked to the operating room in their civilian clothes, climbed onto the operating table without assistance, and were secured to the operating table with safety straps prior to induction.

- Autistic patients were a special challenge because they were invariably frightened and uncooperative. Their parents and caregivers were a great help. In most cases it was necessary to perform mask induction with patients sitting on the side of the operating table with parents and caregivers on either side, and then carefully lowering them to the operating table during mask induction.

- For the sake of safety, wheelchair-bound, mentally handicapped patients were sometimes anesthetized in their wheelchairs and then transferred to the operating table.

- Parents and caregivers routinely accompanied pediatric patients into the operating room and either held children in their lap or placed them on the operating table and comforted them during mask induction.

- EKG, blood pressure cuff, and pulse oximeter were routinely applied prior to induction in adults, but for small children and mentally handicapped patients these monitors were applied as soon as practical during or after mask induction.

- All routine monitoring data, including blood pressure, heart rate, CO_2 capnography readings, and, pulse oximeter readings were automatically captured and recorded in the anesthetic record by AIMS (automated information management system) technology.

- Dr. Lewis Coleman directly supervised the cases managed by Dr. Francisco Jimenez Riera and performed the rest himself to assure that the same technique was employed for all patients.

- All patients were covered with blankets to maintain body warmth, but no heating was applied. Postoperative shivering was seldom problematic.

- Mask induction was employed for all patients.

- Mask induction began with 70% nitrous oxide and 30% oxygen, followed by gradually increasing Sevoflurane ("steal induction") to a maximum of 8%.

- Nitrous oxide was shut off as soon as patients lost consciousness.

- Intravenous access was established by the nurse or the anesthesiologist as soon as patients were adequately anesthetized. Patients were then treated with a bolus dose of 50–250 micrograms Fentanyl IV to optimize nasotracheal intubation, provide pre-emptive analgesia, and suffice for the remainder of the surgery.

- Intravenous catheters were capped with a "pigtail." As noted above, this eliminated hazardous IV tubing in accord with the KISS principle.

- No intravenous fluids were used because they are potentially harmful, expensive, and confer no known benefits in the context of elective surgery.

- Patients stopped breathing as Fentanyl took effect over 3–5 minutes because the Fentanyl temporarily obtunded their respiratory chemoreceptors. They were then ventilated by bag and mask with 8% Sevoflurane overpressure to optimize vocal cord relaxation.

- The induction dose of Fentanyl was estimated to be adequate for intubation and for controlling nociception for the duration of the procedure. The Fentanyl was supplemented when necessary to control respiratory rate within study parameters. The average dose of Fentanyl was about 10mcg/kg, which is the "stress control" dose recommended by the manufacturer.

- The synergistic combination of Fentanyl analgesia and Sevoflurane anesthesia consistently produced vocal cord relaxation equivalent to paralysis but without the safety hazards entailed by paralysis.

- Narcan and Anectine were always readily available but I have never needed either of them in thousands of cases in many years.

- Atropine and epinephrine also were available immediately in case of severe bradycardia and hypotension but atropine was needed only once during the study. See http://www.stressmechanism.com/. On the website, click the link for *50 Years Downloads*, and then click the link for the TcO_2 and $TcCO_2$ anesthetic records. In that PDF, see MRN #24937. In my experience, most patients who were hypersensitive to Fentanyl were bedridden, mentally handicapped individuals. Their chronic inactivity seemed to be the cause of the hypersensitivity.

- Elective nasotracheal intubation was performed as soon as airway relaxation was judged adequate.

- Isoflurane 1% replaced Sevoflurane as soon as patients were successfully intubated.

- The synergistic interaction of anesthesia and analgesia abolished respiratory drive for 5–15 minutes, during which time intubated patients were maintained with apneic oxygenation until rising CO_2 levels in the tissues restored respiratory drive, whereupon spontaneous breathing resumed, and respiratory rate gradually increased as Fentanyl clearance proceeded.

- Maximum end-tidal CO_2 never rose above 116 torr before the resumption of spontaneous breathing, and this was manually recorded in the study spreadsheet because the AIMS software recorded Et CO_2 readings at five-minute intervals and seldom captured the highest readings.

- Peak end-tidal CO_2 levels rapidly declined with the resumption of spontaneous breathing.

- Isoforane was limited to 1% (1/2 MAC) to minimize toxicity and speed emergence.

- The initial dose of Fentanyl was supplemented when necessary to maintain respiratory rate between 8–12 breaths per minute in adults and 10–15 breaths per minute in children to control harmful nociception, prevent CO_2 depletion, and minimize Isoforane toxicity.

- Dentists often injected lidocaine at the surgery site to facilitate analgesia, but their results were inconsistent.

- The Isoforane vaporizer was shut off at or shortly before the conclusion of surgery, and many patients regained consciousness and were extubated soon after the conclusion of surgery.

- All patients were treated with 4mg Ondansetron at the conclusion of surgery, as recommended by the anesthetic gas manufacturer, to minimize postoperative nausea and vomiting. Unlike Decadron, we found this helpful.

- The cessation of surgical nociception often caused temporary apnea, so patients were left intubated and supplemented with oxygen until CO_2 accumulation from normal metabolism restored respiratory drive.

- Patients were turned on their side at the conclusion of surgery to drain blood from the pharynx. Nevertheless, most cases of nausea and vomiting were due to swallowed blood, which is a harmless and unavoidable hazard in all dental and ENT procedures.

- Patients were either recovered in the operating room or transferred to the adjacent recovery room via gurney, still intubated, where they recovered under nursing supervision.

- Narcotized patients tolerated the nasotracheal tube and were left undisturbed until they spontaneously regained consciousness as the effects of Isoflurane dissipated.

- Small infants often slept soundly, which prolonged emergence.

- Patients invariably emerged from anesthesia clear-headed, fully oriented, and free of confusion, delirium, dementia, tachycardia, hypertension, fever, dysrhythmias, and pain. They maintained their airways without help, didn't need pharyngeal suction, and were immediately able to stand and walk.

- Even though patients were able to stand and walk, they were always transported by gurney or wheelchair for the sake of safety.

- Small children typically cried on emergence but regained their composure with the arrival of patients and caregivers, which reflected effective Fentanyl dosage.

- Laryngospasm was rare after extubation and was managed with Ambu bag and mask oxygen.

- Carbogen (a mixture of 95% oxygen and 5% carbon dioxide in a pressurized tank) would have been helpful to supplement CO_2 tissue reserves, stimulate breathing, speed emergence, and treat laryngospasm, but it was not available in the clinic.

- All patients were evaluated by the nurse in accord with established protocols and procedures before being discharged from the clinic.

- Parents and caregivers were instructed to take their children home, allow them to have a nap, watch for any signs of breathing problems, and call ambulance services immediately in the event of problems.

- There were no instances of unexpected postoperative respiratory arrest during the study.

- Parents consistently reported that their children arose from their nap and ran off to play as if they never had surgery.

DATA

Data collected during the study can be viewed and downloaded from my website: http://www.stressmechanism.com/. On the website, click the link for *50 Years Downloads*, and then click the link for the clinical study of anesthesia. The study data is summarized in this PDF spreadsheet:

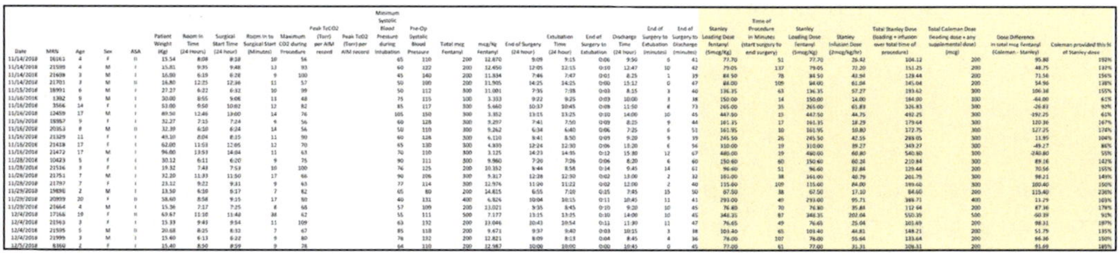

Date	MRN	Age	Sex	ASA	Patient Weight (Kg)	Room In Time (24 Hours)	Surgical Start Time (24 hour)	Room In to Surgical Start (Minutes)	Maximum CO2 during Procedure	Peak TcCO2 (Torr) per AIM record	Peak TcO2 (Torr) per AIM record	Minimum Systolic Blood Pressure during Intubation	Pre-Op Systolic Blood Pressure	Total mcg Fentanyl	mcg/Kg Fentanyl	End of Surgery (24 hour)	Extubation Time (24 hour)	End of Surgery to Extubation	Discharge Time (24 hour)	End of Surgery to Extubation (minutes)	End of Surgery to Discharge (minutes)	Stanley Loading Dose fentanyl (5mcg/Kg)	Time of Procedure in Minutes (start surgery to end surgery)	Stanley Loading Dose Fentanyl (5mcg/Kg)	Stanley Infusion Dose (2mcg/kg/hr)	Total Stanley Dose (loading + infusion over total time of procedure)	Total Coleman Dose (loading dose + any supplemental dose) (mcg)	Dose Difference in total mcg fentanyl (Coleman - Stanley)	Coleman provided this % of Stanley dose
11/14/2018	16161	4	F	II	15.54	8:08	8:18	10	56			65	110	200	12.870	9:09	9:15	0:06	9:50	6	41	77.70	51	77.70	26.42	104.12	200	95.88	192%
11/14/2018	21599	4	M	I	15.81	9:35	9:48	13	93			60	122	200	12.650	12:05	12:15	0:10	12:47	10	42	79.05	137	79.05	72.20	151.25	200	48.75	132%
11/14/2018	21608	3	M	I	16.90	6:19	6:28	9	100			45	140	200	11.834	7:46	7:47	0:01	8:25	1	39	84.50	78	84.50	43.94	128.44	200	71.56	156%
11/14/2018	21701	3	M	II	16.80	12:25	12:36	11	57			50	100	200	11.905	14:25	14:25	0:00	15:12	0	47	84.00	109	84.00	61.04	145.04	200	54.96	138%
11/15/2018	18991	6	M	I	27.27	6:22	6:32	10	99			50	112	300	11.001	7:35	7:38	0:03	8:15	3	40	136.35	63	136.35	57.27	193.62	300	106.38	155%
11/16/2018	1382	9	M	I	30.00	8:55	9:06	11	48			75	115	100	3.333	9:22	9:25	0:03	10:00	3	38	150.00	14	150.00	14.00	164.00	100	-64.00	61%
11/16/2018	3566	14	F	I	53.00	9:50	10:02	12	82			85	117	300	5.660	10:37	10:45	0:08	11:50	8	73	265.00	35	265.00	61.83	326.83	300	-26.83	92%
11/16/2018	12459	17	M	I	89.50	12:46	13:00	14	76			105	150	300	3.352	13:15	13:25	0:10	14:00	10	45	447.50	15	447.50	44.75	492.25	300	-192.25	61%
11/16/2018	19957	9	F	I	32.27	7:15	7:24	9	56			60	128	300	9.297	7:41	7:50	0:09	8:25	9	44	161.35	17	161.35	18.29	179.64	300	120.36	167%
11/16/2018	20353	8	M	II	32.39	6:10	6:24	14	56			50	110	300	9.262	6:34	6:40	0:06	7:25	6	51	161.95	10	161.95	10.80	172.75	300	127.25	174%
11/16/2018	21329	11	F	I	49.10	8:04	8:15	11	90			60	128	300	6.110	8:41	8:50	0:09	9:20	9	39	245.50	26	245.50	42.55	288.05	300	11.95	104%
11/16/2018	21418	17	F	I	62.00	11:53	12:05	12	70			65	130	300	4.839	12:24	12:30	0:06	13:20	6	56	310.00	19	310.00	39.27	349.27	300	-49.27	86%
11/16/2018	21472	17	M	I	96.00	13:53	14:04	11	63			70	110	300	3.125	14:23	14:35	0:12	15:30	12	67	480.00	19	480.00	60.80	540.80	300	-240.80	55%
11/28/2018	10423	5	F	I	30.12	6:11	6:20	9	75			90	111	300	9.960	7:20	7:26	0:06	8:20	6	60	150.60	60	150.60	60.24	210.84	300	89.16	142%
11/28/2018	21536	3	F	I	19.32	7:43	7:53	10	100			76	125	200	10.352	8:44	8:58	0:14	9:45	14	61	96.60	51	96.60	32.84	129.44	200	70.56	155%
11/28/2018	21751	7	M	I	32.20	11:33	11:50	17	66			90	106	300	9.317	12:28	12:30	0:02	13:00	2	32	161.00	38	161.00	40.79	201.79	300	98.21	149%
11/28/2018	21797	7	F	I	23.12	9:22	9:31	9	63			77	114	300	12.976	11:20	11:22	0:02	12:00	2	40	115.60	109	115.60	84.00	199.60	300	100.40	150%
11/29/2018	19898	2	M	I	13.50	6:10	6:17	7	82			65	80	200	14.815	6:55	7:10	0:15	7:45	15	50	67.50	38	67.50	17.10	84.60	200	115.40	236%
11/29/2018	20909	20	F	II	58.60	8:58	9:15	17	80			40	131	400	6.826	10:04	10:15	0:11	10:45	11	41	293.00	49	293.00	95.71	388.71	400	11.29	103%
11/29/2018	21664	4	M	I	15.36	7:17	7:25	8	68			57	109	200	13.021	8:35	8:45	0:10	9:20	10	45	76.80	70	76.80	35.84	112.64	200	87.36	179%
12/4/2018	17166	19	F	II	69.67	11:10	11:48	38	62			55	111	500	7.177	13:15	13:25	0:10	14:00	10	45	348.35	87	348.35	202.04	550.39	500	-50.39	91%
12/4/2018	21563	3	F	I	15.33	9:43	9:54	11	109			63	132	200	13.046	10:43	10:54	0:11	11:30	11	47	76.65	49	76.65	25.04	101.69	200	98.31	197%
12/4/2018	21595	5	M	II	20.68	8:25	8:32	7	67			85	118	200	9.671	9:37	9:40	0:03	10:15	3	38	103.40	65	103.40	44.81	148.21	200	51.79	135%
12/4/2018	21999	3	M	I	15.60	6:13	6:22	9	80			76	132	200	12.821	8:09	8:13	0:04	8:45	4	36	78.00	107	78.00	55.64	133.64	200	66.36	150%
12/5/2018	8360	2	F	I	15.40	8:50	8:59	9	78			64	110	200	12.987	10:00	10:00	0:00	10:45	0	45	77.00	61	77.00	31.31	108.31	200	91.69	185%

Figure J.1: A section of the data spreadsheet.

A PDF of the summary data spreadsheet for over 1500 cases is available for download at http://www.stressmechanism.com/. The yellow columns compare Dr. Coleman's technique to the published technique of Dr. Ted Stanley.

Other data available at Dr. Coleman's website includes:

- Copies of anesthetic records of patients monitored with TcO_2/$TcCO_2$ technology.

- Copies of AIMS anesthetic records from 2007 from Enloe Hospital in Chico, California, documenting my earlier management of hospital cases using similar hypercarbia and narcotic supplementation.
- Three anesthetic records that demonstrate the technique of "Dr. G," who specializes in dental anesthesia. His records illustrate the priority of practicality that dominates ordinary anesthesia practice.

DETAILED DISCUSSION

This section discusses the information derived from the study in the form of graphs and statistics. PDFs of this information and graphs are also available for download from http://www.stressmechanism.com/. On the website, click the link for *50 Years Downloads*, and then click the link for the clinical study of anesthesia.

Overview Statistics

	Age	Kgs	Room In to Incision (mins)	End-Tidal CO2 Max	Systolic Blood Pressure (Min)	Total mcg Fentanyl	mcg/Kg Fentanyl	End of Surgery to Extubation (time in minutes)	End of Surgery to Discharge (time in minutes)	Note
Average	8.88	35.44	11.23	88.40	75.50	274.269	9.997	13.29	50.35	Time of discharge available for 406 cases
Median	6.00	23.77	11.00	88.00	74.00	250.000	10.081	11.00	46.00	

Figure J.2: Overview statistics from the study.

This figure provides averages and medians for study parameters that illustrate the efficiency of the technique. All patients were treated with a loading dose (not shown) of 50–250 mcg Fentanyl to facilitate elective nasotracheal intubation, followed by additional Fentanyl supplementation as needed to maintain respiratory rate within study parameters. The initial bolus of Fentanyl produced a period of 5–15 minutes of apneic oxygenation until metabolic activity elevated CO_2 tissue reserves enough to re-activate respiratory chemoreceptors and restore spontaneous breathing.

Room In to Incision shows the time elapsed from the moment the patients set foot in the operating room until incision. **End-TidalCO_2 Max** is the reading of the capnograph when patients resumed breathing spontaneously. **Systolic Blood Pressure (Min)** represents the average lowest systolic blood pressure during intubation due to the combined effects of Sevoflurane and Fentanyl. **Total mcg Fentanyl** is the dose of Fentanyl treatment per patient. **mcg/Kg Fentanyl** is the dose of Fentanyl per unit weight. Ten mcg/kg is the manufacturer's recommended dose of Fentanyl for effective surgical stress control. **End of Surgery to Extubation** (time in minutes) is the time from the end of surgery to emergence and extubation. **End of Surgery to Discharge** (time in minutes) is the time from the end of surgery until patients were sent home. This latter was recorded for only the first 406 cases because this was deemed adequate to illustrate the efficient discharge of patients. Efficient turnover enabled the clinic to manage up to ten cases per day with only one surgery suite and a single recovery room.

INDUCTION DETAILS

Mask induction is safer than intravenous induction because it takes effect gradually and can be safely aborted, and equally efficient because inhalation agents are quickly absorbed in the lung and transported directly to the brain.[176] It saves time by eliminating pre-oxygenation. It enables intubation without paralysis. It can be made even faster with the addition of narcotic, hypnotic, and CO_2 supplementation.[177, 178, 179, 180] The study used narcotic (Fentanyl) supplementation to optimize vocal cord relaxation.

During this study, mask induction began with 70% nitrous oxide and 30% oxygen, because nitrous oxide is pleasant and devoid of smell. Sevoflurane, which offers the least offensive smell of currently available potent inhalation agents, was gradually introduced until it reached 8%, whereupon nitrous oxide was shut off and the oxygen flow was increased to flush nitrous oxide from the breathing circuit.

Sevoflurane produces toxic combustion products when it reacts with soda lime, but soda lime was removed from the anesthesia machine during this study because it harmfully absorbs carbon dioxide, confers no benefit, and imposes an unnecessary variable on the study.[181, 182]

As soon as patients lost consciousness, the clinic nurse established IV access and administered a front-loading dose of Fentanyl (usually 100 mcg in children and 250 mcg in adults) while bag and mask induction continued. The synergistic interaction of Fentanyl and Sevoflurane accelerated induction, optimized vocal cord relaxation, enabled endotracheal intubation without paralysis, and minimized toxic inhalation agent requirements. If necessary, the Fentanyl could be reversed readily with Narcan, but this was never necessary.

Nasotracheal intubation was performed as soon as the synergistic effects of Sevoflurane anesthesia and Fentanyl analgesia provided adequate vocal cord relaxation, as judged by bag and mask and the decline of systolic blood pressure below 100 torr. The resulting vocal cord relaxation was equivalent to paralysis, but without the danger of "can't intubate, can't ventilate" predicaments caused by sudden paralysis of pharyngeal muscles, especially in the presence of unanticipated anatomical airway difficulties. The stimulation of intubation usually restored blood pressure toward normal, and the narcotic analgesia prevented harmful tachycardia and hypertension. The endotracheal tube was secured with tape, the eyes were taped shut for protection, the patient was covered with a blanket, and the patient's head was wrapped in a towel that was taped in place to complete the induction. In most cases patients were fully prepared for surgery within five minutes after their arrival in the operating room.

The front-loading Fentanyl dose assured control of incision stress and was supplemented as necessary to maintain respiratory rate within study parameters (8–12 breaths/minute in adults, 15–18 breaths/minute in children). This maintained beneficial hypercarbia that opened the capillary gate and released oxygen from the hemoglobin molecule via the Bohr effect, and thereby optimized cardiac efficiency, tissue perfusion, tissue oxygenation, and organ protection.

MASK INDUCTION EFFICIENCY

Overview statistics in Figure J.2 on page 476 show that the average "Room in to Incision" time was just over 11 minutes (the median time was 11 minutes). This time included various delaying factors, such as un-cooperative autistic teenagers and dawdling dentists. In most cases, patients were prepared for surgery within five minutes after they entered the operating room. The speed of mask induction and intubation could be further improved using CO_2 supplementation, but that was judged unnecessary. It is doubtful that any intravenous technique could exceed the safety, simplicity, and efficiency of this approach.

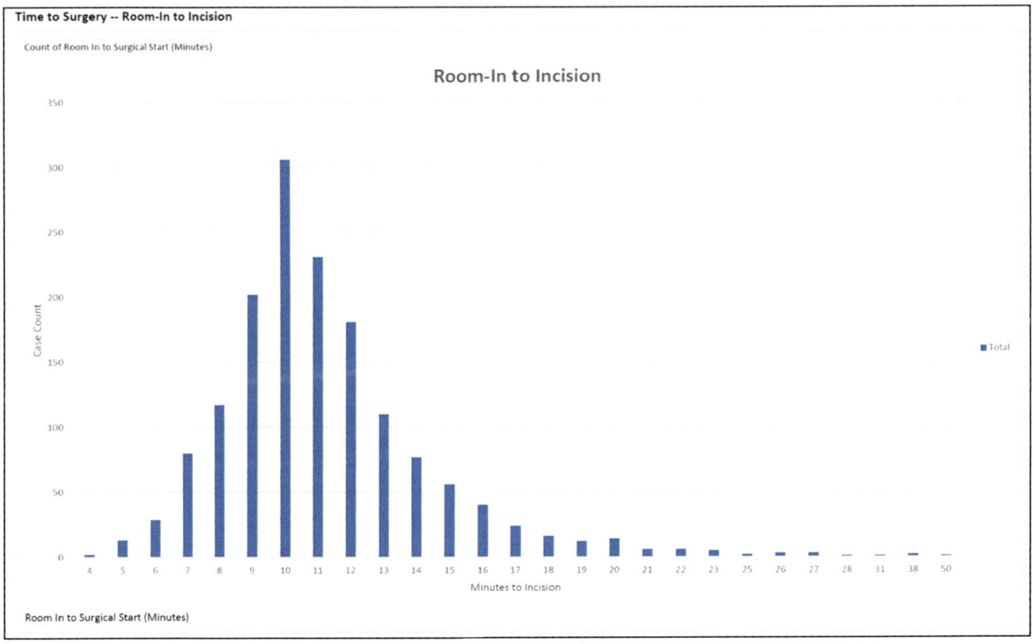

Figure J.3: Patient preparation efficiency
This graph illustrates the efficient preparation of patients for surgery. The average and mean times from room-in to surgical start (incision) hovered around 11 minutes as noted in Figure J.1 on page 475.

An alternative approach is to establish IV access, pre-oxygenate, and treat with 50–100 mcg Fentanyl about three minutes before mask induction to obtund the sense of smell. This approach is useful when Isoflurane, which has an offensive smell, is the only available inhalation agent, or if patients require sedation. With this approach, the pre-treatment with Fentanyl obtunds the offensive smell of Isoflurane even at high concentrations, so that nitrous oxide is unnecessary. Either way, it is safer to secure the airway (intubate patients) before introducing paralysis. CO_2 supplementation accelerates induction with either approach.[183]

Mask Induction Safety

Mask induction and intubation with Fentanyl supplementation offers a practical defense against occult anatomical airway abnormalities, and a simple, safe, efficient, humane alternative to "awake intubation" techniques. It provides warning of the presence of anatomical abnormalities that are difficult to detect with pre- operative inspection. It introduces the effects of toxic inhalation agents gradually, so that induction can be safely aborted in the event of unexpected problems. It avoids the potentially lethal sudden onset of intravenous hypnotic agent toxicity, which is particularly problematic with Propofol.[184, 185, 186] It prevents and controls harmful tachycardia and hypertension during induction and intubation. It does not trigger the collapse of pharyngeal muscle tone that causes soft tissues to unexpectedly obscure the airway and cause the "can't intubate, can't ventilate" dilemma. Also, narcotics can be readily reversed with Narcan, while paralysis cannot. It avoids the postoperative muscle pain, lethal potassium elevations, and malignant hyperthermia caused by succinylcholine. Also, succinylcholine can cause prolonged paralysis in patients with pseudocholinesterase deficiency.

By comparison, bolus intravenous doses of toxic hypnotic agents are dangerous, especially in the case of Propofol, which can cause severe, unexpected, and irreversible cardiac toxicity.[187, 188, 189, 190] Etomidate has little effect on blood pressure but it undermines distant surgical outcome.[191]

Figure J.4 illustrates the mild hypotension produced by the synergistic interaction of Fentanyl and Sevoflurane used to enable mask induction and endotracheal intubation without paralysis during the study. All anesthetic induction approaches reduce blood pressure, but mask induction with Fentanyl supplementation rarely causes severe hypotension or bradycardia.

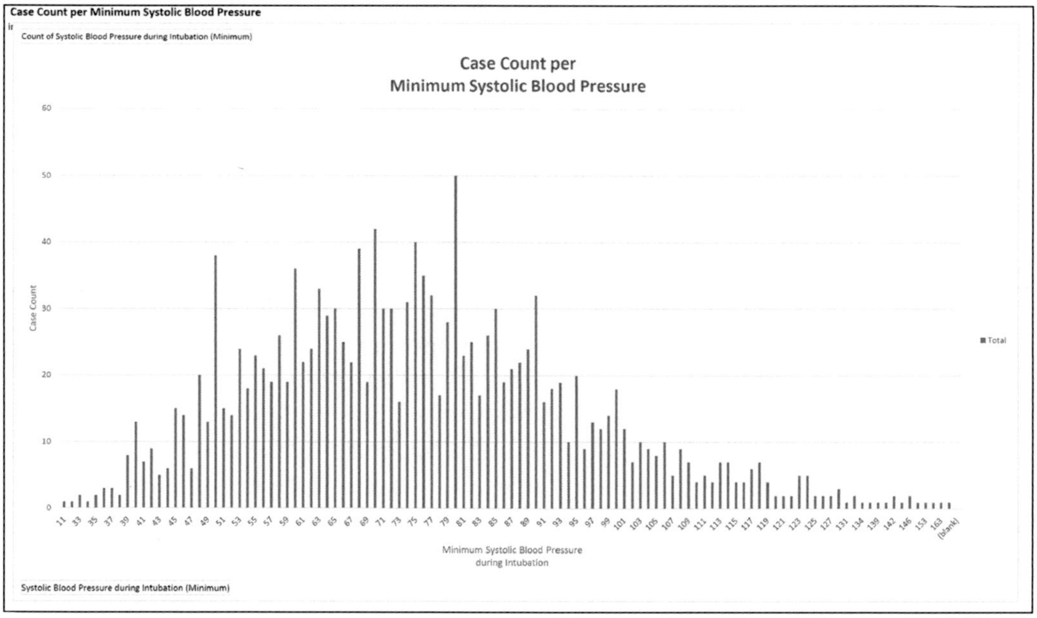

Figure J.4: Hypotension

This graph displays the degree of hypotension that occurred during mask induction. The average minimum systolic blood pressure as shown in Figure J.1 was ~75 torr. In most cases vocal cord relaxation was excellent and intubation was easy when blood pressure fell below 100 torr.

Loose upper incisors and dentures are common hazards with dental cases, but they present no greater hazards with mask induction than paralysis. With either approach dislodged teeth can be swallowed or inhaled into the trachea, which necessitates X-rays to confirm tooth location.

Mask Induction Problems

The fear of vocal cord damage due to inadequate paralysis is greatly exaggerated, and the synergistic interaction of Fentanyl and Sevoflurane relaxes vocal cords as effectively as paralysis. I have never witnessed vocal cord damage with any induction technique or had any patient complain that mask induction was unpleasant in all my years of practice. Laryngospasm occasionally occurs when patients are inadequately anesthetized (with the inhalation agent) or when the narcotic hasn't taken full effect. In such cases it is best to wait patiently until the narcotic effect exerts itself while supplementing the inhalation agent via bag and mask until respiration is easily controlled and blood pressure has declined below 100 torr. Fortunately, laryngospasm is usually mild in such situations. I learned to maneuver the tip of the endotracheal tube into the tracheal orifice and then gently advance it while rotating it counterclockwise. This is the same maneuver used to overcome the *laryngeal shelf syndrome* that plagues fiberoptic equipment. The counterclockwise rotation harmlessly parts the vocal cords, and allows the tip of the tube to slide past them into the trachea.

On perhaps four occasions I have seen hypotension during mask induction that required intervention. Most of these patients were mentally handicapped and wheelchair-bound, and were encountered prior to this study. The only episode of hypotension encountered during the study was MRN#24937, a forty-six-year-old woman with no known medical problems who unexpectedly developed severe bradycardia and hypotension after a 250-mcg bolus dose of Fentanyl. Interestingly, her hemoglobin saturation and transcutaneous oxygen levels remained adequate despite the hypotension, which indicates that effective tissue perfusion was maintained despite the low blood pressure. This is consistent with stress theory, which explains that blood pressure is determined by fluctuating microvascular flow resistance rather than changing cardiac contractility. You can view her anesthetic record at http://www.stressmechanism.com/. On the website, click the link for *50 Years Downloads*, and then click the link for the TcO_2 and $TcCO_2$ anesthetic records. All these hypotensive episodes responded promptly either to atropine or ephedrine. Such severe hypotension is rarely caused by Fentanyl, and is more common with bolus hypnotic intravenous techniques, because hypnotic agent toxicity depresses cardiac contractility, while Fentanyl has negligible effect on the heart. Instead, Fentanyl opens the capillary gate, reduces microvascular flow resistance, lowers blood pressure, and enhances cardiac efficiency and cardiac safety.

MASK INDUCTION AND THE “FULL STOMACH” DILEMMA

There remains no consensus for the management of emergency full stomach management. I can offer no better solution than anyone else. I utilize the conventional “rapid sequence induction” with succinylcholine and Sellick's maneuver to avoid being legally crucified by the “de facto standard of practice” concept that plagues law and medicine, but this approach is far from foolproof. Ironically,

Sellick's maneuver was introduced in the early days of modern anesthesia when needle technology was primitive, intravenous access was impractical, and mask management without intubation prevailed. Perhaps carbon dioxide supplementation facilitated rapid intubation by accelerating ether induction and preventing laryngospasm in those forgotten days. In the absence of full stomach, I believe that the routine use of mask induction and elective intubation without paralysis is the safest induction method. It is safer to introduce paralysis after the airway is secured. Airway security (i.e., elective endotracheal intubation) should be the priority for all situations that necessitate sedation.

MAINTENANCE

The front-loading dose of Fentanyl (250mcg in adults) provided pre-emptive analgesia and minimized the requirement for toxic inhalation agents during surgery as well as speeding induction and facilitating elective endotracheal intubation.[192, 193, 194, 195, 196, 197, 198, 199, 200, 201] Fentanyl supplementation was added to maintain respiratory rate within study parameters (8–12 breaths/ minute in adults, 15–18 breaths/minute in children). This maintained beneficial hypercarbia and optimized safety, simplicity, and outcome.

One percent Isoflurane replaced 8% Sevoflurane after patients were successfully intubated. One percent Isoflurane is approximately half the concentration required to prevent movement due to surgical stimulation. The Fentanyl supplementation enabled this reduction in the inhalation agent concentration, which minimized toxicity and accelerated emergence.[202] Isoflurane offers superior chemical stability, greater potency, and lower cost than Sevoflurane.

The front-loading dose of Fentanyl obtunded respiratory chemoreceptors and halted spontaneous breathing for between 5–15 minutes. Normal cellular metabolism gradually elevated CO_2 tissue reserves during this period of apneic oxygenation until rising levels of CO_2 restored respiratory chemoreceptor function, whereupon spontaneous breathing resumed. The brief period of apneic oxygenation was not problematic because all patients were intubated, connected to 100% oxygen via the anesthesia machine, and monitored with pulse oximetry to assure effective hemoglobin oxygenation. Figures K5 and K6 illustrate the relationships of CO_2 elevations and Fentanyl dosage. I manually recorded these peak end-tidal CO_2 values onto my Excel spreadsheet because the AIMS software recorded data only at five-minute intervals and seldom captured the peak values. The peak range of the clinic capnograph was 116 torr, so it failed to capture a few of the peak CO_2 concentrations. In most cases, surgical nociception restored respiratory drive before CO_2 levels alone were adequate to restore breathing.

DOES HYPERCARBIA EXPONENTIALLY INCREASE RESPIRATORY DRIVE AND/OR NARCOTIC METABOLISM AND CLEARANCE?

Hypocarbia inhibits Fentanyl clearance from the body, while hypercarbia accelerates it.[203] Surprisingly, the graph in Figure J.6 on page 483 shows how spontaneous breathing began before $EtCO_2$ reached 116 torr (the maximum measurement of which the clinic capnograph was capable) in almost all cases. This curious "ceiling" in Et CO_2 suggests the possibility that chemoreceptor sensitivity increases exponentially with increasing CO_2 elevations in the body. The data is confusing because surgical stimulation invariably began before spontaneous breathing commenced. Thus, in most cases surgical stimulation initiated breathing before CO_2 levels had risen high enough to initiate breathing. The average peak end-tidal CO_2 levels would presumably have been higher were it not for the effect of surgical nociception. This observation merits further investigation. The classical apneic oxygenation studies could be adapted to test this hypothesis. Anesthesiology resident volunteers could be anesthetized, intubated, and treated with various mcg/kg doses of Fentanyl, while recording the maximum end-tidal CO_2 concentration when they resumed spontaneous breathing in the absence of surgical stimulation. Such information is pertinent to anesthetic management.

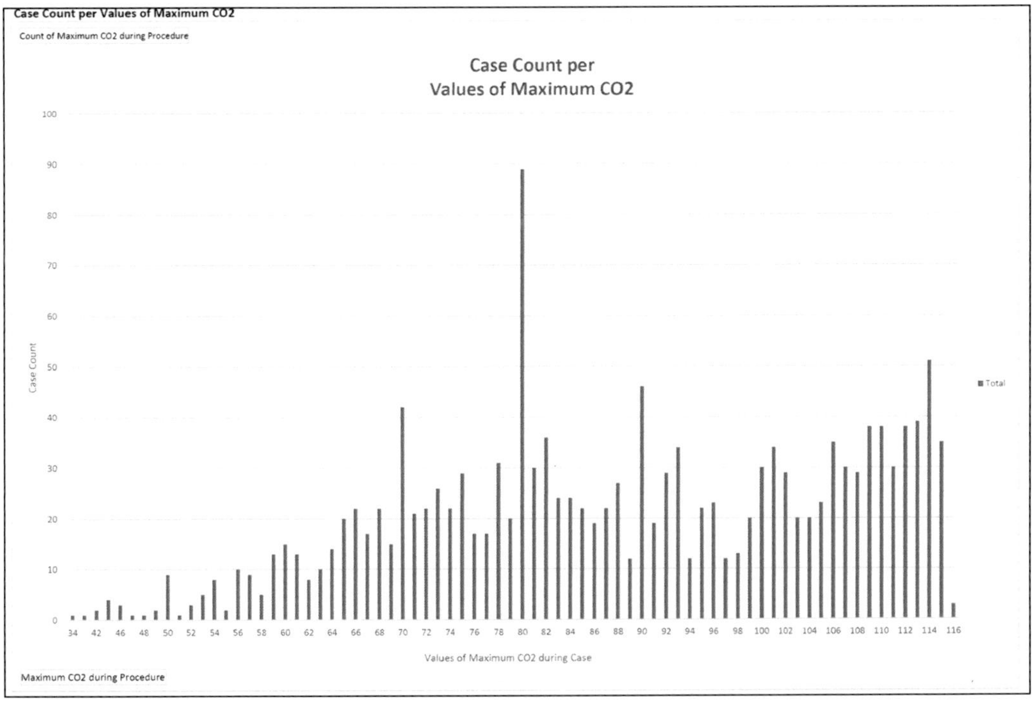

Figure J.5: Values of maximum CO_2

Narcotics inhibit respiratory chemoreceptors that stimulate breathing. This explains why the intravenous front-loading dose of Fentanyl administered during induction temporarily halts breathing. Three factors stimulate respiratory drive and restore breathing: (1) CO_2 production from normal cellular metabolism gradually accumulates in body fluids and tissues, and stimulates respiratory chemoreceptors (2) Rising CO_2 levels accelerate narcotic clearance from the body (3) Surgical nociception also stimulates respiratory drive. Peak end-tidal CO_2 levels were invariably observed when respiratory drive was restored and patients resumed breathing, whereupon end-tidal CO_2 levels progressively fell toward normal. Most patients resumed spontaneous breathing at peak $EtCO_2$ levels below 115 torr regardless of the Fentanyl dose. This striking observation remains unexplained. I hypothesize that the sharp "spike" in patients who resumed breathing at 80 torr was most likely explained by surgical nociception, because surgery typically began about ten minutes after patients entered the operating suite.

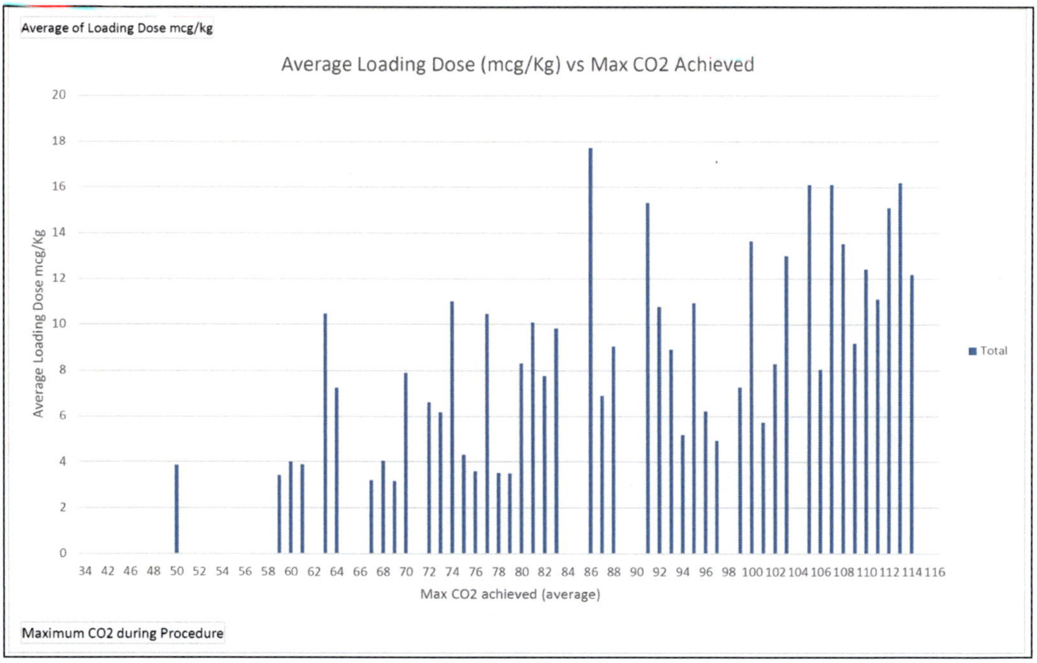

Figure J.6: Average loading dose Fentanyl versus Max $EtCO_2$ when breathing resumed

This graph provides yet another view that illustrates the surprising observation that end-tidal CO_2 rose no higher than 116 torr regardless of the size of the front-loading dose of Fentanyl. I would have expected a bell-shaped curve instead of a sharp cutoff point at 116 torr.

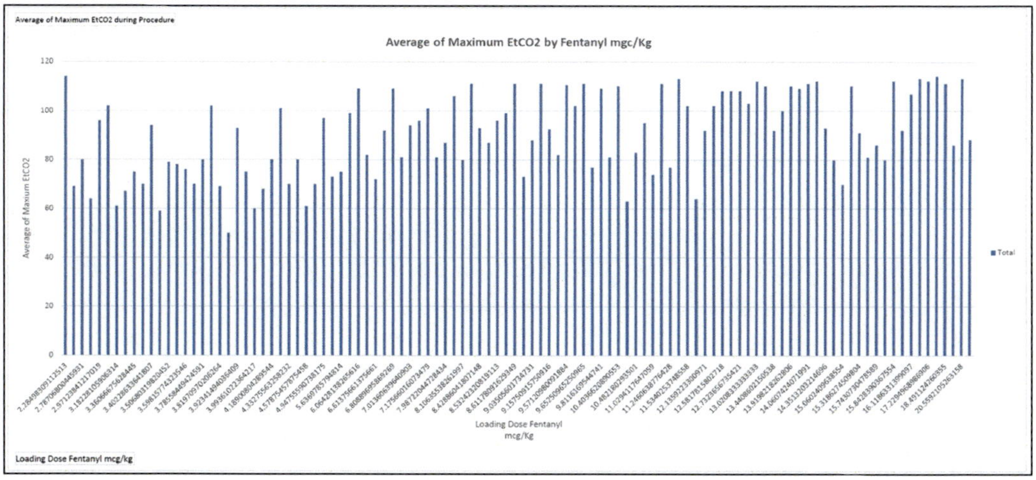

Figure J.7: Average max Et CO_2 by Fentanyl dose

This graph also illustrates the surprising observation that maximum end-tidal CO_2 was 116 Torr for doses ranging from 2–20 mcg/kg. The fact that maximum Et CO_2 remained the same despite a 10-fold variation in mcg/kg dosage suggests an exponential relationship between CO_2 tissue levels and Fentanyl metabolism and clearance. This data was derived from copies of the anesthetic records that were retained to illustrate the measurements produced by the transcutaneous O_2/CO_2 monitor.

Once breathing resumed, the Et CO_2 values declined until the procedure concluded or supplemental Fentanyl was administered. Simultaneously, the transcutaneous oxygen partial pressure climbed progressively because hypercarbia released oxygen into tissues faster than cells consumed it. The increased $TcCO_2$ confirmed that combinations of hypercarbia and analgesia promote tissue oxygenation and organ safety during surgery.

EMERGENCE

All patients were treated with 4mg Ondansetron toward the end of surgery, as recommended by the manufacturer, to minimize postoperative nausea and vomiting. Droperidol would be a better choice now that its availability has been restored.[204] At the conclusion of surgery patients were immediately turned on their side to minimize swallowing of blood, which was the major cause of postoperative nausea and vomiting in these cases.

Figure J.2 on page 476 shows the average and median times for "End of Surgery to Extubation" — the time required for patients to recover from anesthesia (usually less than 15 minutes).

Patients typically exhibited a brief period of respiratory instability or apnea after the cessation of surgical nociception, until normal metabolic CO_2 production restored respiratory chemoreceptor function, whereupon they resumed reliable breathing. As they resumed breathing, they expelled Isoflurane until they regained consciousness (emerged from anesthesia). Figure J.2 shows that the average time required for patients to regain consciousness and be extubated was 13 minutes. The recovery room nurses were instructed to leave the nasotracheal tube in place and allow patients to sleep peacefully until they regained consciousness, whereupon they removed the tube. I was always immediately available in the event of difficulties. The lingering Fentanyl analgesia enabled patients to tolerate the nasotracheal tube. Very young children often slept peacefully for prolonged periods, presumably because they were accustomed to sleeping late in the morning, or because very young children require more sleep than older children and adults.

Once CO_2 accumulation restored respiratory chemoreceptor activity, there were no episodes of unexpected respiratory arrest. In contrast, when mechanically hyperventilated patients emerge from anesthesia, their respiratory drive can remain unstable for hours due to CO_2 depletion, and they remain in danger of respiratory arrest if they fall asleep during this vulnerable period. The danger is exaggerated in the presence of obstructive sleep apnea.

Laryngospasm after extubation was rare, and was treated with bag and mask. The typical pediatric patient would cry immediately after the nasotracheal tube was removed on account of unfamiliar circumstances, but they were soon calmed by the arrival of parents or caregivers, who were called to the recovery room immediately after extubation. Pediatric patients often slept in their parents' arms under nursing supervision while awaiting discharge from the clinic. Parents were instructed to take their children home and allow them to take an observed nap. Children typically woke up after their nap and went off to play as if they never had surgery.

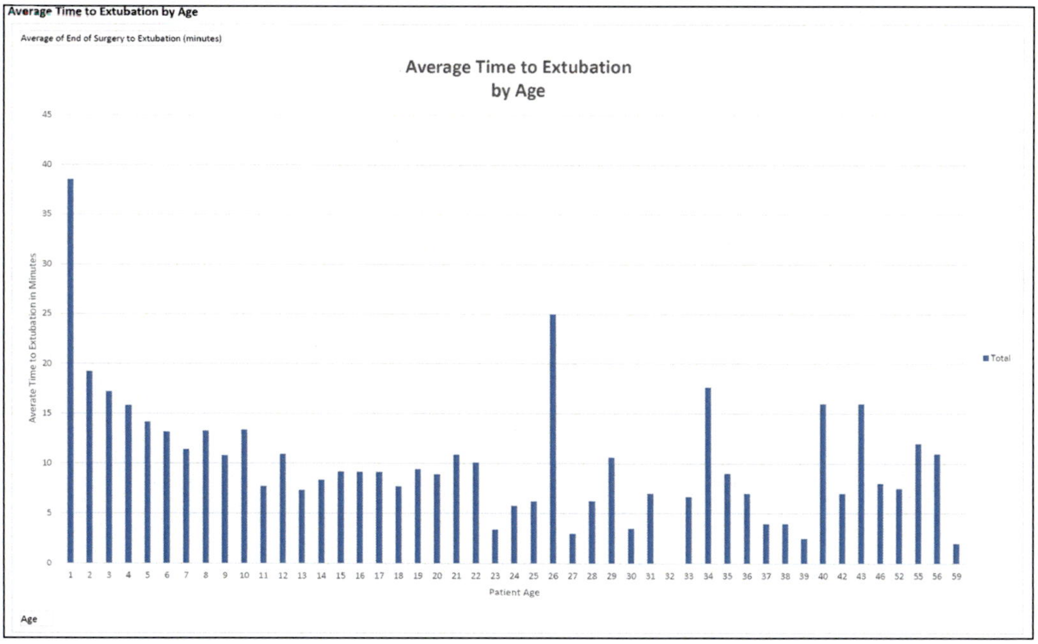

Figure J.8: Average time to extubation by age

The clinic accepted healthy children as young as 2 years old, and this graph illustrates our impression that very young children loved to sleep in the morning. They remained intubated and were allowed to sleep peacefully until they regained consciousness, whereupon the nasotracheal tube was immediately removed, and they would be picked up and comforted. Their mothers or caregivers were invited to the recovery room to hold them as soon as possible to minimize their stress.

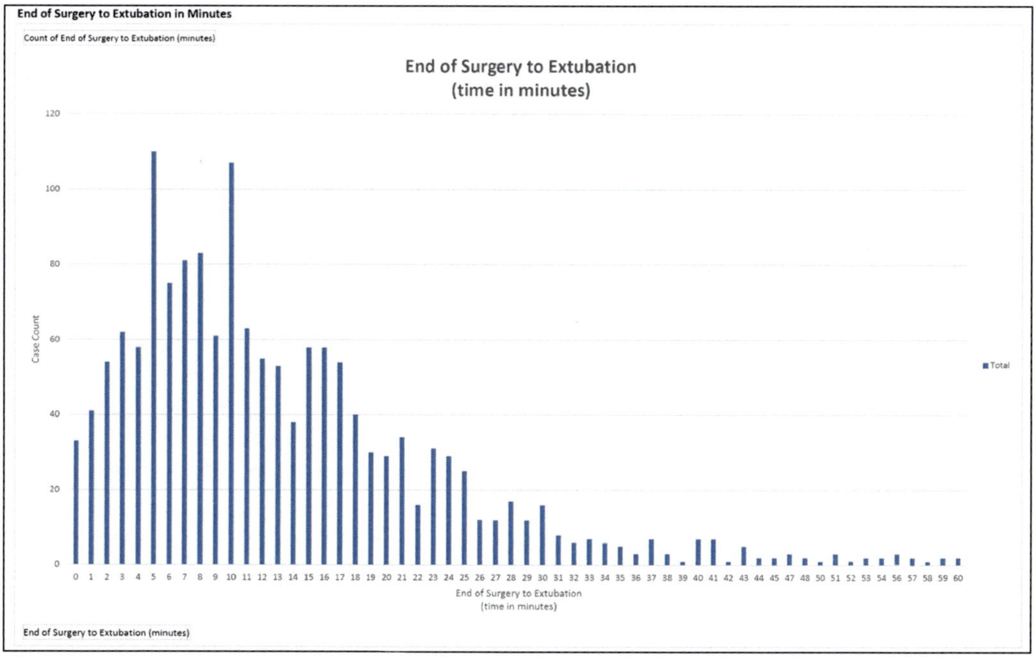

Figure J.9: End of surgery to extubation in minutes

This graph illustrates how most patients emerged from anesthesia and were extubated within 20 minutes of the completion of surgery. Patients would typically be free of pain, calm, clear-headed, and capable of standing and walking if necessary.

There was no significant morbidity or mortality during this study of 1500 cases. Nausea and vomiting were no worse than with other techniques and were mostly related to blood swallowed during the dental procedures. Patients emerged quickly from anesthesia and were clear-headed and free of significant pain. They were capable of walking without assistance, but they were transported to their vehicles via wheelchair as per clinic policy. They were instructed to go home and take a nap to facilitate recovery. Postoperative follow-up indicated that small children typically emerged from their nap and ran off to play as if they never had surgery. There were no instances of unexpected respiratory arrest, such as can occur after patients are mechanically hyperventilated.[205, 206, 207]

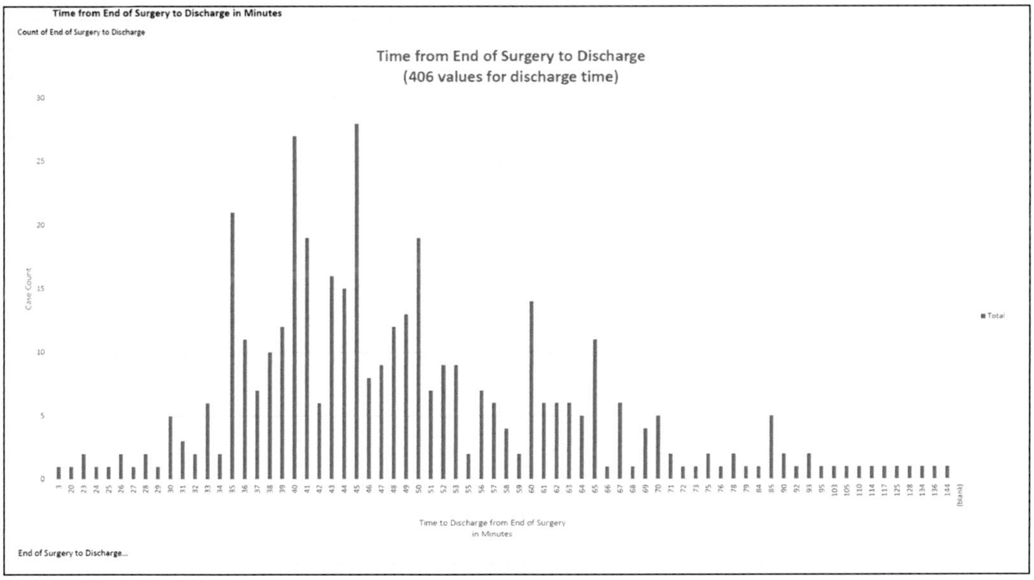

Figure J.10: Time from end of surgery to discharge (406 records)

ADAPTING NARCOTIC ANESTHESIA TO HOSPITAL SURGERY

Today's surgeons usually demand paralysis to facilitate surgery, but this eliminates the ability to regulate narcotic dosage based on respiratory rate during surgery. This presents a problem for narcotic management, because adequate narcotic treatment must be maintained throughout surgery to minimize spinal cord "windup" syndrome that undermines tissue oxygenation and invites myocardial infarction, stroke, bowel ileus, dementia, delirium, and other derangements of the surgical stress syndrome. Inadequate analgesia also exaggerates requirements for paralyzing agents, which invites residual postoperative paralysis and necessitates increased dosage of toxic reversal agents such as neostigmine. Inadequate analgesia becomes especially dangerous if inhalation agent vaporizer supplies inadvertently become depleted, so that patients regain consciousness during surgery and become aware of excruciating surgical pain. Paralysis also necessitates mechanical ventilation, and the customary practice of mechanical hyperventilation depletes CO_2 tissue reserves and exaggerates these problems caused by inadequate analgesia. The CO_2 depletion undermines respiratory drive and invites unexpected postoperative respiratory arrest.[208, 209] Given these considerations, my method of narcotic management in the presence of paralysis is as follows:

- Mask induction with "stress control" narcotic supplementation (10 mcg/kg Fentanyl). This assures effective vocal cord relaxation to optimize elective endotracheal intubation and optimal "pre-emptive" analgesia at the time of surgical incision.

- After the airway is secured, paralysis is introduced using minimal doses of non-depolarizing paralysis agents such as Atracurium or Cis-Atracurium that spontaneously disintegrate in the body, so that their effects dissipate without the necessity of reversal agents. This helps to minimize the toxic effects of reversal agents and the dangerous and unpleasant nuisance of lingering postoperative paralysis.

- Maintain the inhalation agent at ½ MAC during surgery to minimize toxicity and accelerate emergence.

- Adjust mechanical ventilation to maintain permissive hypercarbia with end-tidal CO_2 in the range of 60–100 psi throughout the procedure. This optimizes tissue oxygenation and narcotic metabolism and clearance, and preserves and enhances respiratory chemoreceptor function, which preserves postoperative respiratory drive. Spontaneous respiratory efforts can provide early warning that either narcotic dosage or paralysis requires supplementation. In such circumstances, I re-assess paralysis and observe respiratory rate, and administer small supplementary doses of either narcotics or paralyzing agents as appropriate.

- Watch carefully for any sign of patient movement or surgical complaints of inadequate relaxation and treat with additional narcotics and non-depolarizing agents as necessary, with emphasis on narcotic dosage.

- Minimize both narcotic and relaxant dosage toward the conclusion of surgery to minimize untoward postoperative effects.

- Reverse paralysis at the conclusion of surgery and shut off the inhalation agent.

- Ventilate with 100% oxygen while maintaining hypercarbia to rid inhalation agents from the patient's body.

- Continue mechanical ventilation until the exhaled concentration of inhalation agents falls near the threshold of emergence (about 0.15% in the case of Isoflurane).

- Shut off the ventilator before the patient regains consciousness and allow apneic oxygenation until CO_2 accumulation restores spontaneous breathing. This confirms effective respiratory chemoreceptor function and prevents postoperative respiratory depression and unexpected postoperative respiratory arrest.

- Deflate the ET tube cuff but leave the patient intubated to assure airway safety until anesthetic emergence restores consciousness.

- Either allow the patient to emerge from anesthesia in the operating room or transport him to the recovery room still intubated. Most patients can be safely extubated within five minutes of the conclusion of surgery if they are managed properly.

- Remove the ET tube only after the patient regains consciousness, follows commands, and lifts his head above his pillow to confirm adequate recovery from relaxant effects. Effective narcotization enables patients to comfortably tolerate the ET tube without bucking, coughing, or discomfort.

- Extubate the patient. There should be no need for suction.

- The effective maintenance of hypercarbia and narcotic analgesia should minimize postoperative nausea, vomiting, fever, tachycardia, tachypnea, hypertension, dysrhythmias, dementia, delirium, heart attacks, strokes, and unexpected respiratory arrest.

SUMMARY

This study achieved its objectives, as stated at the outset of this report:

- It refuted the Leake-Waters hoax by demonstrating the harmlessness of hypercarbia — at least as high as 116 mmHg in the presence of oxygen-enriched gas mixtures.

- It demonstrated that the open airway technique offers no advantages in terms of economy, efficiency, or safety as compared to elective endotracheal intubation.

- It demonstrated the safety and benefits of permissive hypercarbia and narcotic supplementation.

- It demonstrated that generous narcotic dosage does not cause respiratory arrest in the presence of hypercarbia.

- It demonstrated the need for professional standards that condemn mechanical hyperventilation and open airway techniques.

- It demonstrated the need for professional standards and guidelines for monitoring and managing beneficial hypercarbia during anesthesia.[210]

- It demonstrated that mask induction and elective endotracheal intubation with narcotic supplementation is as efficient and reliable as with paralysis, and safer.

- It proved that hypercarbia and narcotic supplementation are safe and beneficial.

- It explains how hyperventilation rather than narcotic supplementation causes respiratory depression, unexpected respiratory arrest, and so-called *opioid hypersensitivity.*

- It demonstrated that soda lime is unnecessary and explained why it should be and banned from anesthesia machines.

- It demonstrated that intravenous fluids are not needed for simple surgical procedures. IV fluids are probably best avoided in all types of surgery. Blood loss should be treated with blood transfusion or re-constituted fresh frozen plasma, which was developed by American physicians in France in WWI and proven beneficial, and remains available in France even though it is unheard of in North America.

- It demonstrated a universal anesthetic technique that can be altered to optimize safety, simplicity, efficiency, economy, and outcome in all surgical settings.

- It paved the path for pharmacological development to produce a safe antidote to tissue factor. This would enable control of the tissue disruption pathway, which would abolish the surgical stress syndrome, re-revolutionize surgery, and revolutionize the treatment of critical illnesses.

- It paved the path for combinations of hypercarbia, narcotic nociception control, sedation, and thrombin suppression to provide cures for cancer and critical illnesses.

- It demonstrated that normal cellular CO_2 production is sufficient to enable the benefits of hypercarbia during anesthesia without CO_2 supplementation from external tanks.

- It provided pertinent clinical information for other anesthetists who might be interested in supplementing general anesthesia with narcotic analgesia and permissive hypercarbia.
- It paved the path for improved anesthesia machine and monitoring designs.
- It explained how CO_2 supplementation can facilitate mask induction.
- It justified the use of Carbogen to accelerate anesthetic emergence and recovery and prevent postoperative laryngospasm, nausea, and vomiting.
- It demonstrates how general nasotracheal anesthesia with small portable anesthesia machines can provide safer, simpler, and more efficient anesthesia for dental offices and clinics than the dangerous open airway techniques that presently prevail.
- It suggests fresh studies to clarify the possibility that respiratory chemoreceptors exhibit exponential increases in sensitivity to rising concentrations of carbon dioxide.

CONCLUSION: A CALL TO ACTION

"Medical education does not exist to provide students with a way of making a living, but to ensure the health of the community."
— Rudolf Virchow

Medicine is a social science, and politics is nothing else but medicine on a large scale. Medicine, as a social science, as the science of human beings, has the obligation to point out problems and to attempt their theoretical solution: the politician, the practical anthropologist, must find the means for their actual solution. The physicians are the natural attorneys of the poor, and social problems fall to a large extent within their jurisdiction.
— Rudolf Virchow

Virchow's wisdom runs deep. It is in his spirit that I have endeavored to identify some problems of my profession, explain them, and recommend remedies. Were this anesthetic method to be universally employed, it would improve economy and efficiency and reduce surgical morbidity and mortality. Mask induction and elective endotracheal intubation without paralysis would eliminate the "can't intubate, can't ventilate" dilemma during induction. Combinations of narcotic and carbon dioxide supplementation would optimize tissue oxygenation and organ safety, prevent harmful stress mechanism hyperactivity, reduce paralysis requirements, and minimize postoperative morbidity and mortality.

It has been said that war is the health of the state; but the state depends on the vitality of its citizens, and civilization hangs in the balance. I am no fan of government,[211] which exists for the sake of warfare, and is ill-suited for peace. If civilization is to survive, then government must somehow be altered to encourage health, fertility, intelligence, idealism, integrity, and progress that are essential to attain and maintain public vitality and support. To that end, it is the duty of the state to police medical politics and promote medical integrity and science, for public health is essential for the survival of civilization, which is in the interest of rich and poor alike.

NOTES

1 E. Lowenstein *et al.*, Cardiovascular response to large doses of intravenous morphine in man. *N Engl J Med* 281, 1389-1393 (1969).

2 S. N. Raja, E. Lowenstein, The birth of opioid anesthesia. *Anesthesiology* 100, 1013-1015 (2004).

3 M. G. Davies, P. O. Hagen, Systemic inflammatory response syndrome. *Br J Surg* 84, 920-935 (1997).

4 Y. G. Liu, C. S. Li, [Coagulation dysfunctions in patients with systemic inflammatory response syndrome]. *Zhonghua Nei Ke Za Zhi* 43, 94-97 (2004).

5 K. Takenaka, E. Ogawa, H. Wada, T. Hirata, Systemic inflammatory response syndrome and surgical stress in thoracic surgery. *J Crit Care* 21, 48-53; discussion 53-45 (2006).

6 E. Lowenstein *et al.*, Cardiovascular response to large doses of intravenous morphine in man. *N Engl J Med* 281, 1389-1393 (1969).

7 S. N. Raja, E. Lowenstein, The birth of opioid anesthesia. *Anesthesiology* 100, 1013-1015 (2004).

8 F. J. Overdyk, Postoperative opioids remain a serious patient safety threat. *Anesthesiology* 113, 259-260; author reply 260-251 (2010).

9 L. S. Coleman, in *apsf Newsletter*. (Anesthesia Patient Safety Foundation, Administrator, Deanna Walker Anesthesia Patient Safety Foundation Building One, Suite Two 8007 South Meridian Street Indianapolis, IN 46217-2922 e-mail address: walker@apsf.org FAX: (317) 888-1482, 2010), vol. Winter 2009-2020.

10 F. J. Overdyk, postoperative Opioids Need System-Wide Overhaul. *Anesthesia Patient Safety Foundation Newsletter*, (2010).

11 A. B. Stone, E. Y. Brovman, P. Greenberg, R. D. Urman, A medicolegal analysis of malpractice claims involving anesthesiologists in the gastrointestinal endoscopy suite (2007-2016). *J Clin Anesth* 48, 15-20 (2018).

12 G. Crile, D. H. Dolley, An Experimental Research into the Resuscitation of Dogs Killed by Anesthetics and Asphyxia. *The Journal of experimental medicine* 8, 713-725 (1906).

13 L. W. Crile GW, *Anoci-association*. (Saunders, Philadelphia, 1914), pp. 223-225.

14 Y. Henderson, A Lecture ON RESPIRATION IN ANAESTHESIA: CONTROL BY CARBON DIOXIDE. *Br Med J* 2, 1170-1175 (1925).

15 Y. Henderson, THE PREVENTION AND TREATMENT OF ASPHYXIA IN THE NEW-BORN. *JAMA* JAMA. 1928;90(8):583-586. doi:10.1001/jama.1928.02690350001001, 583-586 (1928).

16 Y. Henderson, The Wesley M. Carpenter Lecture: Atelectasis, Massive Collapse and Related Post-Operative Conditions. *Bull N Y Acad Med* 11, 639-656 (1935).

17 Y. Henderson, Resuscitation with Carbon Dioxide. *Science* 83, 399-402 (1936).

18 Y. Henderson, How Breathing Begins at Birth. *Science* 85, 89-91 (1937).

19 Y. Henderson, in *Cyclopedia of Medicine*. (1940).

20 Y. Henderson, H. W. Haggard, *Noxious gases and the principles of respiration influencing their action*. (Chemical Catalog Co., New York,, 1927), pp. 220 p.

21 A. B. Stone, E. Y. Brovman, P. Greenberg, R. D. Urman, A medicolegal analysis of malpractice claims involving anesthesiologists in the gastrointestinal endoscopy suite (2007-2016). *J Clin Anesth* 48, 15-20 (2018).

22 L. S. Coleman, Routine Mask Induction and Intubation without Paralysis may be Safer. *J Anesth Surg* 2, 1-2 (2015).

23 M. Vlessides, in *Anesthesiology News*. (McMahon, New York, NY, June 4, 2021).

24 N. Anesthesia, N. Anesthesia, Ed. (June 7, 2012).

25 K. Rector, in *The Baltimore Sun*. (December 13, 2011).

26 M. L. Report. (June 7, 2012).

27 M. Santora, in *New York Times*. (The New York Times Company, New York, NY, May 12, 2016).

28 T. D. Company, M. Susan K. Palmer, Ed. (The Doctor's Advocate/Fourth Quarter 2010, 8/24/2011).

29 A. Heller, in *Anesthesiology News*. (McMahon, New York, NY, 2010), vol. 36.

30 A. B. Stone, E. Y. Brovman, P. Greenberg, R. D. Urman, A medicolegal analysis of malpractice claims involving anesthesiologists in the gastrointestinal endoscopy suite (2007-2016). *J Clin Anesth* 48, 15-20 (2018).

31 N. Anesthesia, N. Anesthesia, Ed. (June 7, 2012).

32 K. Rector, in *The Baltimore Sun*. (December 13, 2011).

33 M. L. Report. (June 7, 2012).

34 M. Santora, in *New York Times*. (The New York Times Company, New York, NY, May 12, 2016).

35 D. R. Miller, 2010 Guidelines and international standards to the practice of anesthesia. *Can J Anaesth* 57, 957-960 (2010).

36 S. American Society of Anesthesiologists Task Force on, N.-A. Analgesia by, Practice guidelines for sedation and analgesia by non-anesthesiologists. *Anesthesiology* 96, 1004-1017 (2002).

37 G. R. Dobson, Special Announcement - Guidelines to the Practice of Anesthesia - Revised Edition 2021. *Can J Anaesth* 68, 8-19 (2021).

38 S. A. Guidelines, Practice Guidelines for the Perioperative Management of Patients wth Obstructive Sleep Apnea. *Anesthesiology* Anesthesiology 2006; 104:1081–93, (2006).

39 B. Kronemyer, in *Anesthesiology News*. (McMahon, New York, NY, October 12, 2016).

40 R. N. Merchant, Special announcement: Guidelines to the practice of anesthesia—revised edition 2015. *Can J Anaesth* 62, 8-9 (2015).

41 J. Marquez, J. Magovern, P. Kaplan, T. Sakert, G. P. Gravlee, Cardiac surgery "fast tracking" in an academic hospital. *Journal of cardiothoracic and vascular anesthesia* 9, 34-36 (1995).

42 M. Santora, in *New York Times*. (The New York Times Company, New York, NY, May 12, 2016).

43 C. Alan Duke. (CNN, 2014), vol. 2021, pp. Online report of the cause of death

44 K. Rector, in *The Baltimore Sun*. (December 13, 2011).

45 K. Schultz, in *San Francisco Chronicle*. (San Francisco Chronicle, San Francisco, Ca. , 2016).

46 B. Seidman, in *Sydney Galleger*. (CBS Interactive, September 17, 2015).

47 C. F. c. b. c. B. A. N. Group, In wake of California child's death, dental group releases new guidance on anesthesia. *Enterprise-Record*. 2019.
48 B. Kronemyer, in *Anesthesiology News*. (June 30, 2016).
49 H. Lee *et al.*, Ethics Rounds: Death After Pediatric Dental Anesthesia: An Avoidable Tragedy? *Pediatrics* 140, (2017).
50 B. Seidman, B. Seidman, Ed. (CBS News, 2015), chap. September 17, 2015 4:09PM.
51 T. O. toide@sacbee.com, in *Sacramento Bee*. (June 16 2017).
52 H. Lee et al., Ethics Rounds: Death After Pediatric Dental Anesthesia: An Avoidable Tragedy? *Pediatrics* 140, (2017).
53 E. Lowenstein *et al.*, Cardiovascular response to large doses of intravenous morphine in man. *N Engl J Med* 281, 1389-1393 (1969).
54 S. N. Raja, E. Lowenstein, The birth of opioid anesthesia. *Anesthesiology* 100, 1013-1015 (2004).
55 A. I. McEwan *et al.*, Isoflurane minimum alveolar concentration reduction by fentanyl. *Anesthesiology* 78, 864-869 (1993).
56 G. G. Pacentine, M. Muzi, T. J. Ebert, Effects of fentanyl on sympathetic activation associated with the administration of desflurane. *Anesthesiology* 82, 823-831 (1995).
57 M. J. Laster, E. I. Eger, 2nd, Temperatures in soda lime during degradation of desflurane, isoflurane, and sevoflurane by desiccated soda lime. *Anesth Analg* 101, 753-757, table of contents (2005).
58 S. G. Ainslie, J. H. Eisele, Jr., G. Corkill, Fentanyl concentrations in brain and serum during respiratory acid—base changes in the dog. *Anesthesiology* 51, 293-297 (1979).
59 J. S. Lundy, Carbon Dioxide as an Aid In Anesthesia. *J.A.M.A* 85:1953, 1925, (1925).
60 S. A. Coleman, J. W. McCrory, C. J. Vallis, R. J. Boys, Inhalation induction of anaesthesia with isoflurane: effect of added carbon dioxide. *Br J Anaesth* 67, 257-261 (1991).
61 E. I. Eger, J. W. Severinghaus, The rate of rise of $PaCO_2$ in the apneic anesthetized patient. *Anesthesiology* 22, 419-425 (1961).
62 M. J. Frumin, R. M. Epstein, G. Cohen, Apneic oxygenation in man. *Anesthesiology* 20, 789-798 (1959).
63 K. G. Hickling, Permissive hypercapnia. *Respir Care Clin N Am* 8, 155-169, v (2002).
64 J. G. Laffey, D. O'Croinin, P. McLoughlin, B. P. Kavanagh, Permissive hypercapnia—role in protective lung ventilatory strategies. *Intensive Care Med* 30, 347-356 (2004).
65 H. Morisaki, R. Serita, Y. Innami, Y. Kotake, J. Takeda, Permissive hypercapnia during thoracic anaesthesia. *Acta anaesthesiologica Scandinavica* 43, 845-849 (1999).
66 A. Ohmura, M. Sha, J. Katagiri, How far can we go with permissive hypercapnia? A case presentation and some biased comments with emphasis on maintaining normal haemoglobin level. *Acta Anaesthesiol Scand Suppl* 107, 209-213 (1995).

67 U. H. Thome, W. A. Carlo, Permissive hypercapnia. *Semin Neonatol* 7, 409-419 (2002).

68 Q. Zhou *et al.*, Effects of permissive hypercapnia on transient global cerebral ischemia-reperfusion injury in rats. *Anesthesiology* 112, 288-297 (2010).

69 K. G. Hickling, S. J. Henderson, R. Jackson, Low mortality associated with low volume pressure limited ventilation with permissive hypercapnia in severe adult respiratory distress syndrome. *Intensive Care Med* 16, 372-377 (1990).

70 M. Komori *et al.*, Permissive range of hypercapnia for improved peripheral microcirculation and cardiac output in rabbits. *Critical care medicine* 35, 2171-2175 (2007).

71 E. Lowenstein *et al.*, Cardiovascular response to large doses of intravenous morphine in man. *N Engl J Med* 281, 1389-1393 (1969).

72 S. N. Raja, E. Lowenstein, The birth of opioid anesthesia. *Anesthesiology* 100, 1013-1015 (2004).

73 L. W. Crile GW, *Anoci-association.* (Saunders, Philadelphia, 1914), pp. 223-225.

74 J. G. Bovill, P. S. Sebel, T. H. Stanley, Opioid analgesics in anesthesia: with special reference to their use in cardiovascular anesthesia. *Anesthesiology* 61, 731-755 (1984).

75 T. H. Stanley, Opiate anaesthesia. *Anaesthesia and intensive care* 15, 38-59 (1987).

76 T. H. Stanley, T. D. Egan, H. Van Aken, A tribute to Dr. Paul A. J. Janssen: entrepreneur extraordinaire, innovative scientist, and significant contributor to anesthesiology. *Anesth Analg* 106, 451-462, table of contents (2008).

77 T. H. Stanley, L. R. Webster, Anesthetic requirements and cardiovascular effects of fentanyl-oxygen and fentanyl-diazepam-oxygen anesthesia in man. *Anesth Analg* 57, 411-416 (1978).

78 T. H. E. Stanley, Talmage D., *Anesthesia for the New Millennium.* (Springer-Science + Business Media, 1999), vol. 34, pp. 288.

79 L. S. Coleman, Routine Mask Induction and Intubation without Paralysis may be Safer. *J Anesth Surg* 2, 1-2 (2015).

80 B. G. Goudra, P. M. Singh, Best of both worlds-endotracheal intubation without paralysis. *J Anaesthesiol Clin Pharmacol* 30, 172-173 (2014).

81 M. S. Scheller, M. H. Zornow, L. J. Saidman, Tracheal intubation without the use of muscle relaxants: a technique using propofol and varying doses of alfentanil. *Anesth Analg* 75, 788-793 (1992).

82 J. B. Stevens, L. Wheatley, Tracheal intubation in ambulatory surgery patients: using remifentanil and propofol without muscle relaxants. *Anesth Analg* 86, 45-49 (1998).

83 L. S. Coleman, Routine Mask Induction and Intubation without Paralysis may be Safer. *J Anesth Surg* 2, 1-2 (2015).

84 S. A. Coleman, J. W. McCrory, C. J. Vallis, R. J. Boys, Inhalation induction of anaesthesia with isoflurane: effect of added carbon dioxide. *Br J Anaesth* 67, 257-261 (1991).

85 L. W. Crile GW, *Anoci-association.* (Saunders, Philadelphia, 1914), pp. 223-225.

86 L. S. Coleman, Four Forgotten Giants of Anesthesia History. *Journal of Anesthesia and Surgery* 3, 1-17 (2015).

87 E. Lowenstein *et al.*, Cardiovascular response to large doses of intravenous morphine in man. *N Engl J Med* 281, 1389-1393 (1969).

88 S. N. Raja, E. Lowenstein, The birth of opioid anesthesia. *Anesthesiology* 100, 1013-1015 (2004).

89 J. G. Bovill, P. S. Sebel, T. H. Stanley, Opioid analgesics in anesthesia: with special reference to their use in cardiovascular anesthesia. *Anesthesiology* 61, 731-755 (1984).

90 T. H. Stanley, Opiate anaesthesia. *Anaesthesia and intensive care* 15, 38-59 (1987).

91 T. H. Stanley, T. D. Egan, H. Van Aken, A tribute to Dr. Paul A. J. Janssen: entrepreneur extraordinaire, innovative scientist, and significant contributor to anesthesiology. *Anesth Analg* 106, 451-462, table of contents (2008).

92 T. H. Stanley, L. R. Webster, Anesthetic requirements and cardiovascular effects of fentanyl-oxygen and fentanyl-diazepam-oxygen anesthesia in man. *Anesth Analg* 57, 411-416 (1978).

93 T. H. E. Stanley, Talmage D., *Anesthesia for the New Millennium.* (Springer-Science + Business Media, 1999), vol. 34, pp. 288.

94 W. Liu, A. V. Bidwai, T. H. Stanley, J. Isern-Amaral, Cardiovascular dynamics after large doses of fentanyl and fentanyl plus N2O in the dog. *Anesth Analg* 55, 168-172 (1976).

95 M. J. Frumin, R. M. Epstein, G. Cohen, Apneic oxygenation in man. *Anesthesiology* 20, 789-798 (1959).

96 M. J. Frumin, N. A. Bergman, D. A. Holaday, Carbon dioxide and oxygen blood levels with a carbon dioxide controlled artificial respirator. *Anesthesiology* 20, 313-320 (1959).

97 E. I. Eger, J. W. Severinghaus, The rate of rise of $PaCO_2$ in the apneic anesthetized patient. *Anesthesiology* 22, 419-425 (1961).

98 P. Slinger, P. E. Blundell, I. R. Metcalf, Management of massive grain aspiration. *Anesthesiology* 87, 993-995 (1997).

99 L. Litt *et al.*, Cerebral intracellular changes during supercarbia: an in vivo 31P nuclear magnetic resonance study in rats. *J Cereb Blood Flow Metab* 5, 537-544 (1985).

100 Y. Cohen *et al.*, Stability of brain intracellular lactate and 31P-metabolite levels at reduced intracellular pH during prolonged hypercapnia in rats. *J Cereb Blood Flow Metab* 10, 277-284 (1990).

101 Y. Xu, Y. Cohen, L. Litt, L. H. Chang, T. L. James, Tolerance of low cerebral intracellular pH in rats during hyperbaric hypercapnia. *Stroke* 22, 1303-1308 (1991).

102 C. Prys-Roberts, W. D. Smith, J. F. Nunn, Accidental severe hypercapnia during anaesthesia. A case report and review of some physiological effects. *Br J Anaesth* 39, 257-267 (1967).

103 K. G. Hickling, Permissive hypercapnia. *Respir Care Clin N Am* 8, 155-169, v (2002).

104 J. G. Laffey, D. O'Croinin, P. McLoughlin, B. P. Kavanagh, Permissive hypercapnia—role in protective lung ventilatory strategies. *Intensive Care Med* 30, 347-356 (2004).

105 H. Morisaki, R. Serita, Y. Innami, Y. Kotake, J. Takeda, Permissive hypercapnia during thoracic anaesthesia. *Acta anaesthesiologica Scandinavica* 43, 845-849 (1999).
106 K. G. Hickling, S. J. Henderson, R. Jackson, Low mortality associated with low volume pressure limited ventilation with permissive hypercapnia in severe adult respiratory distress syndrome. *Intensive Care Med* 16, 372-377 (1990).
107 C. R. Carvalho *et al.*, Temporal hemodynamic effects of permissive hypercapnia associated with ideal PEEP in ARDS. *American journal of respiratory and critical care medicine* 156, 1458-1466 (1997).
108 H. Hager *et al.*, Hypercapnia improves tissue oxygenation in morbidly obese surgical patients. *Anesth Analg* 103, 677-681 (2006).
109 G. M. Hare *et al.*, Hypercapnia increases cerebral tissue oxygen tension in anesthetized rats. *Can J Anaesth* 50, 1061-1068 (2003).
110 J. K. Hino, B. L. Short, K. Rais-Bahrami, W. R. Seale, Cerebral blood flow and metabolism during and after prolonged hypercapnia in newborn lambs. *Critical care medicine* 28, 3505-3510 (2000).
111 J. Ramirez *et al.*, Oxygen-carrying capacity during 10 hours of hypercapnia in ventilated dogs. *Critical care medicine* 28, 1918-1923 (2000).
112 O. Akca, Carbon dioxide and tissue oxygenation: is there sufficient evidence to support application of hypercapnia for hemodynamic stability and better tissue perfusion in sepsis? *Intensive Care Med* 34, 1752-1754 (2008).
113 O. Akca *et al.*, Hypercapnia and surgical site infection: a randomized trial. *Br J Anaesth* 111, 759-767 (2013).
114 O. J. Kirkeby, S. Kutzsche, C. Risoe, I. R. Rise, Cerebral nitric oxide concentration and microcirculation during hypercapnia, hypoxia, and high intracranial pressure in pigs. *J Clin Neurosci* 7, 531-538 (2000).
115 A. T. Mazzeo, A. Spada, C. Pratico, T. Lucanto, L. B. Santamaria, Hypercapnia: what is the limit in paediatric patients? A case of near-fatal asthma successfully treated by multipharmacological approach. *Paediatric anaesthesia* 14, 596-603 (2004).
116 Y. Henderson, A Lecture ON RESPIRATION IN ANAESTHESIA: CONTROL BY CARBON DIOXIDE. *Br Med J* 2, 1170-1175 (1925).
117 Y. Henderson, Resuscitation with Carbon Dioxide. *Science* 83, 399-402 (1936).
118 Y. Henderson, in *Cyclopedia of Medicine*. (1940).
119 L. S. Coleman, Four Forgotten Giants of Anesthesia History. *Journal of Anesthesia and Surgery* 3, 1-17 (2015).
120 Y. Henderson, Resuscitation with Carbon Dioxide. *Science* 83, 399-402 (1936).
121 A. J. C. DeSilva, Normocapnic Ventilation Using The Circle System. *Canad. Anaesth.Soc. J* 23, (1976).
122 E. I. Eger, J. W. Severinghaus, The rate of rise of $PaCO_2$ in the apneic anesthetized patient. *Anesthesiology* 22, 419-425 (1961).
123 M. J. Frumin, R. M. Epstein, G. Cohen, Apneic oxygenation in man. *Anesthesiology* 20, 789-798 (1959).
124 M. J. Frumin, N. A. Bergman, D. A. Holaday, Carbon dioxide and oxygen blood levels with a carbon dioxide controlled artificial respirator. *Anesthesiology* 20, 313-320 (1959).

125 A. I. McEwan *et al.*, Isoflurane minimum alveolar concentration reduction by fentanyl. *Anesthesiology* 78, 864-869 (1993).
126 G. G. Pacentine, M. Muzi, T. J. Ebert, Effects of fentanyl on sympathetic activation associated with the administration of desflurane. *Anesthesiology* 82, 823-831 (1995).
127 M. R. Murphy, C. C. Hug, Jr., The anesthetic potency of fentanyl in terms of its reduction of enflurane MAC. *Anesthesiology* 57, 485-488 (1982).
128 C. L. Westmoreland, P. S. Sebel, A. Gropper, Fentanyl or alfentanil decreases the minimum alveolar anesthetic concentration of isoflurane in surgical patients. *Anesth Analg* 78, 23-28 (1994).
129 L. S. Coleman, 30 Years Lost in Anesthesia Theory. *Cardiovasc Hematol Agents Med Chem* 10, 31-49 (2012).
130 Y. Henderson, The Wesley M. Carpenter Lecture: Atelectasis, Massive Collapse and Related Post-Operative Conditions. *Bull N Y Acad Med* 11, 639-656 (1935).
131 Y. Henderson, Resuscitation with Carbon Dioxide. *Science* 83, 399-402 (1936).
132 Y. Henderson, How Breathing Begins at Birth. *Science* 85, 89-91 (1937).
133 Y. Henderson, in *Cyclopedia of Medicine*. (1940).
134 Y. Henderson, H. W. Haggard, *Noxious gases and the principles of respiration influencing their action.* (Chemical Catalog Co., New York,, 1927), pp. 220 p.
135 A. Campbell, E. P. Poulton, *Oxygen and carbon dioxide therapy.* Oxford medical publications (Oxford University Press, H. Milford, London,, 1934), pp. xv, 179 p.
136 E. Lowenstein *et al.*, Cardiovascular response to large doses of intravenous morphine in man. *N Engl J Med* 281, 1389-1393 (1969).
137 S. N. Raja, E. Lowenstein, The birth of opioid anesthesia. *Anesthesiology* 100, 1013-1015 (2004).
138 K. G. Hickling, Permissive hypercapnia. *Respir Care Clin N Am* 8, 155-169, v (2002).
139 K. G. Hickling, S. J. Henderson, R. Jackson, Low mortality associated with low volume pressure limited ventilation with permissive hypercapnia in severe adult respiratory distress syndrome. *Intensive Care Med* 16, 372-377 (1990).
140 H. Hager *et al.*, Hypercapnia improves tissue oxygenation in morbidly obese surgical patients. *Anesth Analg* 103, 677-681 (2006).
141 G. M. Hare *et al.*, Hypercapnia increases cerebral tissue oxygen tension in anesthetized rats. *Can J Anaesth* 50, 1061-1068 (2003).
142 O. Akca, Carbon dioxide and tissue oxygenation: is there sufficient evidence to support application of hypercapnia for hemodynamic stability and better tissue perfusion in sepsis? *Intensive Care Med* 34, 1752-1754 (2008).
143 O. Akca *et al.*, Hypercapnia and surgical site infection: a randomized trial. *Br J Anaesth* 111, 759-767 (2013).
144 O. J. Kirkeby, S. Kutzsche, C. Risoe, I. R. Rise, Cerebral nitric oxide concentration and microcirculation during hypercapnia, hypoxia, and high intracranial pressure in pigs. *J Clin Neurosci* 7, 531-538 (2000).
145 A. T. Mazzeo, A. Spada, C. Pratico, T. Lucanto, L. B. Santamaria, Hypercapnia: what is the limit in paediatric patients? A case of near-fatal asthma successfully treated by multipharmacological approach. *Paediatric anaesthesia* 14, 596-603 (2004).

146 O. Akca, Hypercapnia: what is the upper limit? *Paediatric anaesthesia* 15, 80-81 (2005).

147 O. Akca, Optimizing the intraoperative management of carbon dioxide concentration. *Current opinion in anaesthesiology* 19, 19-25 (2006).

148 O. Akca, Lower tidal volumes and positive end-expiratory pressure prevents alveolar coagulation in patients without lung injury. *Anesthesiology* 106, 1065-1066 (2007).

149 O. Akca *et al.*, Hypercapnia improves tissue oxygenation. *Anesthesiology* 97, 801-806 (2002).

150 O. Akca *et al.*, Effect of intra-operative end-tidal carbon dioxide partial pressure on tissue oxygenation. *Anaesthesia* 58, 536-542 (2003).

151 O. Akca *et al.*, Tissue oxygenation response to mild hypercapnia during cardiopulmonary bypass with constant pump output. *Br J Anaesth* 96, 708-714 (2006).

152 A. M. Mukhtar, G. M. Obayah, A. Elmasry, N. M. Dessouky, The therapeutic potential of intraoperative hypercapnia during video-assisted thoracoscopy in pediatric patients. *Anesth Analg* 106, 84-88, table of contents (2008).

153 M. C. Seevers, WH; et al, The Role of Hypercapnia in the Production of "Ether Convulsions". *J Pharmacol Exp Ther* 63, (1938).

154 L. S. Coleman, Four Forgotten Giants of Anesthesia History. *Journal of Anesthesia and Surgery* 3, 1-17 (2015).

155 P. Slinger, P. E. Blundell, I. R. Metcalf, Management of massive grain aspiration. *Anesthesiology* 87, 993-995 (1997).

156 L. Litt *et al.*, Cerebral intracellular changes during supercarbia: an in vivo 31P nuclear magnetic resonance study in rats. *J Cereb Blood Flow Metab* 5, 537-544 (1985).

157 Y. Xu, Y. Cohen, L. Litt, L. H. Chang, T. L. James, Tolerance of low cerebral intracellular pH in rats during hyperbaric hypercapnia. *Stroke* 22, 1303-1308 (1991).

158 C. Prys-Roberts, W. D. Smith, J. F. Nunn, Accidental severe hypercapnia during anaesthesia. A case report and review of some physiological effects. *Br J Anaesth* 39, 257-267 (1967).

159 S. L. Klein, J. K. Lilburn, An unusual case of hypercarbia during general anesthesia. *Anesthesiology* 53, 248-250 (1980).

160 A. J. C. DeSilva, Normocapnic Ventilation Using The Circle System. *Canad. Anaesth.Soc. J* 23, (1976).

161 D. J. Cullen, E. I. Eger, 2nd, The effect of extreme hypocapnia on the anaesthetic requirement (MAC) of dogs. *Br J Anaesth* 43, 339-343 (1971).

162 H. Hager *et al.*, Hypercapnia improves tissue oxygenation in morbidly obese surgical patients. *Anesth Analg* 103, 677-681 (2006).

163 G. M. Hare *et al.*, Hypercapnia increases cerebral tissue oxygen tension in anesthetized rats. *Can J Anaesth* 50, 1061-1068 (2003).

164 O. Akca, Carbon dioxide and tissue oxygenation: is there sufficient evidence to support application of hypercapnia for hemodynamic stability and better tissue perfusion in sepsis? *Intensive Care Med* 34, 1752-1754 (2008).

165 O. Akca *et al.*, Effect of intra-operative end-tidal carbon dioxide partial pressure on tissue oxygenation. *Anaesthesia* 58, 536-542 (2003).

166 Y. Henderson, Resuscitation with Carbon Dioxide. *Science* 83, 399-402 (1936).
167 Y. Henderson, in *Cyclopedia of Medicine*. (1940).
168 M. J. Laster, E. I. Eger, 2nd, Temperatures in soda lime during degradation of desflurane, isoflurane, and sevoflurane by desiccated soda lime. *Anesth Analg* 101, 753-757, table of contents (2005).
169 T. C. Harvey *et al.*, Effect of carbon dioxide in acute mountain sickness: a rediscovery. *Lancet* 2, 639-641 (1988).
170 L. S. Coleman, Should soda lime be abolished? *Anesth Analg* 102, 1290-1291 (2006).
171 L. S. Coleman, Soda lime is a dangerous relic of a forgotten past. *Anaesthesia* 70, (2016).
172 S. Reitsma, D. W. Slaaf, H. Vink, M. A. van Zandvoort, M. G. oude Egbrink, The endothelial glycocalyx: composition, functions, and visualization. *Pflugers Arch* 454, 345-359 (2007).
173 T. H. E. Stanley, Talmage D., *Anesthesia for the New Millennium*. (Springer-Science + Business Media, 1999), vol. 34, pp. 288.
174 P. L. Bailey, The use of opioids in anesthesia is not especially associated with nor predictive of postoperative hypoxemia. *Anesthesiology* 77, 1235 (1992).
175 P. L. Bailey *et al.*, Frequent hypoxemia and apnea after sedation with midazolam and fentanyl. *Anesthesiology* 73, 826-830 (1990).
176 L. S. Coleman, Routine Mask Induction and Intubation without Paralysis may be Safer. *J Anesth Surg* 2, 1-2 (2015).
177 S. A. Coleman, J. W. McCrory, C. J. Vallis, R. J. Boys, Inhalation induction of anaesthesia with isoflurane: effect of added carbon dioxide. *Br J Anaesth* 67, 257-261 (1991).
178 M. S. Scheller, M. H. Zornow, L. J. Saidman, Tracheal intubation without the use of muscle relaxants: a technique using propofol and varying doses of alfentanil. *Anesth Analg* 75, 788-793 (1992).
179 J. B. Stevens, L. Wheatley, Tracheal intubation in ambulatory surgery patients: using remifentanil and propofol without muscle relaxants. *Anesth Analg* 86, 45-49 (1998).
180 M. V. J. K. Matthias W. Koenig, MD, in *Anesthesiology News*. (McMahon, New York, NY, 2016).
181 T. C. Harvey *et al.*, Effect of carbon dioxide in acute mountain sickness: a rediscovery. *Lancet* 2, 639-641 (1988).
182 L. S. Coleman, Should soda lime be abolished? *Anesth Analg* 102, 1290-1291 (2006).
183 S. A. Coleman, J. W. McCrory, C. J. Vallis, R. J. Boys, Inhalation induction of anaesthesia with isoflurane: effect of added carbon dioxide. *Br J Anaesth* 67, 257-261 (1991).
184 B. Vasile, F. Rasulo, A. Candiani, N. Latronico, The pathophysiology of propofol infusion syndrome: a simple name for a complex syndrome. *Intensive Care Med* 29, 1417-1425 (2003).
185 F. J. Overdyk, Lack of tolerance with propofol in radiation therapy needs patience. *Anesthesiology* 101, 1044; author reply 1044 (2004).

186 E. A. Bermudez, M. H. Chen, Cardiac arrest associated with intravenous propofol during transesophageal echocardiography before DC cardioversion. *Heart Dis* 4, 355-357 (2002).

187 B. Vasile, F. Rasulo, A. Candiani, N. Latronico, The pathophysiology of propofol infusion syndrome: a simple name for a complex syndrome. *Intensive Care Med* 29, 1417-1425 (2003).

188 F. J. Overdyk, Lack of tolerance with propofol in radiation therapy needs patience. *Anesthesiology* 101, 1044; author reply 1044 (2004).

189 E. A. Bermudez, M. H. Chen, Cardiac arrest associated with intravenous propofol during transesophageal echocardiography before DC cardioversion. *Heart Dis* 4, 355-357 (2002).

190 A. R. Painter. (Painter Law Firm, March 30, 2015), pp. Propofol is the dangerous anesthetic drug involved in the deaths of Michael Jackson and Joan Rivers.

191 S. A. Forman, Clinical and molecular pharmacology of etomidate. *Anesthesiology* 114, 695-707 (2011).

192 A. I. McEwan *et al.*, Isoflurane minimum alveolar concentration reduction by fentanyl. *Anesthesiology* 78, 864-869 (1993).

193 M. R. Murphy, C. C. Hug, Jr., The anesthetic potency of fentanyl in terms of its reduction of enflurane MAC. *Anesthesiology* 57, 485-488 (1982).

194 C. L. Westmoreland, P. S. Sebel, A. Gropper, Fentanyl or alfentanil decreases the minimum alveolar anesthetic concentration of isoflurane in surgical patients. *Anesth Analg* 78, 23-28 (1994).

195 J. E. Ilkiw, P. J. Pascoe, S. C. Haskins, J. D. Patz, R. Jaffe, The cardiovascular sparing effect of fentanyl and atropine, administered to enflurane anesthetized dogs. *Canadian journal of veterinary research = Revue canadienne de recherche veterinaire* 58, 248-253 (1994).

196 G. S. Murphy, J. W. Szokol, J. H. Marymont, M. J. Avram, J. S. Vender, The effects of morphine and fentanyl on the inflammatory response to cardiopulmonary bypass in patients undergoing elective coronary artery bypass graft surgery. *Anesth Analg* 104, 1334-1342, table of contents (2007).

197 K. J. Anand, M. Maze, Fetuses, fentanyl, and the stress response: signals from the beginnings of pain? *Anesthesiology* 95, 823-825 (2001).

198 K. J. Anand, W. G. Sippell, A. Aynsley-Green, Randomised trial of fentanyl anaesthesia in preterm babies undergoing surgery: effects on the stress response. *Lancet* 1, 243-248 (1987).

199 J. P. Cravero, M. Beach, B. Thyr, K. Whalen, The effect of small dose fentanyl on the emergence characteristics of pediatric patients after sevoflurane anesthesia without surgery. *Anesth Analg* 97, 364-367, table of contents (2003).

200 D. M. Philbin, C. E. Rosow, Fentanyl and sufentanil anesthesia revisited. *Journal of cardiothoracic and vascular anesthesia* 5, 651-652 (1991).

201 M. Tverskoy, S. Gelman, K. C. Fowler, E. L. Bradley, Influence of fentanyl and morphine on intestinal circulation. *Anesth Analg* 64, 577-584 (1985).

202 A. I. McEwan *et al.*, Isoflurane minimum alveolar concentration reduction by fentanyl. *Anesthesiology* 78, 864-869 (1993).

203 S. G. Ainslie, J. H. Eisele, Jr., G. Corkill, Fentanyl concentrations in brain and serum during respiratory acid—base changes in the dog. *Anesthesiology* 51, 293-297 (1979).

204 T. J. Gan, P. F. White, P. E. Scuderi, M. F. Watcha, A. Kovac, FDA "black box" warning regarding use of droperidol for postoperative nausea and vomiting: is it justified? *Anesthesiology* 97, 287 (2002).

205 L. S. Coleman, in *apsf Newsletter*. (Anesthesia Patient Safety Foundation, Administrator, Deanna Walker Anesthesia Patient Safety Foundation Building One, Suite Two 8007 South Meridian Street Indianapolis, IN 46217-2922 e-mail address: walker@apsf.org FAX: (317) 888-1482, 2010), vol. Winter 2009-2020.

206 L. S. Coleman, A call for standards on perioperative CO(2) regulation. *Can J Anaesth*, (2011).

207 A. J. Salvatore, S. F. Sullivan, E. M. Papper, Postoperative hypoventilation and hypoxemia in man after hyperventilation. *N Engl J Med* 280, 467-470 (1969).

208 L. S. Coleman, in *apsf Newsletter*. (Anesthesia Patient Safety Foundation, Administrator, Deanna Walker Anesthesia Patient Safety Foundation Building One, Suite Two 8007 South Meridian Street Indianapolis, IN 46217-2922 e-mail address: walker@apsf.org FAX: (317) 888-1482, 2010), vol. Winter 2009-2020.

209 L. S. Coleman, A call for standards on perioperative CO(2) regulation. *Can J Anaesth*, (2011).

210 L. S. Coleman, A call for standards on perioperative CO(2) regulation. *Can J Anaesth*, (2011).

211 L. S. Coleman, Take a Walk on the Wild Side. *Voluntaryist*. 1995.

INDEX

Symbols

A

B

C

D

F

G

H

I

J

N

O

P

Q

R

T

U

V

W

X

Y

Z

About the Author

Lewis S. Coleman is Chair of the Science and Education Board of The American Institute of Stress. He is a board-certified American anesthesiologist whose discovery of Selye's long-sought stress mechanism is a classic adventure of amateur scientific curiosity fueled by the fickle finger of fate. It began when he employed carbon dioxide to anesthetize cockroaches during his undergraduate research at the Ohio State University. Knowing that all animal cells continuously produce CO_2 as a product of normal metabolism, he suspected that CO_2 asphyxiation (i.e., suffocation) mimics anesthesia by causing brain hypoxia, and worried that harmful hypoxia might perturb his research results. His suspicions were subsequently confirmed by CO_2 pathophysiology lectures at New York Medical College that explained how breathing excessive concentrations of CO_2 and other atmospheric gases can mimic general anesthesia by causing brain oxygen starvation. By great luck, these lectures took place during the two years when NYMC retained Dr. Johannes Rhodin, a famous Swedish researcher, to revise and reform the basic sciences curriculum. Dr. Rhodin was an expert in the stress theory of Hans Selye, and he provided spellbinding stress theory lectures that left Dr. Coleman convinced that stress theory represented the future of medicine. Then, after a surgical internship he chose a career in anesthesiology, which is the essence of stress control. Anesthesia research during that time sought to identify ways to alter anesthetic technique to control the "surgical stress syndrome" and optimize surgical outcome. This resulted in the rapid introduction of new machines, medications, monitors, ideas, and techniques. Several years later he discovered fresh research information that coagulation factor VIII is a "chimera" consisting of two seemingly unrelated proteins. He knew that biological chimeras are exceedingly rare, and suspected that factor VIII must be part of an undiscovered physiological mechanism. His curiosity inspired an intense review of published medical literature using computer techniques that hadn't existed during his training. Factor VIII proved to be a "Rosetta Stone" that gradually led to the discovery of the stress mechanism that Dr. Rhodin had described.

COLOPHON

This book was designed, edited, and composed by Jeff Walden using Adobe InDesign, Photoshop, and Acrobat DC. It is set in Work Sans and Open Sans, both fonts publicly available from Google. Illustrations are from the author.

Made in United States
Orlando, FL
05 November 2026

90240245R00343